W9-AFS-622

BEHOLD MAN

BEHOLD MAN

A Photographic Journey of Discovery inside the Body

Lennart Nilsson

in collaboration with

Jan Lindberg

Text by **David H. Ingvar** **Stig Nordfeldt** **Rune Pettersson**

Translator **Ilona Munck**

Drawings by **Bernt Forsblad**

Little, Brown and Company Boston Toronto

Design by Håkan Lindström

ISBN: 91-0-038093-8

LC: 73-14087

First American Edition

Lithography by Angsö Lito AB
Printed at AB Åetåtryck, Stockholm
Bound at Bonniers Bookbindery
Printed in Sweden

Published simultaneously in Canada by Little, Brown & Company (Canada) Limited

Preface

When we began work on this book we knew that we would never be able to describe the human body in complete detail. It would take more than a lifetime to draw a full portrait of man using the variety of electronic and optical instruments currently available. Our bodies are too complex, many processes are not yet understood, and others can not yet be reproduced photographically. We have worked with the pictures in this book for many years and are pleased to be able to present them now as a kind of first report.

New optical systems, lenses with extremely short focal lengths and wide visual angles, which we were able to develop with the aid of international experts, have made it possible to observe and film parts of the body never before photographed. We hope that in some instances this technique has provided new information. For example, our photographs of the interior of various blood vessels, which, among other things, show the changes associated with arteriosclerosis, drew attention to our technique when they were published in *Life* magazine and won awards from the American Heart Association.

What we have done could not have been done without the collaboration of experts in many different fields. We are grateful to all who have helped us and their names, individuals as well as institutions and organizations, are listed at the end of the book. In particular, Dr. Lena Lagerström has played a most active role. Her contribution is exceptional, and we realize that without her, many of the photographs appearing in the book would never have been taken. Our especially warm thanks go to her.

Stockholm, August 1973

Lennart Nilsson
Jan Lindberg

Contents

Preface 5

Exploring Man 8

The Cell 10

Cellular Organization
Tissue and Organs 21

Digestion and Excretion 21 Respiration 23
Blood and Lymph 23
Skin, Skeleton, and Muscles 24
The Sense Organs 26 The Nervous System 26
Reproduction 28 The Parts and the Whole 30

Reproduction 33

The Male Reproductive System 34
Male Sex Hormones 35 The Testes 36
Sperm Cells 39
The Female Reproductive System 40
Female Sex Hormones 40 The Ovaries 42
The Fallopian Tube 44 The Ovum 46
The Female Sexual Cycle 48
The Development of the Fetus 49
Egg and Sperm Cells 50 Fertilization 51
The Fertilized Egg 52 Fetal Development 53

Blood
The Body's Fluid Tissue 57

The Body's Transport System 57
Defense Against Infection 59 Blood Cells 60
Blood Heals Wounds 62 Blood Vessels 63
The Circulation of the Blood 64
What Is an EKG? 64 The Heart 65
The Aorta 66 Intestinal Blood Vessels 69
Circulation through the Lungs 71 Blood Pressure 72
Inside the Heart 72 Blood Types 74
Fetal Circulation 76

The Lymph Vessels
The Body's Drainage System 78

Respiration 81

Breathing in and out 81 The Voice 86
The Trachea and Bronchi 88
Pulmonary Alveoles 89
The Lungs and the Air We Breathe 90 Smoking 91

Body Temperature 94

The Kidneys
The Body's Cleansing System 96

The Urinary System 96 The Renal Cortex 96
The Renal Pelvis 103
The Ureter and the Bladder 104

Skin 105

The Functions of the Skin 105 Perspiration 110
The Color of Skin and Hair 110

Muscles 115

The Living Skeleton 120

Skeletal Bones 120 Bone Formation 121

Solid and Spongy Bone 122
The Human Fetus Has a Tail 124 Joints 125

Teeth 126

What Happens to Food 129

The Digestive Tract 129
The Mouth and the Pharynx 130 The Stomach 132
The Small Intestine 136 The Pancreas 142
The Liver 143 Bile 144 The Colon 148
The Rectum 150 Metabolism 150

The Nervous System 153

The Central Nervous System 153
The Spinal Cord 154 Nerves 156
The Development of the Nervous System 160
The Brain 162 Nerve Cells 166
The Pineal Body and Corpora Quadrigemina 171
The Pituitary Gland 172 The Cerebellum 174
Glial Cells 174 The Nerve Impulse 176

The Senses 177

Sensory Receptor Cells 178
Evolution of the Sense Organs 179

Vision 180

Light 180 The Eye 180
The Anterior Part of the Eye 180
Inside the Eye 182 The Cornea 184
The Pupil 185 The Iris 186 The Lens 188
Eye Movements 190 The Fundus of the Eye 191
The Macula Lutea 192 Rods and Cones 196
Tears 198 The Retina 198

Hearing 201

The External Ear 202 The Middle Ear 204
Inside the Tympanic Cavity 208 The Inner Ear 210
Auditory Receptor Cells 216
The Development of the Ear 218

The Organs of Equilibrium 220

Touch and Pain 225

Olfaction 231

The Nose 232 The Olfactory Mucosa 235
Olfactory Receptor Cells 238

Taste 240

The Tongue 241 The Papillae of the Tongue 244

Some Concluding Words 248

Index 251

Exploring Man

The human body is marvelous. It can move freely, act deliberately, and survive under the most variable conditions. Its construction is complex and its requirements many. Compared with other species, man can exercise greater choice in meeting these demands. Nature has endowed him with a brain that enables him to think abstractly, using symbols. This ability underlies man's intellectual life, allowing a wealth of perceptions and memories, reasoned thoughts and actions. In particular, man's capacity for symboling has its greatest application in language — the uniquely human vehicle for thought and communication.

Spermatozoa with long tails swarm around the ovum's surface. Magnified about 2000 times.

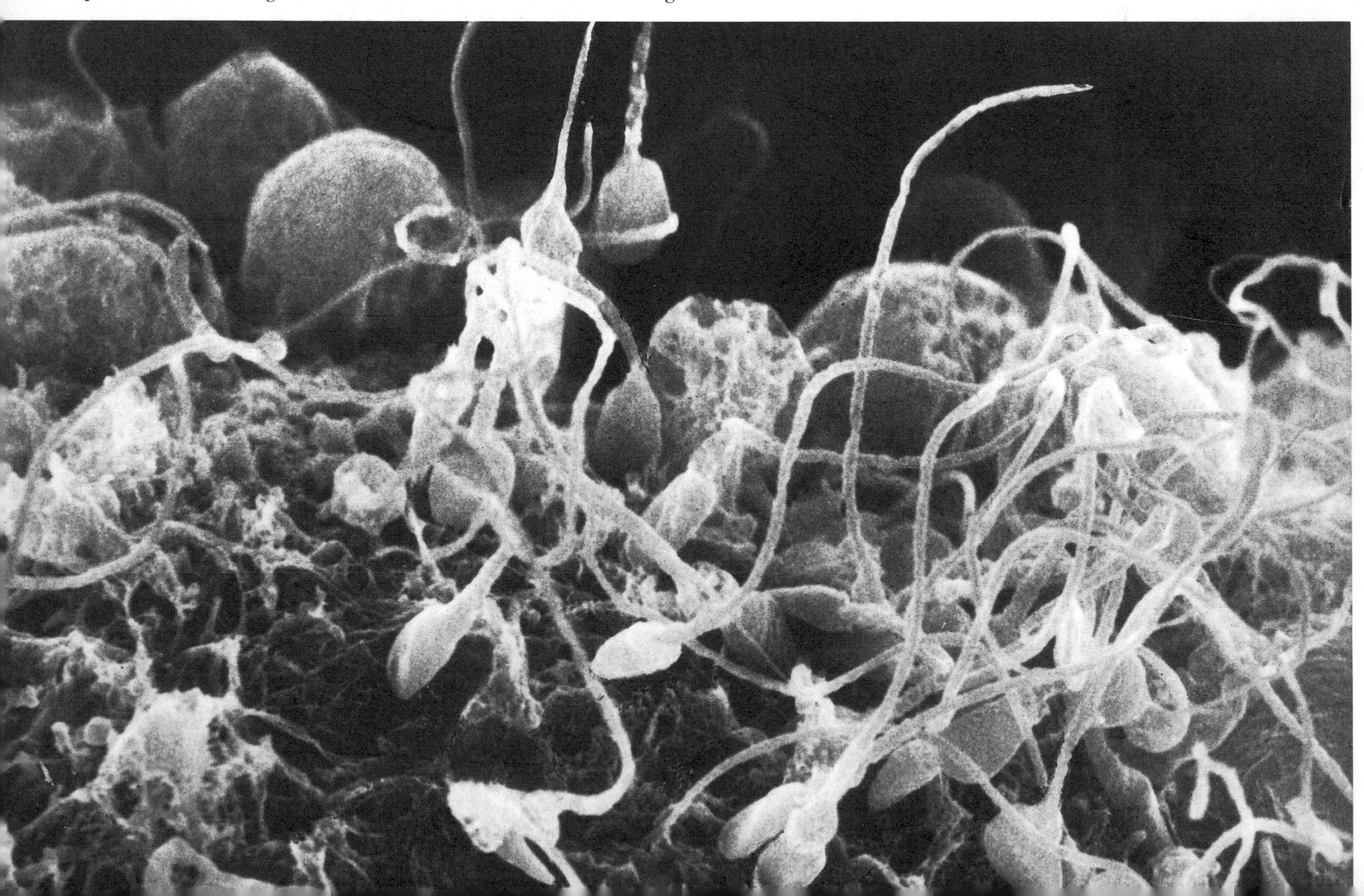

Spermatozoa meet egg
and fertilization is about to take place. The photo on the left-hand page was taken with a scanning electron microscope with a magnification of about 2 000 times. A portion of the surface of the ovum is shown with sperm cells trying to penetrate the wall. The human ovum, which is a single cell, is between 0.1 and 0.2 mm in diameter. The tadpole-shaped sperm cell, also a single cell, is much smaller. Its head is about one-sixteenth the size of the ovum.
The photo on the right was taken through a light microscope and also shows an ovum surrounded by sperm cells. Here the magnification is only about 500 times, and the cells are not as clearly distinguished. Movements of the sperm cells in trying to penetrate the ovum may make it rotate slowly.

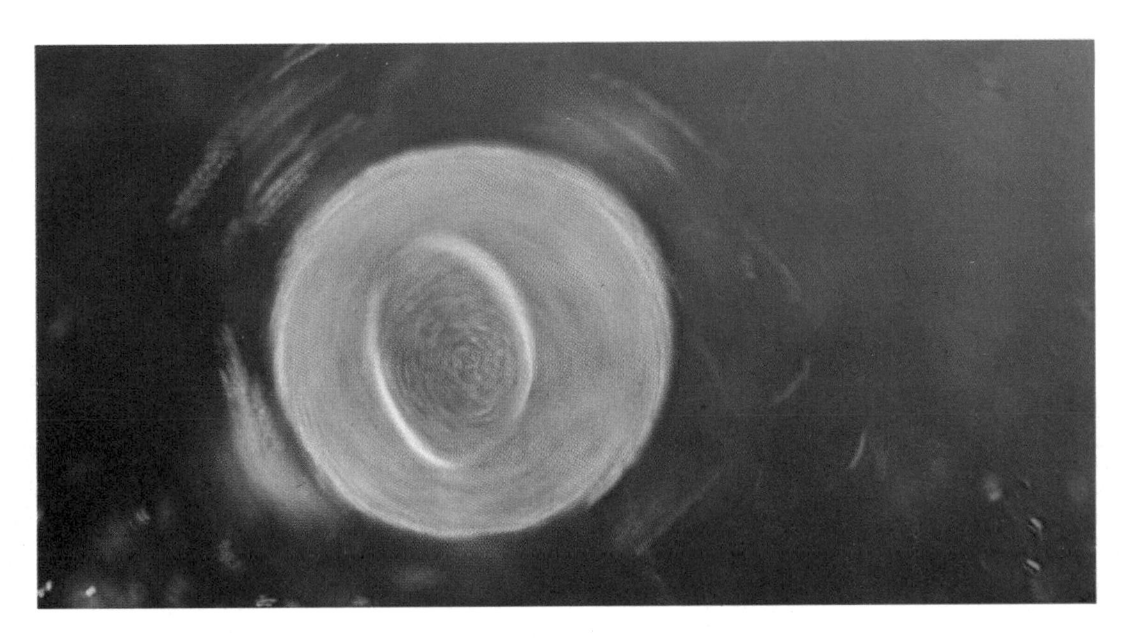

The human body is an exciting object to study, to learn about in minute detail. We know from biology that all living matter is made up of small units, cells, which cooperate with each other. So to understand how the human body functions, it makes sense to begin with a study of cells. A knowledge of how they work will make the function of the human body as a whole easier to grasp.

The first cell in the series which leads to the creation of a new individual is the fertilized egg or ovum. It splits into two daughter cells which continue to divide, becoming organized and specialized. Groups of cells form organs, and eventually a complete organism develops. All growth and everything else that happens inside the body is directed toward that end.

It was remarkably late in human history that scientists began to study the human body in a relatively unprejudiced and objective way. Research has intensified over the years and has come to involve more and more people. An enormous amount of knowledge has been amassed and published in books and journals, scientists all the while engaging in a lively exchange of ideas. The medical literature published over the last decade alone exceeds in quality and quantity the total published in the field up to that time.

It was not until the eighteenth century that the attitude toward medical research that is so natural today was first formulated — to find out as accurately as possible "how things really are." During the latter part of the nineteenth century, advances in techniques and equipment, particularly in the microscope, created exciting new possibilities. Thinner and thinner tissue sections could be prepared, and better staining techniques revealed new details under the lens. It was possible to study the structure of the minutest components of tissue, and this led to a better understanding of their function. Microscopes continued to be improved, with greater enlargements possible, while at the same time comparable advances were made in tissue preparation techniques.

Several special types of light microscope have been constructed, including the interference microscope and the phase-contrast microscope. In recent years electron microscopes of various types have been increasingly used. These can give far greater detail because they are based on the use of electron beams striking a target rather than on visible light rays. Objects smaller than the smallest wavelength of visible light cannot be resolved in a light microscope. Since electron wavelengths are much smaller than this, however, an electron microscope can be used to observe objects a hundred times smaller than those visible with light microscopes.

Photography was used early in the study of anatomy, and many fine pictures were taken by pioneers at the end of the nineteenth century. The technique of black-and-white photography has greatly advanced as both cameras and films have reached new high standards. Color photography has also been developed. Scientists have been quick to adopt all these innovations. This book presents a variety of examples of how advanced techniques of microscopy combined with extensive knowledge of the latest achievements of photography can yield a new and fascinating picture of the body and its functions. It is a book that explores man from a new angle.

The Cell

All living matter is made up of small units called cells. The cell has a number of properties which distinguish it from inert or dead matter. For example, it can breathe, it can take in nutrients, and it has the capacity to reproduce.

There are a great many kinds of cells. The human body is composed of approximately 200 billion, all collaborating in the common purpose of survival and reproduction.

A cell can be thought of as an independent individual belonging to the larger organization which the body represents. It is a being with a life of its own, equipped with organs that supply nourishment, and handle metabolism and excretion. Like the body as a whole, the individual cell can respond to external stimulation, some — nerve and muscle cells, for example — more than others. Within limits, the individual cell can determine its own activity, its biochemical behavior. And just as individuals influence other individuals, cells influence each other in a variety of ways, most often chemically, but also by means of special communications systems, such as the neural pathways.

Most cells can only be seen with the aid of a microscope. The human egg cell is comparatively large, up to 0.2 mm across, about the size of a pinpoint. Nature's largest cells are the eggs of birds. A red blood cell is .007 mm in diameter. Most of the body's cells are between .02 and .08 mm in diameter.

The cell is surrounded by a fine membrane between .00001 and .00002 mm thick. The membrane is composed of rows of fat molecules sandwiched between layers of carbohydrates and proteins. It can be thought of as the cell's "skin": it holds the cell together, helps preserve a constant internal environment, and protects the cell from the actions of chemicals or other agents. The cell membrane is so constructed that certain substances can

Cells

are the building blocks of the body. They can breathe, take in nourishment, and multiply. Some can even move freely. Inside the cell membrane (1), which surrounds the cell, are the cytoplasm (2) and the nucleus (5). The cytoplasm is the viscid ground substance of all cells. It is composed chiefly of water, proteins, fats, and carbohydrates. Small particles (13) or fat droplets (10) contain food reserves. The vacuoles (11) are cavities in the cytoplasm. The cytoplasm also contains structures with specific functions, cell organelles. Mitochondria (9) generate the energy necessary for cellular activities. Ribosomes (4) lie in rows of small granules embedded in a complex network of thin membranes, the endoplasmic reticulum (3). They are the site of protein synthesis. This process is under control of the genes, the units of heredity, which are located in the cell nucleus. A centrosome with one or two centrioles (8) is also present in the cytoplasm. The centrioles are involved in cell division. The function of the Golgi apparatus (12) is not clearly understood. The nucleus (5) with its nucleolus (6) is generally spherically shaped and enclosed by a nuclear membrane (7).

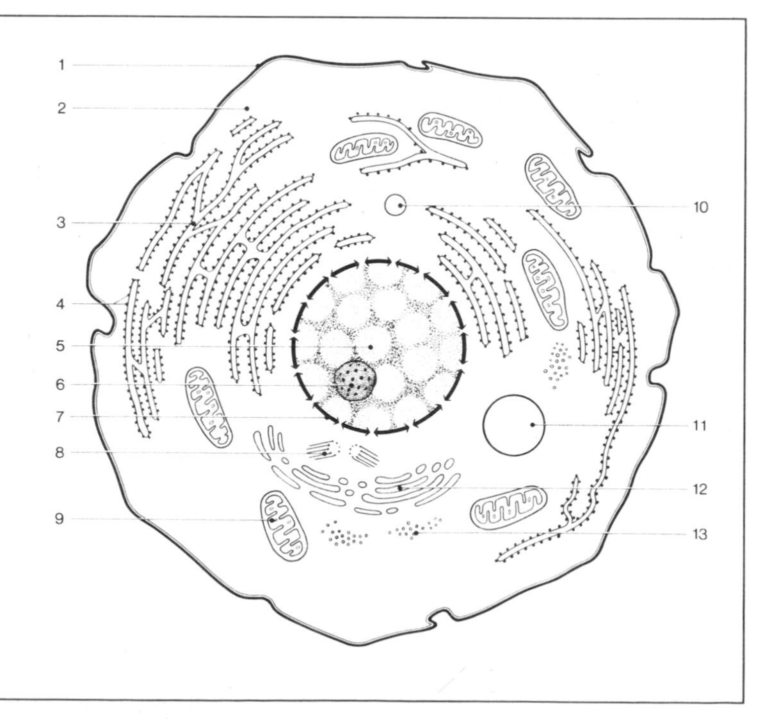

pass across it. These include nutrients and water, as well as waste products. The permeability of the membrane can be adjusted to suit the needs of the cell, as, for example, in regulating the uptake of nutrients.

The interior of the cell consists of the cell body, the cytoplasm or protoplasm, and the nucleus. About 70 or 80 percent of the cytoplasm consists of water; the remainder is made up of proteins, fats, carbohydrates, salts, and so on. Traversing the cytoplasm is a fine fibrous network, or reticulum, and embedded throughout it are the cell's "organs" — organelles — the most important of which are the mitochondria. These are the cell's power stations, generating the energy the cell uses to carry out its vital processes. Energy is produced through the breakdown of certain organic molecules, mostly by the process of oxidation (or burning), but sometimes through a process of fermentation. Enzymes direct the chemical reactions involved in the storage or release of energy, while high- and low-energy phosphate bonds play the major roles. Ribosomes — organelles which are only visible with an electron microscope — are the cell bodies responsible for the synthesis of specific proteins required by the cell. The chemical reactions involved here are ultimately controlled by the chromosomes.

The nucleus of the cell is generally spherical and enclosed by a nuclear membrane approximately .00003 mm thick. The chromosomes, each containing a set of genes, which are the individual units of heredity, appear as extremely thin threadlike bodies in the nucleus. The number of chromosomes for each species is constant. In man the total is forty-six, of which two are the sex chromosomes. This pair determines whether an individual is genetically a male or a female.

Cells reproduce by dividing, a process in which each cell gives rise to two new cells with the same properties as the parent. When the cell divides, the chromosomes become shorter and thickened and align themselves in pairs along a plane parallel to the cell's "equator." Each chromosome then splits along its length into identical halves. One set of new chromosomes now migrates toward one side or pole of the cell; the other set migrates toward the opposite pole. The cell body then begins to fold in toward the equator. When this process is completed, the cell membrane completely separates the two halves, so that two new cells come into existence.

Cell division goes on continually, making growth possible and sustaining life by replacing worn-out cells. Rates of renewal vary for different parts of the body. A skin cell may die and be sloughed off in less than a week, and the skeleton is renewed every twenty to twenty-four months. But the ability of nerve cells to divide comes to an end in childhood. The longevity of individual nerve cells varies. Some may live for twenty or thirty years; others may last a lifetime. Some scientists believe that the constant loss of nerve cells from maturity until the end of life may in part explain the changes that occur in aging.

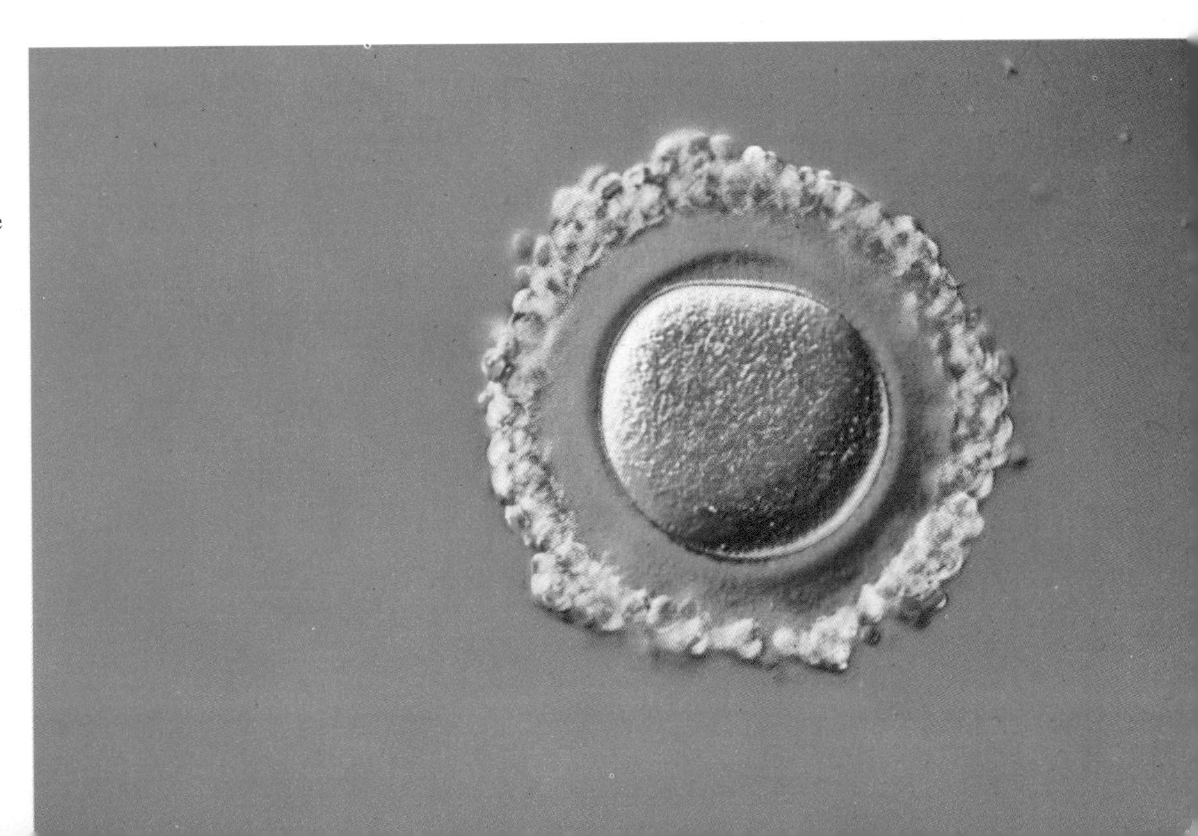

The egg cell
is the largest cell in the human body. Other cells are much smaller, the majority being only one-tenth the size of the ovum, which is between 0.1 and 0.2 mm across. The photo was taken through an interference microscope. Surrounding the ovum is a protective membrane. A number of yolk granules are visible in the cytoplasm.

A nerve cell from the cerebellum magnified approximately 8,000 times ▷

Connective tissue cells ▷
have a fairly simple structure in comparison to other kinds of cells. They can easily be made to grow and multiply outside the body, given adequate nutrient media. The picture was taken with a phase-contrast microscope. The granular nature of the nucleus in each of the elongated cell bodies shows up very clearly. The photo also illustrates how connective tissue cells unite into bundles or bands. Such connective bands are the wrappings that enclose our interior organs or parts of organs. Connective tissue cells also figure in wound-healing, multiplying and growing into the break to bind up the edges.

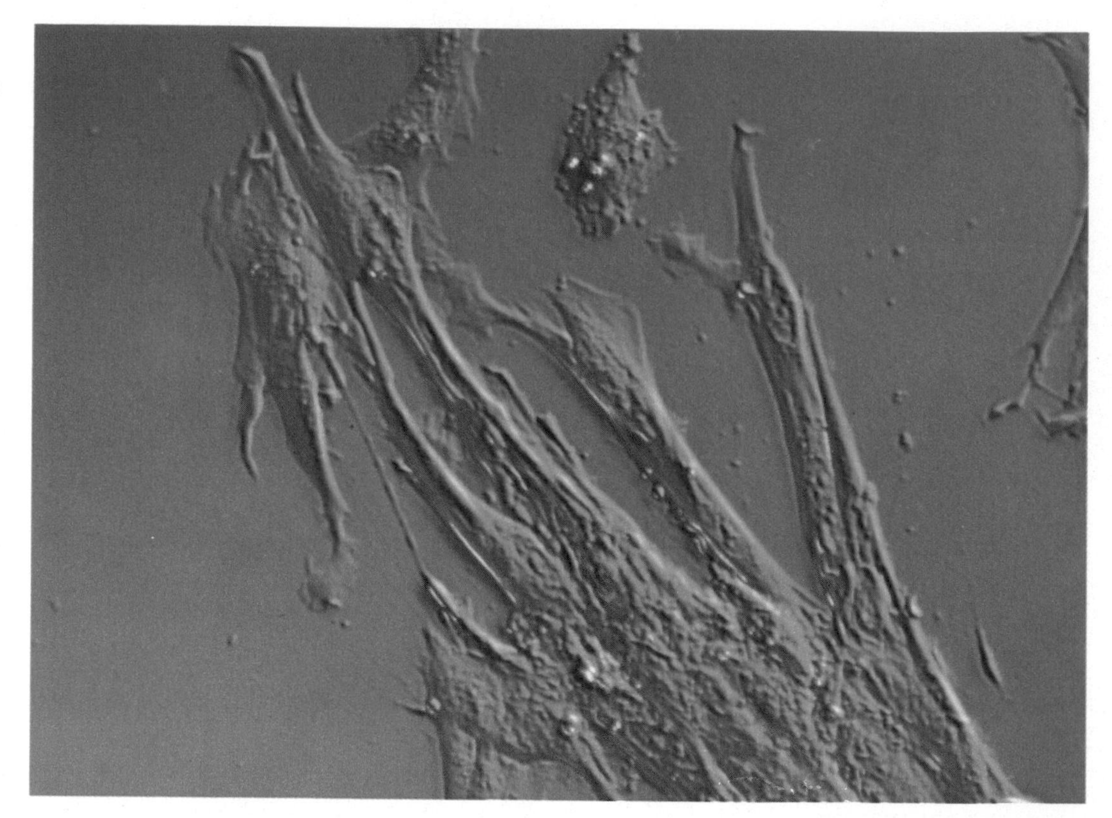

Fat cells ▷
can store fat as a food reserve. In a state of starvation, the fat content of fat cells decreases. Fat cells also insulate against cold and provide some mechanical protection against pressure: the cells act like small cushions which distribute the load. The fat cells in this preparation are round and yellow. They are lying in muscle tissue, the cells of which appear red in the photo.

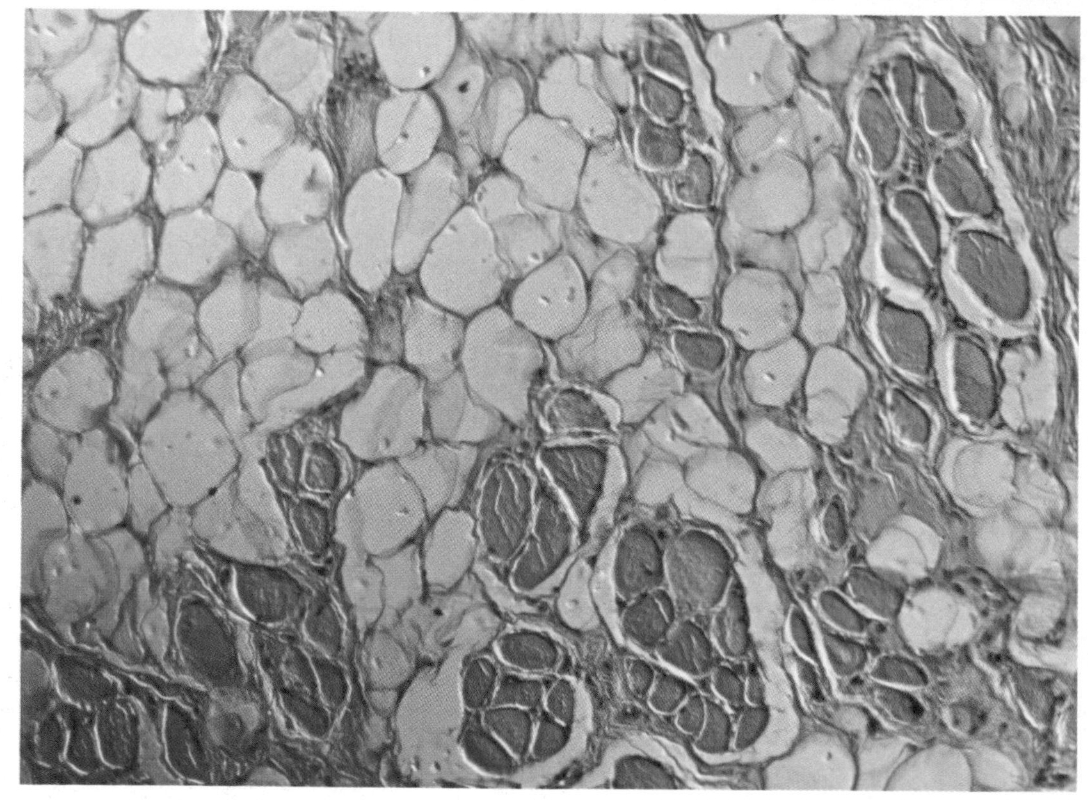

A nerve cell
The photo on the right-hand page shows a nerve cell magnified 8,000 times in a scanning electron microscope. This particular nerve cell is a Purkinje cell from the cortex of the cerebellum. A thick extension from the cell body, called a process, divides into finer branches called dendrites. The dendrites are in contact with other nerve cells. Stimulation reaching the nerve cell body through the dendritic network can in turn activate the cell to transmit a nerve impulse along another elongated extension, its axon (not visible in this picture).

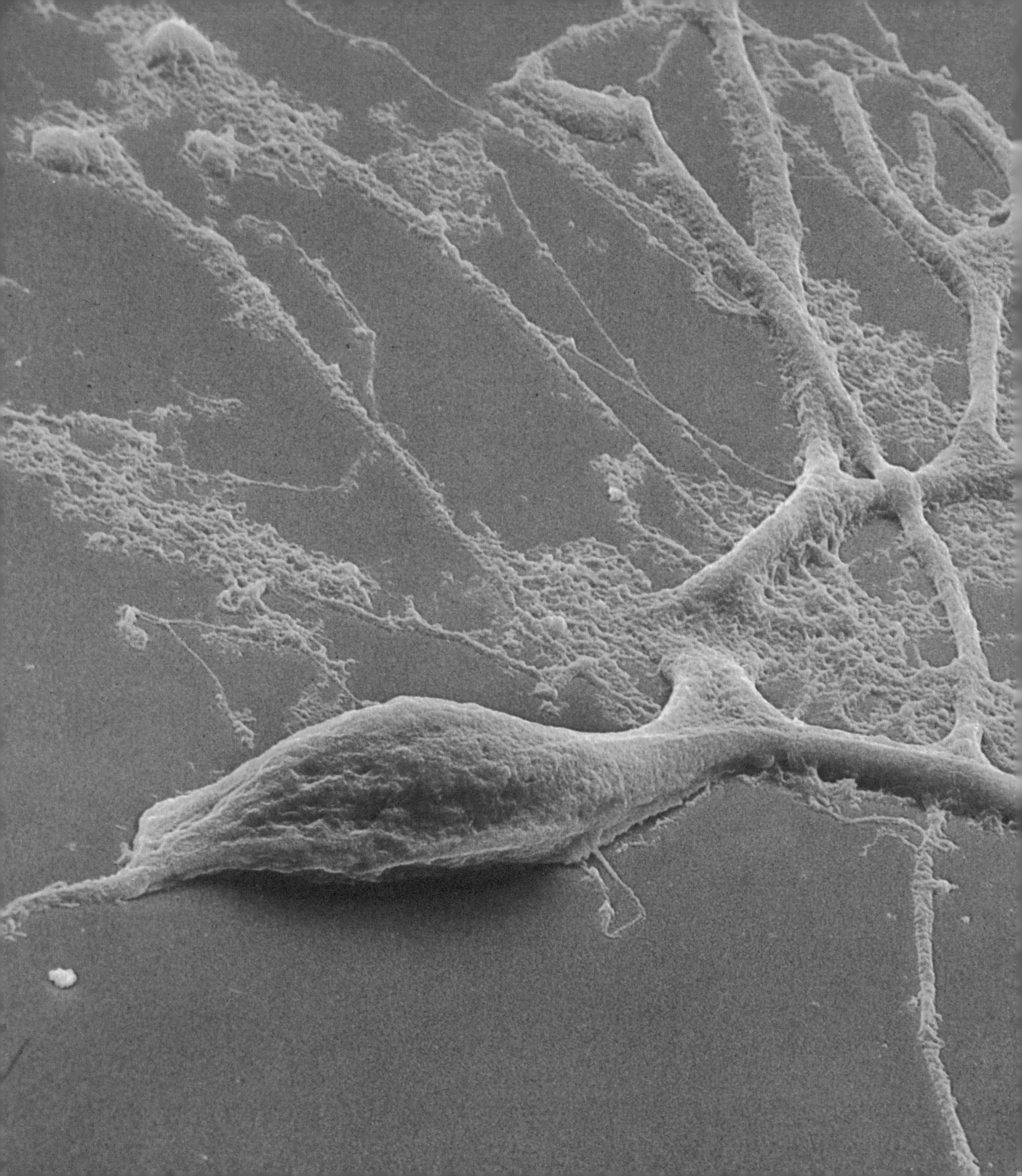

Cartilage cells
In this cross section of cartilage tissue, the nuclei of the cartilage cells appear as dark areas. They are surrounded by a whitish ground substance which contains tensile and elastic fibers. These characteristics make cartilage a major supporting tissue of the body. Cartilage acts as a shock absorber between vertebrae, a friction-bearing surface over joints, and as flexible supporting tissue in the external ears and the nose, for example.

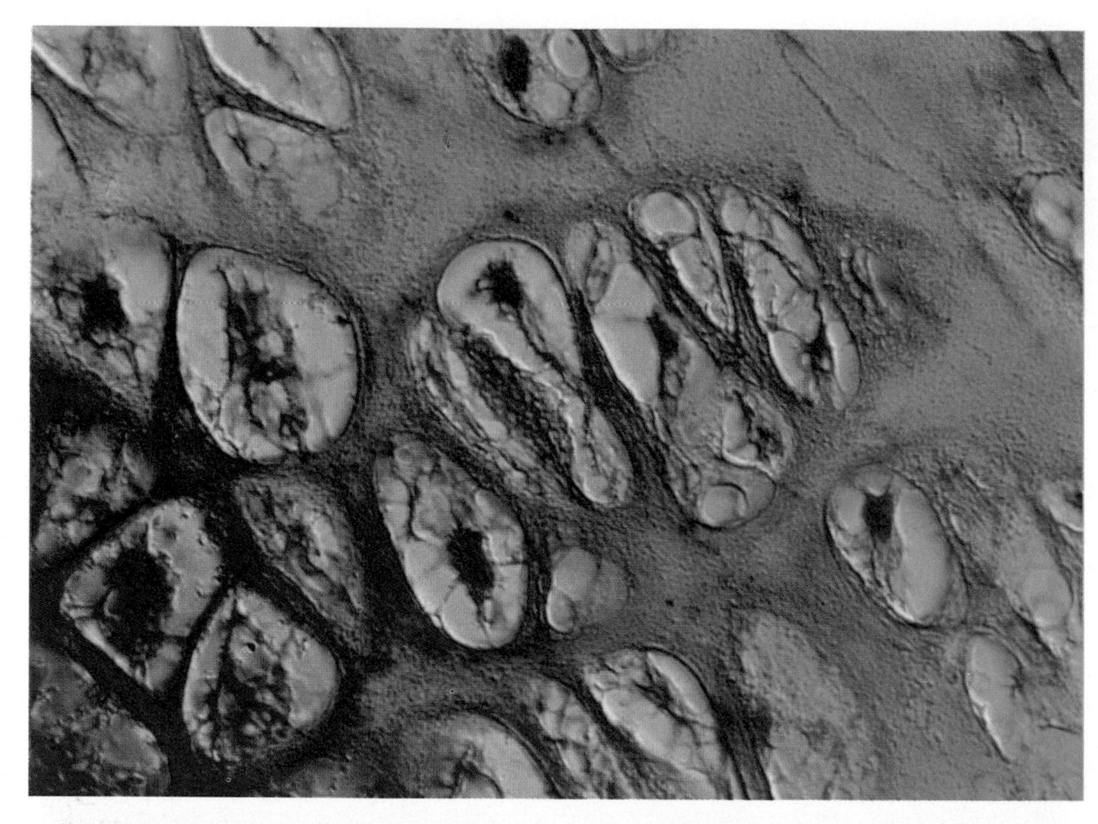

Bone cells
Bone tissue is not dead. It contains cells whose nuclei control the process of calcification in skeletal bones. The cell body is richly branched and, in this cross section through the bone, the cells look like spiders. At the bottom right is a small canal for blood vessels and nerves.

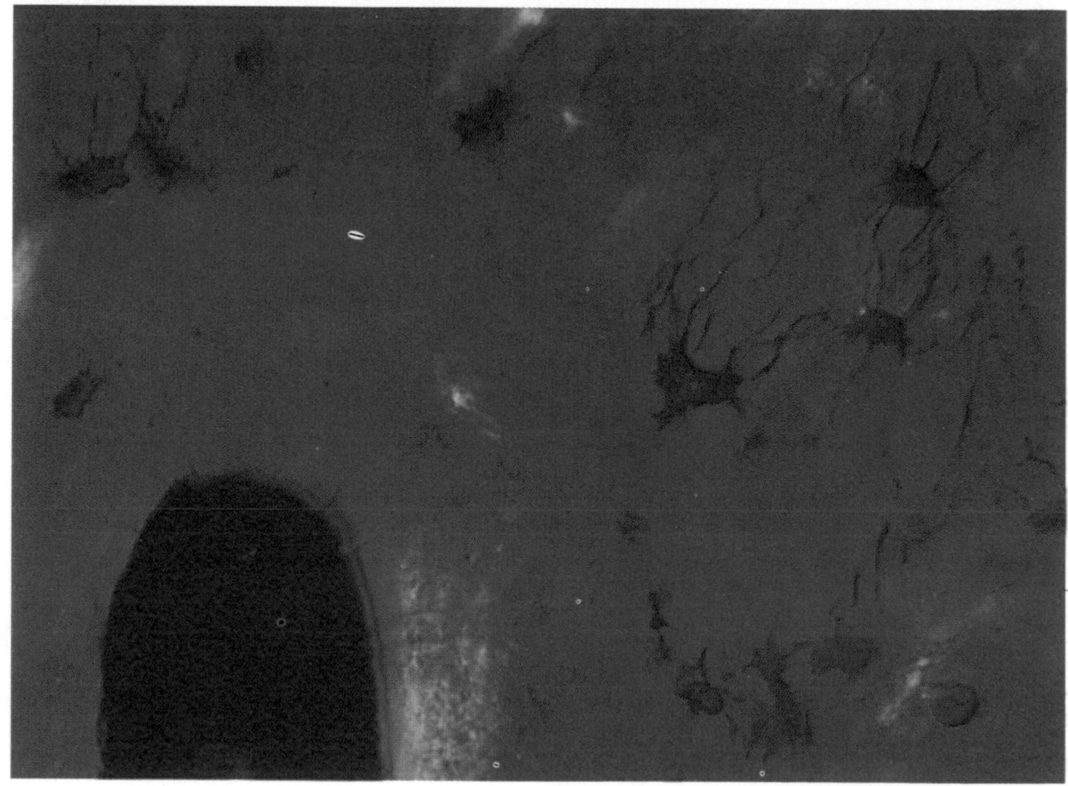

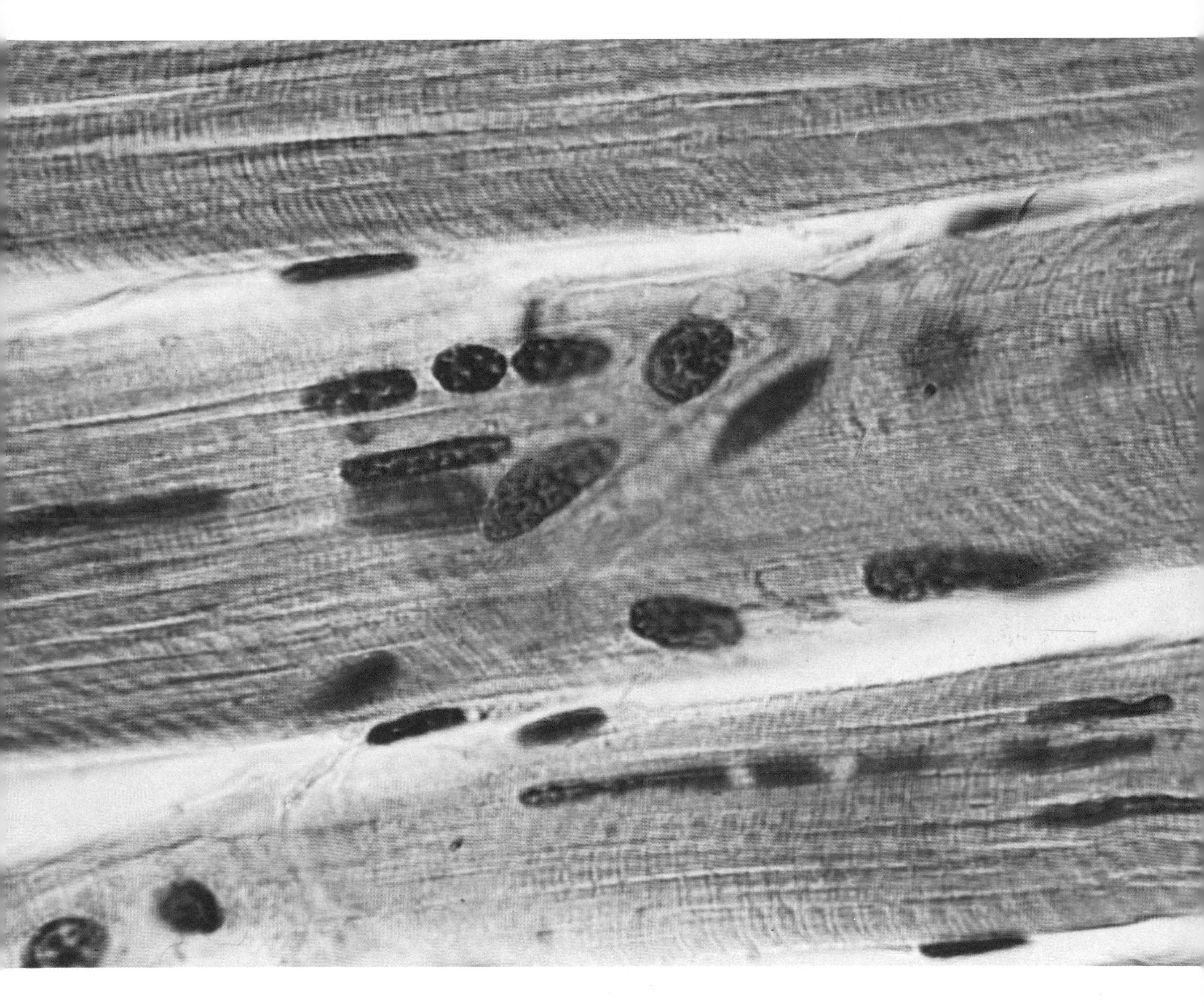

Muscle cells
in a longitudinal section through skeletal muscle. The dark oval structures are cell nuclei, whose granular nature can be seen in a few cells. Muscle cells contain special proteins which can be extended or shortened when the muscle relaxes or contracts — a process controlled by the central nervous system. The orderly arrangement of the protein molecules accounts for the muscle's striped or striated appearance. Muscle cells in interior organs — those lining the walls of the intestines or blood vessels, for example — lack these striations and are called smooth muscle cells.

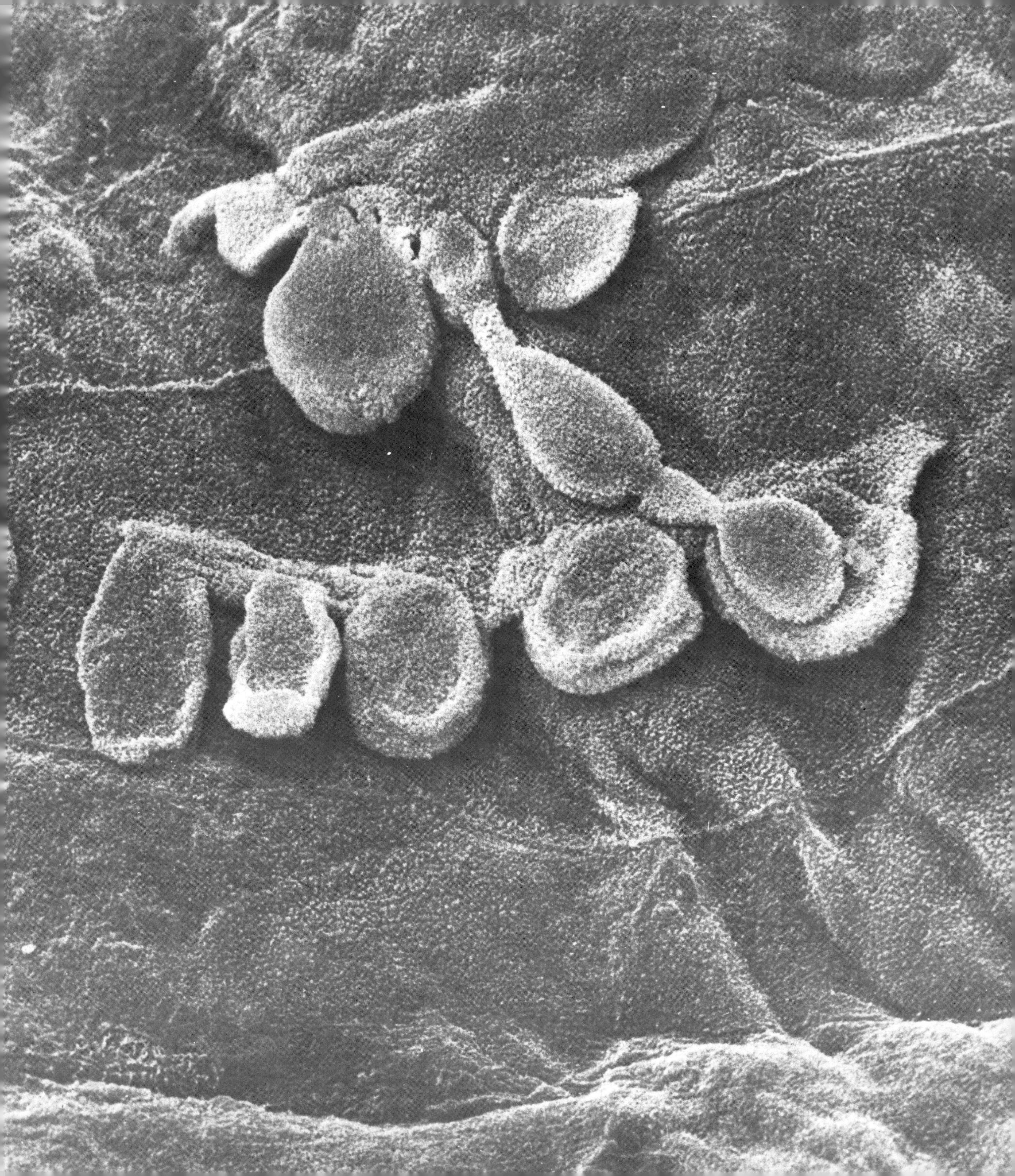

Epithelial cells
The outer surface of the skin, as well as mucous membranes in the body, consists of one or more layers of regularly arranged cells called epithelial cells. All epithelial cells are nucleated, and inside the nucleus are granular structures containing molecules of deoxyribonucleic acid (DNA) and ribonucleic acid (RNA). Epithelial cells are generally strongly attached to one another so that epithelial tissue forms a soft, smooth, and protective covering.

The photo on the left shows the skin of a four-month-old fetus, magnified 11,000 times in a scanning electron microscope. At this stage of development the outermost layer of epidermis is still alive. The cells have not hardened to form the keratinized dead layer characteristic of skin later in life. The photo shows the cells are still in close connection.

An enlargement of an epithelial cell from the skin of the same fetus is shown at the top right. The nucleus appears as a small elevation in the center of the cell. The surface of the cell is covered with small extensions, microvilli. These can be seen in living cells.

Epithelial-lined tissue from an adult is shown at bottom right, magnified 1,600 times in a light microscope. The cell membranes and nuclei are clearly visible.

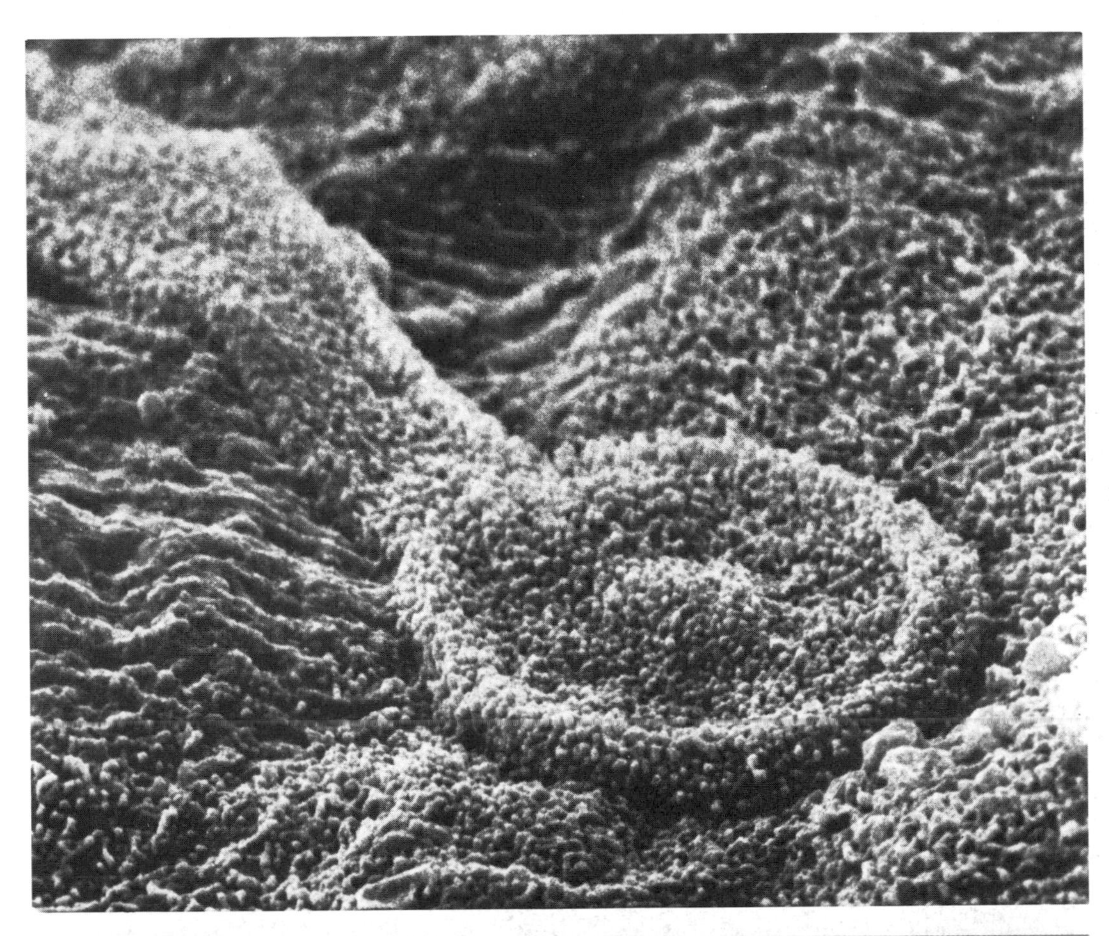

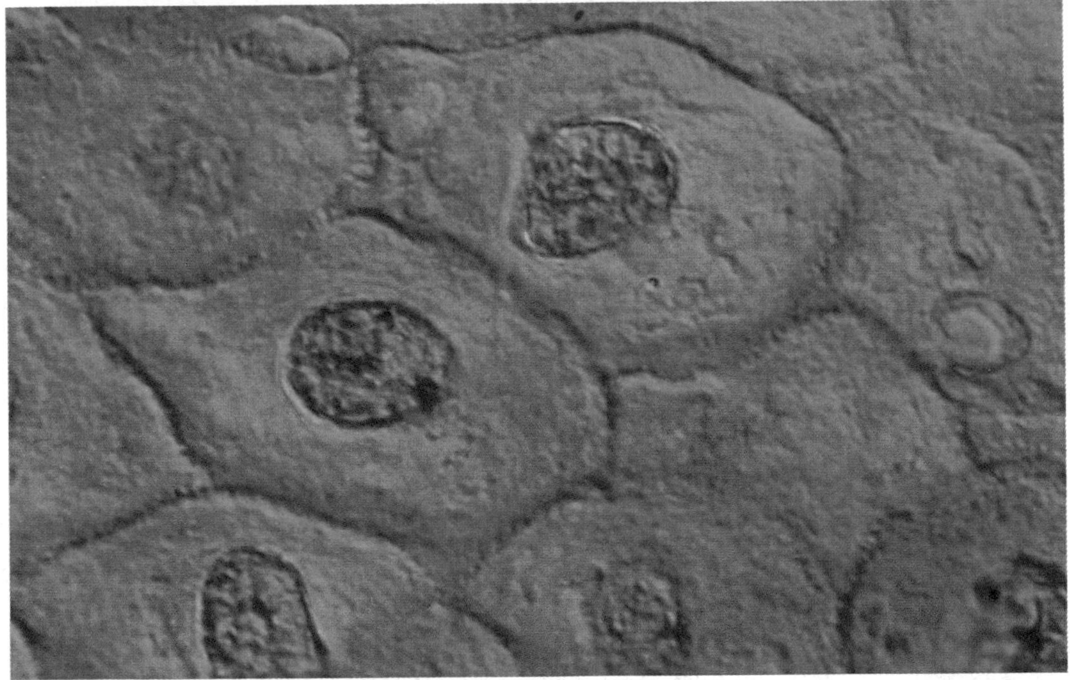

Ciliated cells from the fallopian tube
Many cells in the body are highly specialized. Shown here are cells lining the inner surface of the fallopian tube in an adult female. Hair-like extensions or processes, cilia, project from the cell surfaces and face inward toward the center of the tube. These cilia, moving in sweeping motions, can transport fluid or mucus down the tube. In particular, the fallopian cilia move the ovum through the tube toward the uterus, where development of a new individual takes place.

A sperm cell in the fallopian tube
in precisely the region where fertilization most often occurs. The number of sperm cells ejaculated at coitus is extremely high. Very few of them manage to pass from the vagina through the cervix to reach this point. As the passage is so long and difficult, it is extraordinary to observe a sperm cell at this point. Its tail is making the swimming motions that move the cell forward against the downstream flow of mucus, which is produced by the cilia in the epithelium of the fallopian tube. Some cilia are visible behind the cell.

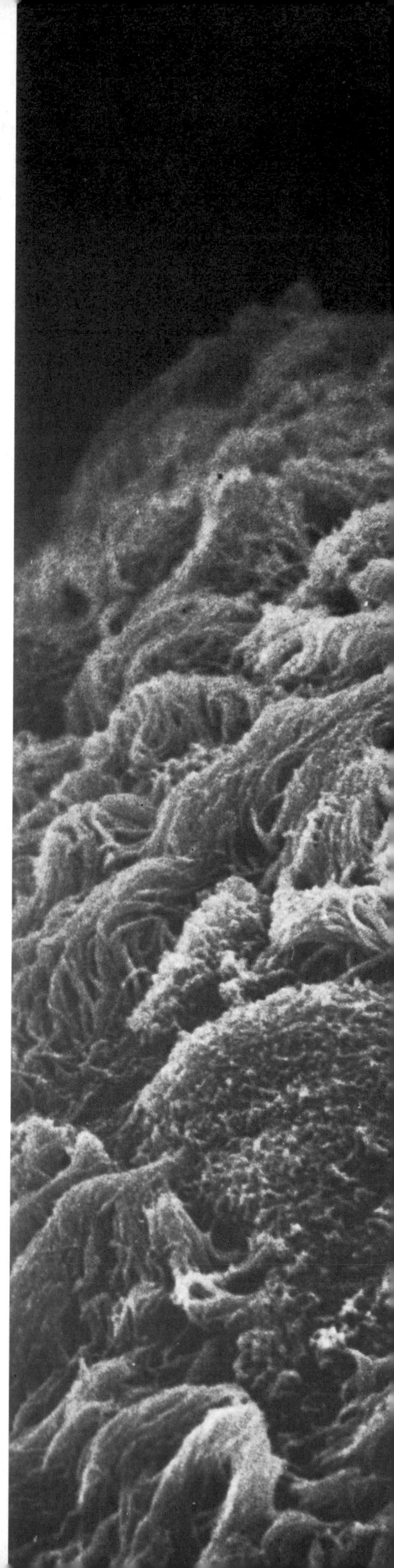

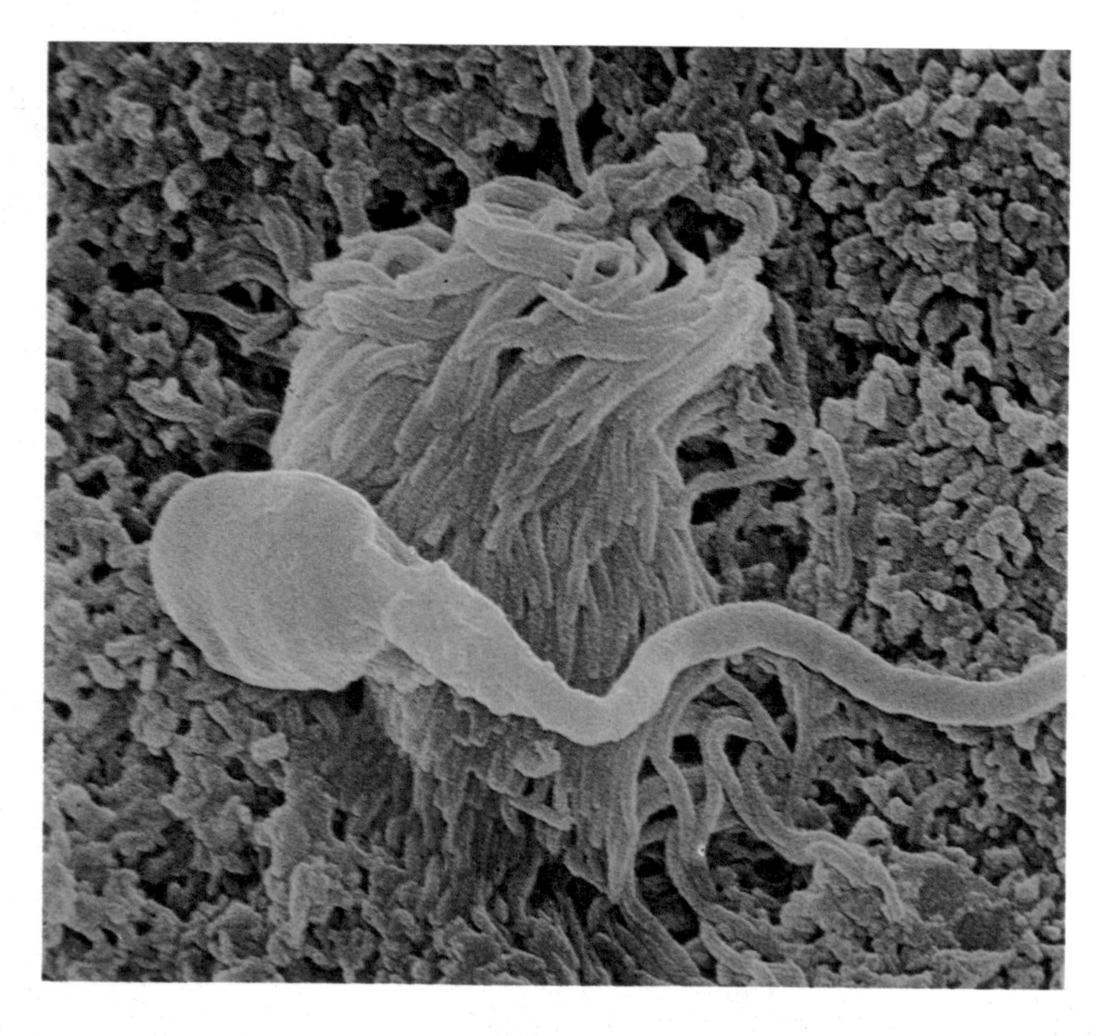

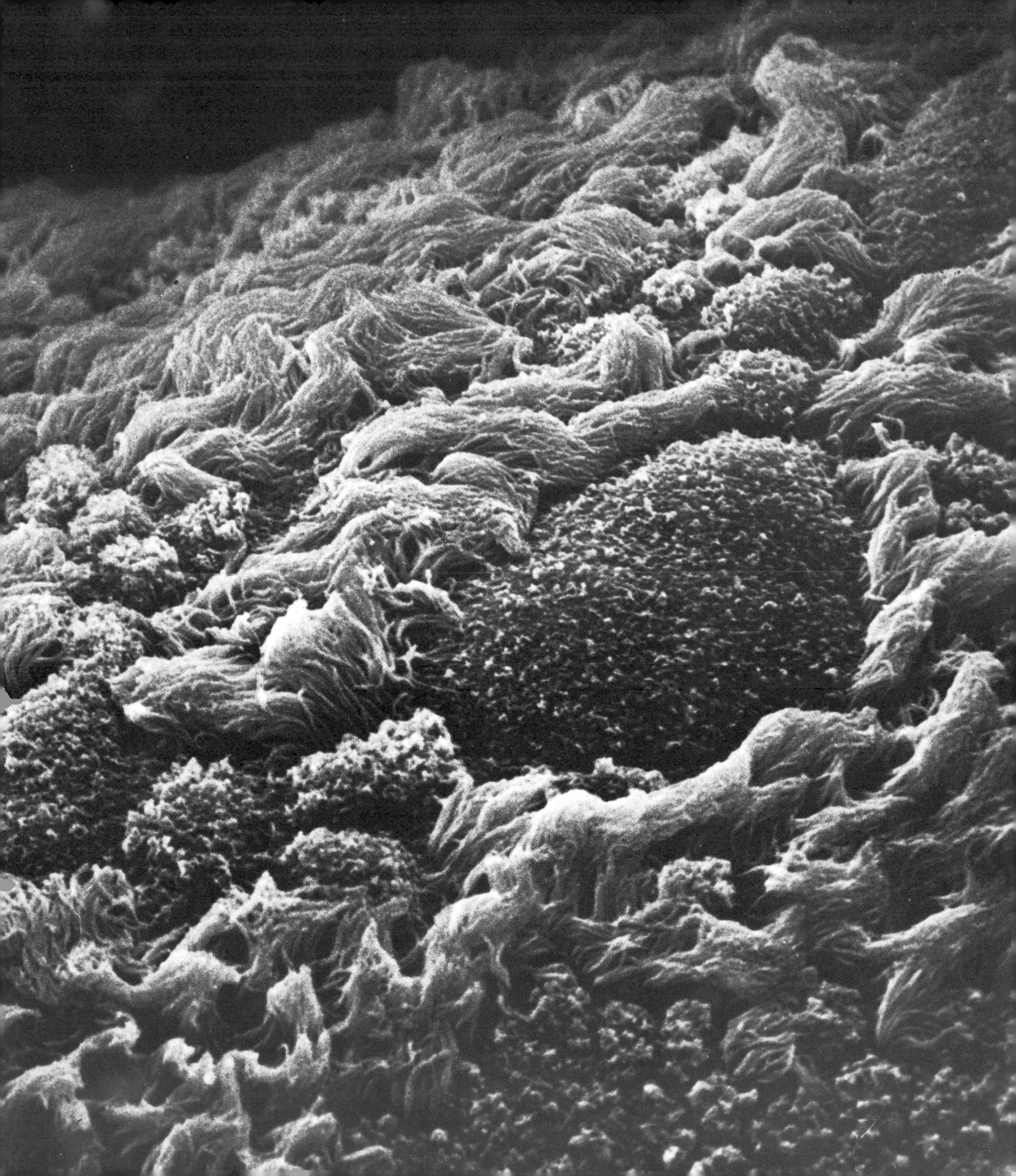

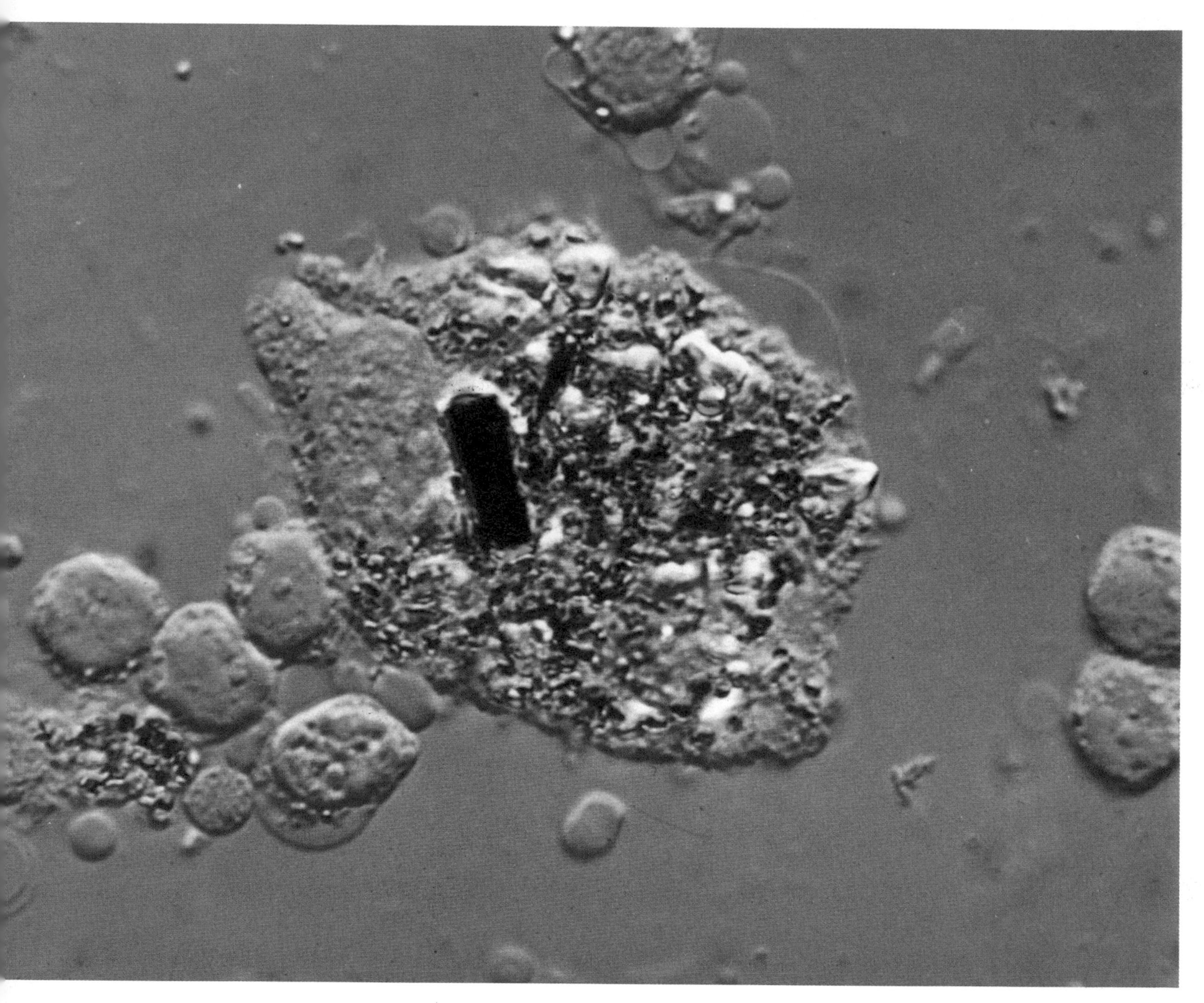

A macrophage
is a large connective tissue cell. This type of cell is present everywhere in the body and is concerned with sanitation or refuse removal. The macrophage in the center is surrounded by white blood cells. A second macrophage is visible at bottom left. Macrophages and white blood cells ingest foreign matter. In both macrophages such "dirt" appears as black dots. The large black rectangular structure is probably a silicon fragment. The macrophages shown here were found in lung tissue. One of their functions is to take care of the dirt in the air we breathe.

Cellular Organization

Tissues and Organs

The human body exhibits a variety of functions, in which different organs participate. If we examine the nature of these functions and note the chief organs or organ systems involved, we obtain a list something like the following:

the intake of nourishment and metabolism — the digestive apparatus (but also every cell)
respiration — the lungs
excretion — the kidneys and liver
circulation — the heart and blood vessels, the lymph system
supporting apparatus — skeleton, cartilage, connective tissue, skin
movement — muscles
reactive capacity, reasoning, regulation, and coordination — the brain, other parts of the nervous system, the sense organs
regulation and coordination — the endocrine glands
growth and reproduction — internal and external genital system (but also individual cells)

These different organs and systems are integrated into the "cell society" that is represented by the body as a whole. The functioning of the body depends on the individual parts, and the parts in turn are constructed so that they work together to serve the whole body. By belonging to a body — with all the duties this involves — the organs are assured of living conditions which will promote their survival. Single cells and organs have therefore become specialized and organized into systems.

The analogy between human society and the human body can be extended further to clarify some current theories on body function. As individuals, we work together in groups, in firms, and in other organizations, and these organizations in turn cooperate within the framework of society, becoming responsible for society's many different activities. Society as such makes it possible for individuals and groups to survive, thus ensuring that society itself can continue. Cells are comparable to individuals, organs to organizations, and the human body as a whole to human society.

Fetal development provides a fine illustration of the way individual cells unite to form tissues and organs — how they "organize." One fertilized cell, the ovum, divides into two, four, eight, sixteen, thirty-two, eventually into thousands of millions of cells. Even when the total cell number is only a few hundred thousand, the primordial extremities and organs of the microscopically small embryo can be seen.

The body's main organs and systems develop from one or more of the three primitive "germ layers" which make up embryonic tissue. The outer germ layer on the back of the embryo, the ectoderm, contains groups of cells which fold in toward each other down the center to form a neural tube. At the upper or head end of the tube, the brain begins to take shape, above the primitive face of the embryo. The inner germ layer at the front of the embryo is the entoderm. From here arise the primordia of the digestive tract and the respiratory organs. The mesoderm, the germ layer between the ectoderm and the entoderm, is the site of development for bones, cartilage, muscles, and also the vascular system and the heart.

Our knowledge of the forces that govern the development of a single fertilized ovum into the millions of cells distributed throughout the body and performing a variety of functions is still incomplete. We know that the program of development is contained in the molecules of the genes. The information coded there determines whether a newly formed life is going to be male or female, short or tall, what shape the head and face will be, whether hair will be blond or dark, eyes blue or brown, and so on. The program also determines the development of the interior organs, indicating where in the chest the heart will be, what curves the stomach will have, how large the lobes of the liver will be, and so on. The genes determine too the size of the brain and its basic structure, that is, how the nerve cells are arranged to communicate with each other. This structure in turn forms the basis for the psychological life of the individual.

Digestion and Excretion

Monocellular organisms live in fluids from which they derive the substances they need. Cells organized into larger aggregates or organs likewise live in fluid — the extracellular fluid. Most cells are stationary and function in close collaboration with the cells surrounding them. A large number are movable, however, such as cells circulating in the blood or lymph vessels. Body cells

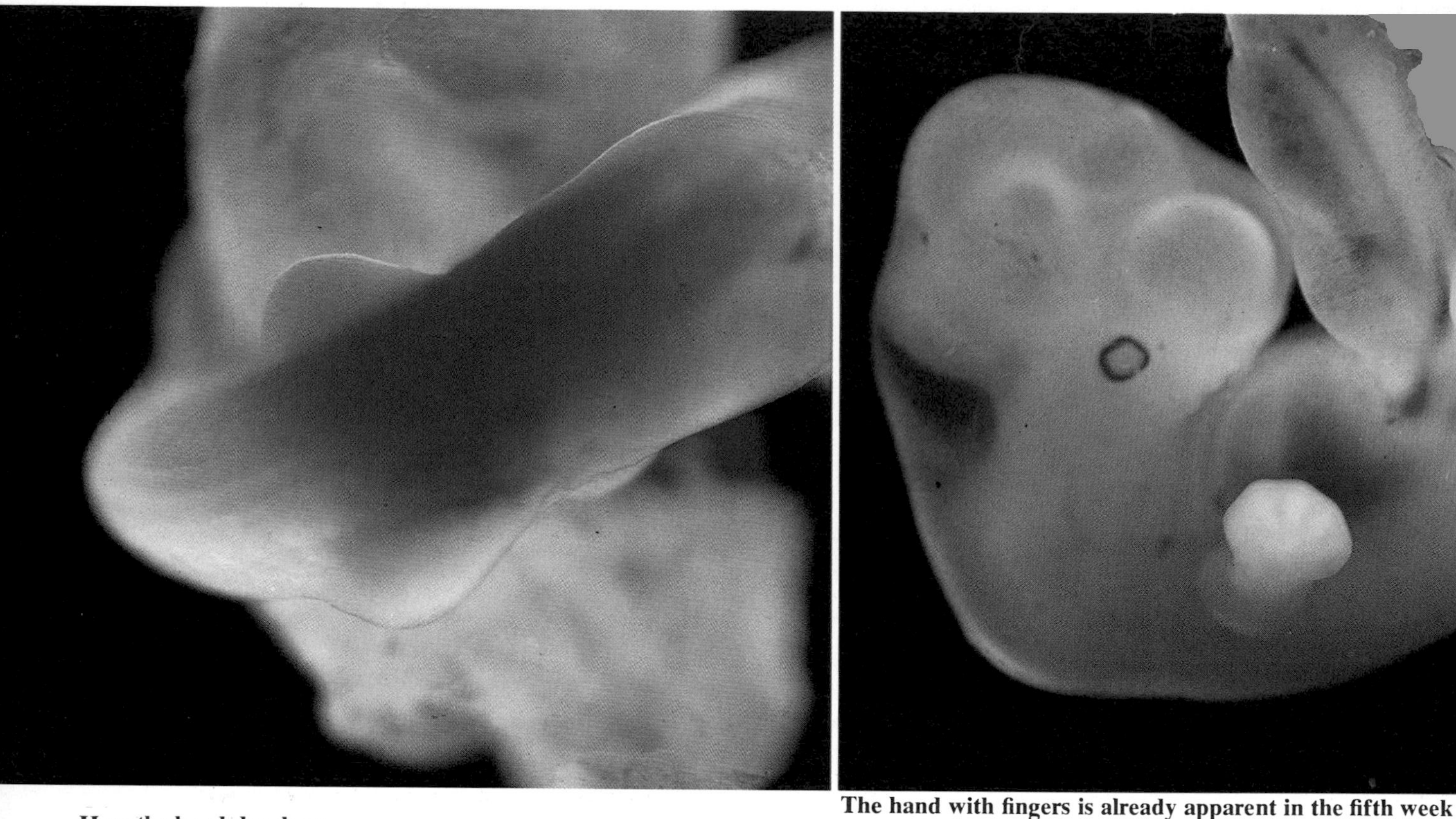
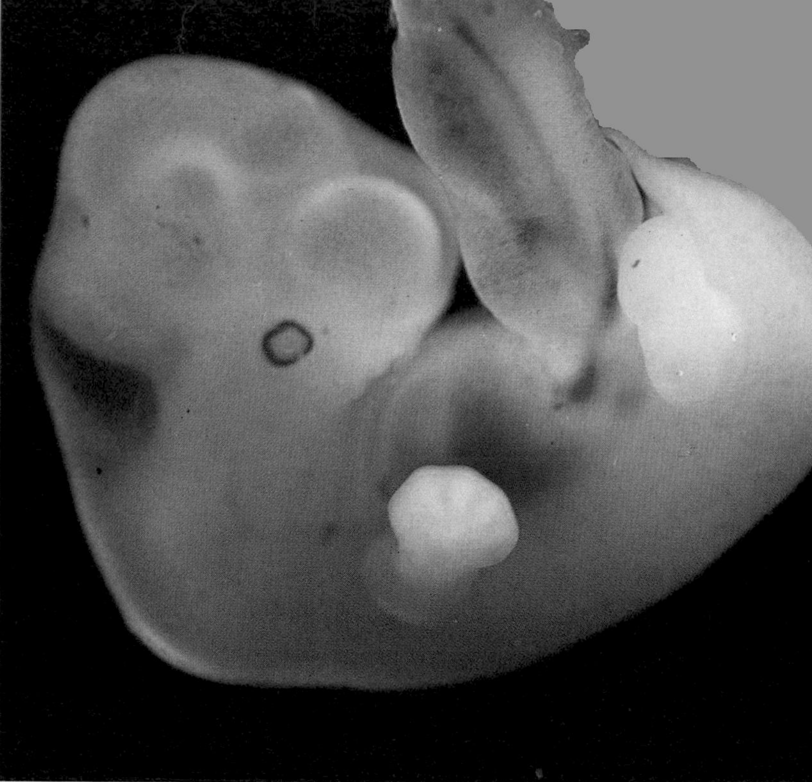

The hand with fingers is already apparent in the fifth week

How the hand develops
The development of the various parts of the body and the organs is programmed in the genes. This series of pictures illustrates how the hand and the arm develop from a small protuberance on the side of the embryo during the fourth week of gestation. At this stage, the somites (small masses of mesoderm) on the back of the embryo are visible. During the fifth week, a tiny hand with primordial fingers appears. Development continues, and by the eighteenth week the human hand exhibits its characteristic opposable thumb. Primordial fingernails are clearly visible. By the twentieth week the hand and the arm are well developed, and the blood vessels supplying the new extremity can also be seen.

normally exist in an adequate environment which supplies appropriate nourishment. The digestive apparatus breaks down the food we eat so that it can be readily absorbed by body tissue. The circulating blood then transports these products of digestion to cells to serve the cells' metabolic needs.

In its passage from mouth through pharynx, stomach, and intestines, food is broken down into simple chemical compounds that the body can use for growth, maintenance, or repair. Only simple carbohydrates and salts can be taken up directly. Digestive enzymes — complex protein molecules — are needed to break down the larger molecules of fat or protein. An adequate diet must contain salts and vitamins. Digestion also requires a great quantity of fluid. This is supplied in part by what we drink, in part by salivary, gastric, pancreatic, and intestinal secretions, and bile from the liver. Most of this fluid is reabsorbed in the colon (the lower intestine) prior to the exit of indigestible residues from the body through the anus.

After the digestive apparatus has prepared the nutrients, they are distributed to the body's cells via the blood and lymph systems. Nutrients supply the cell with the fuel and building blocks it needs to carry on its own metabolic processes. Oxidation, or burning of the fuel, provides the energy for cellular movement, growth, or reproduction. These vital processes themselves result in waste products which must be removed. The wastes pass into the bloodstream or lymph channels and eventually reach the kidneys, which are the main filtering stations in the body. Here the often poisonous residues are removed from the blood and leave the body with the urine.

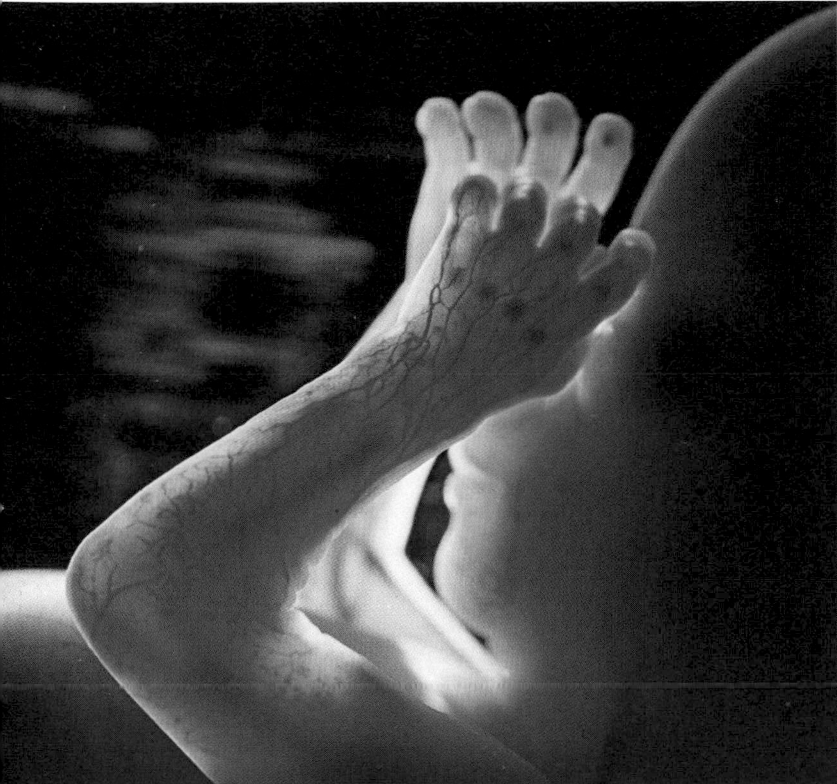

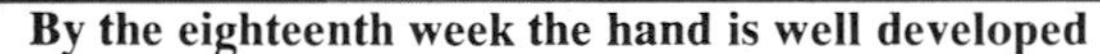

By the eighteenth week the hand is well developed

In the twentieth week the thumb can oppose the other fingers

Respiration

Oxidative processes in the cell require the presence of oxygen. The source for this oxygen is ultimately the air we breathe. Red blood cells coming into contact with air in the lungs are able to take up the oxygen molecules. Only two cell membranes separate the cells from air, and the oxygen molecules diffuse across these membranes, become bound to the red cells, and are thus transported throughout the body. Red cells also bind and transport carbon dioxide (CO_2), the waste product of cellular oxidation. The CO_2 is returned to the lungs, where it is excreted in exhaled air.

Blood and Lymph

Blood circulating in the body comes into contact with most cells. It is kept in motion by that reliable pump, the heart. In order to reach individual cells, blood is conveyed in very small, thin blood vessels, the capillaries. Some blood plasma (the fluid part of the blood) seeps out of the vessels to become part of the fluid which fills the spaces, or interstices, between cells in any tissue. This interstitial fluid is present everywhere in the body. The greater part of the fluid moves back into the veins and becomes part of the circulatory system again. If too much blood is pooled somewhere, it is drained off by lymphatic vessels. The fluid portion enters their widespread network and is gradually returned to the blood. In its circuit through the body, lymph passes through lymph nodes (often erroneously called lymph glands). There, special white blood cells ("lymphocytes" are one such type) remove bacteria or other harmful substances.

Skin, Skeleton, and Muscles

Skin, connective tissue, cartilage, and bones form the supporting apparatus of the body.

Connective tissue, because it is soft and stretchable but also durable, serves many vital functions in the body. It forms a strong and protective layer under the skin. The skin covering the body protects and preserves the internal environment and helps to keep it constant. Connective tissue also contains fat cells, which, besides being food depositories, insulate against cold, and, because of their softness, function as shock absorbers.

The "connective" nature of connective tissue becomes evident under the microscope. The wrappings or "capsules" that surround the body's organs consist of connective tissue. Similarly, the walls enclosing blood vessels are composed of connective tissue, and it is connective tissue that serves to attach the vessels to the tissue they pass through. Many parts of organs, such as the lobes of the liver or the lung, are encased in layers of connective tissue, and connective tissue also surrounds bundles of individual nerve fibers, making them into smaller or larger "cables." Muscle cells in the limbs are similarly divided into

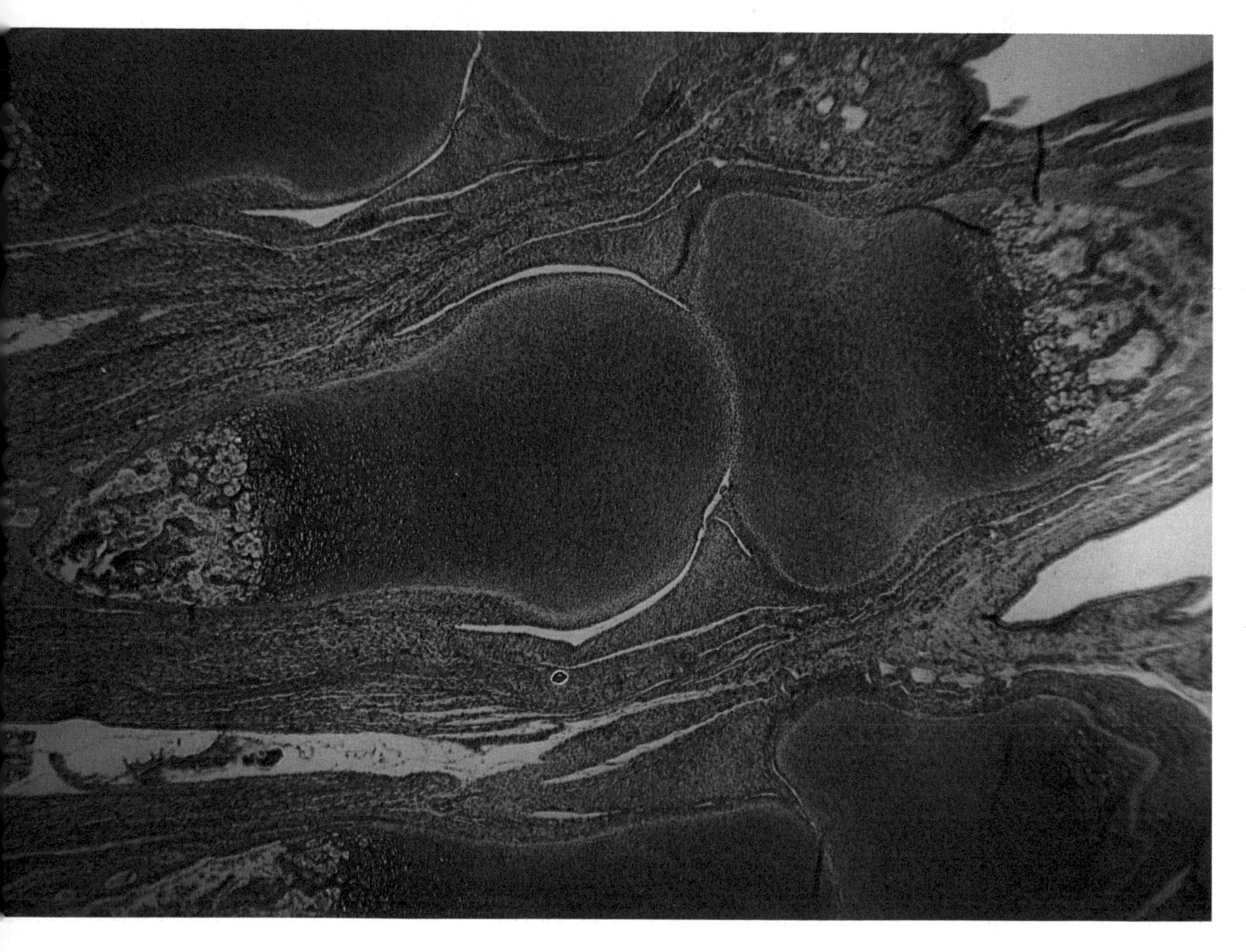

bundles that are covered by elastic connective tissue. At the ends of the muscles, these coatings come together to form tough tendons, which are attached to the bone shaft by means of a capsule of connective tissue which covers the bone, the periosteum.

Cartilage and bone cells are specialized cells in the connective tissue system. They can form molecules capable of enduring compressive forces or mechanical strain. The hardness of bones comes about through the deposit of calcium salts in bone cells. This hardness enables the bones of the skeleton to bear the weight of the body and more, whether at rest or in motion.

Under the microscope, the majority of muscle cells look striped or striated. The striations are due to the regular arrangement of the protein molecules in the cells. These molecules move closer together and thus shorten the muscle's length when the muscle contracts. Other types of muscles, those in the interior organs, for example, lack these striations and are called smooth muscles. Striated muscles are also called skeletal muscles. They are subject to voluntary control, while smooth muscles generally are not.

Bone formation
The development of the skeleton is a process in which connective tissue cells, cartilage cells, and bone cells themselves collaborate. Skeletal bones are first indicated in connective tissue, which is subsequently converted into cartilage. Bone cells then appear in the cartilage and mediate the deposition of calcium salts in particular growing zones. Generally, these zones are located near the ends of tubular bones. This photo of a bone from the hand shows the growth zones along with connective tissue, cartilage, and bone cells.

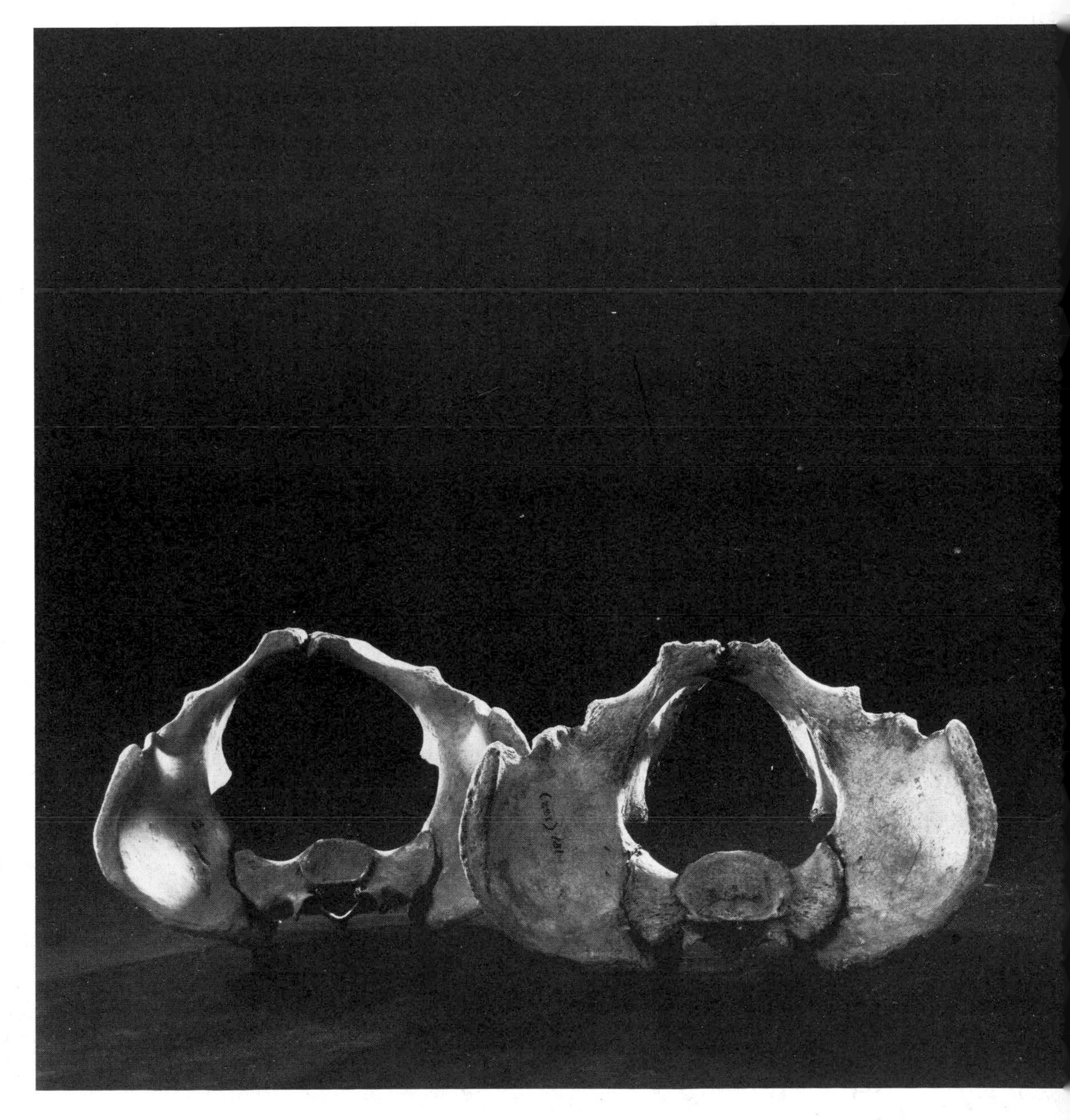

The pelvis in an adult man and woman
Among the differences between males and females are certain characteristics of the skeleton. The female pelvis is constructed to accommodate the birth process. For this reason, the pelvic opening in females is wider than in males. This permits the normal passage of the baby during delivery.

These two pelvises were found at excavations at a medieval gravesite on the island of Frösön near Östersund in Sweden. They are about a thousand years old. In archaeologic excavations the structure of preserved pelvic bones is used as one of the criteria to determine the sex of the remains.

Human skeletal remains over nine thousand years old have been found that are equally well preserved. This shows how resistant to decomposition bone tissue can be.

The Sense Organs

Observable activities, such as movements, facial gestures, and speech, together with internal functions, such as endocrine secretions, heart rate, and respiratory volume, are part of the behavioral repertoire of the individual. Behavior in this sense can be considered the end-product of a number of complex events that are set in motion when sensory input is received by the sense organs and coded according to the nature of the nervous system. The sense organs translate a number of the physicochemical properties of the environment into nerve impulses. The impulses constitute the language of the nerve cells, or neurons, which are the basic elements of the nervous system.

The receptor cells that respond to pressure, touch, vibrations, muscle stretch, sound waves, or postural stimulations are called mechanoreceptors. They are found throughout the body in skin, muscles, and tendons, but also at specific sites in the inner ear which contain the centers for the senses of hearing and balance. Visible light, that is, electromagnetic waves with a wavelength between four thousand and seven thousand ångstrom, provides the stimulation for the visual receptor cells, the cones and rods in the retina. The chemical senses, taste and smell, translate the properties of various molecules into sweet, sour, salt, and bitter if they are present in the mouth, or into a variety of smells if they are present in the surrounding air. The taste receptors are located in the taste buds on the tongue; the receptors for smell lie in the olfactory mucous membrane inside the upper part of the nose. Special sensory receptors inside the body respond to certain internal changes. Receptors in the walls of certain blood vessels, for example, are stimulated by changes in blood pressure or in the carbon dioxide content of the blood.

Much of what we see, hear, smell, feel, or experience in one way or another through our senses, especially through the receptors in the interior organs, may not be consciously perceived. What reaches consciousness is often something new or different, something that presents a problem, that tempts us, or that gives rise to needs or desires. Pain sensations, however, are of a special nature. They appear to be generated in nonspecific nerve endings or in specific receptors which are usually stimulated by touch and pressure. Pain receptors are present all over the body, in the skin as well as in the interior organs, in muscles, and around joints. They are guards that report when the body or a part of it is injured by pressure or piercing, by blows, by chemical agents, by a lack of oxygen, or by extremes in temperature.

The Nervous System

The nervous system controls and integrates the entire activity of the body. In man, the nervous system consists of a peripheral division, with nerve fiber and groups of neurons (ganglia) located in different parts of the body, and the central nervous system, which includes the spinal cord, the brain stem (with the cerebellum), and the cerebrum. Sensory impulses from peripheral receptors are conducted to appropriate central nervous system centers in a few thousandths of a second. Some impulses may lead to muscle contractions that are triggered at the spinal cord level. Other impulses are conveyed to the higher areas in the brain stem and the cerebrum. There the decision may be made to ignore or to respond to the incoming information by a change in behavior. Once received, sensations can be stored in the central nervous system as memory, and recollected when needed. The nervous system also generates needs, or feelings that the body requires something — food, water, more heat or cold, the desire to defecate or micturate, or to have sexual contact. At a higher level, personal and social needs can also be developed — the child's need for his parents, the adult's need for emotional and social contact, for self-realization, achievement, success, competition, and so forth. The question of which of these and other needs are inborn in man and which derive from his cultural upbringing remains open.

Nerve cells could be described as having their own special need — a need for stimulation. In the course of satisfying this need the nervous system is supplied with a constant stream of information about the nature of the environment. The state of consciousness, while itself still little understood, involves perceiving oneself as a separate being, and this requires a continuous supply of sensory input. Such input makes us aware of our surroundings, even in its most trivial aspects. Nerve endings in the interior organs also send a constant stream of impulses to the brain, more or less informing it that the body is still alive and that everything is in order. A marked reduction in this stream of background impulses, this background "noise," seriously affects cerebral functioning. Self-consciousness, or a sense of identity, may begin to fade, and the individual may feel threatened or anxious. Exactly how or where consciousness arises in the nervous system is not understood, although the phenomenon appears to involve interactions of the cortex with nerve cells lower down in the brain stem.

Sense organs
of all shapes and variety are present throughout the body. They are concentrated on the surface, but also numerous in the interior organs. The bulb-shaped structures in this papilla (literally, a nipple-shaped protuberance) from the tongue are the sensory receptors for taste, called taste buds. The papillary mounds give the tongue its rough surface. This helps to mix food in the mouth and carry it back toward the pharynx.

Cellular Organization

Reproduction

The survival of a species depends on the ability of its members to reproduce successfully. Procreative powers are reflected in the division of individual cells, in the fertilization of an ovum, in a variety of growth processes, and in the development of the reproductive organs along with appropriate sexual behavior.

Male and female sex organs develop from identical structures, and at an early embryonic stage there is no observable differentiation into female or male organs. The embryo normally develops later into either a male or a female according to the sex chromosomes it has inherited. All cells in the body contain the two sex chromosomes, and in cases where physical development gives rise to uncertainty, genetic sex is determined by a chromosomal analysis of cells which may conveniently be taken from the oral cavity.

The process of reproduction begins with the fertilization of an ovum, when the female cell is united with a male sperm cell. Soon after fertilization, the ovum begins to divide and moves down the fallopian tube. Subsequently, it attaches to the wall of the uterus, where the process of growth and division will continue to the point where the fetus is ready to be delivered.

The unfertilized ovum and the sperm cell contain the chromosomes, which carry the genes which will program fetal development as well as exert powerful influences on the adult. During normal cell division, where one cell gives rise to two new ones, all the chromosome pairs are duplicated and each new cell receives a full complement of chromosomes. The unfertilized ovum

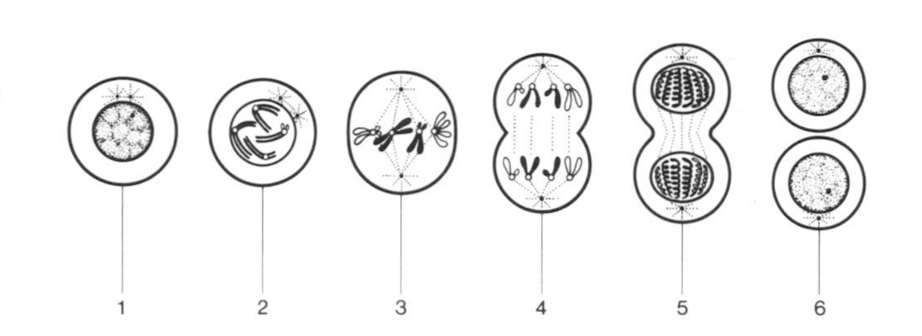

Diagram of normal cell division (mitosis)

1 resting stage
2 the chromosomes in the nucleus double lengthwise
3 the chromosomal pairs are aligned along a plane
4 the pairs separate
5 the new sets of chromosomes gather at opposite poles
6 resting stage: two new cells have formed

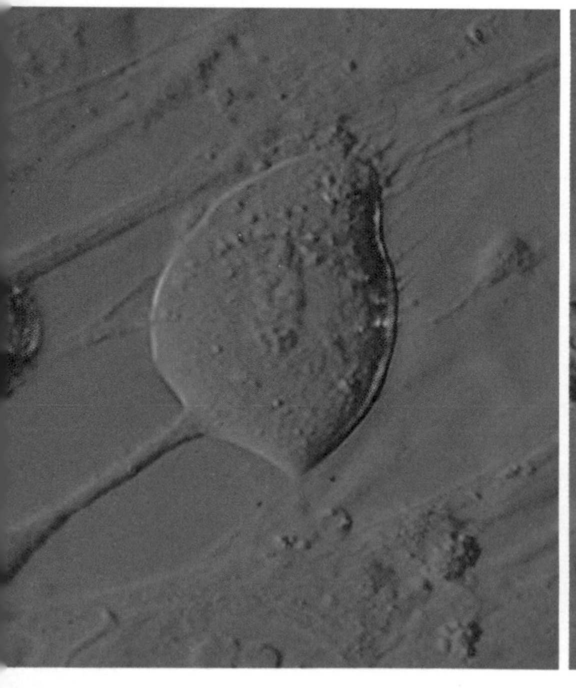

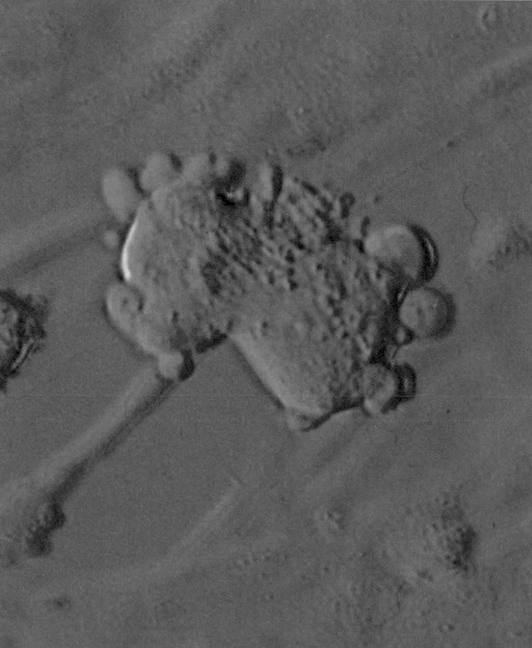

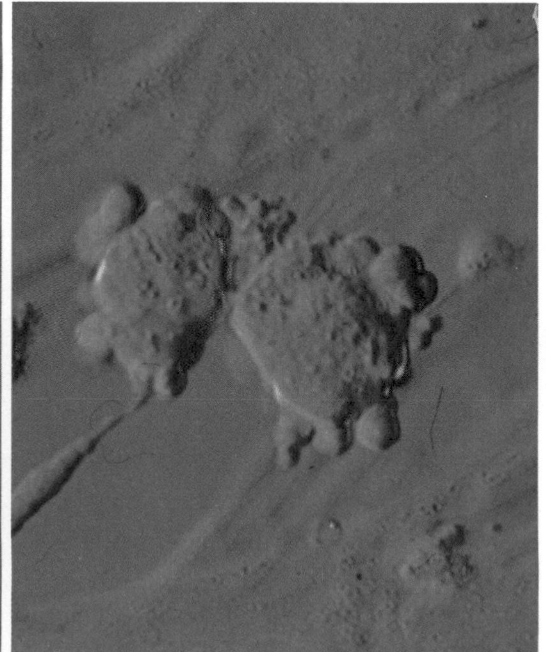

Meiosis ▷
is one of the most important events in sexual reproduction. When ordinary cells in the body reproduce, each of the chromosomes in the parent cell is duplicated and each new cell gets the same number of chromosomes as the original. In sexual reproduction, two cells unite, the ovum and the sperm cell. If both had the full complement of chromosomes, the new individual would have a double number of chromosomes. This does not happen. During the development of ova and sperm cells, the number of chromosomes is halved. This is called meiosis, or meiotic division. The diagrams on the right show the principal stages in the development of sperm cells and ova. The halving of the chromosomes takes place at stages 2 and 8.

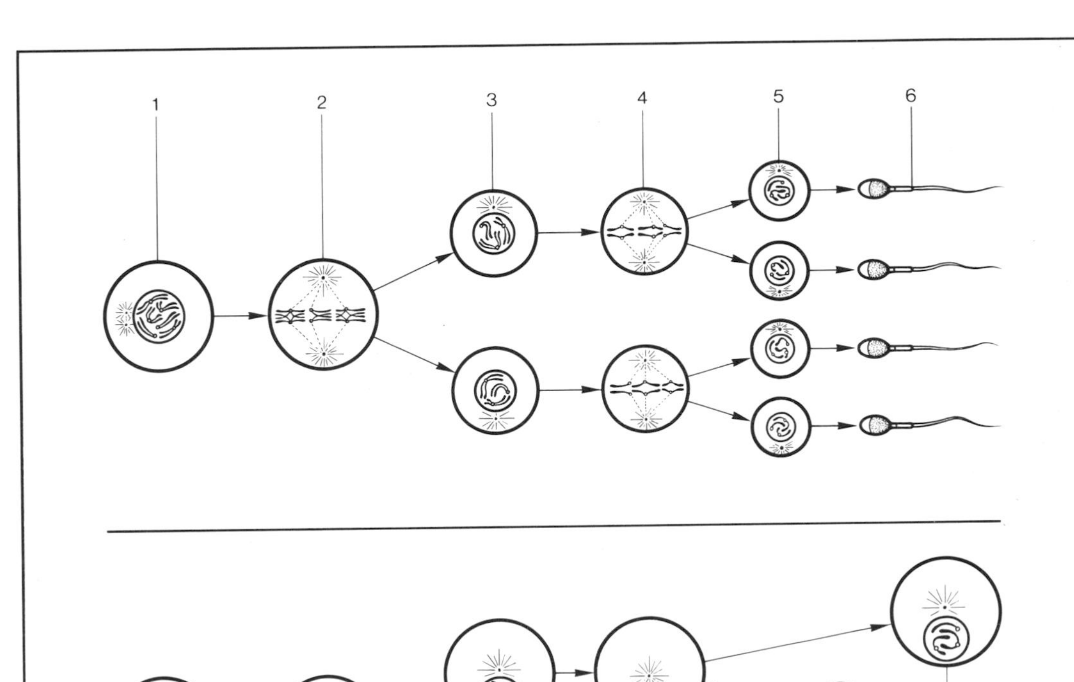

Development of sperm cells (upper diagram)

Stages 1 and 2: Primitive germ cell no. 1, the primary spermatocyte. At stage 1, the chromosomes in the nucleus double lengthwise. At stage 2, the chromosomal pairs are arranged in a plane for meiotic division. This will give rise to two new cells each with half the normal complement of chromosomes.
Stages 3 and 4: Primitive germ cell no. 2, the secondary spermatocyte. At stage 3, the chromosomes in the nucleus again double longitudinally. At stage 4, the chromosomes are aligned in a plane for ordinary cell division, not meiosis.
Stage 5: Four sperm cell precursors, or spermatids, are produced from the original spermatocyte.
Stage 6: Mature sperm cells.

Development of an ovum (lower diagram)

Stages 7 and 8: The primitive germ cell, or the primary oocyte. At stage 7, the chromosomes in the nucleus double lengthwise. At stage 8, the chromosomes are aligned in a plane for meiotic division. This gives rise to a secondary oocyte and the first polar body, each with half the number of chromosomes.
Stages 10 and 12: The second primitive germ cell, or the secondary oocyte. At stage 10, the chromosomes in the nucleus double longitudinally. At stage 12, the chromosomes are aligned in a plane for ordinary cell division, not meiosis.
Stages 9 and 11: First polar bodies. At stage 11, the chromosomes of the first polar body are arranged for ordinary cell division.
Stage 13: Three secondary polar bodies are produced.
Stage 14: Mature ovum.

◁ **One cell becomes two**
The series of photos opposite corresponds to the last two stages outlined in the diagram above the pictures. At far left, the nucleus has already divided into two new nuclei. Next, the newly forming cells bubble and wriggle in order to get free of each other. Finally, they achieve complete separation.

Cellular Organization

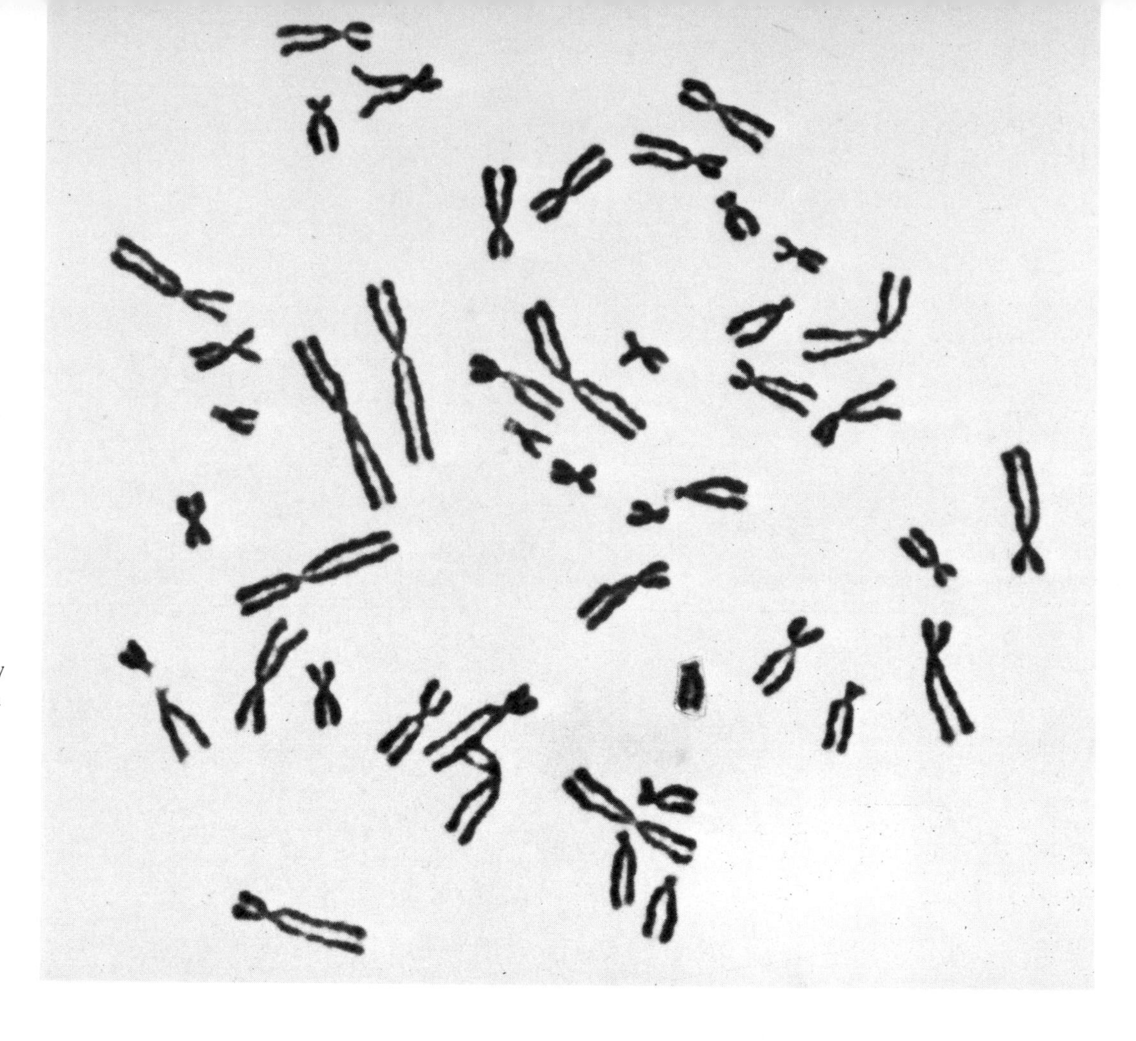

Human chromosomes magnified about 500 times in a light microscope. The molecular structure of the chromosomes determines the properties we inherit. Each body cell has its full complement of chromosomes. When the cell is engaged in its normal activity, the chromosomes lie like long threads in the nucleus. When the cell prepares to divide, the chromosomes assume the shape of short, thick rods. This is the stage illustrated here.

In the process of division, each chromosome divides lengthwise into a pair of identical chromosomes, one for each new cell. The total number of chromosomes in man is forty-six. The individual units of heredity, the genes, are located at fixed sites along the chromosome. Certain genes along the sex chromosomes determine the specific sexual characteristics of the individual.

and the sperm cell each have only half the full number, however, one from each pair. (Otherwise, the fertilized ovum would have a double complement of chromosomes.) In the process of formation of ova and sperm cells, the chromosomes are reduced to half their usual number. This process is called meiosis. In the primitive germ cells — the cells that give rise to an ovum or to a sperm cell — the chromosomes are arranged in pairs. One chromosome in each pair originated from the mother of the individual, the other from the father. In subsequent divisions, an ovum or a sperm cell is produced which contains only one chromosome from each of the pairs. In this way the number of chromosomes in mature ova and sperm cells is reduced to half that in other body cells. Since the chromosomes in each sex cell represent a random selection of the heritable properties contained in the primitive germ cell (because sometimes the chromosome came from the mother, sometimes from the father), there is enormous potential for variation in the offspring.

The Parts and the Whole

The different organs in the body have specialized in order to meet different demands. While the function of individual cells differs, and they are constructed to meet different demands, their basic structure is similar. Cells organize into organs and groups of organs, and here the main characteristic is specialization for the different tasks required. Finally, the organs form an organism, a whole, where the most important feature is not specialization, but coordination. While their versatility is obvious, the collaboration among organs for the benefit of the whole is much more important. The various parts in the body must be strongly united to assure survival.

The nervous system is mainly responsible for the coordination of activities in the body. Hormones, which circulate in the bloodstream, also play an important role by adjusting the body to the various demands of the environment. Hormones are chiefly produced in the endocrine glands — that is, glands that secrete directly into the blood. Some hormones exert specific effects on particular organs — their "target" organs — while others have more generalized effects. Hormones are numerous. Among the most important is adrenaline (epinephrine), which is produced in the adrenal glands. It has a powerful, generally stimulating effect, mobilizing the body's powers of defense. Sex hormones regulate sexual functions. A very complex hormonal interaction governs the menstrual cycle.

These extremely complex reactions and interactions strive to bring about a state of balance among the billions of cells that make up a human being — an individual who has the ability to act freely and independently, and to influence the world about him.

Hormone crystals
The endocrine glands in the body secrete chemical substances called hormones. These enter the bloodstream directly, where at very low concentrations they exert a profound influence on other organs or on the body as a whole. Today, many hormones can be produced synthetically in crystalline form. Hormone crystals often have the special property that they can refract light, producing beautiful patterns of color. Shown here are crystals of two important hormones, cortisone, at the top, and insulin. Cortisone is produced in the adrenal cortex. It is secreted when the body is exposed to stress and strain of various kinds.

Insulin, produced in the pancreas, is necessary for sugar metabolism. Disturbances in the production of insulin are characteristic of diabetes.

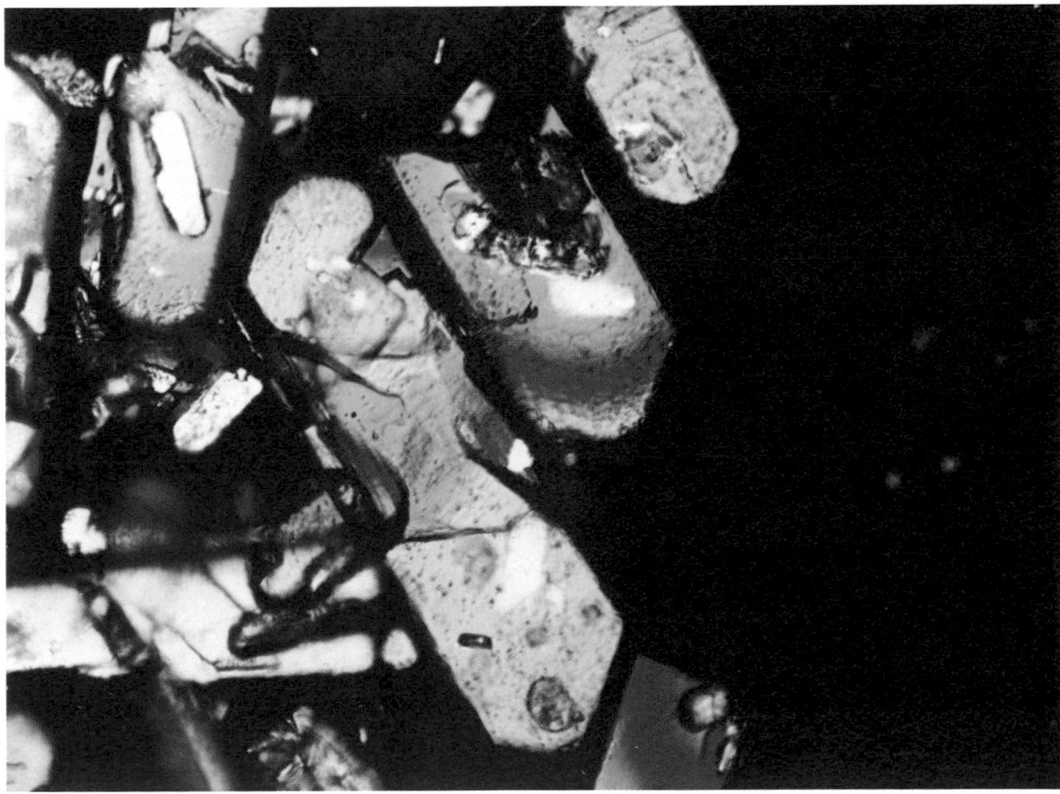

Two important hormones
one from the thyroid gland, the other from the adrenal medulla (the central part of the gland). Thyroxine, the thyroid hormone, participates in the regulation of metabolism and of growth. Thyroxine increases oxidation, promotes protein, carbohydrate, and fat metabolism, and influences the regulation of body temperature. Diminished function of the thyroid leads to a general reduction in the metabolic rate, which can affect mental functioning. An excessively active thyroid, on the other hand, leads to an abnormally high rate of metabolism.

Noradrenalin and adrenaline (the latter shown on page 34) are produced in the adrenal medulla. Noradrenalin is also produced and secreted at certain nerve endings, in particular ones in the sympathetic nervous system. These two hormones collaborate and complement each other. Adrenaline acts as an "emergency" hormone. It stimulates the heart, raises the metabolism rate, mobilizes sugar from the liver, counteracts muscular exhaustion, dilates the vessels in the heart and in working musculature, dilates the pupils of the eyes and the bronchi of the lungs, and promotes blood coagulation. Noradrenalin stimulates the heart, contracts the walls of blood vessels (thus raising blood pressure), inhibits intestinal movements, and has a small but noticeable stimulating effect on metabolism rate.

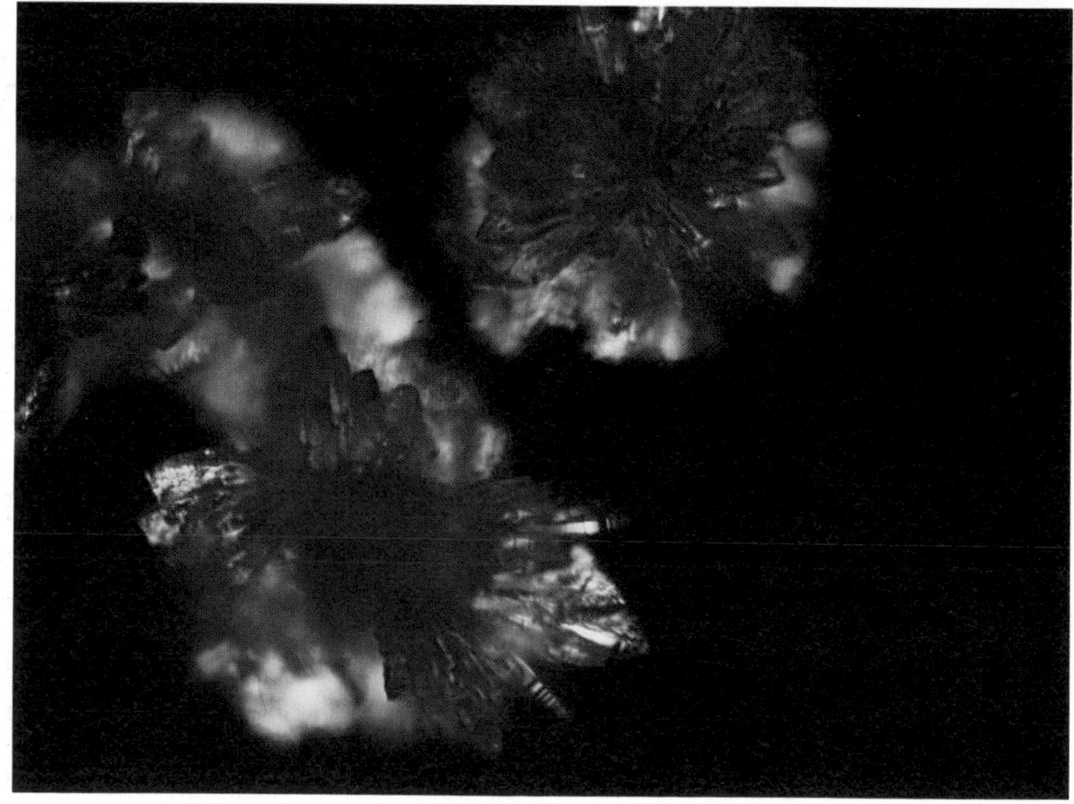

Reproduction

Life began in the primordial sea. Animate beings were washed ashore and eventually spread over the earth and into the air. The course of evolution was marked by divergence. Animals and plants fought for room, and won out or perished. Ever since the beginning, over three billion years ago, life forms have been changing, and the abundance of types which have emerged is extraordinary. Yet the fundamental principle remains unchanged: life is reproduced from generation to generation. All the information necessary for the development of a new human being is stored in the sex cells. This solitary pair of cells determines the new individual.

The Male Reproductive System

A new human life arises from the union of a female ovum and a male sperm cell. At the very moment the chromosomes of the ovum and the sperm cell come together, a number of the offspring's properties are determined, among them its sex. This depends on whether the sperm cell contains a female (X) chromosome or a male (Y) chromosome. The ovum always contains an X chromosome. About half of the 200 to 500 million sperm cells present in a single ejaculation carry the X chromosome that will result in a girl, the other half the Y chromosome that will result in a boy.

The Y chromosome of the sperm cell exerts its influence in many ways. It affects the development of the skeletal structure and the muscles, for example, so that these are generally larger in a man than in a woman. Its primary function, of course, is to steer the course of development of the reproductive system and secondary sexual characteristics into those appropriate to the male.

The male reproductive organs function to produce sperm cells and transport them to the female. Sperm cells are formed in the testes, the male sexual glands, which lie in the scrotum, a fold of peritoneal tissue which hangs outside the pelvis. If the prevailing temperature is cold the scrotum can be contracted upward toward the pelvis by special muscles. When warm, the scrotum sinks downward. The normal temperature of the scrotum is a little lower than that of the abdominal cavity, where the corresponding female organs, the ovaries, are located. Theoretically, the placement of the male reproductive organs outside the pelvis cavity allows for better temperature control. It has been shown experimentally, for example, that a long-term increase in scrotal temperature can lead to a reduction in sperm production.

Sperm cells are formed in the seminiferous tubules, very long but microscopically thin and narrow canals. The tubules are coiled and grouped into larger units, the lobules. These make up approximately 80 percent of the mass of the testes. Seminiferous tissue contains cells with high rates of division. These cells mature at puberty and remain active throughout life. Male and female reproductive systems differ in this respect, for the production of ova comes to an end after the woman reaches menopause. The testes continue to produce millions of sperm cells every day, although the count is lower in older men than in the young.

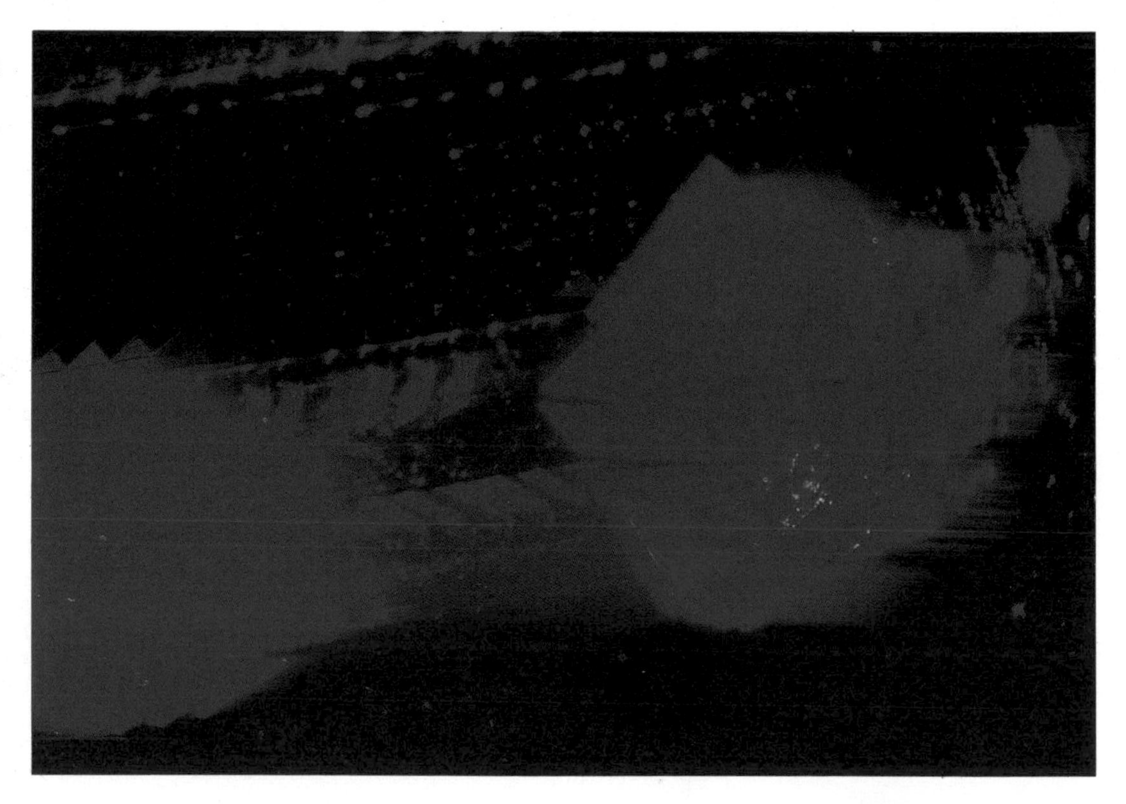

Adrenaline in crystalline form
This hormone is produced in the adrenal medulla (the central part of the gland). One of its effects is to increase pulse rate.

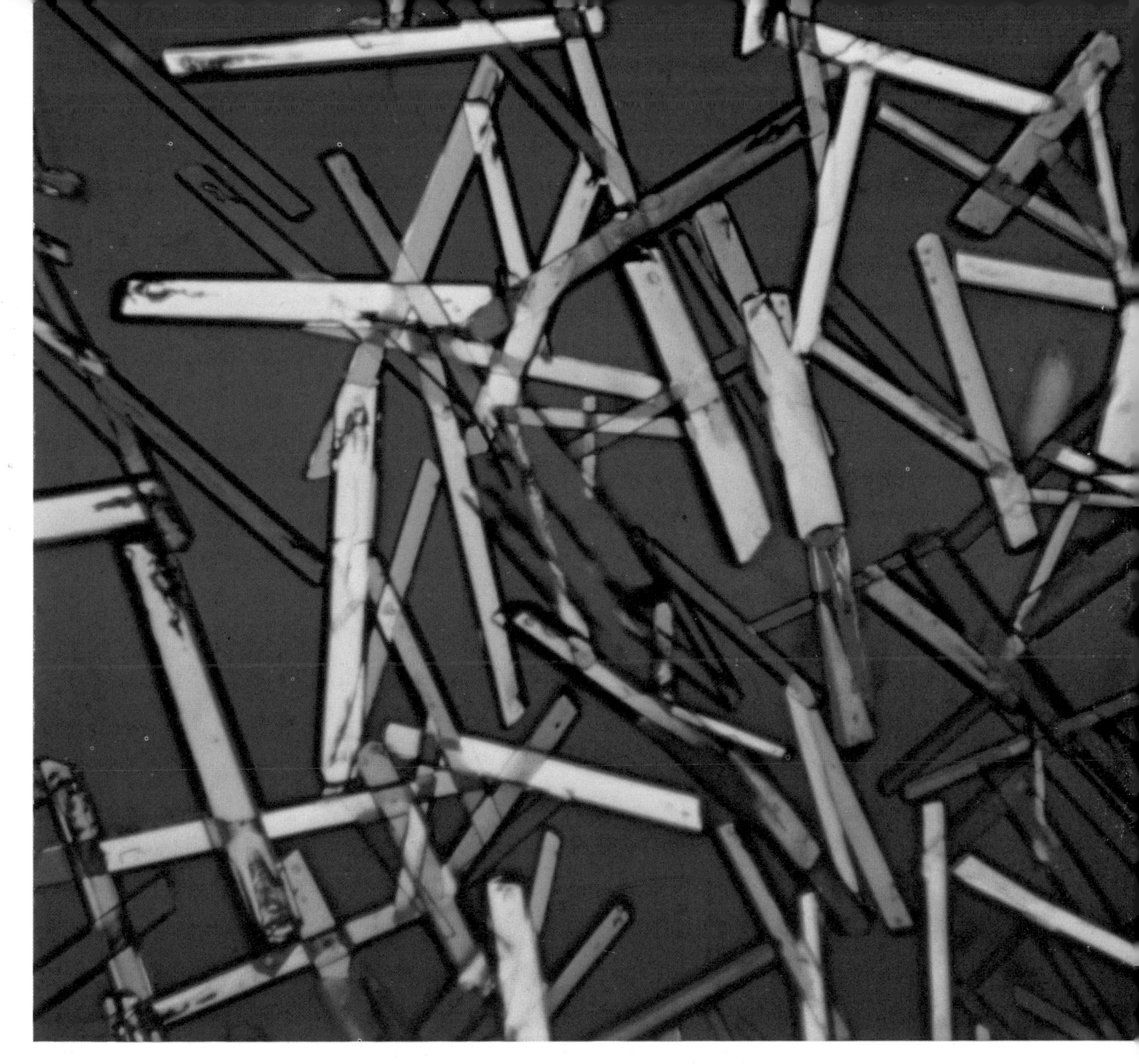

Testosterone crystals under high magnification
These rods may not look very impressive — more like a handful of multicolored matches against a fuchsia tablecloth — but in a dissolved and biologically active state, the male hormone is extremely potent.

Testosterone is produced at sexual maturity. Its synthesis is stimulated by certain pituitary "gonadotropic" hormones, hormones which after release in the bloodstream exert their influence on the sexual organs, or gonads. After synthesis, testosterone undergoes further changes in the liver, where it is converted to the hormone androsterone and other substances.

Strangely enough, the testes also produce female sex hormones, estrogens, though in small amounts. It is testosterone and androsterone that determine the development of the male secondary sex characteristics such as body configuration, hair distribution, a large larynx, and deep voice. They also affect the development of the penis, scrotum, epididymis, spermatic ducts, seminal vesicles, and prostate. Through their influence on metabolism, the sex hormones also inhibit the growth of the long tubular bones during puberty.

The sperm cells produced in the tubules are immature and lack motility (they are unable to move on their own). They are transported to the epididymis, where they can be stored and complete their growth. The epididymis consists of a network of channels which collect into the spermatic ducts. During sexual arousal, mature sperm cells leave the ducts and pass through the seminal vesicles and the prostate gland, where various secretions are added. These secretions contain substances which neutralize the acidity of the male urethra and the female vagina (which would otherwise have lethal effects on the sperm). At the same time, the secretions increase the motility of sperm cells.

The male organ, the penis, consists of three parts: glans, body, and root. The glans is particularly rich in nerve endings. In some species, including one as closely related to man as the chimpanzee, there is a penis bone. The human male has no such bone. Penile erection is due to the presence of erectile tissue — cavernous or spongy tissue well supplied with blood vessels. In response to nervous system signals associated with sexual excitement, the erectile tissue becomes engorged with blood. At the same time, the veins which would normally drain the tissue become constricted, blocking the outflow. The penis becomes rigid and expands by about one-third its length. When the erection subsides, the veins open again, and the penis resumes its usual size. At ejaculation, the pelvic muscles contract rhythmically. This allows semen to be deposited in the upper part of the vagina.

The testes are formed in the abdominal cavity ▷
They are shown here in a four-month-old fetus, surrounded by a loop of intestine. When a boy is born, the testes have usually descended into the scrotum.

Testis with spermatic cord and epididymis ▽
The testis is roughly egg-shaped, about an inch and a half long. Spermatozoa are formed in the tubules, the small tubes in the interior of the testes.

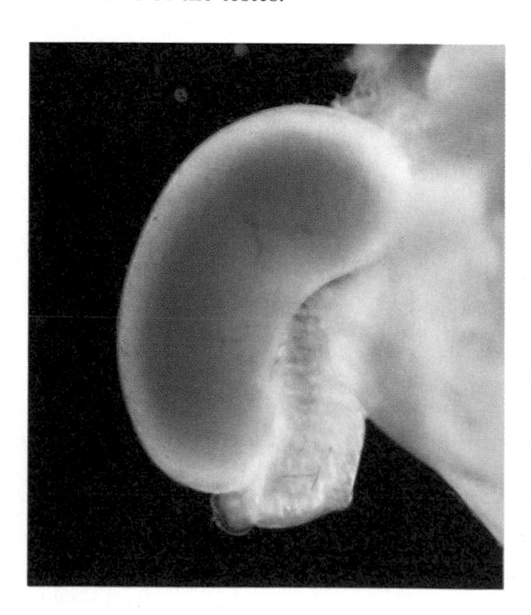

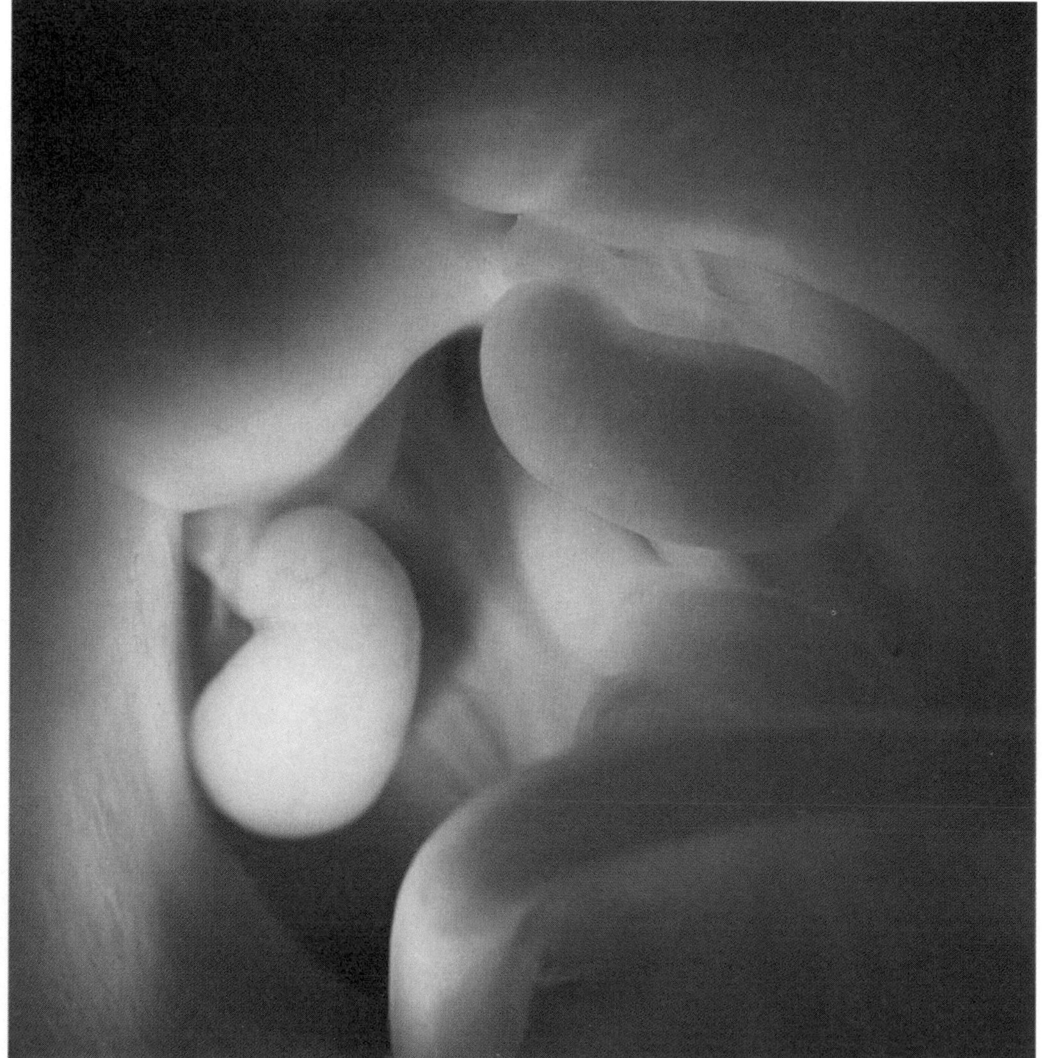

The testes in a twenty-year-old man, isolated here with the crescent-shaped epididymis above. Sperm cells are transported from the testis to the epididymis, which consists of a network of coiled ducts. Here the sperm cells complete their growth and are stored. The tube visible in the upper right part of the picture is the spermatic duct.

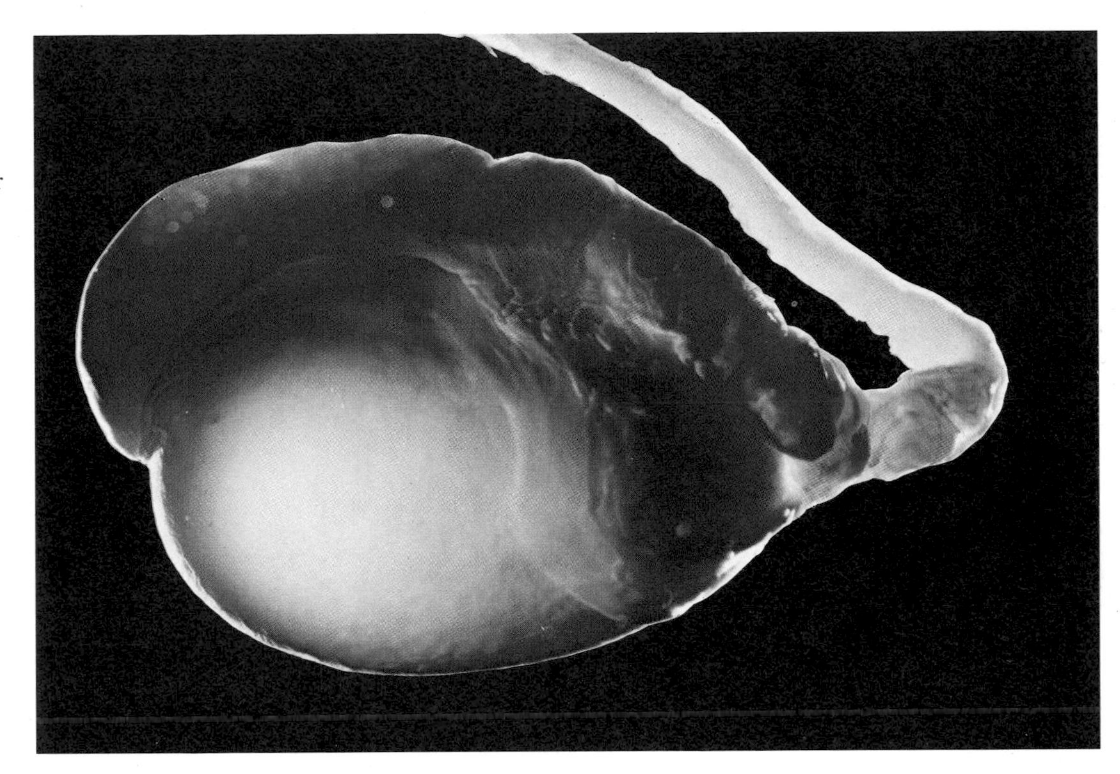

In the interior of the testis are hundreds of lobules. They lie like a ball of wool under the solid wrapping of connective tissue that surrounds the testis but that is removed here. These ducts in turn are composed of microscopic tubes, the tubules. The total length of the tubules of a testis is several hundred yards. The formation of sperm cells, spermatogenesis, takes place in the tissue lining the tubules, under the influence of a pituitary hormone. Newly formed sperm cells accumulate in the tubular lumen (the hollow center of the tubule) and collect in still wider tubes. Spermatogenesis begins at puberty and continues throughout adult life. Normal production in mature males amounts to tens of millions of sperm cells per day.

The epididymis,
seen on the right under the testis, is the place where sperm cells are stored while they complete their growth. Immature sperm arrive at the epididymis in a never-ending stream. The epididymis is made up of soft tissue arranged in a network of coiled channels feeding into the spermatic duct. The photo also shows how the testis is suspended in connective tissue. The suspending ligament contains spermatic ducts, blood vessels, and nerves.

Penis in a four-month-old fetus
The male and the female reproductive organs have the same genetic origin and early in development cannot be distinguished. Even though the child's sex is determined at the moment of fertilization, it is not until the end of the eighth week of gestation that the sex chromosomes begin to exert their effects. In the girl, the primordial structure that resembles a small bud develops into the clitoris; in the boy, the penis. The tissue destined to become the labia in the female becomes the scrotum in the male.

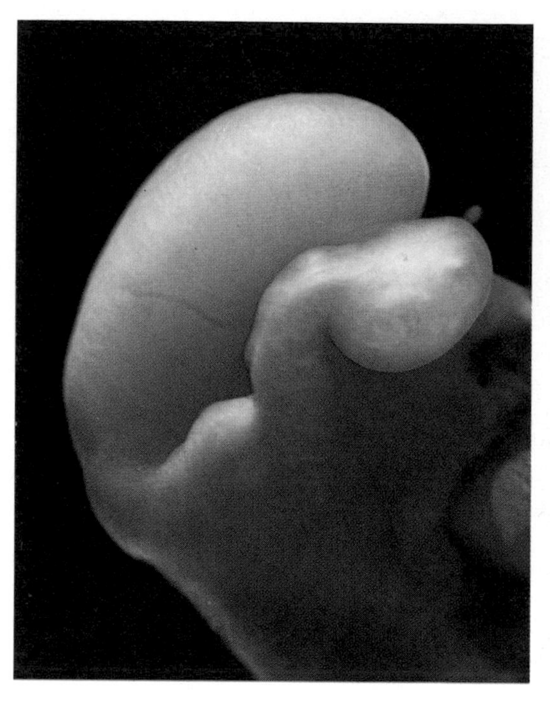

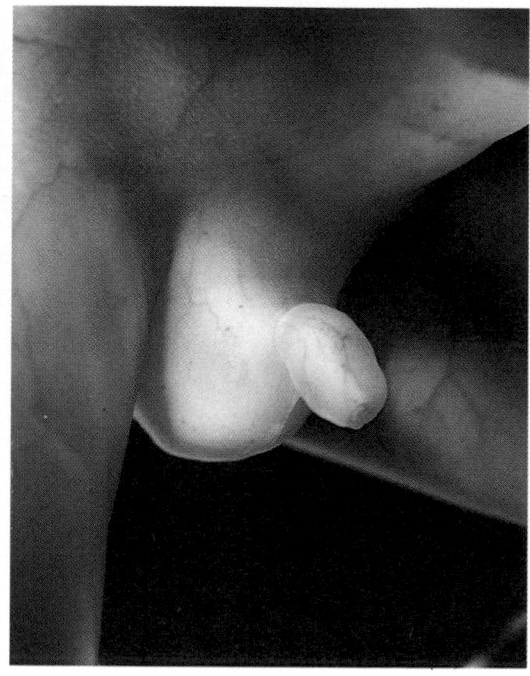

Sperm cells are produced in the testes and stored in the epididymis together with semen. At ejaculation, spermatozoa from the epididymis pass through the spermatic ducts and secretions are added from the seminal vesicles and the prostate. During orgasm, the muscles in the walls of the urethra contract and the ejaculate is expelled through the opening. The penis is composed of three kinds of erectile tissue, richly supplied with blood. Erection results when blood is pumped to the tissue and fills the dilated arterial vessels. At the same time the arteries become engorged, the veins are constricted, maintaining the erection. The penis lengthens and broadens, becomes warm and rigid. The foreskin, which normally covers the sensitive glans, is drawn back. In volume the ejaculate amounts to about 3 to 5 ml of semen, which contains between 200 and 500 million sperm cells.

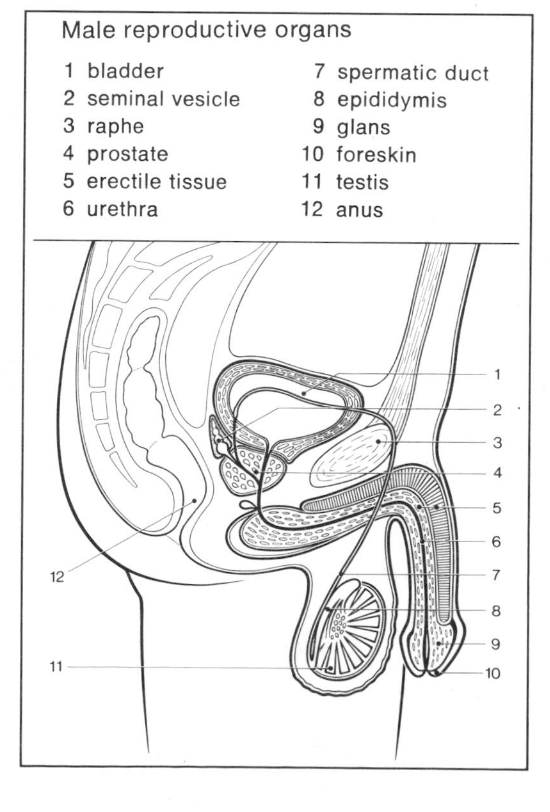

Sperm cells, magnified about 10,000 times ▷
in a scanning electron microscope. The actual size of the sperm is about .06 mm, making it one of the smallest cells in the body. In contrast, the ovum is one of the largest — its diameter may be as much as 0.2 mm. A sperm cell, also called a spermatozoon, consists of a head, a midsection, and a tail, which is comparatively long. The head of the sperm contains the genetic information which is transmitted from father to child.

Sperm Cells

Spermatozoa are formed ▷
in microscopic, coiled ducts, the seminiferous tubules. Extended full length, a single tubule is over twenty inches long. Two to four tubules form a lobule. The lobules lie with their bases against the wall of the testis like small balls of wool. Newly formed sperm cells pass into the center of the testis from openings in the tubules. In this cross section of a tubule, the large, round, brown-spotted structures are the nuclei of the primary spermatocytes (see diagram). Nuclei at different stages of development are also visible. At right is a group of almost fully developed sperm cells lying with their tails toward the tubular lumen. At this stage they cannot use their tails for swimming, but are passively transported to the epididymis. Fructose, a sugar present in semen, appears to be essential for the motility of sperm cells. One of the hormones regulating the supply of fructose is testosterone.

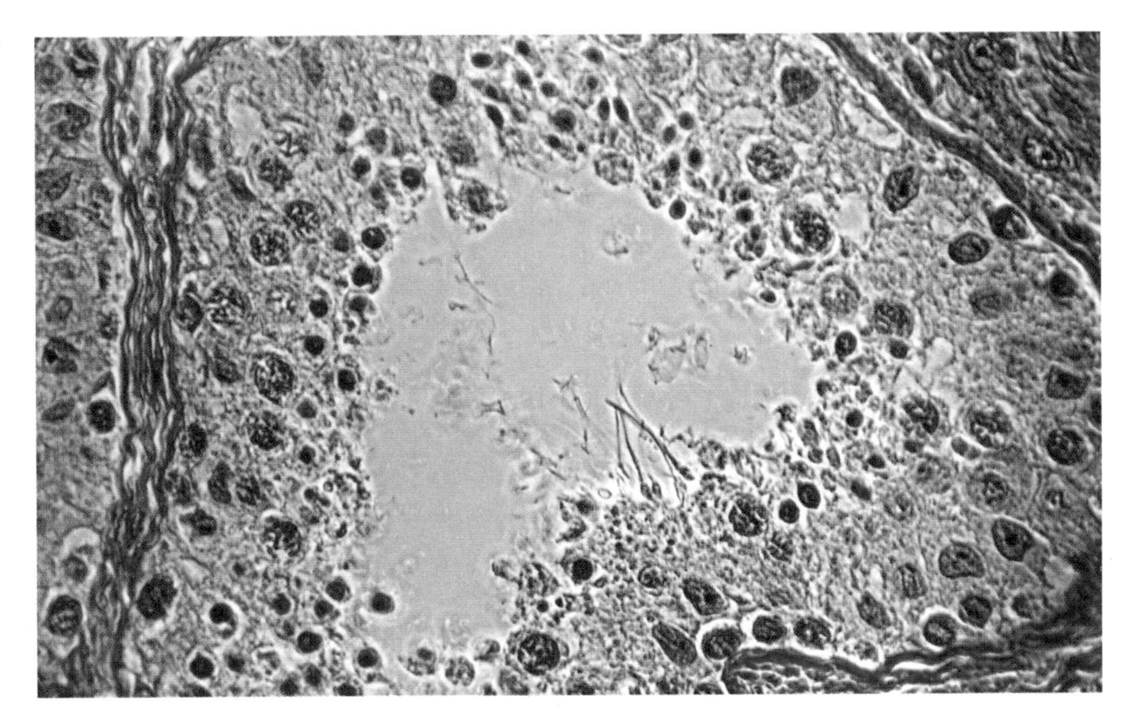

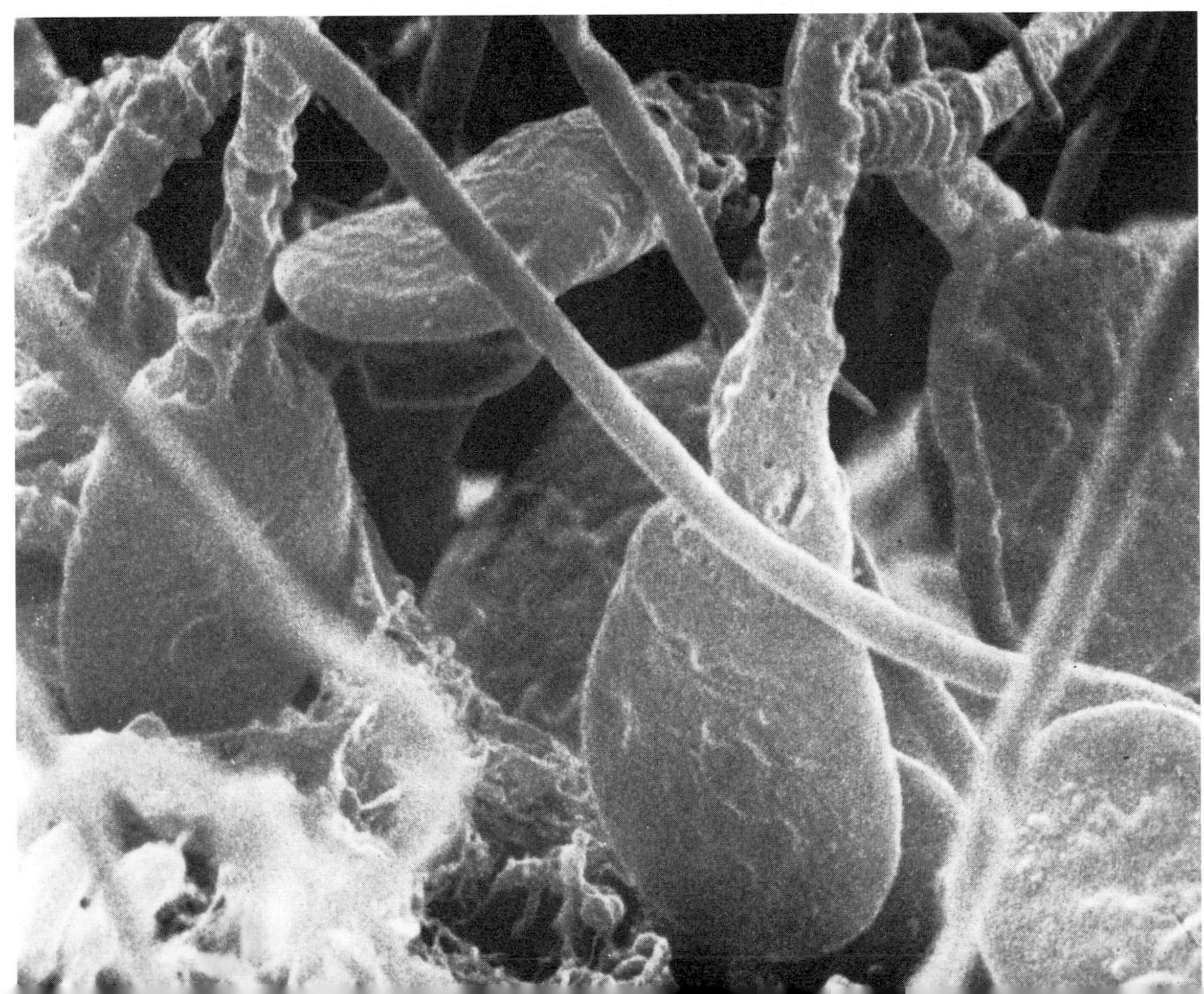

The Female Reproductive System

The female reproductive organs have three main functions: to produce ova which can be fertilized, to receive the sperm cells from the male, and to provide space and nourishment for the fetus during its development. The X chromosome in the female determines such female characteristics as a more delicate body structure and broader hips as well as the development of ovaries, fallopian tubes, uterus, vagina, and enlarged breasts with milk-producing glands. Each cell in the body of a normal female contains two X chromosomes, while male cells contain one X and one Y chromosome. Half the sperm cells produced contain an X chromosome, the other half a Y chromosome.

The ovaries and testes are analogous organs. At birth each ovary already contains its full complement of about two hundred thousand ova in primordial form. Only a fraction of these (about one in a thousand) will have a chance to mature, and still fewer the chance to be fertilized. Most will gradually degenerate over a woman's lifetime. The ovaries remain small until puberty. In the mature woman, they are almond-shaped organs between one and two inches long on either side of the body. Their size reduces after the menopause, and in older women they are quite small.

Some ova are encapsulated in follicles in the outer layer of the ovary, the cortex. These follicles vary from pinhead- to pea-size. The larger size represents a more mature stage of development. The midpoint of the menstrual cycle is marked by the rupture of a mature follicle and the release of a ripe ovum. This phenomenon is stimulated by a complex sequence of hormonal events which will be described in greater detail in the section on menstruation (page 48). The development of a mature follicle may occur at any point in the cortex of the ovary. Since the ovum is not mobile, some means must be provided to transport it to a place where it can be fertilized. This is the function of the two fallopian tubes. Each tube is about four or five inches long and a few millimeters in diameter. One end opens into the uterus, and the other, which is fringed and funnel-shaped, rests near the ovary that it serves. When a follicle is almost mature, the fallopian tube moves toward it and picks up the egg when it is released. As a rule, fertilization takes place in the upper part of the tube. Cell division starts at once, and the fertilized ovum proceeds down toward the uterus, whose internal opening is very narrow — between 0.5 and 1 millimeter in diameter. The mucous membrane lining (the mucosa) of the fallopian tube is extremely convoluted and is fringed with cilia, whose motions carry the ovum and mucus down toward the uterus. Peristaltic movements of the tube (similiar to intestinal contractions) accelerate the journey.

Crystals of estriol, a female sex hormone (1), magnified approximately 200 times. Estriol is excreted in urine during the final stages of pregnancy.

Crystals of estradiol (2), biologically the most potent of the estrogens. Estradiol is formed in the ovaries. The female sex hormones govern the development of such female sexual characteristics as the broadening of the pelvis, the enlargement of the mammary glands, the distribution of fat, and certain phases of the menstrual cycle. Ovarian hormone production is stimulated by interactions between pituitary hormones and the various sex hormones.

1

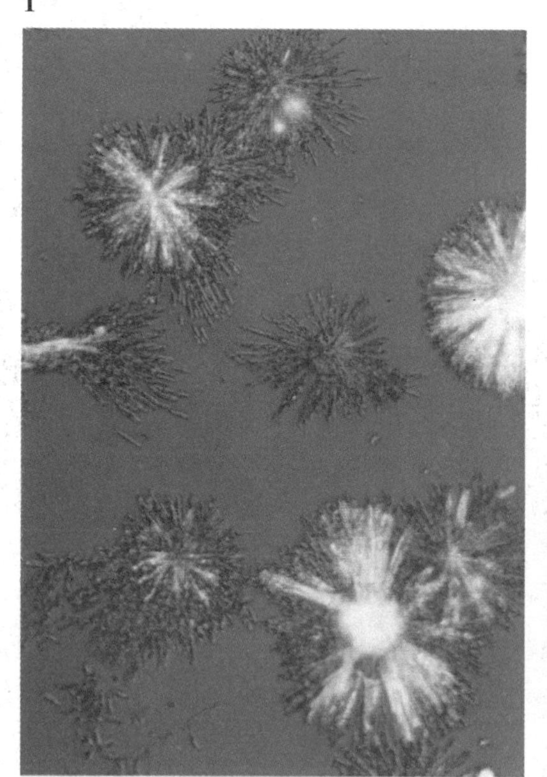

2

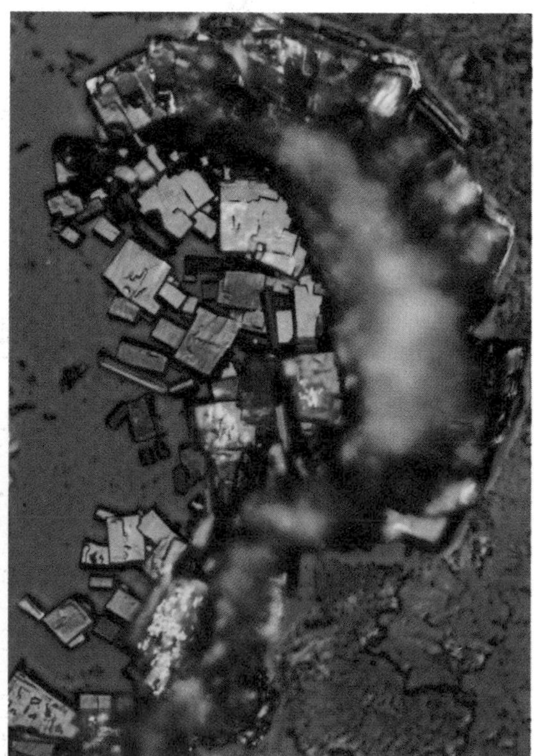

Crystals of estrone (3)
Physicians measuring the relative amounts of estrone, estriol, and estradiol in the urine can determine if there are hormonal imbalances in a female patient.

Progesterone (4)
is an ovarian hormone produced in the latter half of the menstrual cycle when the follicle changes into a gland. Progesterone stimulates the growth and thickening of the uterine mucosa.

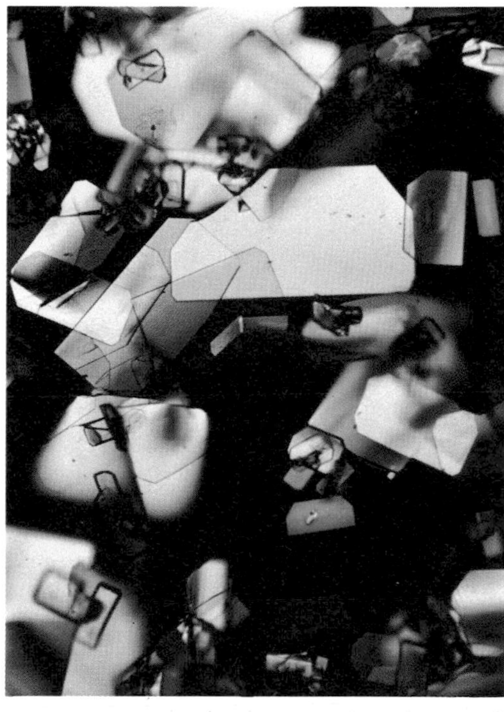
3

4

Progesterone in crystalline form
sometimes looks like a fantastic landscape. Together with estradiol, progesterone determines the phases of the menstrual cycle.

While the ovum matures, the ovary produces estriol, estradiol, and estrone. After ovulation, its main production is progesterone. Progesterone stimulates the changes designed to protect a fertilized ovum. The mucosa of the uterus becomes more "egg-friendly" — it thickens and its blood supply increases. The motility of the uterine musculature is reduced too, because progesterone inhibits the action of oxytocin, the hormone which stimulates the contractions associated with labor. If the ovum is fertilized, the ovary continues producing progesterone for about three months. Then the placenta takes over this function.

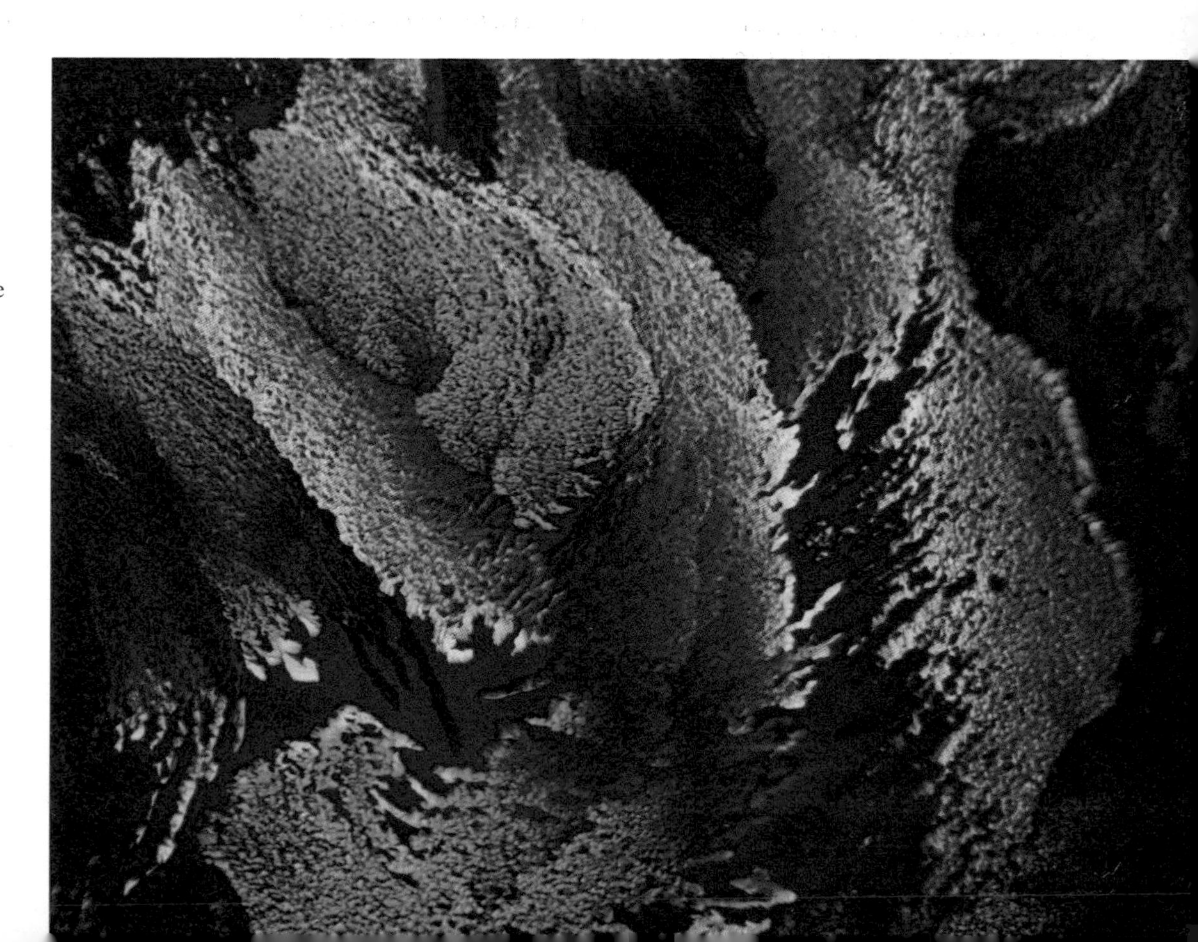

The clitoris in a three-month-old female fetus
This organ in the female corresponds to the penis in the male. It is prominent during the early stages of development of the reproductive system because of hormonal influence. Notice the rudimentary tail of the fetus also present at this stage.

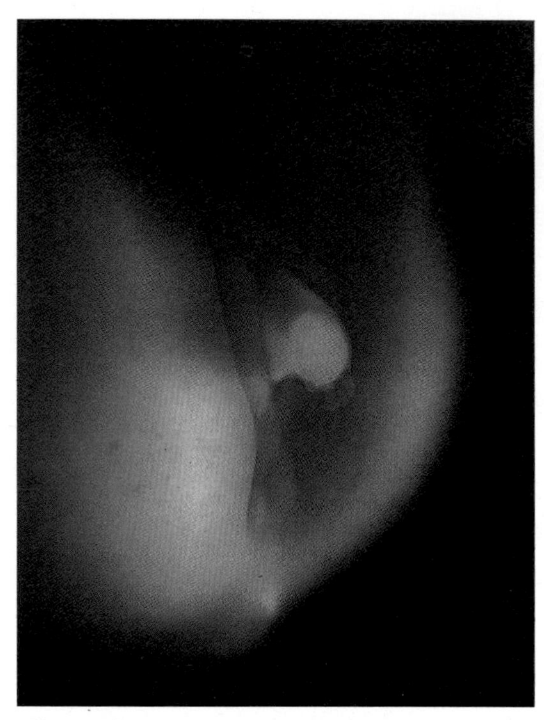

External sex organs in a female fetus

Female reproductive organs

1 fallopian tube
2 ovary
3 uterus
4 symphysis pubis
5 bladder
6 mons veneris
7 clitoris
8 labium minus
9 labium majus
10 hymen
11 vagina
12 anus

The ovary in a three-year-old girl, with the fallopian tube and its flower-like, funnel-shaped end. At the time of birth, each ovary contains about two hundred thousand primordial ova. Only about four hundred will develop into mature ova later in life.

The ovary of a fourteen-year-old girl is larger
The function of the funnel-shaped end of the fallopian tube, the infundibulum, is to move toward the follicle ready to rupture and pick up the ovum.

The ovary of a twenty-seven-year-old woman
By and large, the ovary looks the same in a fertile woman as in a small child or a fourteen-year-old girl.

The ovary of a fifty-six-year-old woman is scarred from hundreds of ovulations. Follicles with undeveloped ova are also seen.

Corpus hemorrhagium,
the blood-filled follicle after rupture, is the red elevation seen at top right. Subsequently it turns yellow in color and becomes the corpus luteum.

Ovary in a three-year-old girl

In a fourteen-year-old girl

The Ovaries

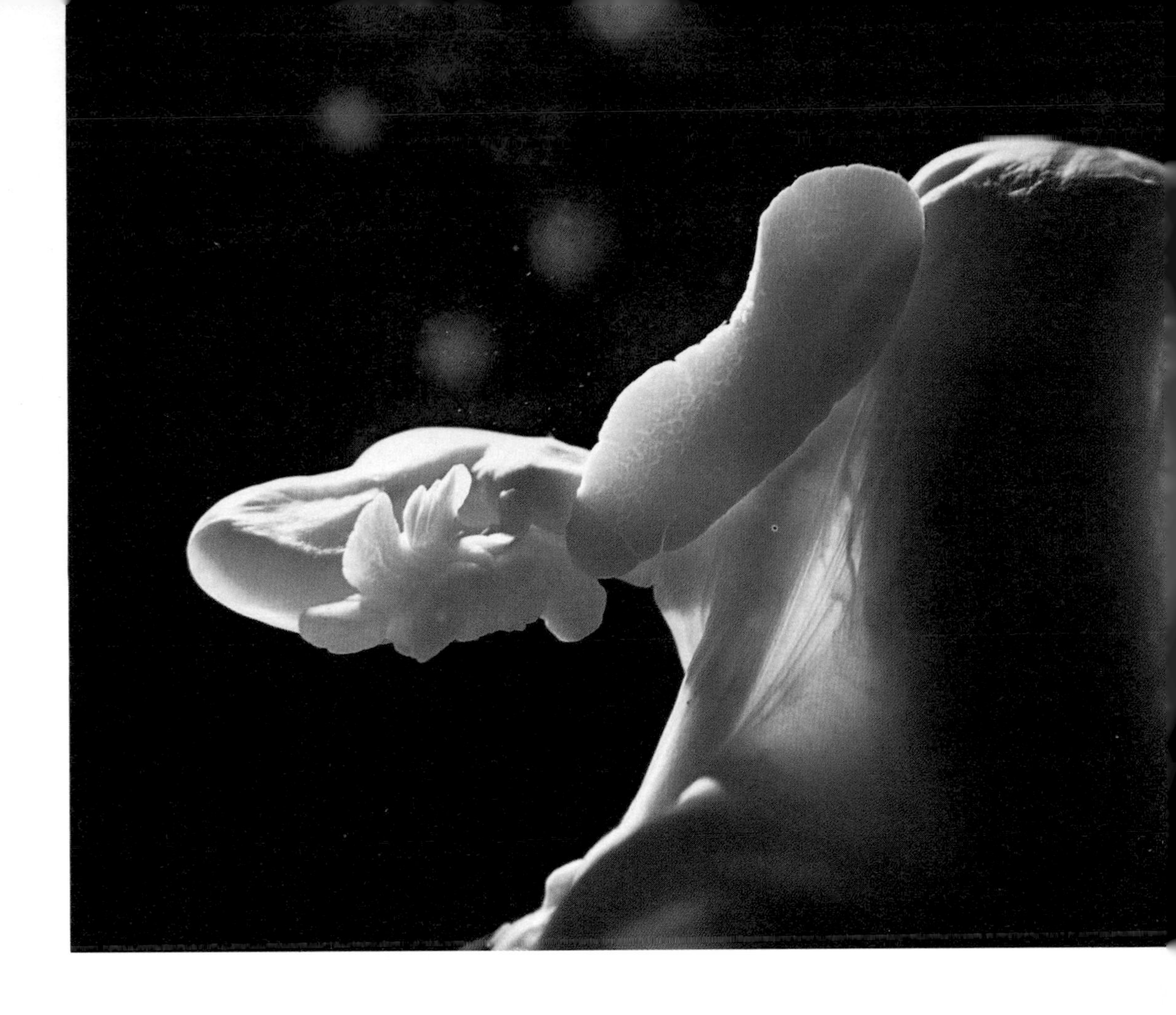

Right ovary, fallopian tube, and infundibulum
in a young fetus. The uterus is at the right.

In a twenty-seven-year-old woman

In a fifty-six-year-old woman

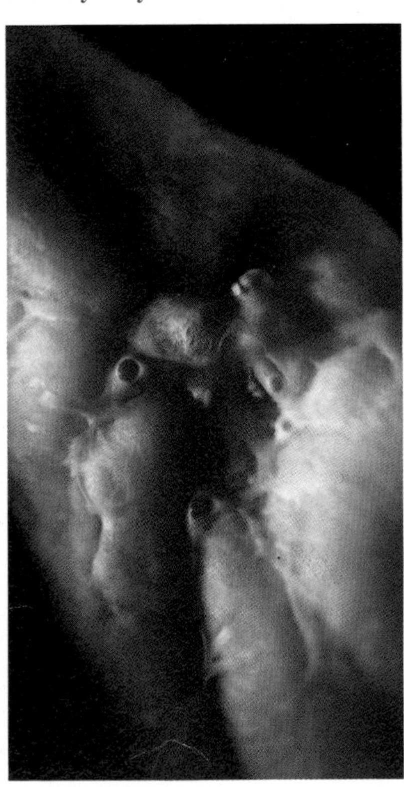

Corpus hemorrhagicum

The Fallopian Tubes

Once settled in the uterus, the fetus will continue its growth and development for almost nine months. This process requires that the uterus itself change and expand to accommodate its contents, the fetus and the placenta. The uterus, shaped like a flattened pear, is about four inches long in a normal nonpregnant woman, and weighs only two ounces. The wall itself consists of three layers: mucosa, muscles, and outer capsule. During pregnancy, the uterus expands to about a foot in length, and weighs more than two pounds at the time of delivery.

The uterus narrows toward its lower end and opens into the uterine neck, the cervix, which extends into the upper part of the vagina. The vagina is a tube three to four inches long, which slants up toward the cervix. The vaginal wall is muscular and extremely elastic to allow for the passage of the head of a baby during delivery. There are no mucous glands in the upper part of the vagina. The secretions present there, which are acidic and bactericidal, come from the cervix. These surroundings, clearly uncongenial to sperm cells, are nevertheless the place where semen is usually deposited at coitus. Many sperm cells die, but some of the vigorous ones manage to swim on.

The outer vaginal opening, or orifice, is protected by two pairs of lips, the minor and the major labia. Slightly forward of this orifice is the clitoris, the female counterpart of the male penis. In sexual excitement, the clitoris too becomes engorged with blood and enlarges. The glans area at the tip is similarly rich in nerve endings. Excitation of the clitoris may trigger orgasm in the female.

◁ **The ovum outside the fallopian tube**
The egg cell, only a fraction of a millimeter in size, is the light blue spot at the left of the fallopian tube. At ovulation, the fimbriae forming the fringe around the end of the fallopian tube fill with blood and are activated for their important function of picking up the ovum. If the ovum comes to rest outside the fallopian tube or remains in the tube, and is fertilized, an ectopic or extrauterine pregnancy occurs. This is a rare and serious complication, which usually results in a miscarriage. The ovum must reach the uterus for the fetus to develop properly. The fallopian tube is about five inches long and is lined with smooth muscles. Its interior is covered by mucosa. The fallopian tube provides the means for elaborate transport of the ovum: cilia, mucus, and gentle peristaltic movements conduct the ovum toward the uterus.

The infundibulum ▷
in a mature female. The fimbriae, or fringed surface, provide a very large surface area. When ovulation is near, the infundibulum — the end of the fallopian tube — moves toward the follicle to pick up the ovum when it is released. A left fallopian tube can even move toward the right ovary if both the left ovary and the right fallopian tube have been removed. The mucosal lining of the fallopian tube is convoluted and covered with cilia.

The mucosal folds in the fallopian tube resemble an airy veil. The mucosa is extremely convoluted and contains a large number of mucous glands. Convolution is also the solution found in other regions of the body where a large surface is required within a limited space (for example, in the intestines). When the fallopian tube is active, the ridges and valleys are smoothed out by mucus. It is in this rough terrain that the sperm cell finds the ovum. Examination of the photo makes it clear why so many sperm cells are required for successful fertilization.

The ovum
surrounded by a protective membrane, traveling through the fallopian tube. The three small round bodies to the right of the ovum are the polar bodies, formed when the primary and secondary oocytes divided. The ovum is sensitive to pressure and friction. One of the reasons why the ovum is so much bigger than the sperm cell is that it provides nourishment for the first cell divisions. After the growing cell mass has settled in the uterus by imbedding itself in the mucosa, it is nourished by the many blood vessels there.

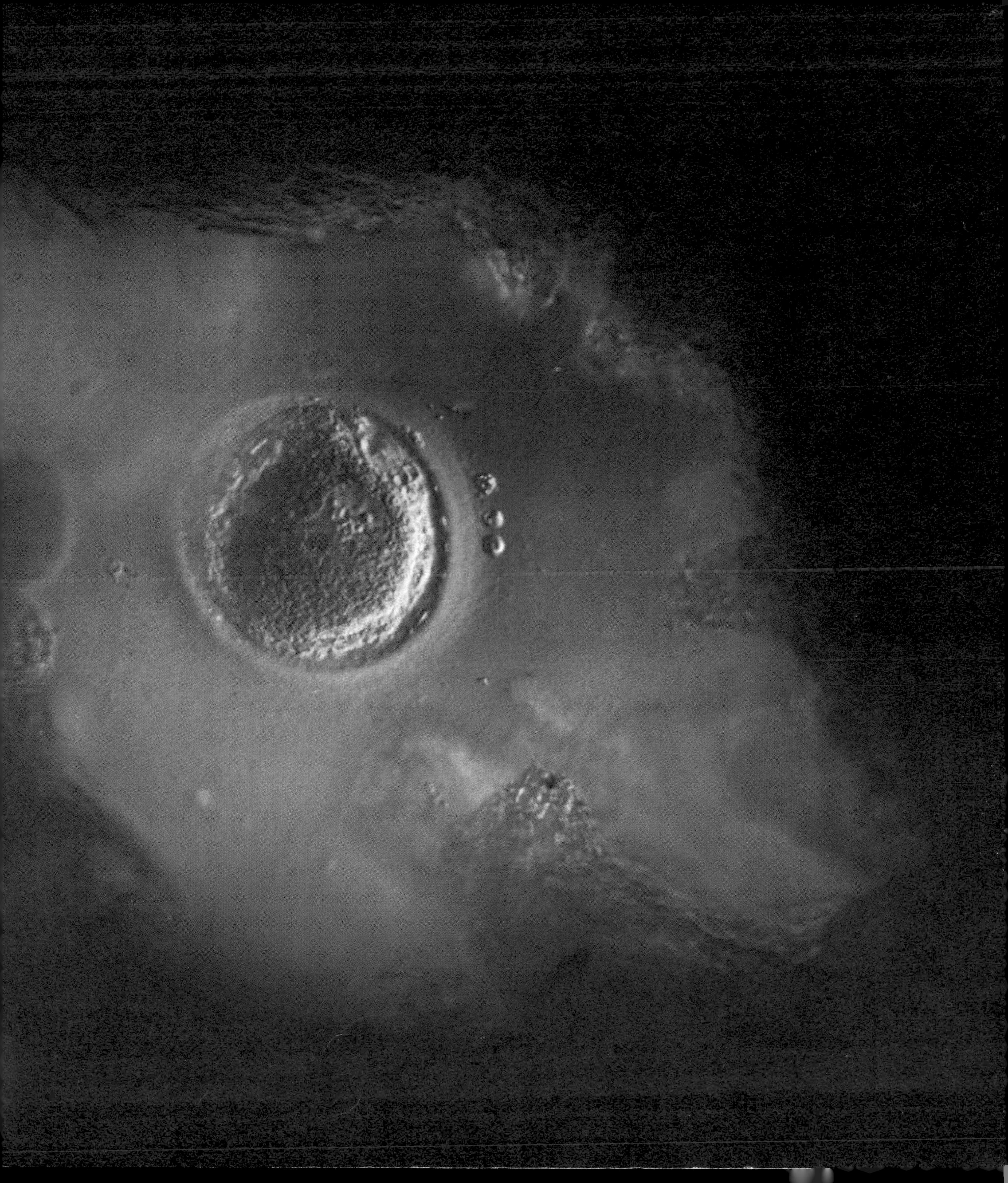

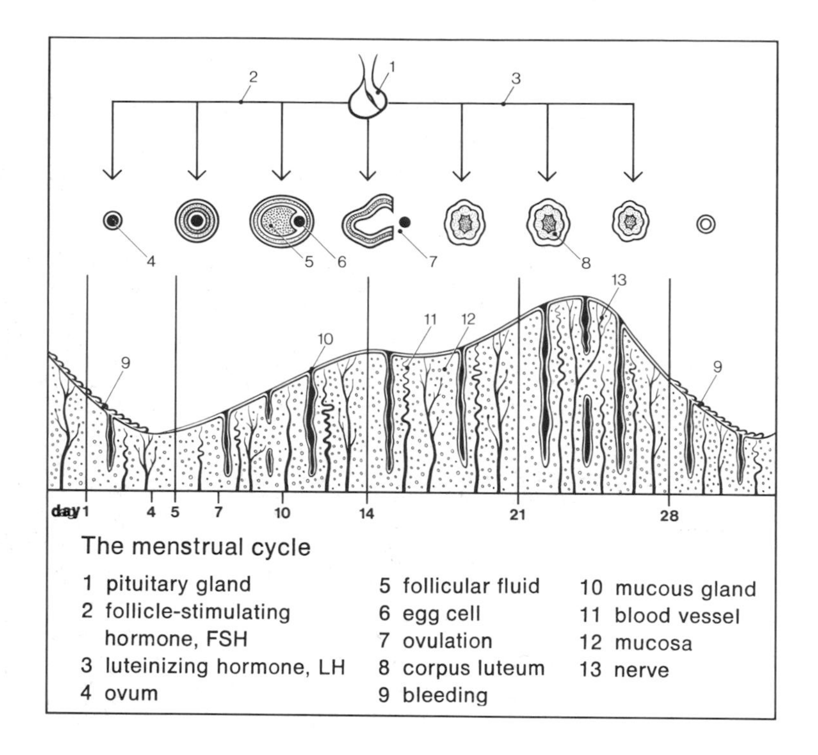

The menstrual cycle

1 pituitary gland
2 follicle-stimulating hormone, FSH
3 luteinizing hormone, LH
4 ovum
5 follicular fluid
6 egg cell
7 ovulation
8 corpus luteum
9 bleeding
10 mucous gland
11 blood vessel
12 mucosa
13 nerve

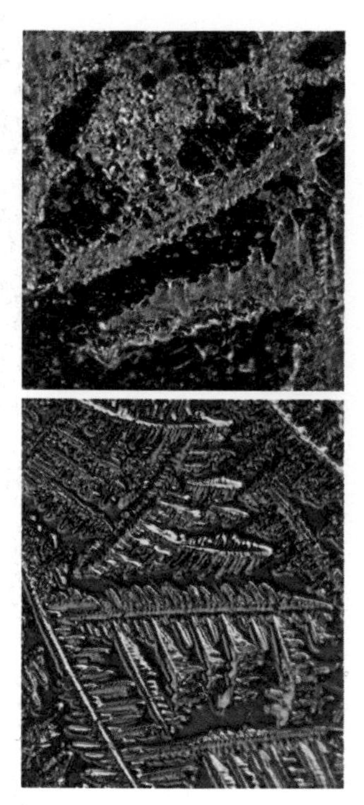

A plug of mucus in the uterine neck (the cervix) protects the uterus from bacteria and other microorganisms. During the greater part of the menstrual cycle the mucus plug is almost impervious to contact — even by sperm cells. But at ovulation it changes its character for two or three days. The photos show the difference: under the influence of an ovarian hormone, the mucous glands gradually produce a thinner and more fibrous mucus. On the fourteenth day of the cycle, the mucus is as "friendly" toward sperm cells as possible — millions of sperm cells with whipping tails can clear the hurdle without difficulty. The upper photo shows crystallized mucus from the first days of the menstrual cycle. The lower photo shows the characteristic fern-like pattern of crystallized mucus on the day of ovulation.

The Female Sexual Cycle

There is no process in the man that corresponds to the menstrual cycle in the woman. During puberty the testes begin producing sperm cells, and continue doing so throughout life. There is no clear-cut stage comparable to menopause, although many authorities acknowledge that there is a male "climacteric" in middle age and that it is associated with psychological and physiological changes.

The first menstruation in girls usually begins between the ages of eleven and thirteen. At first the cycles may be irregular, but in time they are usually established at intervals of about twenty-eight days. Ovulation marks the midpoint climax of each period. This pattern continues until the onset of menopause at some time in the forties or fifties. The fertile period for a woman thus amounts to about thirty-five or forty years, with approximately four hundred ovulations. (This means that only about one out of a thousand primordial ova matures.)

Menstrual cycles are found only among the primates — man and his closest relations among the animals, monkeys, and apes. Nor is there any special mating period, although a mature ovum can only be fertilized over a short time period (a few days) each month. Normally, one ovum matures every month except during pregnancy. Then the body adjusts itself to develop the fertilized ovum.

The menstrual cycle in a woman is the result of complex interactions among hormones released from the anterior lobe of the pituitary gland and the ovaries. The hormonal pendulum moves to and fro between ovulation and menstrual bleeding. Normally, the only thing that alters this rhythm is pregnancy itself. Once pregnancy is over, the hormonal pendulum starts again. Its course is intricate and largely automatic, set by controls and feedback mechanisms that carefully monitor the hormone levels in the blood and in the organs involved.

The anterior lobe of the pituitary gland secretes a follicle-stimulating hormone (FSH), presumably under the control of a center in the hypothalamus, one of the great relay stations in the brain. (It has been shown that the hypothalamus also plays a large role in regulating hunger, thirst, and body temperature, as well as the sexual functions.) FSH, which is a water-soluble protein, circulates in the blood to reach the ovaries. FSH is responsible for the early growth of the follicles. Subsequently, one follicle is singled out; it enlarges and forms a bulge on the ovary. This is the first part of the cycle.

During the next stage, the pituitary gland secretes another hormone, called luteinizing hormone (LH). This accelerates the maturation of the egg in the follicle. FSH and LH together stimulate the production of estrogens in the follicle. These are the female hormones associated with the development of the female reproductive system as well as the secondary sex characteristics. Estriol, estradiol, and estrone are estrogenic hormones. At this stage of the cycle, the mucosa in the uterus is also stimulated so that it grows and thickens, to provide a proper nesting site for the ovum. As estrogen production proceeds, some estrogens cir-

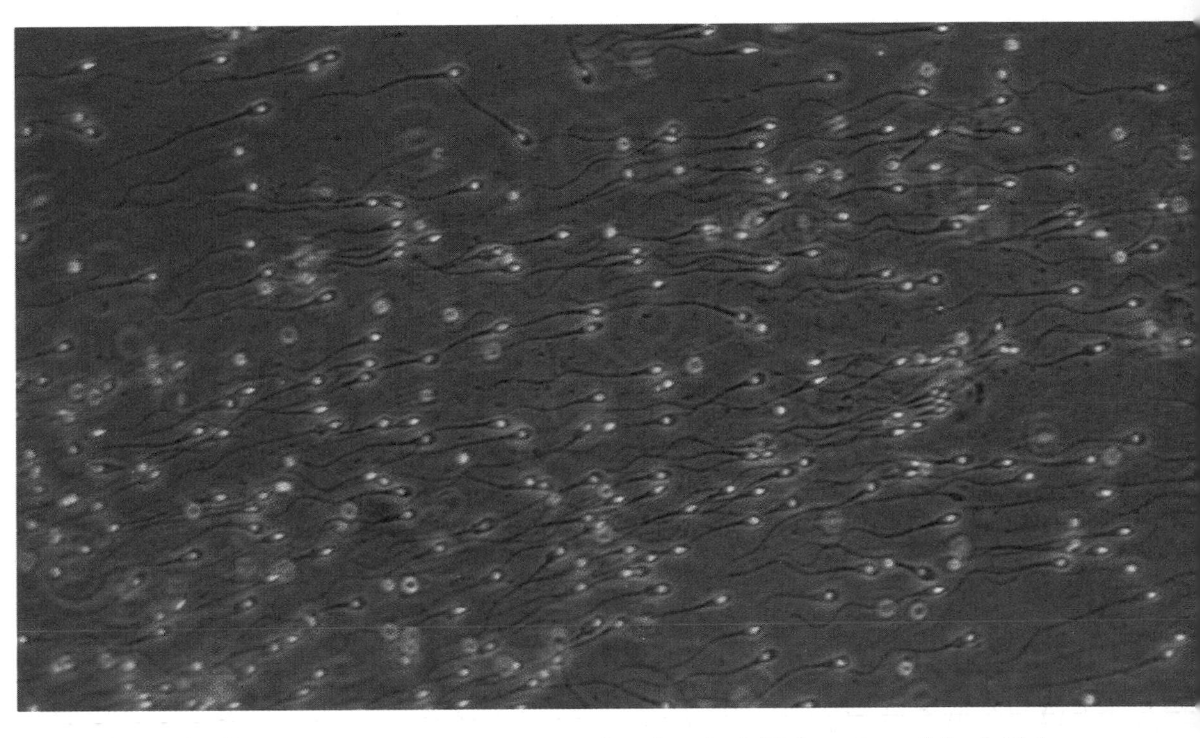

Several million sperm cells moving through the mucus in the cervix. A follicle has ruptured and a mature ovum moves down the fallopian tube. Millions of sperm are destroyed en route to or in the tube; other millions swim up the wrong one. But all that is necessary is that one sperm cell find the right way at the right time.

culate in the bloodstream and reach the pituitary gland. Figuratively speaking, the pituitary gland now "knows" its secretions have had the desired results and now changes its hormone production schedule. This is an example of "feedback" control: the pituitary gland is able to regulate its present behavior by reacting to the very hormones it had earlier stimulated other organs to produce. Now the secretion of FSH becomes inhibited, while that of LH increases. As the estrogen level in the blood continues to increase, the maturation hormone LH eventually becomes dominant. By about the fourteenth day of the cycle this will result in the rupture of the follicle. The egg is now pushed out and picked up by the nearby fallopian tube. Following this, the follicle undergoes further changes into a hormonal gland, the corpus luteum (yellow body). The production of estrogens now decreases rapidly and progesterone, the hormone produced by the corpus luteum, takes over. When progesterone reaches the uterus, it exerts an almost explosive effect on the cells. The mucosa grows rapidly and prepares itself to protect the ovum. Progesterone also inhibits the pituitary production of FSH so that no new follicle is developed. (Birth control pills also make use of this inhibitory effect.) Now the system waits to see what will happen to the ovum in the fallopian tube. And as long as the corpus luteum continues to secrete progesterone, the ovum will be protected.

If the ovum is not fertilized, the corpus luteum gradually degenerates. The uterine mucosa stops growing, and as it begins to be sloughed off and discharged, the woman menstruates. This occurs some two weeks after the initial rupture of the follicle, on about the twenty-eighth day of the cycle. Now the pituitary gland, which has monitored the reduced secretion of progesterone, again begins to secrete FSH and a new follicle starts developing.

On the other hand, if the ovum had been fertilized, the corpus luteum would have continued to produce progesterone for about another three months. Further production would then have been taken over by the placenta. The ovaries, having ceased their monthly activities of ovum production, would not resume them until some months after the child was born, again as a result of stimulation by the pituitary hormones.

The Development of the Fetus

Fetal development is still little understood. In nine months' time, a single fertilized ovum grows into the 6,000 billion cells which constitute a human being. How does the ovum "know" that it will become a human being? What mechanism governs the development of the various organs? How do the organs "know" when to stop growing? The questions concerning fetal development far exceed the answers, despite many years of intensive study by embryologists, geneticists, and cell biologists. However, they have succeeded in elucidating the broad outlines of the course of development and the principles governing it.

The nucleus of the ovum and the sperm cell embody the genetic inheritance from the mother and the father respectively. The individual hereditary units, the genes, arranged in orderly

Ovum and Sperm Cells

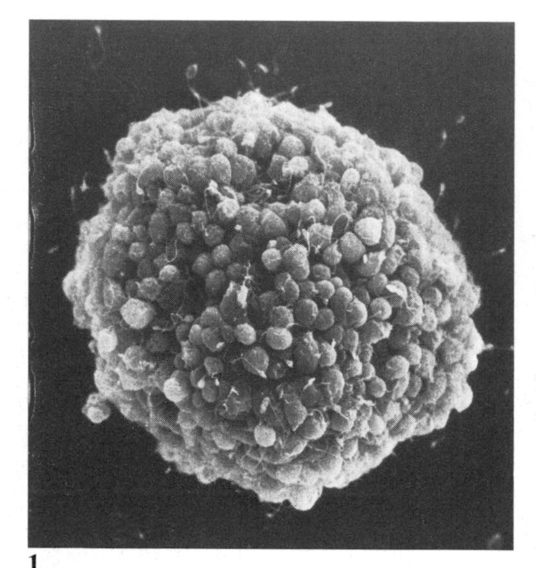

1

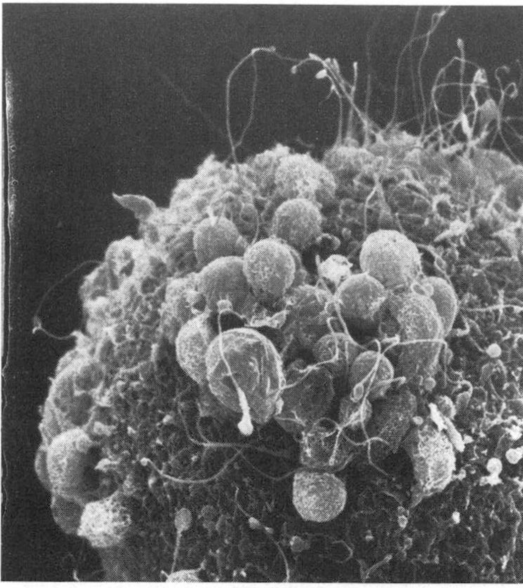

2 Ovum and sperm cells

Fertilization (4)
Here, the head of one of the many millions of sperm cells has succeeded in penetrating the cell membrane of the ovum. The nucleus of the sperm cell enters the ovum and merges with the egg nucleus. As soon as a sperm cell has penetrated the membrane, an impervious "fertilization membrane" develops which makes it virtually impossible for any other sperm cells to penetrate the ovum.

3 Sperm cell

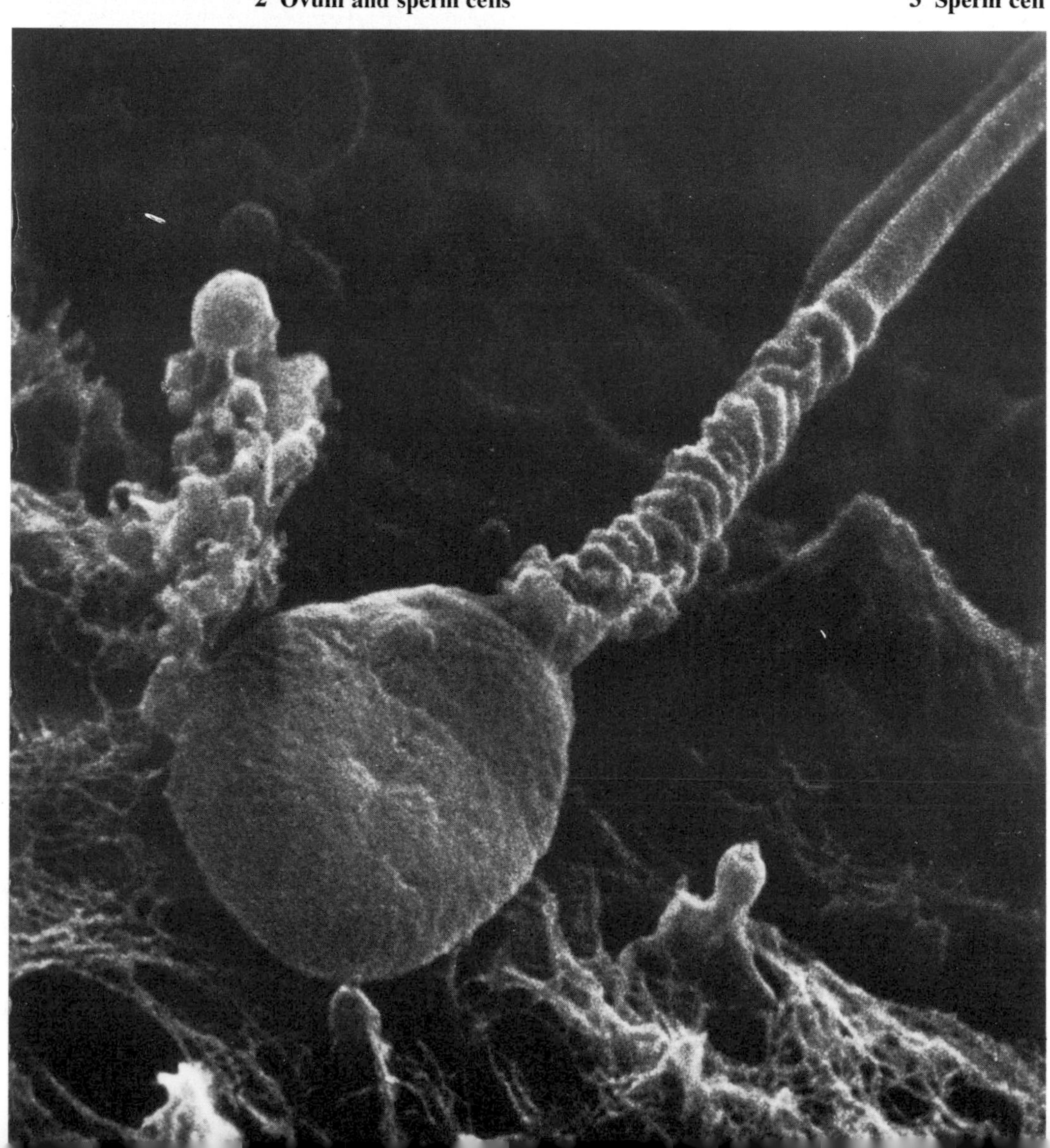

The primary oocyte surrounded by nutritional cells (1)
During its formation in the ovary, the egg cell undergoes a meiotic division. The primary oocyte divides into a secondary oocyte with half the full number of chromosomes and a much smaller cell, the polar body. The secondary oocyte is expelled from the ovary when the follicle (called a graafian follicle) ruptures. It is surrounded by a layer of follicular cells, which supply it with nourishment. In the fallopian tube the secondary oocyte divides, like the primary one, into a larger cell and a smaller polar body. The larger cell becomes the female reproductive cell, the ovum. At the same time, the first polar body also divides into two polar bodies. Eventually all three polar bodies die. The photo shows the density of follicular cells around the primary oocyte.

Sperm cells approaching the ovum (2)
The ovum matures in the fallopian tube. The nourishing follicular cells gradually diminish, but still surround the ovum in this picture.

A sperm cell (3)
When a sperm cell is formed, the primary spermatocyte divides into two secondary spermatocytes, each with half the full number of chromosomes. Each of these also develops into sperm cells (see drawing, page 29). A sperm cell is only .06 mm long, but it contains the complete hereditary information from the father.

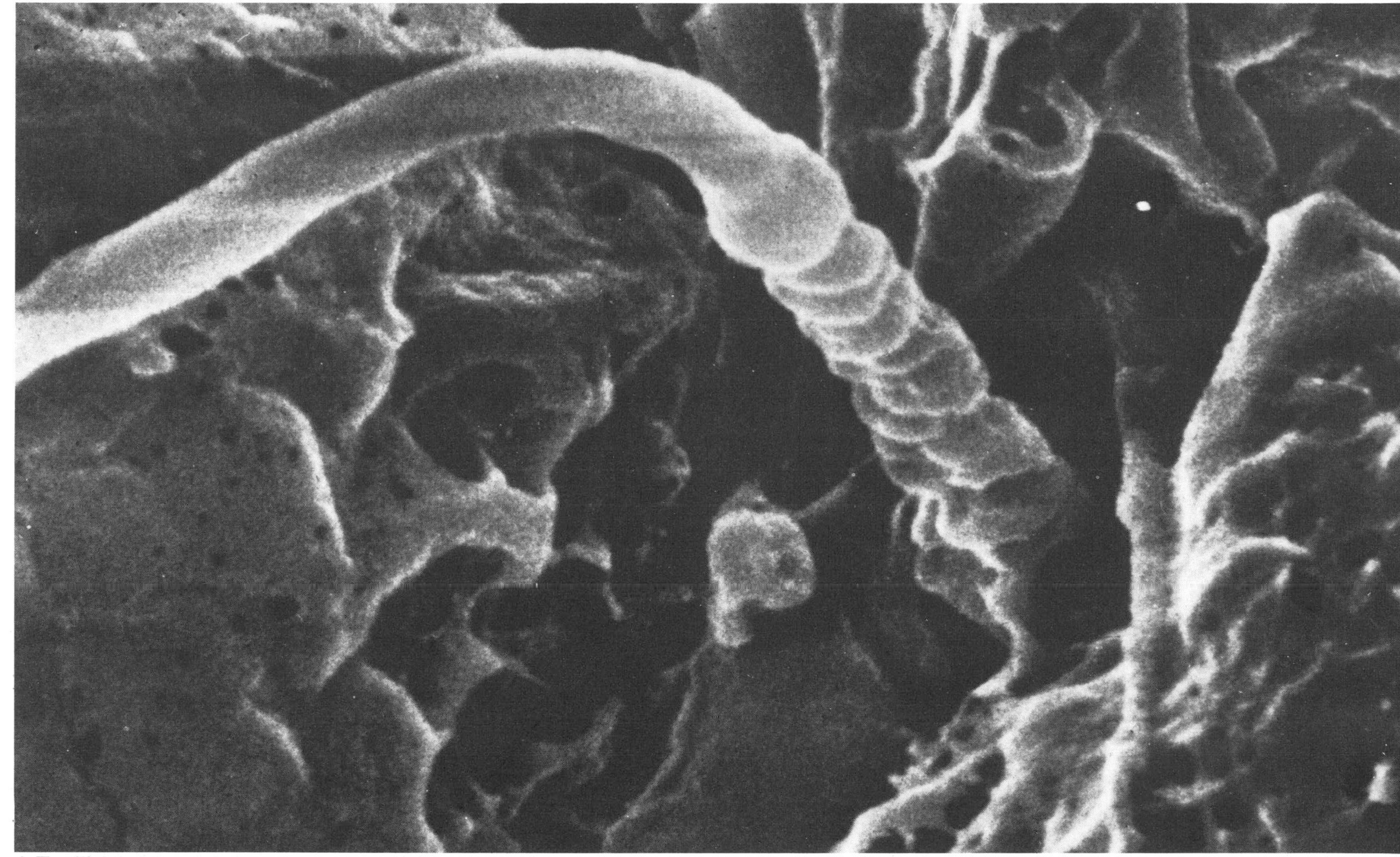

4 Fertilization

sequences along the chromosome, are the blueprints for the new being. In the fertilized ovum the twenty-three chromosomes originating from the mother and the twenty-three from the father are paired up. One of these pairs consists of the sex chromosomes, an X-X combination if the sperm cell fertilizing the ovum happened to contain an X chromosome, or an X-Y pair if the sperm carried a Y chromosome. This chromosomal pattern will appear in all the cells of the new individual except the sex cells (also called gametes), which have only half the number of chromosomes. The hereditary combinations possible given the hundreds of thousands of genes and the total of forty-six chromosomes ensure that no two human beings are ever likely to be exactly alike — except in cases of identical twins, where a single fertilized ovum splits in two early in development.

As a general rule, fertilization takes place in the upper part of the fallopian tube, and the first cell division usually occurs within the first day or so. During the week or so it takes the ovum to travel to the uterus, the cells divide once or twice a day. The cells become smaller at each division, so that the total cluster of cells stays about the same size as the original fertilized ovum. The rounded cell cluster at the time it is ready to pass through the small (0.5 to 1 mm) opening of the fallopian tube into the uterus is called the morula and may number thirty-two cells.

For almost three weeks the uterus has been preparing itself to receive the morula. During the first half of the cycle, estrogens have been stimulating it to build up its mucosa. During the third week, progesterone from the corpus luteum stimulates the development of blood vessels and glands in the uterine mucosa. When the morula leaves the fallopian tube, the mucosa is already soft and its increased blood supply will ensure adequate nourishment for the new life. During the last part of its journey in the uterus, the compact cell cluster changes into a blastocyst, a hollow ball of cells that burrows into the mucosa. Small extensions from its outer cellular layer seek out blood vessels which will provide nourishment. This is the beginning of the placenta.

During the second week of development, the cells forming the inner layer of the blastocyst arrange themselves into an embryonic disk which has two layers, the ectoderm and the entoderm. A third layer, the mesoderm, will develop from the so-called primitive streak, which appears as a lengthwise line of division on the upper side of the disk. These three primitive germ layers give rise to all the organs and tissues of the body. The orderly sequence of development is controlled by an organizing center in the anterior part of the primitive streak.

How cells are differentiated into layers, tissues, and organs is not known in detail. But it is known that a particular cell's destiny is laid out in advance and that its growth toward that end must occur at some appropriate time, usually within a relatively short period. Disturbances in the process involve the risk of malformation. A special phenomenon called induction has been described in the germ layers. In this process, a group or a layer of cells which comes into contact with another cell group, even for only a short time, can "induce" the development of a special kind of tissue, for example, nervous tissue. The rate of development at this stage is also remarkable. During the first eight weeks of life almost all the organs are formed. In the eighth week, the embryo is only about two inches long, but could be recognized as human.

The outer germ layer, the ectoderm, gives rise to the nervous system, skin, nails, hair, dental enamel, the mucous membranes in the mouth and the throat, and parts of the eye.

The mesoderm gives rise to muscles and connective tissue, cartilage and bones, the vascular system and the heart, the bony parts of the teeth, and the urogenital organs.

The inner layer of entoderm is the source for lung tissue, the digestive tract, and the glands.

The ovum has divided for the first time (1)
The chromosomes from the father and the mother have united. The composition of genes in all new cells formed at subsequent cell divisions will be identical. Cell division starts at once, and the developing cluster of cells moves slowly down the fallopian tube toward the uterus. After a week or so the blastocyst moves to the soft uterine mucosa.

The growing embryo obtains nourishment (2)
from maternal blood circulated through a network of fine blood vessels. But the maternal and the fetal blood circulations are completely separated by a thin membrane. The photo shows the attachment of the umbilical cord to the placenta.

Three-week-old embryo (3)
At this stage, the mother may not know that she is pregnant. Blood from the umbilical cord travels to the liver and from there to the heart. The large white protuberance is a primordial leg; arms are seen at the gill slits. Note segmentation: the orderly arrangement of vertebrae in the embryo has its evolutionary heritage in the worm. Eye development is also well under way.

Four weeks old and one-fifth of an inch long (4)
Note the characteristic bending of the embryo. The neural tube, the beginning of the nervous system, is still open. The protuberances at the top are the future arms. The legs, which develop more slowly, are visible at the bottom. In the background is the placenta, the depot for exchange of maternal and fetal blood.

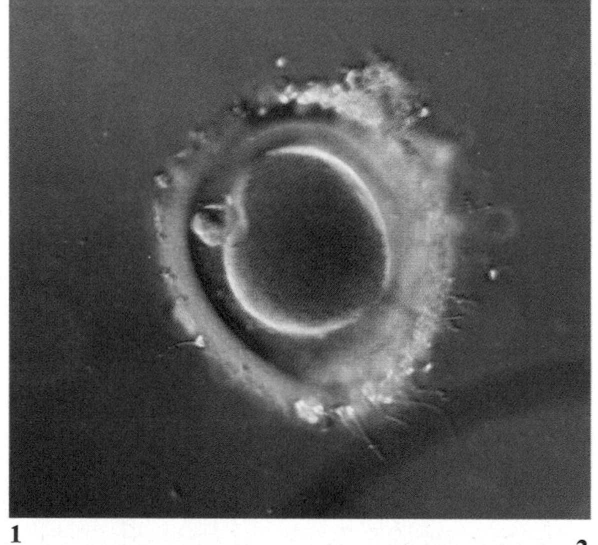

1

2

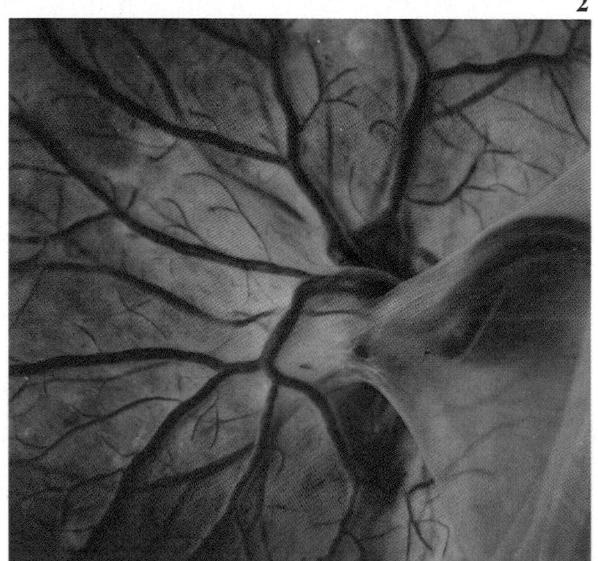

Five weeks old, and two-fifths of an inch long (5)
The embryo floats in amniotic fluid, anchored to the placenta by the umbilical cord. At the left is the yolk sac, the main producer of blood corpuscles at this time. Now the major divisions of the brain can be seen, an eye, the hands, the arms, and a long tail, and it is obvious that this is a human embryo. At earlier stages, the embryos of different animals can hardly be distinguished from a human embryo. The upper part of the body develops more rapidly than the lower one — development takes place from the top down.

Six weeks old and three-fifths of an inch long (6)
The embryo rests securely in its shock-absorbing amniotic sac. The heart beats rapidly. The brain is growing and the eyes are taking shape. The dark red swelling at the level of the stomach is the liver. The external ears are developing from skin folds. Throughout fetal life, the developing organism is very sensitive to virus infections or toxins which, unfortunately, may cross the placenta.

Seven weeks old, nearly an inch long, and weighing about two grams (7)
A tiny human being with all its outer and inner organs formed is now developing rapidly. It has a face with eyes, nose, lips, tongue, and even primordial milk teeth. The hands have the beginning of bones, though they are not more than a few millimeters long. There are muscles covered by a thin skin. The fontanel of the skull is seen on the crown.

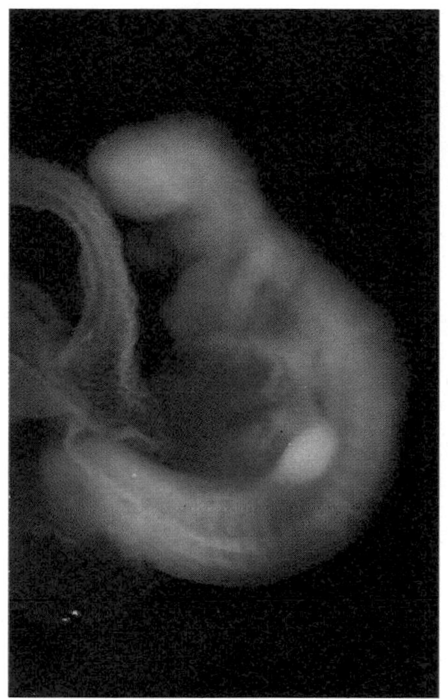
3 Three-week-old embryo

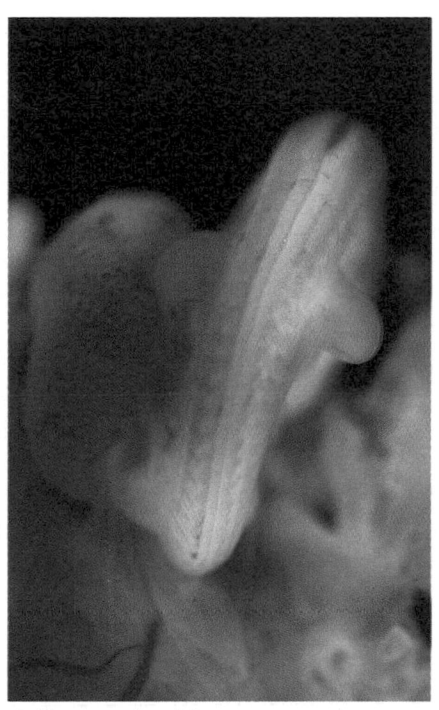
4 Four weeks old, one-fifth of an inch long

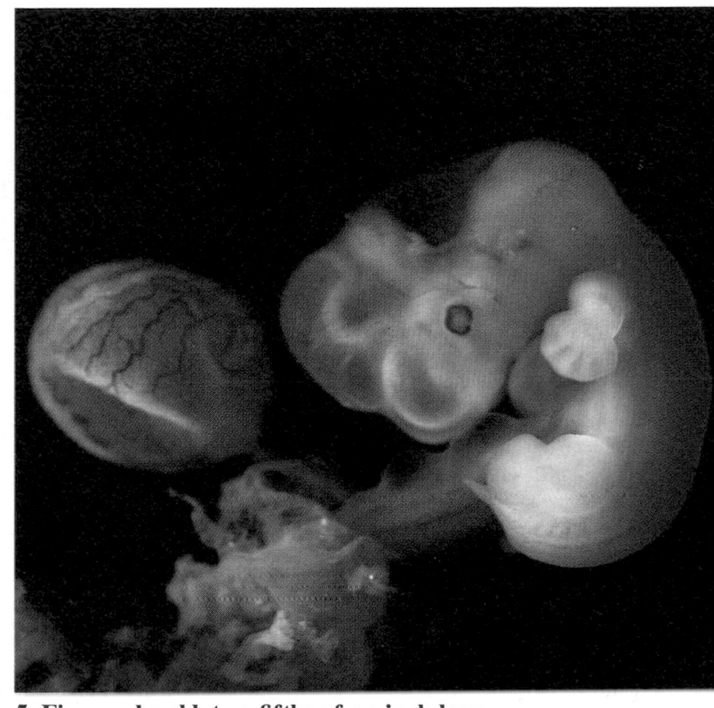
5 Five weeks old, two-fifths of an inch long

6 Six weeks old, three-fifths of an inch long

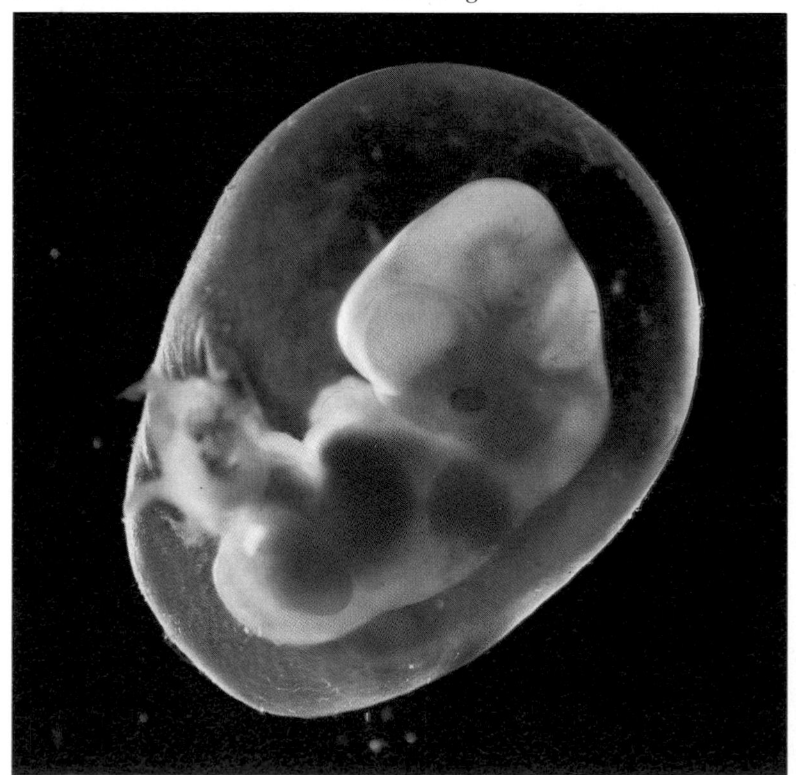

7 Seven weeks old, nearly an inch long, and weighing about two grams

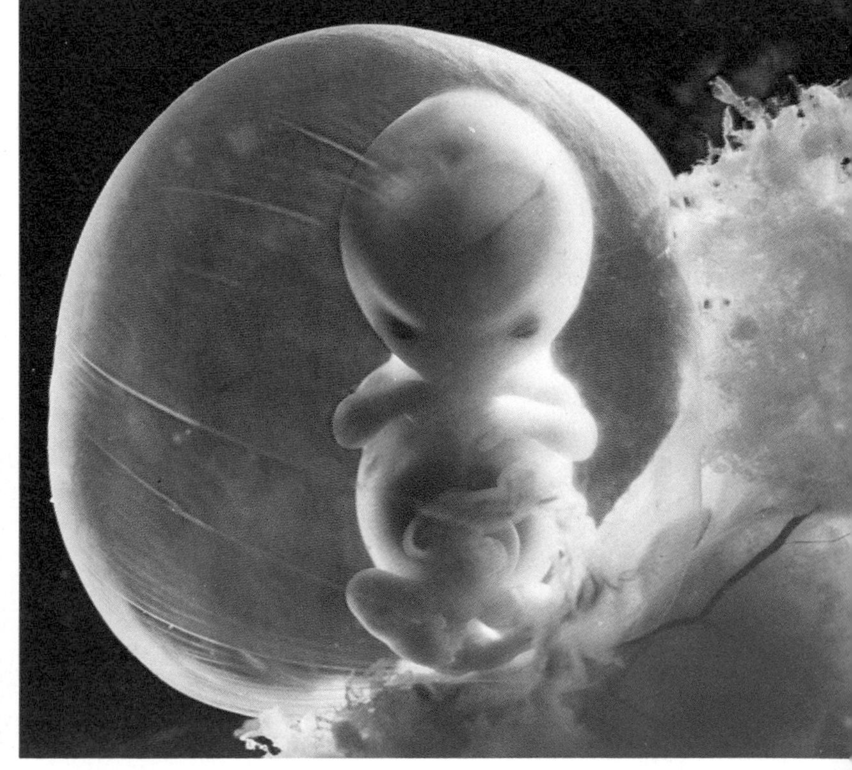

The Growth of the Fetus

Four months old, more than six inches long, and weighing about seven ounces

Eight weeks old, a little more than an inch long, and weighing about two and a half grams
At this time the mother may feel she is pregnant. The baby is now growing by a few millimeters a day; the two umbilical veins carry oxygenated blood from the placenta to the embryo — now called the fetus. The umbilical arteries remove waste products.

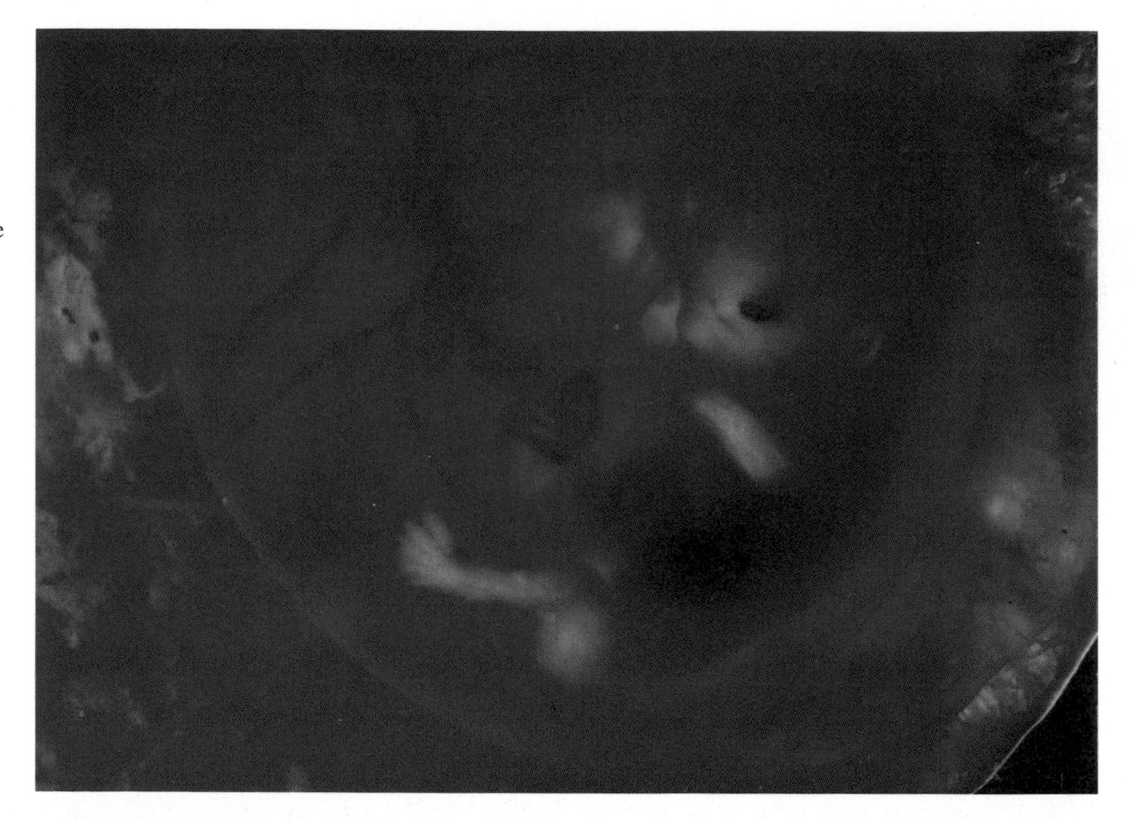

Three months old, over three inches long, and weighing almost an ounce
The head of the fetus is disproportionally large, but its features are clearly human. The fingers and toes are fully developed and external ears and eyelids have formed. The umbilical cord grows longer. Now comes a period when the fetus moves freely in its capsule — weightless as an astronaut in space.

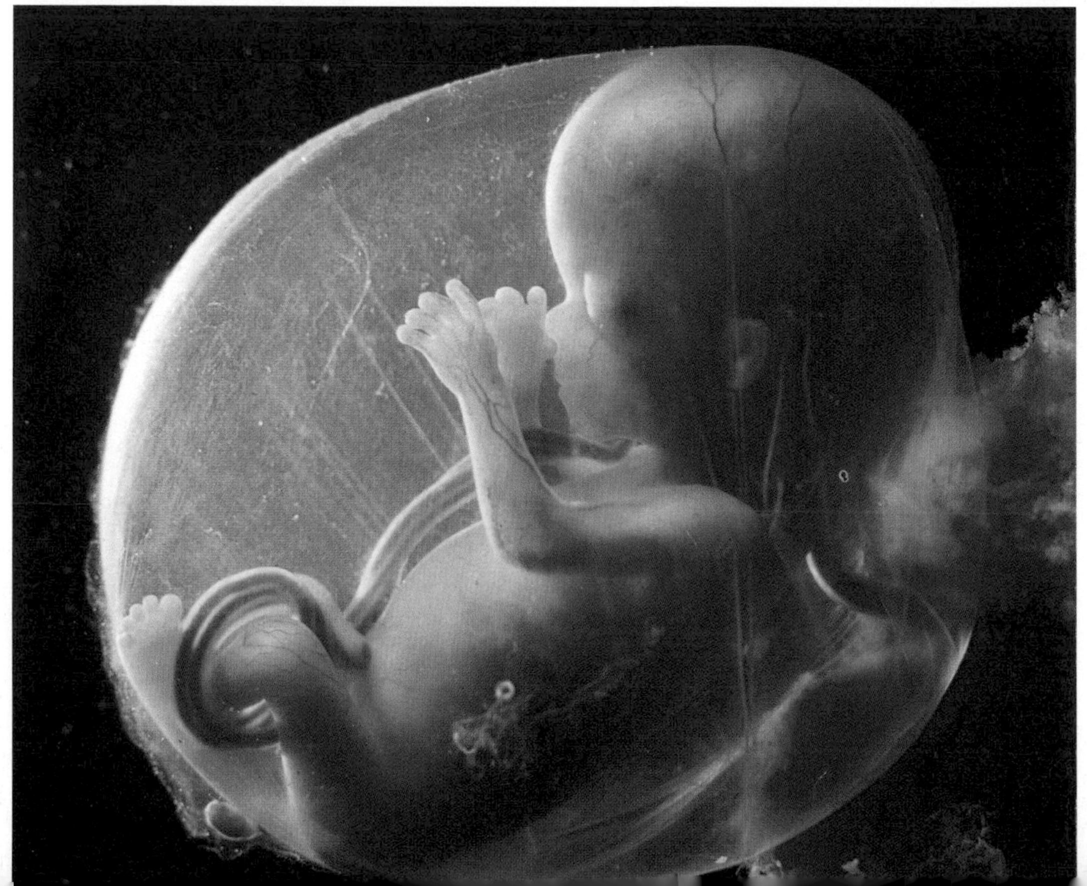

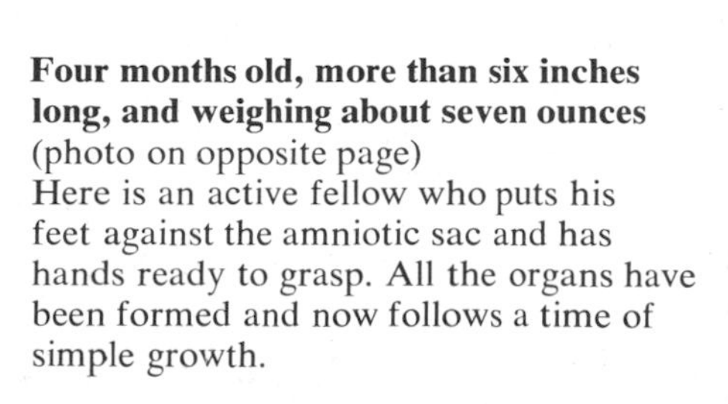

Four months old, more than six inches long, and weighing about seven ounces
(photo on opposite page)
Here is an active fellow who puts his feet against the amniotic sac and has hands ready to grasp. All the organs have been formed and now follows a time of simple growth.

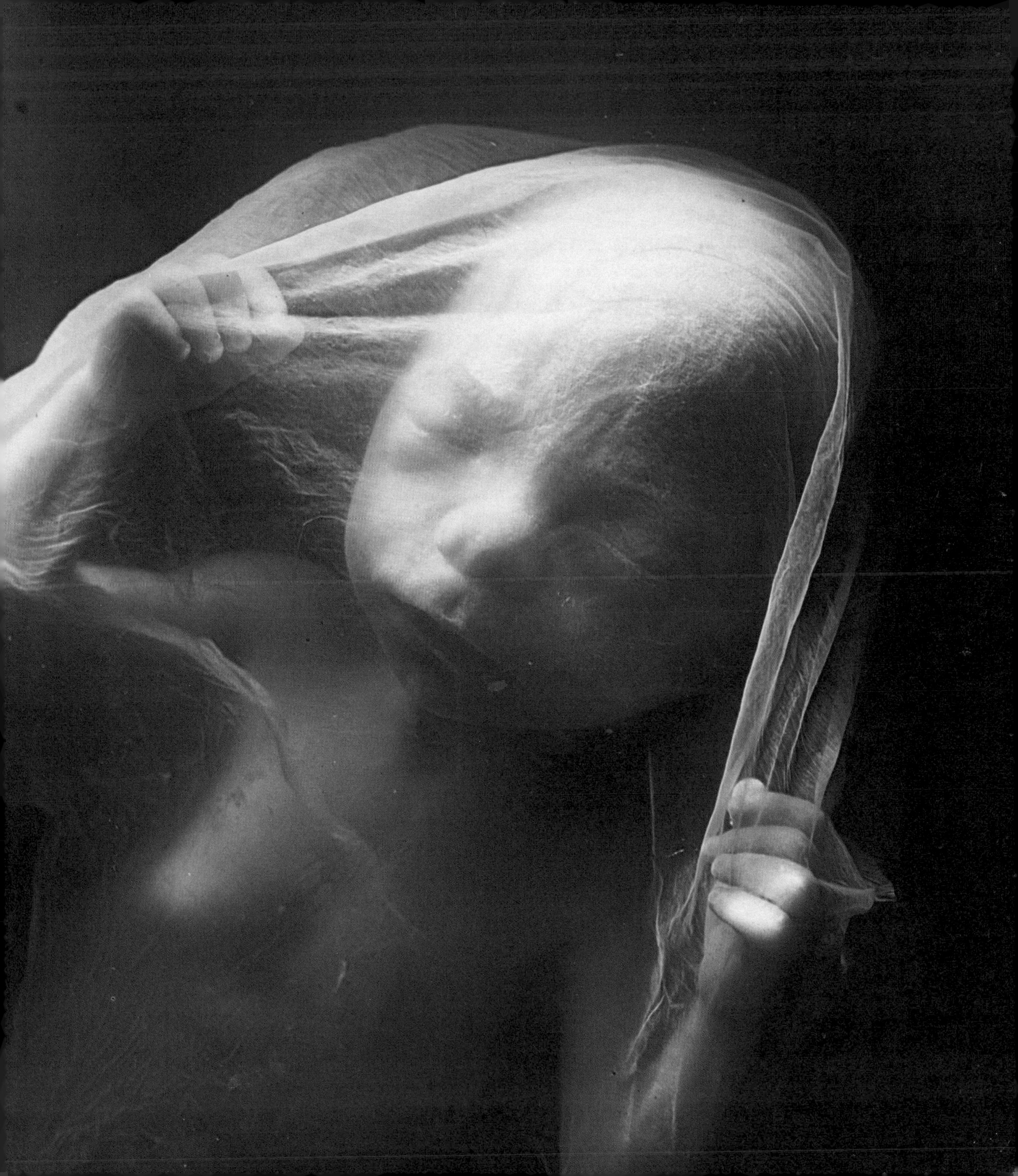

Early Stages of Fetal Development

- First week: the ovum is fertilized, divides, and is implanted into the uterus.
- Second week: the ectoderm, entoderm, primitive streak, and mesoderm develop.
- Third week: appearance of the first somites, or body segments, which eventually form the permanent spine, and the beginning of the brain and the spinal cord.
- Fourth week: heart, blood circulation, and digestive tract form.
- Fifth week: arms and legs begin to take shape; the heart pumps blood.
- Sixth week: formation of the eyes and ears.

Though the baby lives and develops inside its mother, the two are effectively separated. This is necessary in order that fetal tissue not be rejected as foreign and incompatible with the tissue of the mother. Contact between mother and child takes place through the placenta. Normally, the placenta serves as a biochemical barrier, preventing the exchange of blood corpuscles or large protein molecules between mother and child, but letting dissolved substances pass. The total area of contact the placenta establishes between the fetal and the maternal circulation is remarkably large — about twelve square yards.

The child lives inside its mother for nine months, floating weightlessly in a dark wet world of amniotic fluid. At delivery, it will literally be pressed and pushed out into a very different world. Not even the pearl diver returning to the surface experiences such a dramatic change. At the moment the umbilical cord is cut, the lungs must function, and the foramen ovale, the opening between the two upper chambers of the heart, is closed.

An independent life has begun.

The Blood
The Body's Fluid Tissue

One-celled animals and plants have no circulatory system. They do not need any, since each cell is in immediate contact with its surroundings and the vital substances to be found there. Multicellular organisms require some means of transport so that substances can pass into and out of the various cells.

Blood is a fluid tissue that carries oxygen and nourishment to cells and removes waste products. Blood also carries hormones and chemicals that regulate acidity so that cells have a suitable chemical environment. In addition to these vital activities, blood also participates in the control of body temperature and provides protection against infections.

The volume of blood in an adult is over six quarts. Half this volume consists of liquid plasma, yellowish in color. Blood corpuscles and platelets, small bodies important in the coagulation process, make up the other half. The red color of the blood comes from hemoglobin, a pigment containing iron, found in red blood cells.

The Body's Transport System

All cells in the body need oxygen to function. Brain cells in particular are very dependent on oxygen, and may be seriously damaged after only a few moments' deprivation. Other cells are less sensitive. Oxygen transport is the primary function of red blood cells (erythrocytes), which are manufactured in bone

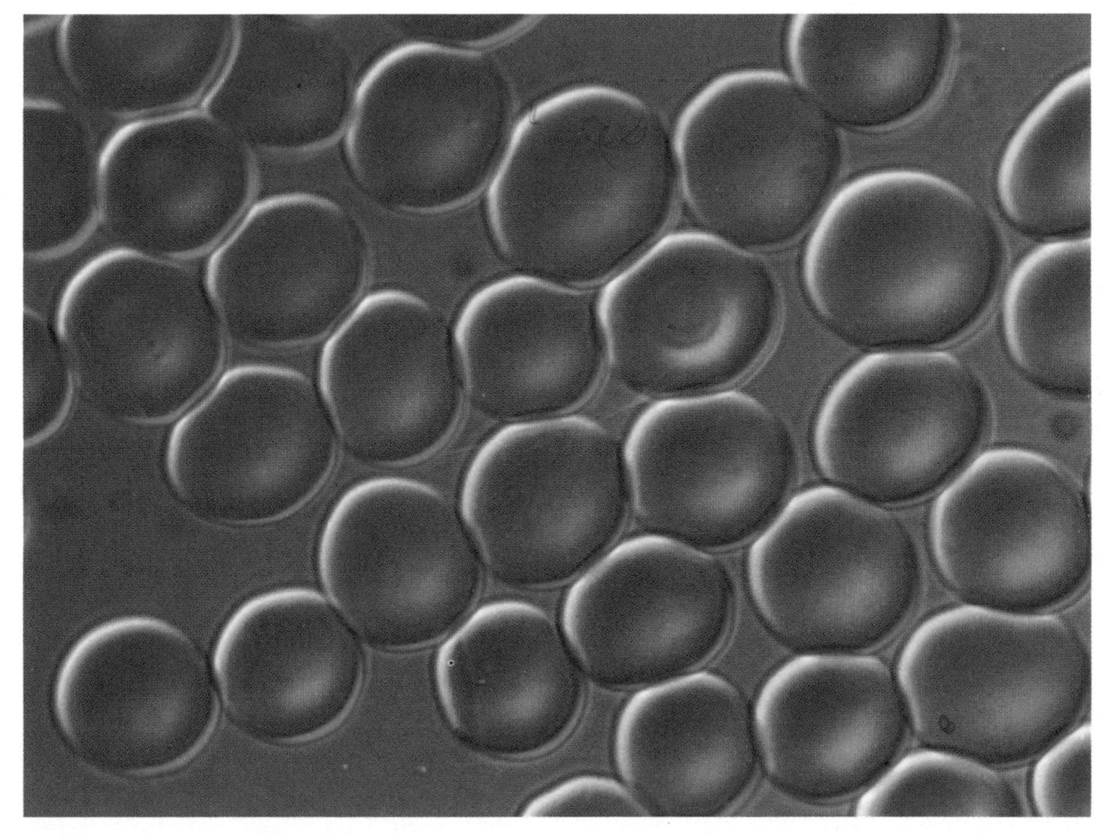

Red blood cells floating freely in blood plasma. In endless procession the disk-shaped, oxygen-carrying cells flow through blood vessels. They are small: a row of 150 would be about 1 mm long. There are so many of them, however, that if they were arranged in a line they would extend four times around the equator.

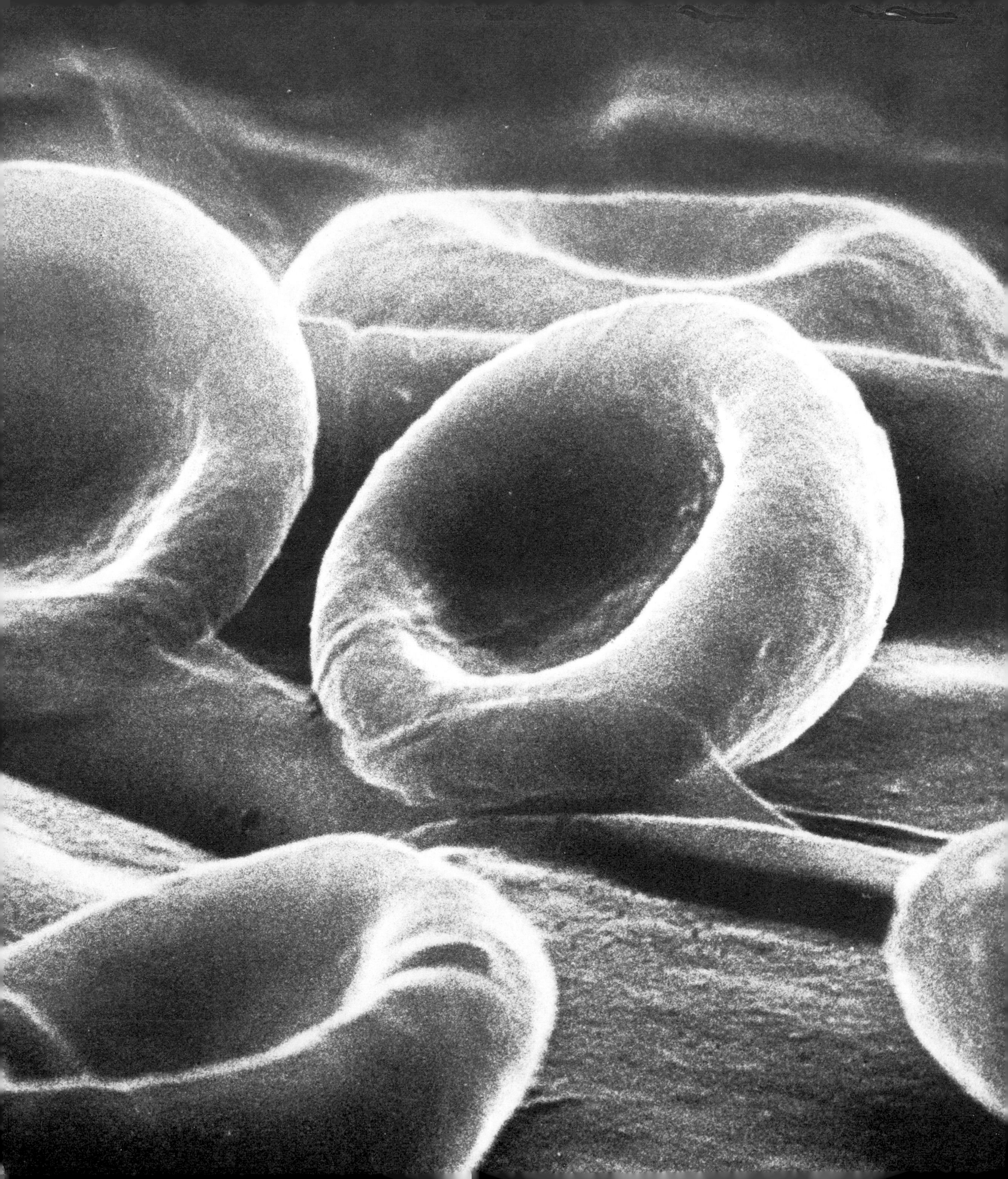

marrow. Oxygen becomes bound to the hemoglobin of red blood cells circulating in the lungs and is subsequently delivered to body tissue. Red blood cells are disk-shaped with slight depressions at the center, like two saucers placed back to back, or a solid doughnut.

Blood plasma transports fats, carbohydrates, and proteins from the digestive tract and liver to body cells. Plasma also contains salts, vitamins, and other substances necessary for cell metabolism.

Carbon dioxide is one of the waste products of metabolism. It is circulated to the lungs and leaves the body in exhaled air. Other products of metabolism circulate in the blood to the kidneys, where they are filtered and excreted in urine. Some excretion of urea also takes place through the sweat glands.

When body temperature rises, blood vessels in the skin dilate. This permits the movement of a large volume of warm blood from the body's interior to the surface where heat may be lost through radiation. The result is a drop in body temperature. Under conditions of cold just the opposite happens: surface blood vessels constrict to retain as much heat as possible.

Defense against Infection

In defending the body against bacterial infection, the circulatory system recruits a variety of troops and weapons. These include white blood cells and antibodies, a kind of self-governing army of robots. White blood cells (leukocytes) can leave blood vessels and move freely among body cells to seek out the enemy and destroy him.

Antibodies are proteins that are produced by lymphocytes, a type of white blood cell. Particular antibodies are produced in response to particular types of infectious matter: the antibody is said to "recognize" the invader on the basis of previous contacts. The portion of the blood called gamma globulin is the source of antibodies against a variety of infectious diseases.

When blood vessels are damaged, a series of events takes place to stop bleeding, reduce leakage, and repair the damage. The coagulation process involves the formation of a network of fibers which traps blood platelets (thrombocytes) and blood corpuscles, preventing further blood loss.

◁ **Red blood cells magnified 20,000 times** in a scanning electron microscope. Under the same magnification, a man would be twenty-two miles tall. Red blood cells are very elastic. They can easily squeeze through the narrow capillaries. This elasticity is aided by their biconclave (doughnut-like) shape which makes them thinnest at the center. Their diameter varies between .007 and .008 mm, and they are approximately .002 mm thick. Their water content is about 60 percent. Red blood cells contain hemoglobin, an iron-containing protein which can bind oxygen. Hemoglobin is a pigment, which gives blood its red color. Oxygenated arterial blood is a brighter red than venous blood, which is low in oxygen.

Blood cells in a scanning electron microscope ▷
The acidity of blood (measured in pH) and its salt composition are kept very constant. Even small changes in the balance lead to disturbances in many organs. The red cells in the photo are floating in a lower than normal salt solution. Under these conditions the cells absorb water, swell, and may even burst.

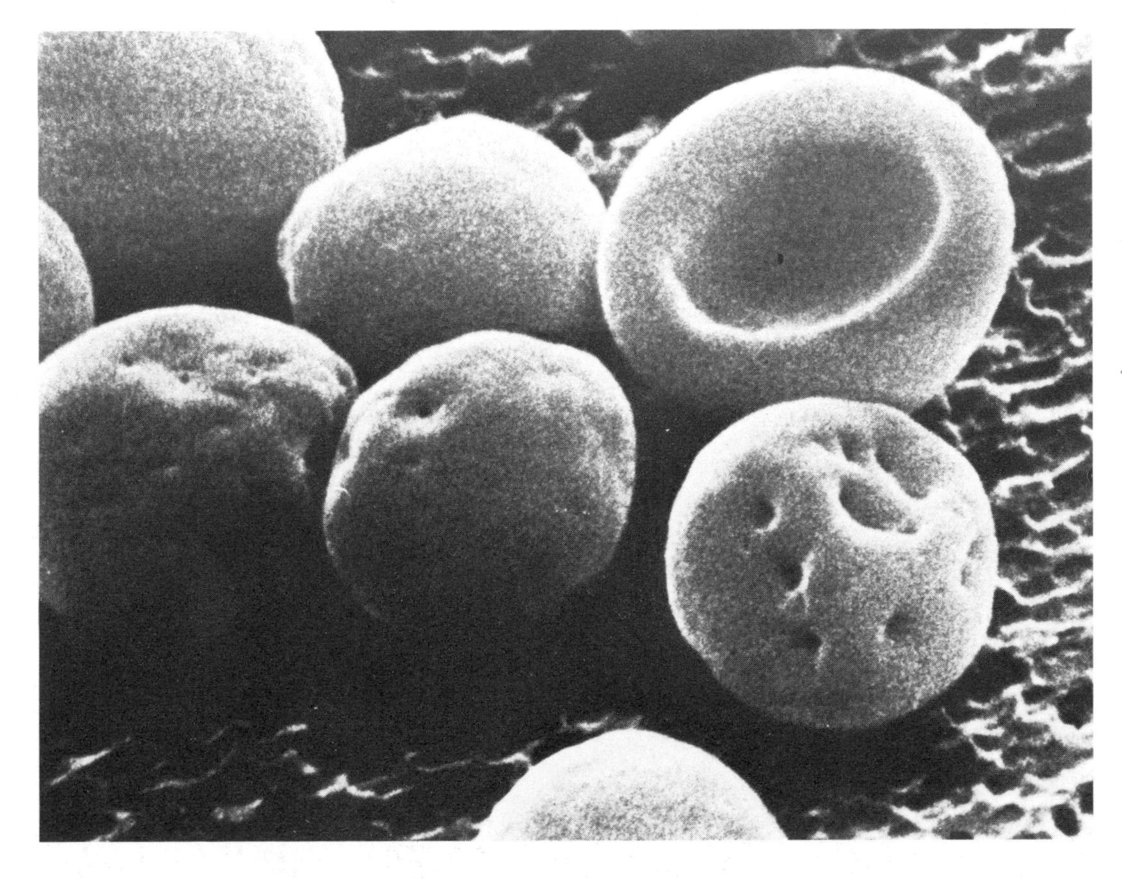

The Formation of Blood Cells

The site where blood cells are formed (1)
Red blood cells have no nucleus and cannot reproduce. New cells are formed in red bone marrow. Every second between two and three million are launched into the bloodstream. An equal number is rejected every second, having lived a lifetime of about four months. The old cells are disposed of in the spleen, the liver, and in bone marrow. Reusable parts, in particular iron, are recycled for use in new cells. The remaining substances are excreted in urine and feces. The cavities in the vertebra shown in this photo contain bone marrow.

1 Cavities in a vertebra

Newly produced red blood cells (2)
have nuclei, as seen in this stained microscope slide. The nuclei disappear just before the cells enter the bloodstream. The absence of nuclei in red blood cells is a mammalian characteristic. All other vertebrates have nucleated blood cells. The normal red blood cell count in man is between 4 and 6 million per cubic millimeter of blood — more in men than in women.

More mature red blood cells (3),
soon ready to enter the vascular bed. Some still have a nucleus, but most have not. The function of red blood cells as carriers of carbon dioxide to and oxygen from the lungs is made possible by the iron-containing pigment hemoglobin. Most of the iron comes from old, destroyed blood cells. However, we normally need an additional .01 gram of iron a day, which is obtained from food. We also need a small amount of copper, since this forms part of one of the blood-cell-producing enzymes.

White blood cells (4)
differ from red ones in that they have a nucleus. There are three main types of white blood cells: granulocytes, lymphocytes and monocytes. Their number varies between 4,000 and 9,000 per cubic millimeter of blood. They are formed in bone marrow, in lymphatic tissues, and in the liver and spleen.

2 New red blood cells

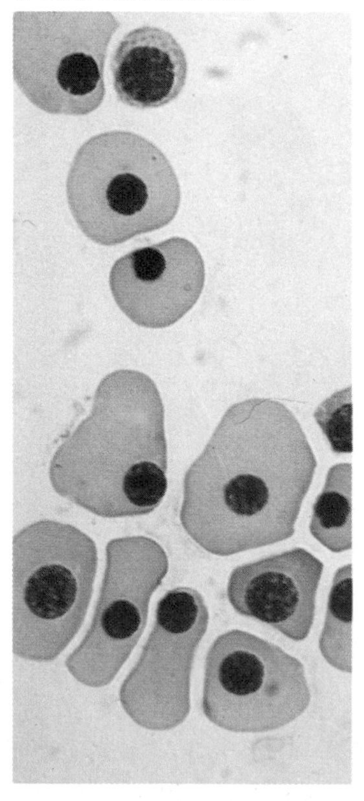

3 More mature red blood cells

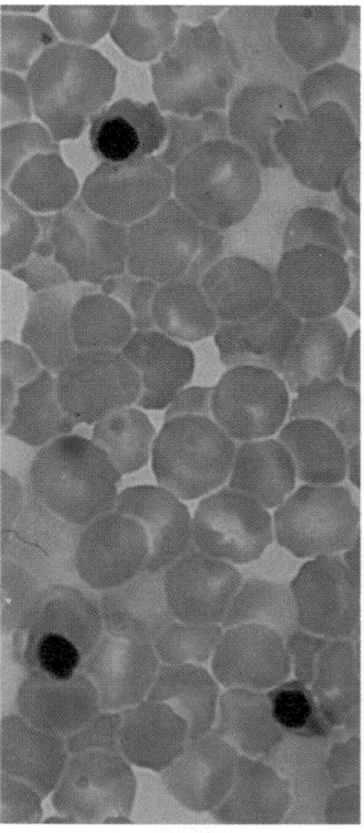

4 White blood cells

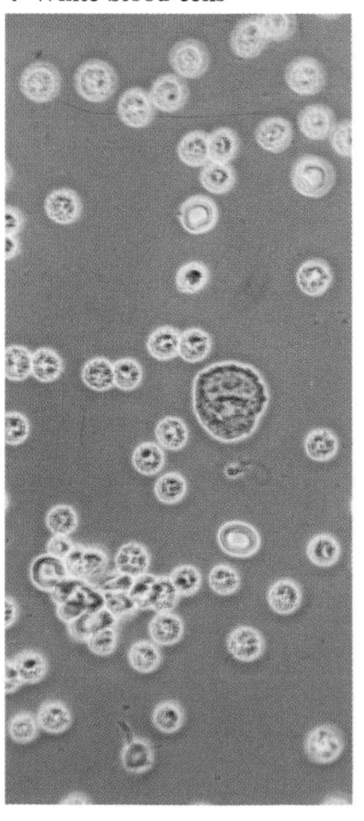

White Blood Cells in Action

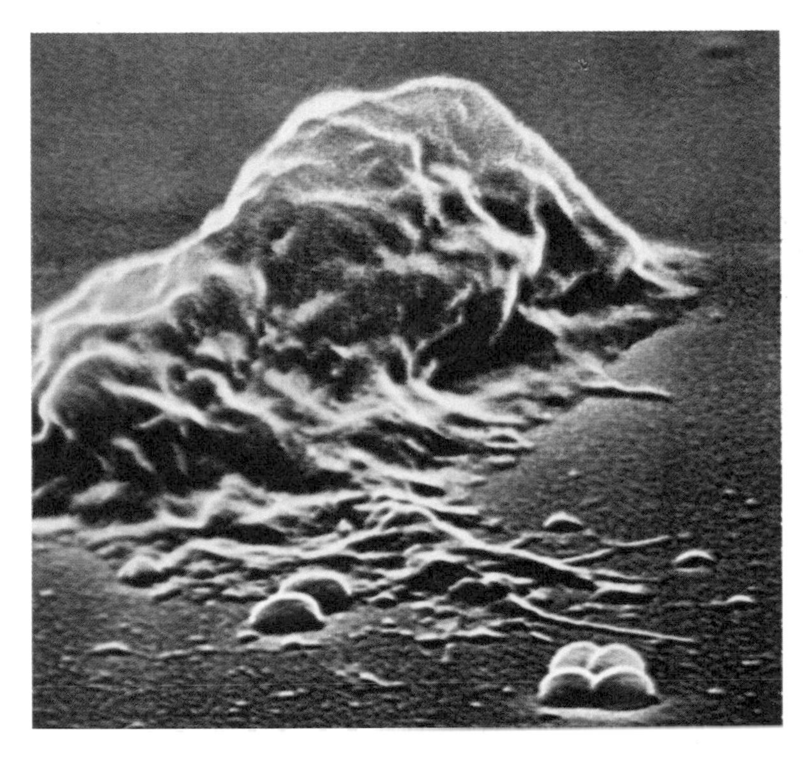

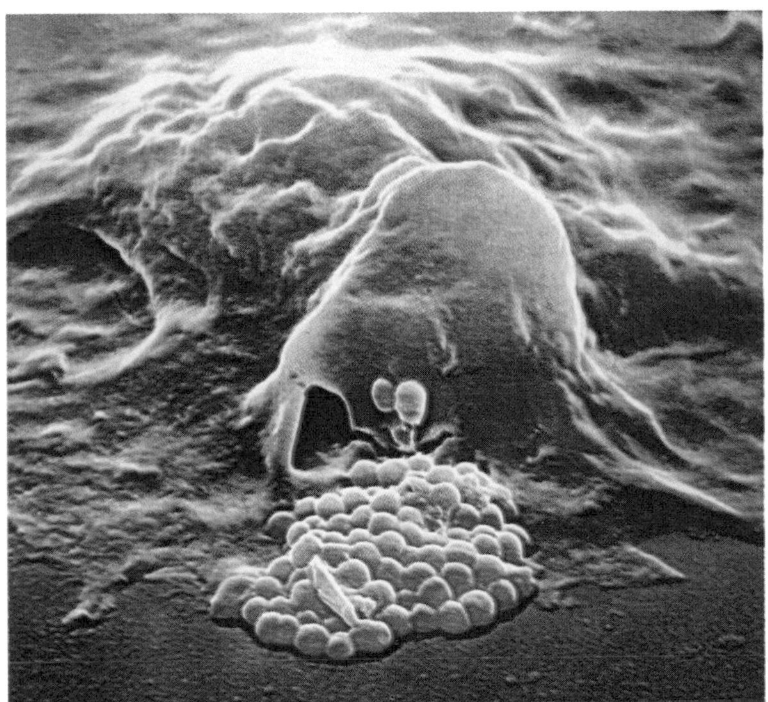

White blood cells in defense of the body This series of photos shows the contest between white blood cells and invaders of various kinds — bacteria in this case. The struggle rages from the first day of life to the last. Above left, a white blood cell, a lymphocyte, approaches a cluster of bacteria (green spheres). Next, the lymphocyte surrounds, encapsulates, and ingests the bacteria, taking the solid matter into its own protoplasm. The process is called phagocytosis. Ingesting bacteria is not always beneficial to the cell. It sometimes dies after having performed its duty. The pus in wounds consists of active and dead white blood cells, bacteria, and tissue remnants. During an infection, the number of white blood cells increases rapidly. After recovery, it soon normalizes again. Each cell in the body must have the proper chemical identity, reveal the "password," so to speak. If not, it may be attacked by lymphocytes and be destroyed. An example of such a process is the rejection reaction that is frequently seen in organ transplants.

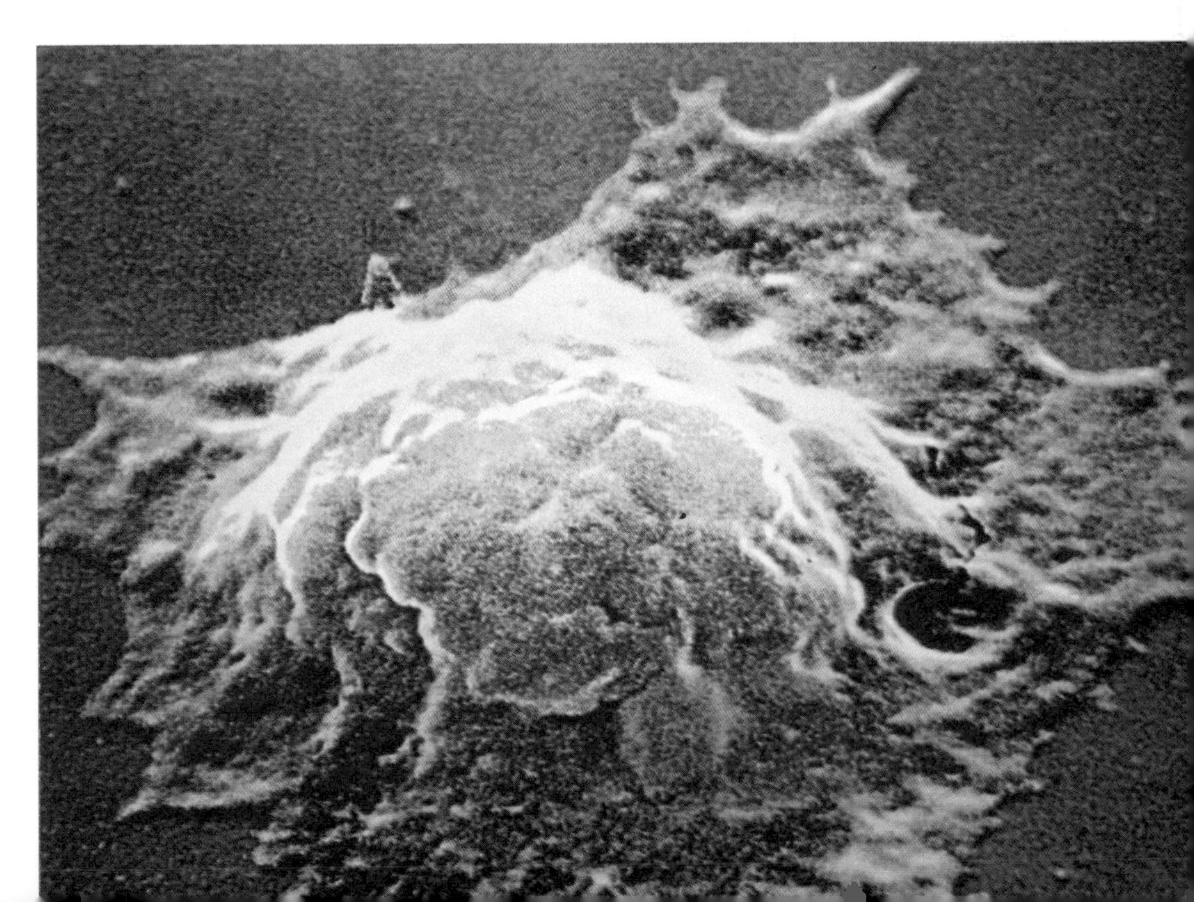

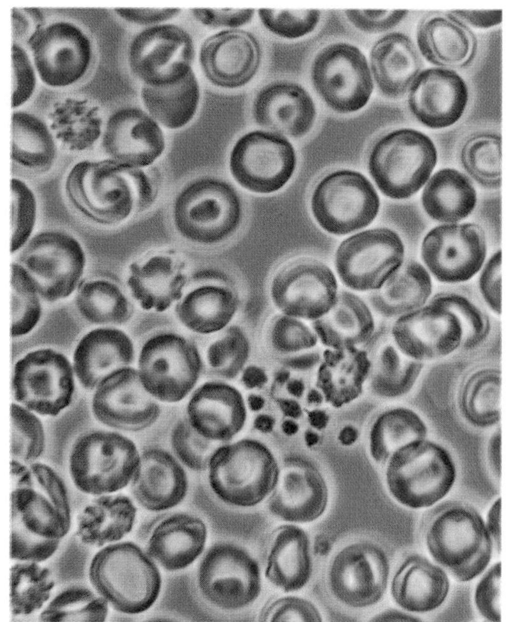

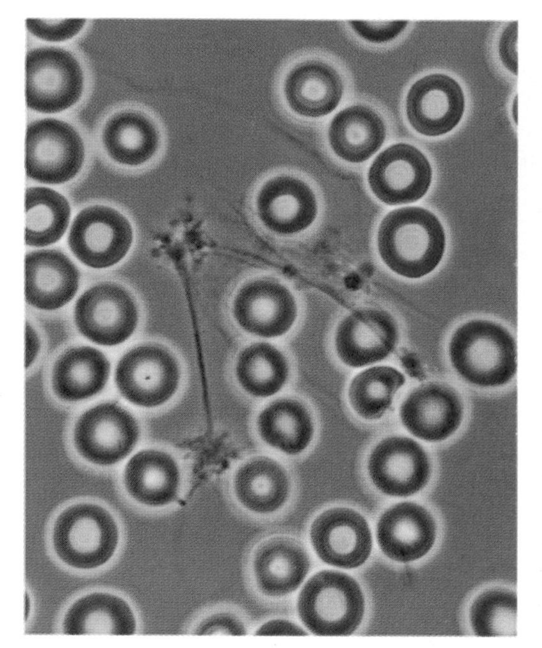

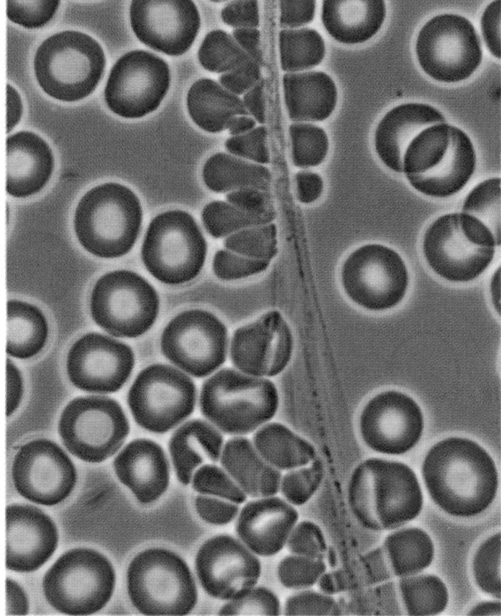

How blood clots
The capacity for clotting or coagulating, for stopping bleeding from a wound and starting healing, is one of the most mysterious and important properties of blood. The details of the process are not fully understood, but it is known that direct contact with air is not necessary. Even internal leaks of blood can heal rapidly in the absence of air. More than ten different chemical substances appear to be necessary in the clotting process.

The photo series illustrates how a clot is formed. The first picture shows small black bodies near the center. These are blood platelets (thrombocytes), small unnucleated bodies which are formed by special giant cells in bone marrow. One cubic millimeter of blood contains between 200,000 and 300,000 thrombocytes. If a blood vessel is injured, the platelets rapidly collect. The middle picture above shows how, simultaneously, or moments before, a protein in the plasma, fibrinogen, is converted into long fibers, fibrin. These fibers rapidly weave a net that traps platelets and blood cells. The remaining pictures show later stages of the process up to the completed clot, which later becomes the scab of the wound.

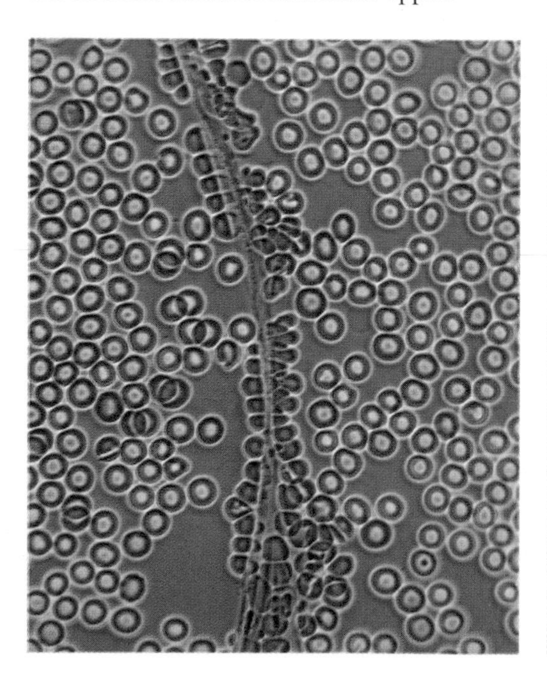

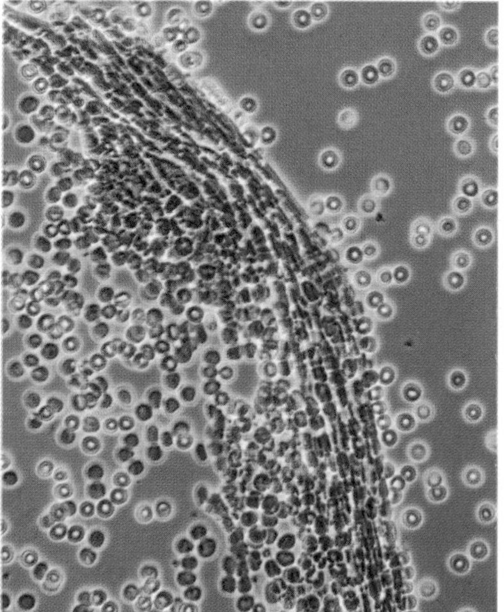

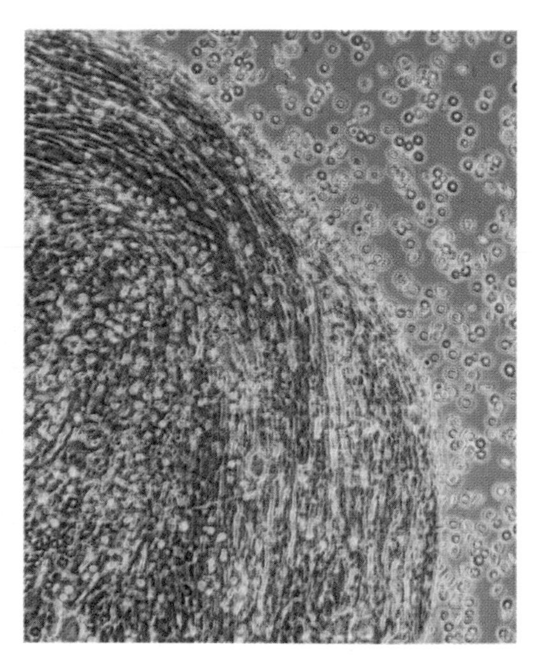

Blood Vessels

Almost all cells in the body are dependent upon an adequate blood supply. Without it, they die from a lack of oxygen and nourishment, or are poisoned by their own waste products. This happens to brain cells in the case of a cerebrovascular accident, or stroke, for example, when a blood clot in some part of the brain cuts off circulation in the vicinity. If such a clot occurs in the heart, the condition is called a myocardial infarction.

Arteries, veins, and capillaries are the main types of blood vessel. The capillaries are the fine blood vessels which wind among individual cells and provide the points of contact for the exchange of gases or dissolved substances. The wall of a capillary consists of a single layer of cells, and its narrow lumen — it is only a few thousandths of a millimeter in diameter — makes it necessary for blood corpuscles to move in single file as they pass from capillaries into the venous system.

If capillaries are comparable to bicycle paths or footpaths, then arteries are major highways. The largest artery is the aorta, which arises from the left ventricle of the heart. Arteries carry oxygenated blood to all parts of the body. Their walls are composed of smooth muscles and elastic connective tissue. When the heart pumps blood into the aorta, its wall dilates and a wave or surge of blood moves through the arterial system. The rhythmic movements of the pulse reflect these pressure waves moving down the vessels, creating successions of dilations and contractions which travel at a rate of 23 feet a second.

Blood which has circulated in the capillaries has given up most of its oxygen to body cells and is dark in color as it trickles into the smaller veins (venules). These in turn drain into larger and larger veins as they approach the heart. The largest vein in the body is the vena cava, which returns blood to the upper right chamber (atrium) of the heart. Blood pressure in veins is much lower than in arteries. It is generated not by an active pumping arrangement, but by the pressure of surrounding tissues, applied to the walls of veins. Contractions of limb muscles are especially important in the return of venous blood to the heart because they apply force directly on vein walls, which are much thinner than arterial walls. The proper direction of the venous blood flow (toward the heart) is ensured by one-way valves along the inner walls. These valves close immediately in the case of reverse flow.

The skin, lungs, and spleen are among the major blood reser-

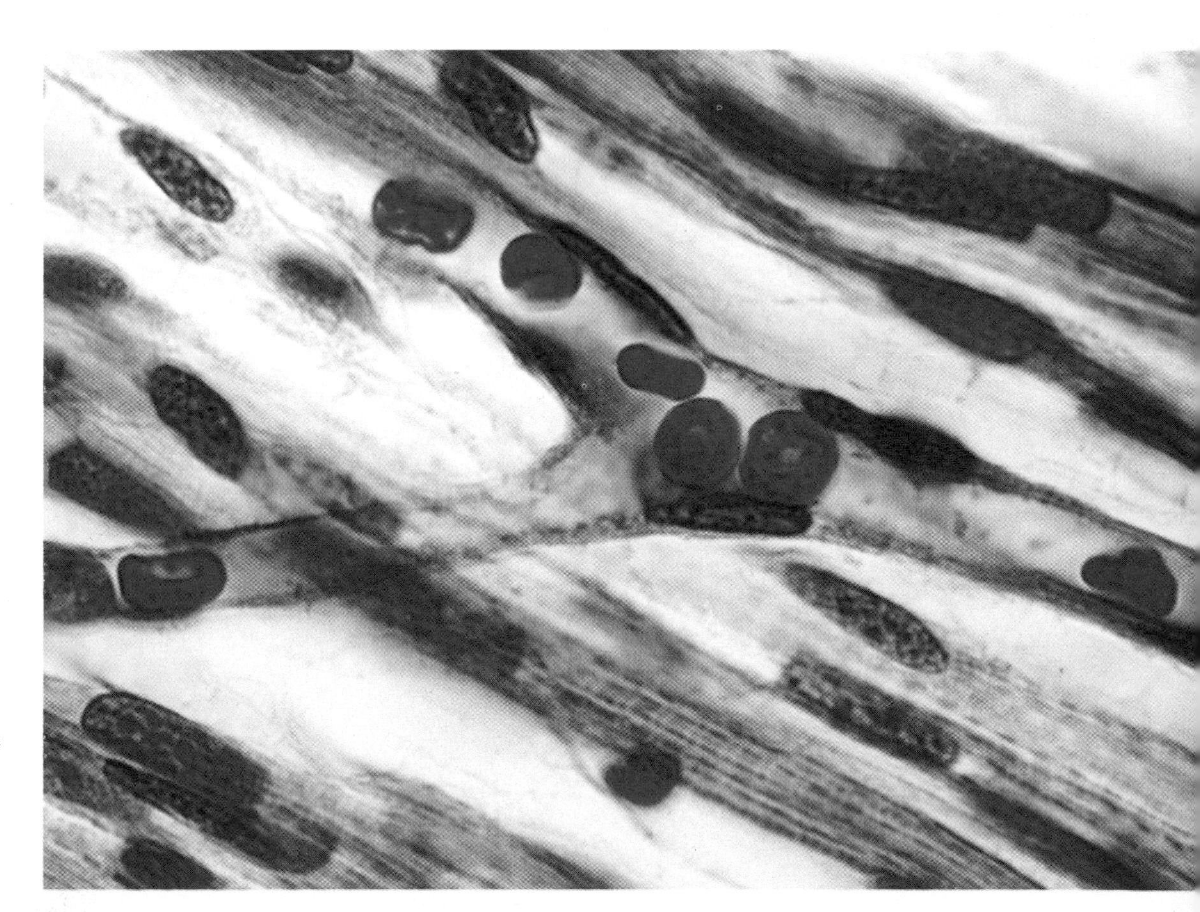

Red blood cells in capillaries, the smallest blood vessels in the body, through which blood reaches the body cells. Capillaries are also the connecting links between arteries and veins. Because the capillary diameter is only between .007 and .008 mm, red blood cells sometimes have difficulty in proceeding and may have to move sidewise, as shown here.

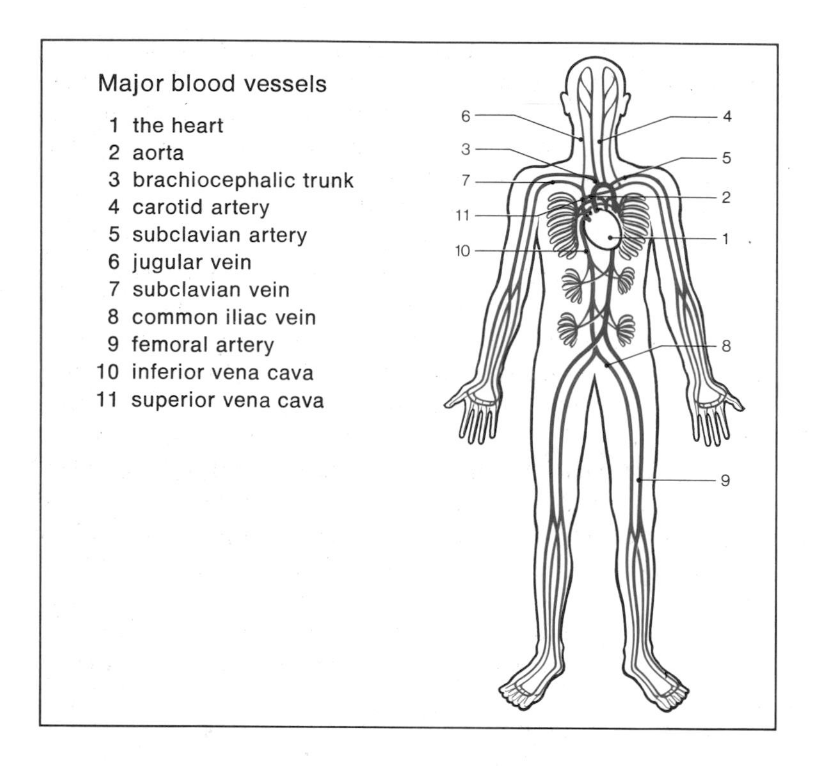

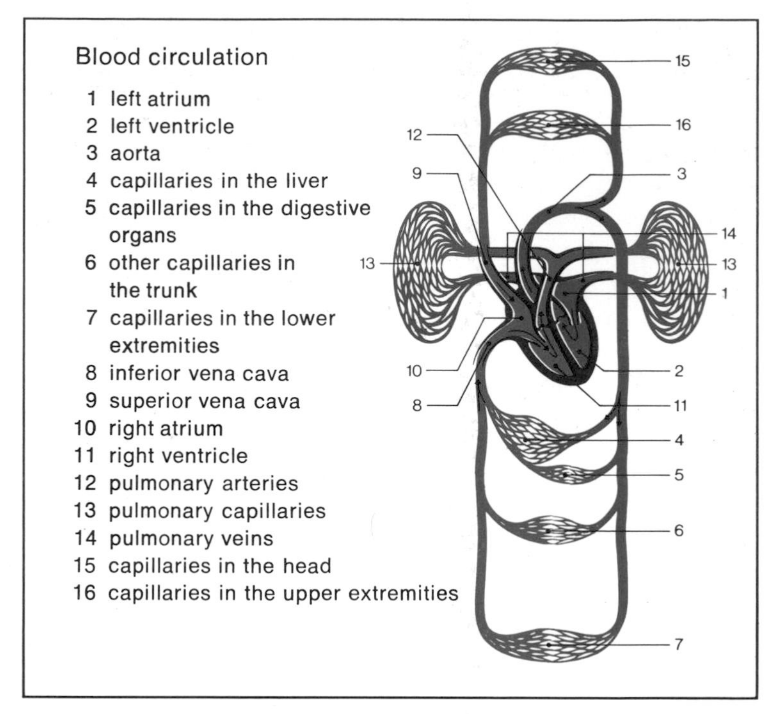

voirs in the body. When the vessels in these organs dilate, they can hold more blood; when they constrict, blood is made available to vital parts in an emergency. This happens, for instance, to counteract the effects of massive bleeding.

The Circulation of the Blood

The heart's pumping keeps blood in circulation. As it courses through the body, blood passes through the heart twice. The contractions of the left ventricle, the more powerful pump, sends blood through the body. This pumping motion can easily be felt as the pulse at the wrist, for example. Blood is distributed to body tissue through the capillaries and drains into the veins, which gradually increase in size. Venous blood finally enters the right atrium from the vena cava and flows from there into the right ventricle. This is the route of the systemic or general circulation.

The circuit of blood through the heart and lungs is the pulmonary circulation. Greater speed is necessary here so that the blood can pass through the lungs efficiently. This is the work of the right ventricle, the pump that forces blood through the pulmonary arteries to the lungs, where it gives off carbon dioxide and becomes oxygenated. The large pulmonary veins then carry oxygenated blood to the left atrium, from which it enters the left ventricle. It is now ready for systemic circulation.

The circulation of blood through the body was unknown until about 1600, when William Harvey, an English physician, demonstrated that two pumps in the heart were responsible for its movement. He was unable to explain how arterial blood passed into veins, however. It took another fifty years before capillaries were discovered.

What Is an EKG?

The electrocardiogram, or EKG, was invented by Willem Einthoven, a Dutch physiologist. It is based on the principle that as heart muscle contracts it produces small differences in electrical potential (on the order of a thousandth of a volt). These can be amplified, and visualized on an oscilloscope or recorded on film. The potential differences are picked up by electrodes placed at the wrist and the ankle as well as over the heart. The amplifier is coupled to a recorder that plots the EKG curve. The EKG is an important indicator of how well the heart is functioning.

The heart in a five-week-old embryo is already visible as a red spot next to the liver. The circulation of blood in the embryo is completely separate from the maternal circulation, yet in contact with it. The umbilical cord contains blood vessels which transport oxygen to the embryo and carry waste products away. Oxygenated blood in the cord empties into organs close to the fetal heart. Returning blood, low in oxygen, passes through vessels in the cord to the placenta. Here maternal and embryonic blood systems meet, but they do not intermingle, and here the exchange of waste products and oxygen takes place.

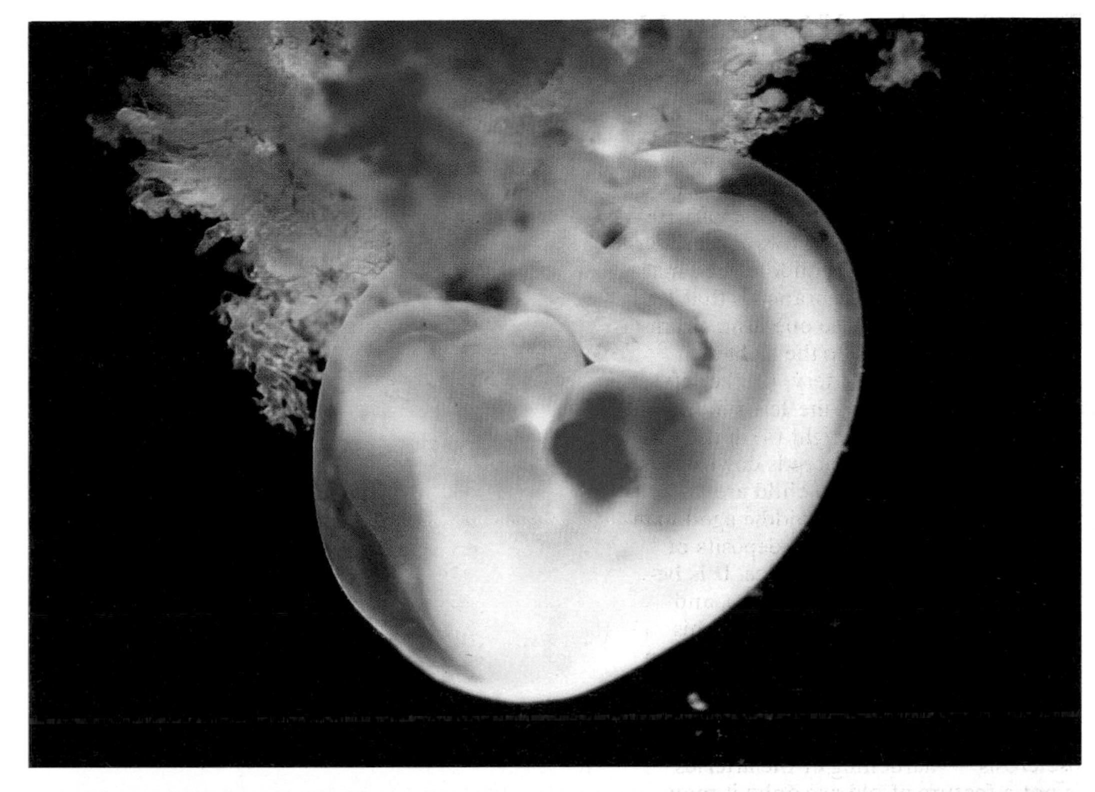

The size of your heart is about the size of your fist, the saying goes. In an adult it weighs between 8 and 12 ounces. The heart is located between the lungs in the middle of the chest, pointing forward and downward to the left. Its tip can be felt to the left of the breastbone. The heart is a hollow muscle. It functions as a pair of pumps arranged in series.

The heart is supplied with oxygenated blood through the coronary arteries, which are branches of the aorta. This front view of the heart shows how the major coronary arteries twist around the surface of the cardiac muscle.

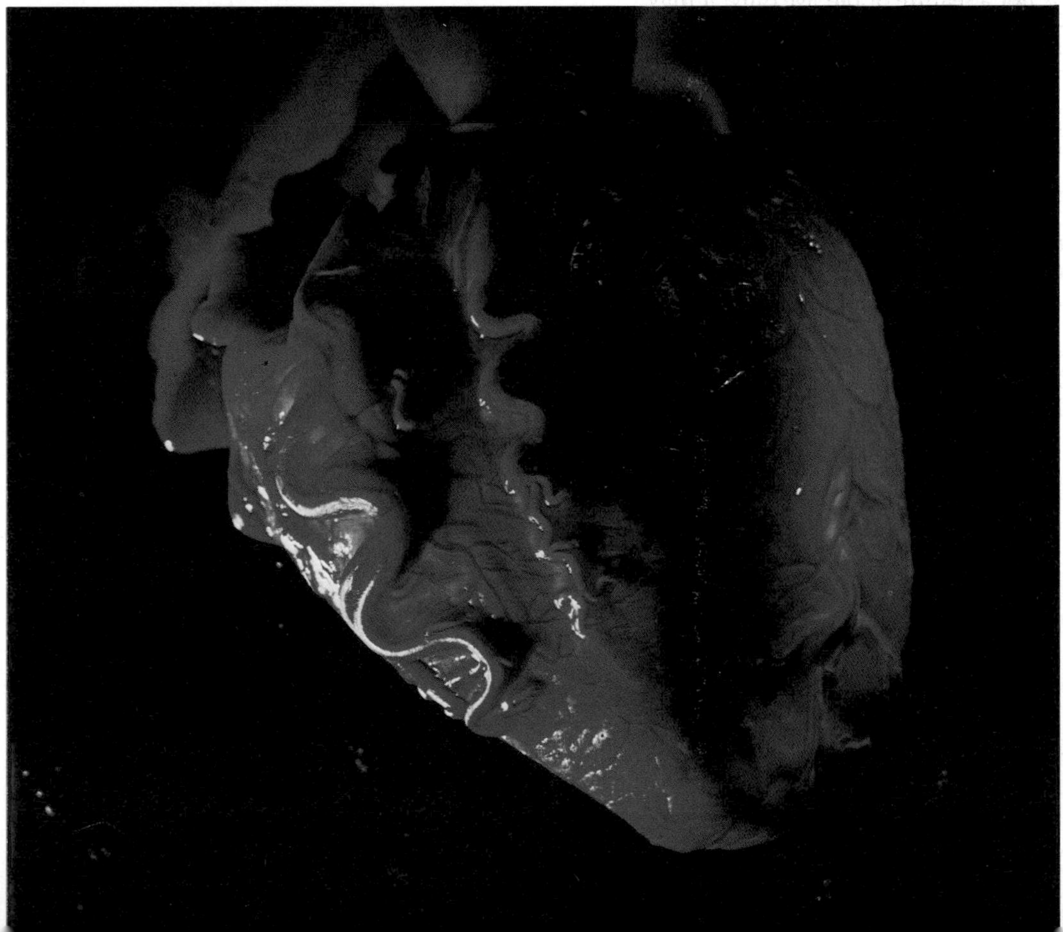

Inside the Aortic Arch

The aortic arch in a child and in an adult
The aorta is the major artery in the body. It arises from the upper part of the left ventricle, arches over and passes downward behind the heart. These photos show the aortic arch from inside.

The large picture shows the arch in a three-year-old child, magnified about 100 times. The small picture shows a sclerotic, calcified aortic arch in a middle-aged man. The three openings are branches that lead blood to the head and to one arm. From the left: (1) the artery to the right arm and the right carotid artery, (2) the left carotid artery, and (3) the left subclavian artery. Farthest to the right (4) in the dark area, the aorta proceeds downward. The vascular walls in the child are smooth and fine. In the middle-aged man they are often infiltrated by deposits of cholesterol, a fat-like substance. It is believed that lack of exercise, stress, and excess smoking may contribute to this condition. The deposits gradually harden so that the walls become rigid. Blood may clot on their rough surface, giving rise to the danger of thrombosis. Arteriosclerosis — hardening of the arteries — is not a feature of old age only; it may start in youth.

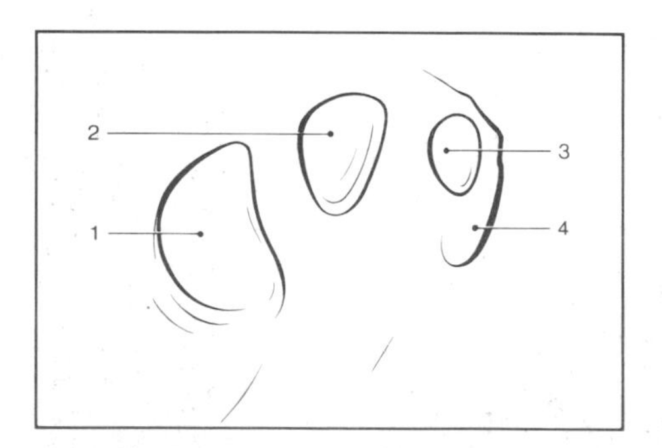

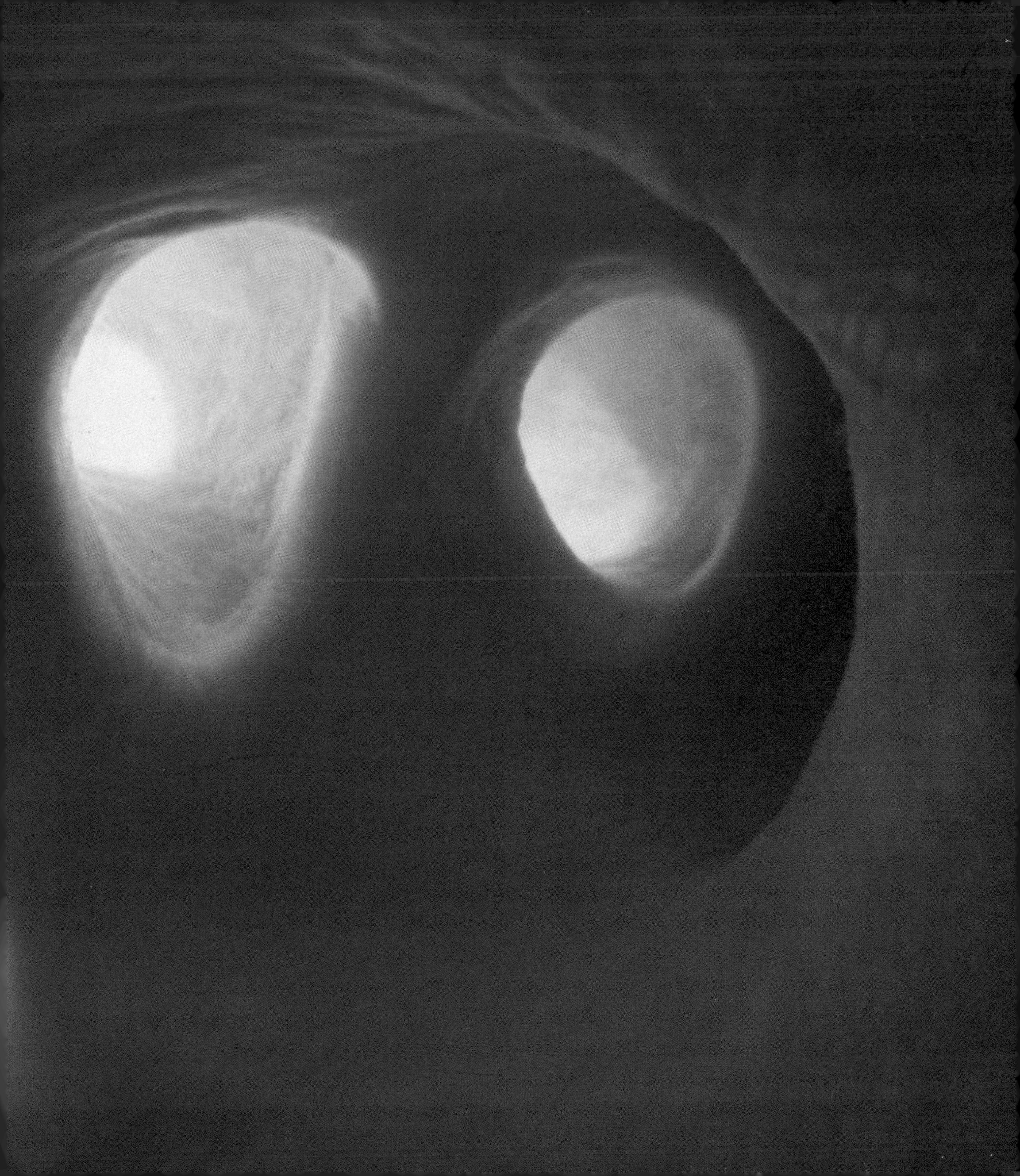

The Aorta

The aortic valve (1, 2)
The heart has four valve systems which assure that blood will flow in the right direction. Between the atria and the ventricles are cuspid valves (shaped like the points of a tooth). There are also valves in the aorta at its point of exit from the left ventricle, and in the pulmonary artery where it leaves the right ventricle. These are the semilunar (crescent-shaped) valves. Photos 1 and 2 show the aortic valve viewed from below and above. It consists of three segments of thin connective tissue. When the left ventricle contracts and presses blood against the aortic valve, the valve opens and a wave of blood is ejected into the aorta. We can feel this wave as a pulse, easy to detect at the neck, wrist, or groin. When the ventricle then pauses for a moment, the aortic valve shuts with a thud, preventing blood from flowing back into the heart. What we perceive as heartbeats are the sounds of both the aortic and pulmonary valves. These sounds are among those the physician listens to with his stethoscope.

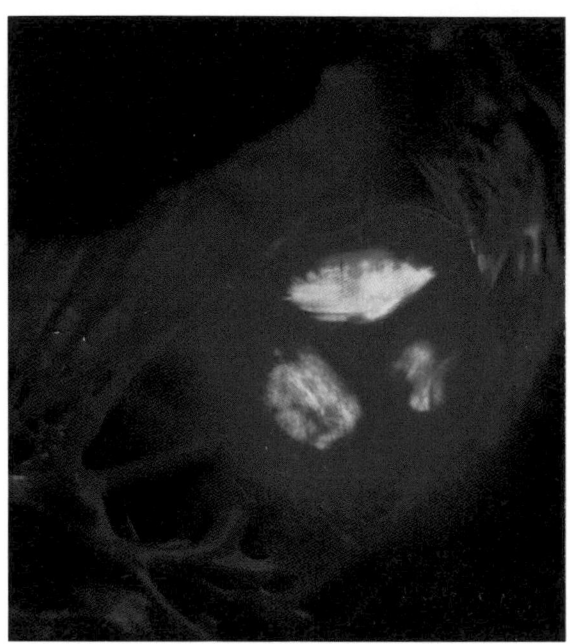

1 The aortic valve in the heart from below

2 The aortic valve in the heart from above

The aorta,
describing a wide backward and downward curve, takes blood from the heart and proceeds down through the body. This view looking down into the aorta makes its farther reaches look like a deep pit. The two bright points at the bottom are openings into the common iliac arteries. Just above them arise the renal arteries. One of them can be seen as a bright red spot. The wall of the aorta consists of several concentric layers of elastic tissue and smooth musculature, a necessary structure if the arteries are to fulfill their function. The heart does not propel the whole mass of blood forward in one motion at a time, but the vessels expand and contract elastically, driving a wave of blood onward. When arteries become sclerotic, they are less elastic, so the heart has to work harder. Moreover, they are less able to adapt themselves to the various demands made on the vascular system. Such vascular deterioration is often combined with high blood pressure. If the aorta is severed, as may happen in an automobile accident if the wheel is pressed hard against the chest of the driver, blood flows into the chest or abdominal cavity, and death is instantaneous.

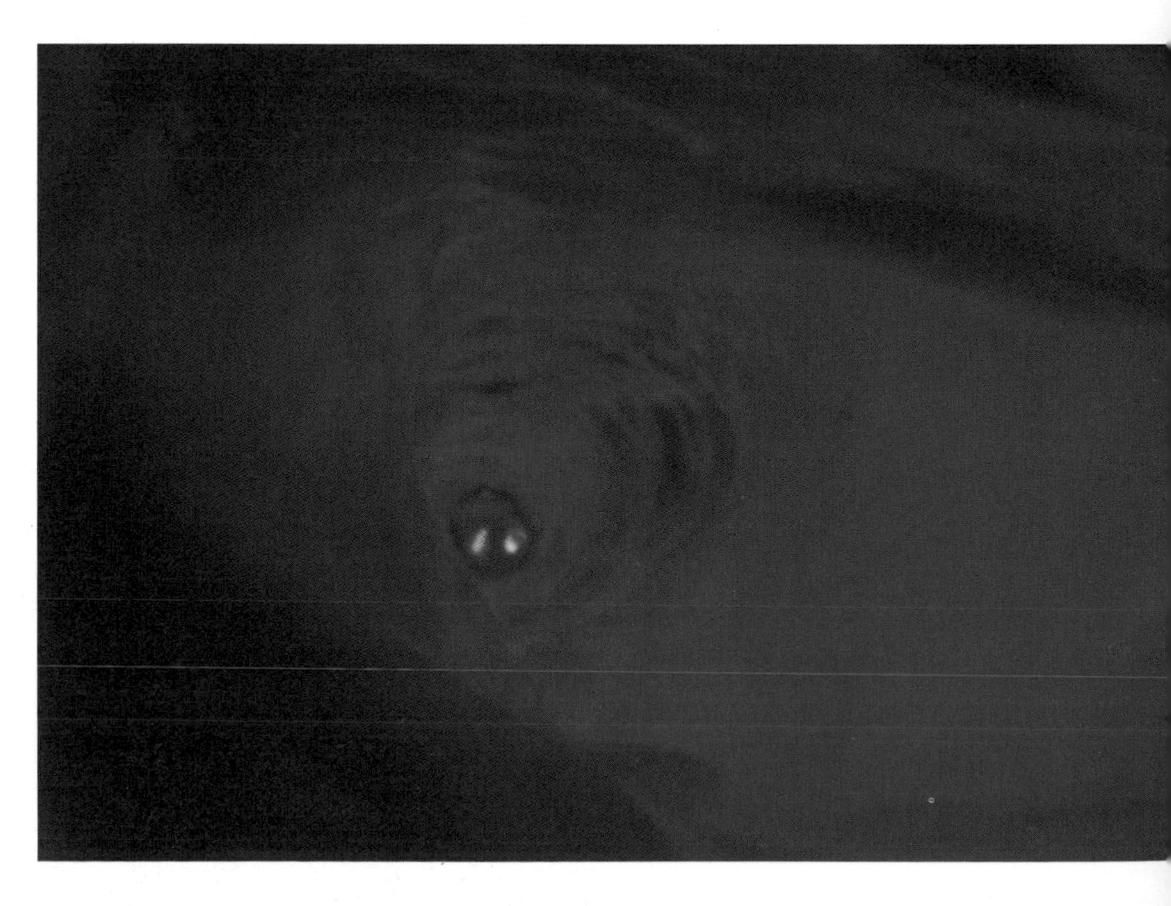

Blood vessels in the intestines ▷
Blood takes up nourishment in the intestines through a widely branched system of vessels and carries the nutritive substances to the various parts of the body. The vessels can be seen clearly in the photo because they have been injected with fluorescent dye. The nutritive substances are present in a dissolved state and form part of the plasma. If necessary, the blood can also pick up sugar stored in the liver. Blood also carries hormones from the many endocrine glands in the body. As hormones stimulate, accelerate, and regulate a number of chemical processes in the body, blood plays an important role as a carrier of signals.

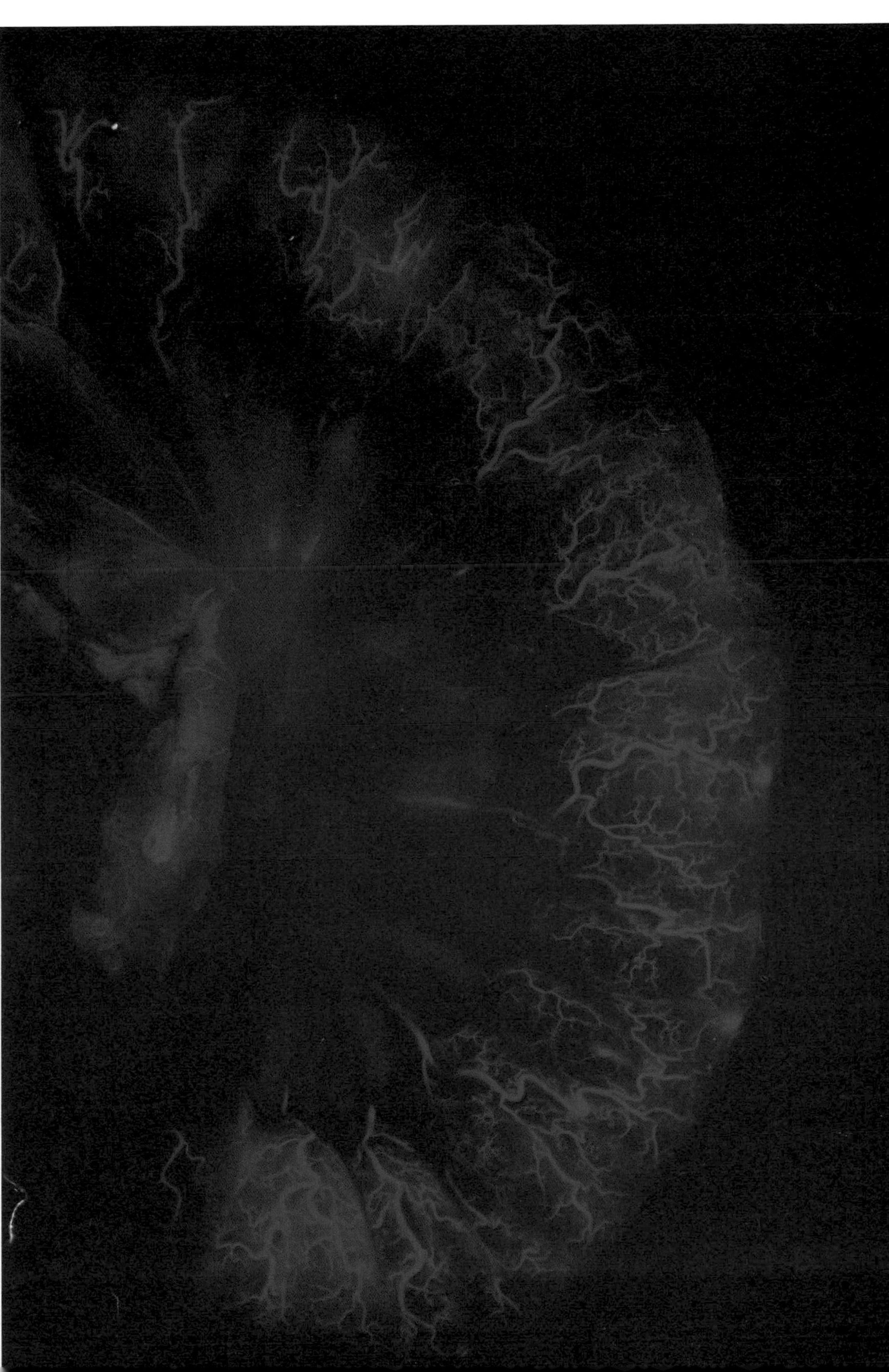

The interior part of the heart
We are at the base of the right ventricle of the heart, looking up toward the right atrium. We can see the tricuspid valve, which prevents blood from returning to the atrium when the ventricle is contracting. The three flaps then extend out to cover the opening. In the same way, blood is directed through the pulmonary valve to the pulmonary artery. The right side of the heart drives the lesser, or pulmonary, circulation; that is, it pumps returning venous blood, low in oxygen, to the lungs for oxygenation. The cuspid valves are controlled by fibers of connective tissue that arise from the papillary muscles in the ventricular wall and form a flexible framework. The contractions of the heart are controlled by parts of the nervous system, one of which is the sinoatrial node, a body of nerve cells located at the junction of the superior vena cava with the right atrium. The signals to contract are given by the sinoatrial node; from there they pass to nerve centers in the ventricles and finally to centers in the cardiac musculature. However, the heart is also stimulated by nerves from the brain and by hormones such as adrenaline in the circulating blood. In some cases of abnormal rhythm, heart rate can be returned to normal use by use of a battery-operated pacemaker, to replace the weak or faulty natural pacemaking system.

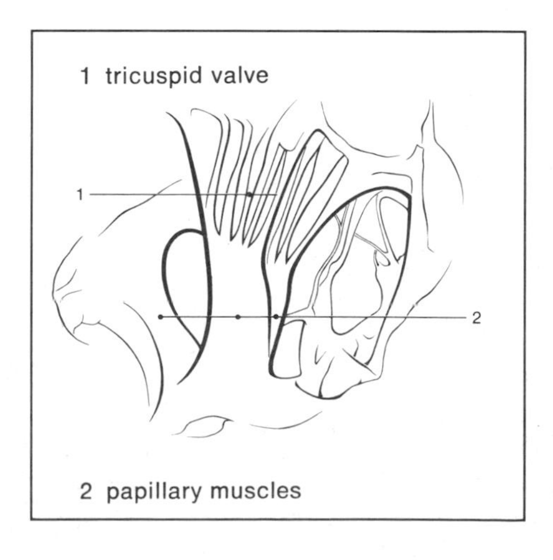

Circulation in the Lungs

The small air sacs of the lungs
Pulmonary alveoli are clusters of small, thin-walled sacs that form the tiny end-branches of the trachea, or windpipe. Here blood deposits carbon dioxide and takes up oxygen. The exchange of gases can occur because the pressures of oxygen and carbon dioxide differ in blood and in inhaled air. When the blood in the capillaries passes between the alveoli, surplus carbon dioxide passes into the air while oxygen becomes bound to the hemoglobin of red blood cells. Dark, venous blood has an oxygen pressure of about 14 mm Hg (mercury); oxygen pressure in bright arterial blood is over seven times higher. The pulmonary capillaries drain into larger vessels, and finally into the four pulmonary veins that empty into the left atrium. The accepted term for all vessels carrying blood *from* the heart is arteries, and for those carrying blood *to* the heart, veins. Thus the pulmonary "artery" actually transports venous blood, and the pulmonary vein, arterial blood.

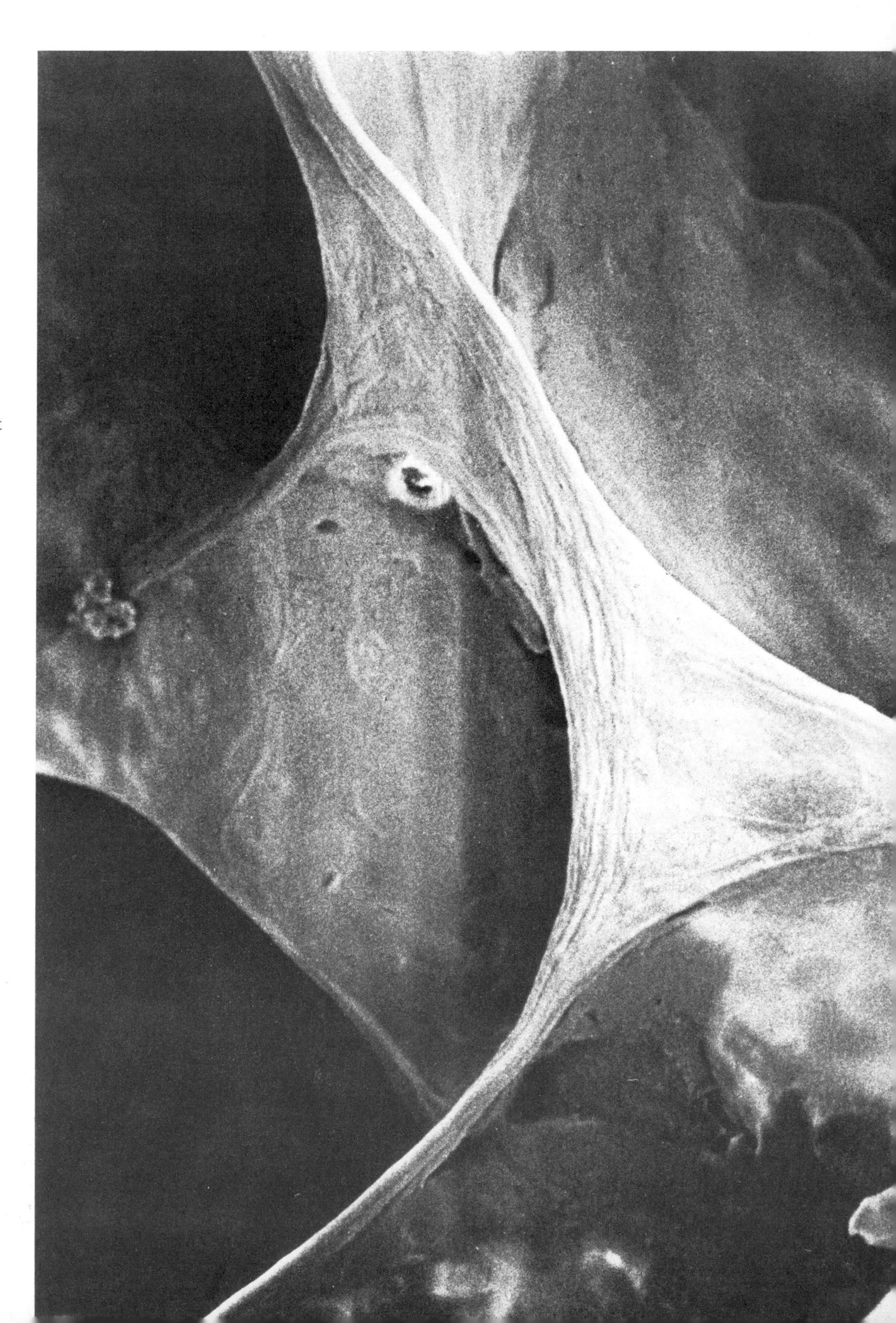

Blood Pressure

The adaptability of the circulatory system is remarkable. At rest, the heart pumps between four and five quarts of blood a minute. During great exertions, heart output may go up to forty quarts a minute. This variation makes heavy demands not only on cardiac muscles but also on the vascular system and respiration. It takes some time after stress for the supply of oxygen and the removal of waste products from heart muscles to be in complete balance. The intimate byplay between the cardiac and the circulatory systems is one of many examples of the fine way in which the various organ systems of the body cooperate.

Like water in a pipe, blood must be under a certain pressure in order to flow. If water pressure becomes too low in a high-rise building, for example, the upper floors will have no water. If blood pressure falls below a certain level, the brain is affected first, and we may lose consciousness in a faint. When we are running or working hard, blood pressure rises. This allows more blood to pass more quickly through dilated vessels to supply skeletal muscles, for example. After a period of rest, blood pressure falls, and blood circulation adjusts to the new situation.

Blood pressure is regulated by a vasomotor center in the lower part of the brain. Nerve impulses from pressure-sensitive receptors in the carotid arteries and in the aorta relay information to the center, which is also sensitive to variations in the acidity and carbon dioxide content of the blood. The vasomotor center can relay nerve signals directly to muscles to constrict or dilate blood vessels, or to glands which secrete vasoconstricting hormones such as adrenaline and noradrenalin.

Blood pressure is measured in two figures, for example 125/80. The first figure denotes blood pressure during the working phase of the heart, the systole; the second measures pressure during the resting phase, the diastole. Systolic pressure is higher, as it is recorded during the contraction of the heart. The measurement is made by means of an inflatable cuff that is wrapped around the upper arm and connected to a manometer. The cuff is rapidly inflated until the pressure in it exceeds that in the artery in the arm. Then air is slowly let out until a pulse can first be heard with a stethoscope placed against the artery at the inside of the elbow. This is the systolic measurement. The diastolic pressure is read on the manometer when pulse sounds are no longer heard.

The left atrium
A pair of veins from each lung carries oxygenated blood to the left atrium. The opening between the left atrium and ventricle is closed by the mitral valve, two thin flaps of connective tissue that are operated by the papillary muscles, which are connected to the valve by fibrous cords. The point of entry for the four pulmonary veins is shown in the diagram below.

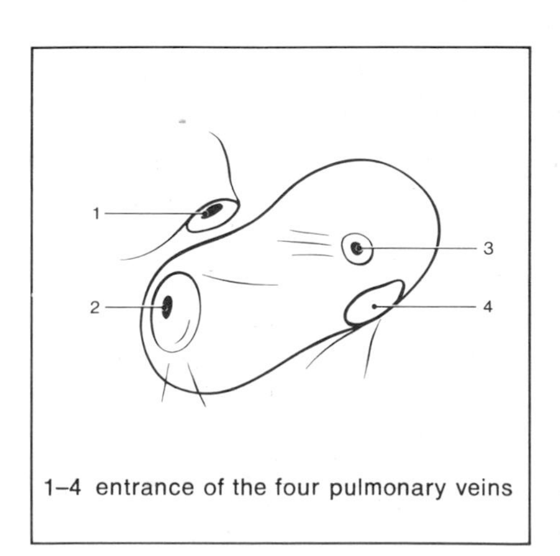

1–4 entrance of the four pulmonary veins

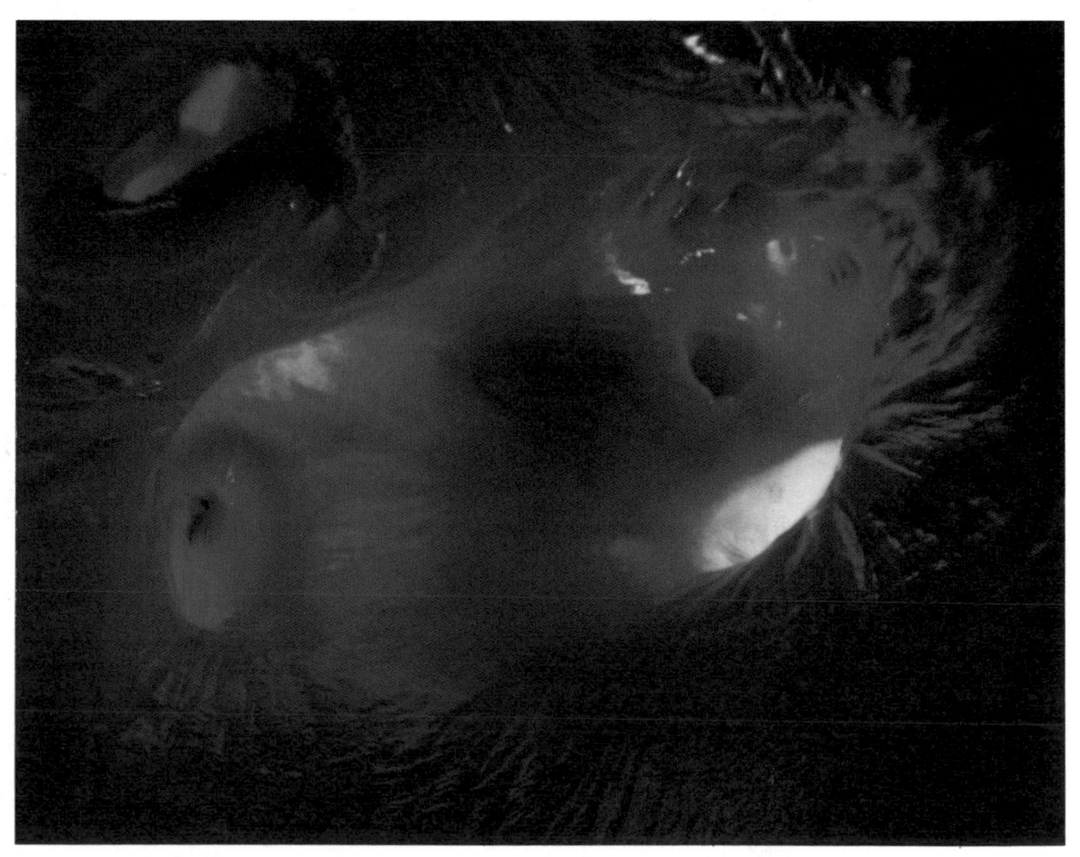

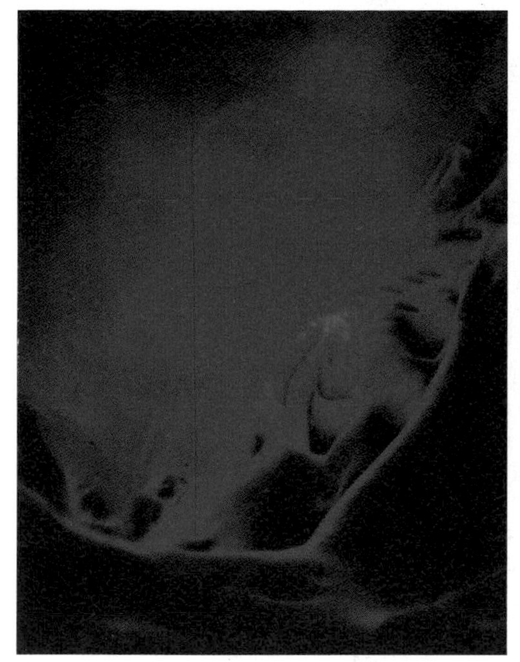

Left ventricle filled with blood

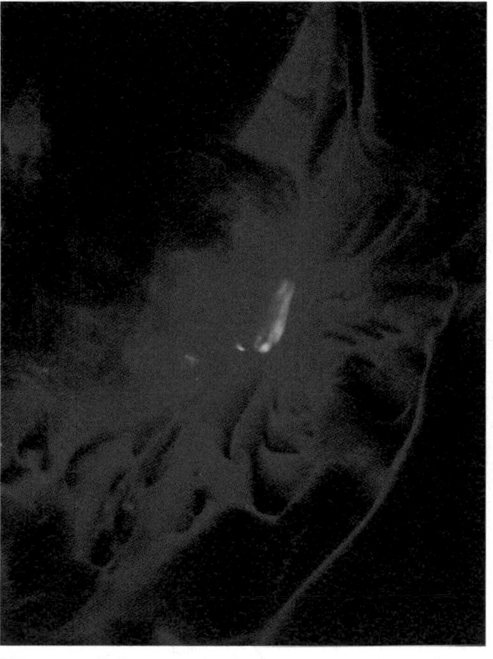

The ventricle contracting
The aortic valve can be seen behind the blood

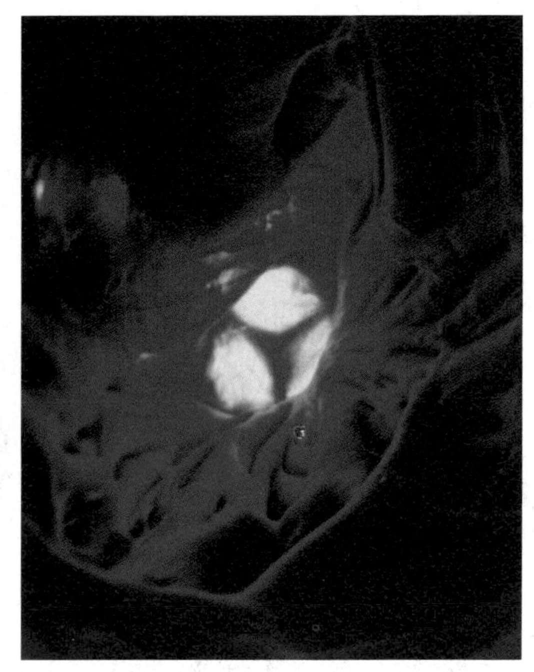

The ventricle, almost empty

A self-lubricating, high-capacity pump

The design specifications for the heart considered as a mechanical pump might read as follows:

Type: blood pump for human bodies.

General description: The heart consists of two pumps arranged in serial order. It is self-lubricating and self-regulating. The model has been tested over a very long time and under extreme conditions. It is insensitive to changes in external temperature. Under normal conditions it will work continuously for seventy-five to eighty-six years, but in certain cases may work for one hundred years or longer.

Construction: Each pump consists of two chambers, an atrium and a ventricle, separated by membranous valves. There is a conduction organ, or pacemaker, the right atrium, and flap valves between chambers and connecting pipes.

Flow diagram: Fluid first fills both atria and flows from there into the ventricles. On signals from the pacemaker, the pumps contract, and fluid is pumped into the system. During the process, the valves between atria and ventricles close automatically, while those between ventricles and the connected pipes open. Normal pump frequency during rest is 50 to 70 beats per minute, during extreme load, about 200.

Dimensions: Shape: conical. Width at base: approximately 100 mm. Height: approximately 155 mm. Weight: approximately 300 grams (10.5 ounces).

For its size, the heart pump has unrivaled capacity, as shown by the following performance data:

Stroke volume: 80–100 ml per beat, or 5–6 liters a minute. Normal performance per day: 100,000 beats and 8,000 liters. Converted into energy, each day the heart performs work corresponding to lifting a 154-pound person almost 1,000 feet.

Total output: During its lifetime, the pump makes 2,500,000,000 beats and could fill a vessel with a capacity of 240 million liters.

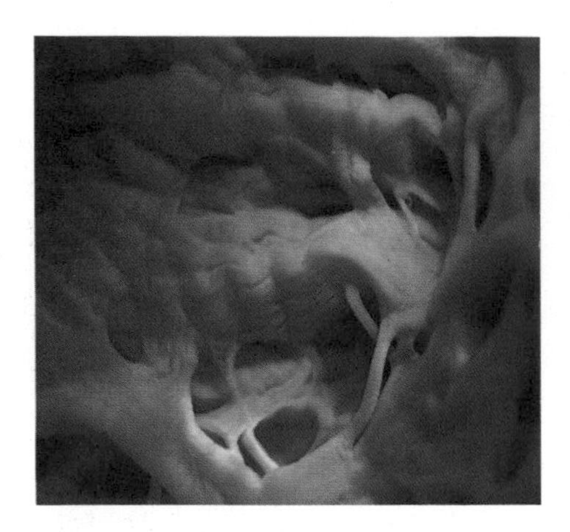

The strong wall of the left ventricle, △ which performs the heaviest labor in the heart, is composed of muscles that pump blood to all parts of the body.

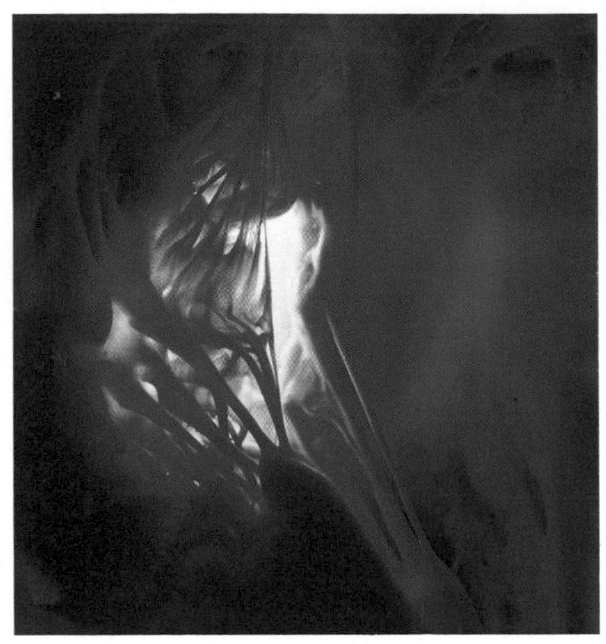

The valves in the heart open and close between four and five thousand times an hour. They are very strong despite their tender, semi-transparent structure. The photo on the left shows the location of the two cusps of the mitral valve between the left atrium and ventricle when the ventricle is distended. The photo at right justifies the assertion that the mitral valve is one of the most beautiful tissues in the body, resembling a sail catching the wind. The sheets (or lines) are fibers of connective tissue which control the position of the valve.

Blood Types

All human beings have one of four blood types: A, B, AB, or O in the ABO blood system. The system is based on the presence or absence of certain complex molecules called agglutinogens in red blood cells. An individual who has only agglutinogen A has blood type A, and someone who has only agglutinogen B has blood type B. If both agglutinogen A and agglutinogen B are present, the individual has type AB. If neither A nor B is present, the blood type is O. The mixture of certain blood types can be disastrous. Red blood cells will become clumped, or agglutinated, and will choke the capillaries, a condition which can be fatal. This is because blood serum contains specific proteins called antibodies. The antibody against type A blood is called anti-A, and against type B blood, anti-B. As seen in the diagram, an individual with blood type A has anti-B antibody in his blood serum. Should he receive a transfusion of B type blood, the foreign B-agglutinogens will react with his native anti-B antibodies and cause the red cells to agglutinate. The blood from individuals with blood type O can be given to anyone; such persons are "universal donors," because their blood does not contain any agglutinogens to arouse antibodies in the recipient's blood. Consequently the blood will not be considered "hostile" or incompatible. On the other hand, a person with type O blood can only receive blood from other individuals with this type of blood, because his blood serum contains antibodies against both A and B. Persons with AB blood are "universal recipients," since their serum contains both A and B agglutinogens and no antibodies against either type A or type B blood.

Blood type	A	B	AB	O
Agglutinogen in blood cells	A	B	AB	none
Antibodies in blood serum	anti-B	anti-A	none	anti-A + anti-B

In addition to the agglutinogens of the ABO system, another blood factor of practical importance is the Rh factor, an antigen present in 85 percent of the Western European population. Such people are termed Rh-positive. An Rh-negative individual does not have this antigen, but may develop Rh-antibodies following a blood transfusion with Rh-positive blood. A later transfusion with Rh-positive blood will stimulate the production of Rh-antibodies, which will then attack the blood cells supplied. If an Rh-negative woman is pregnant and the fetus has inherited the Rh factor from the father, antibodies produced in the mother may pass into the blood of the fetus and destroy red blood cells. Nowadays it is possible to counteract this mortal danger by a major transfusion of blood in the fetus.

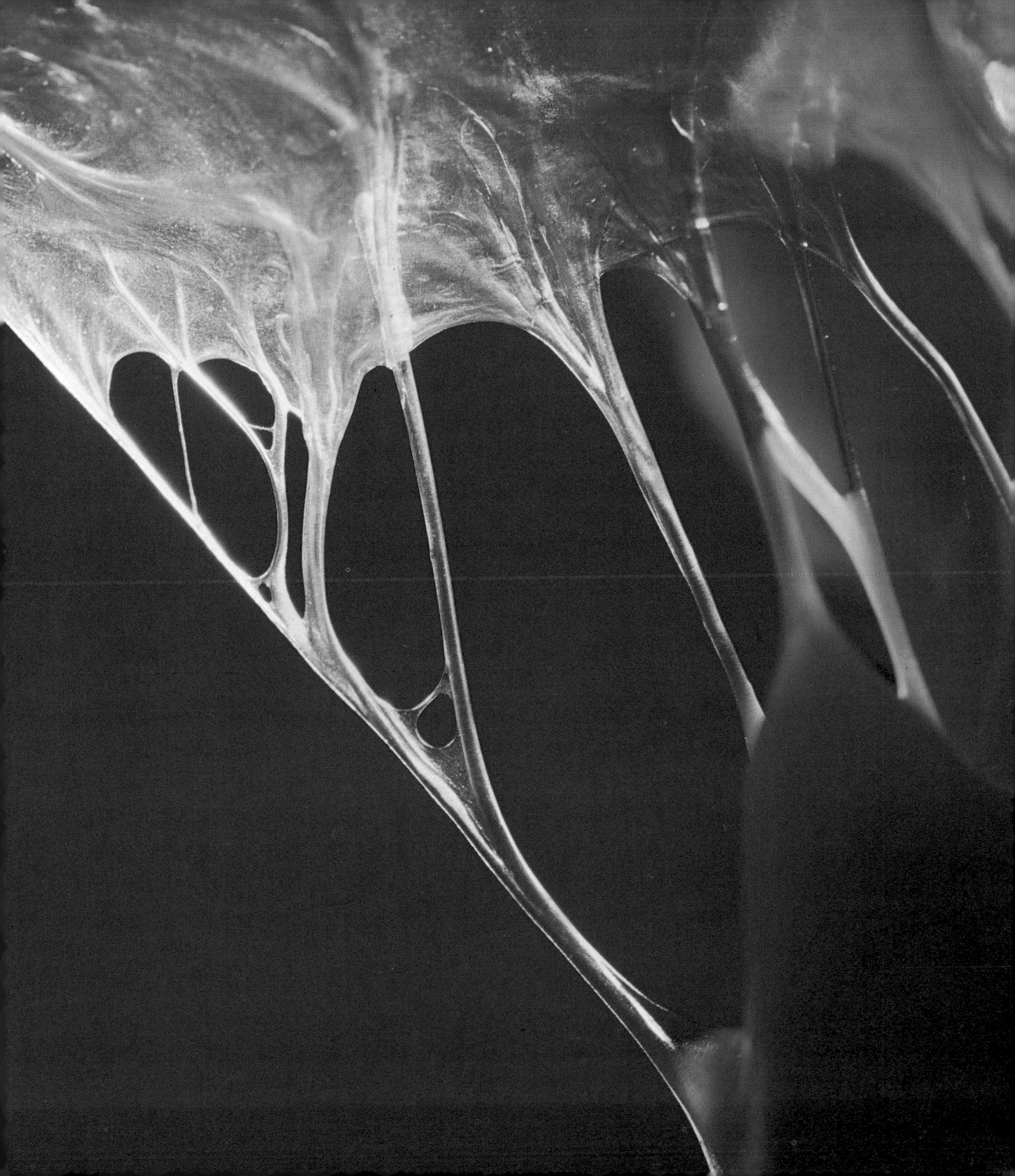

Fetal Circulation

The blood vessels in a four-month-old fetus are continuously developing

The fetus develops in liquid surroundings. During an early phase of its development it has a number of gill slits in the neck region. However, the human fetus never breathes through gills nor through its lungs. Oxygen is supplied through the maternal placenta. There, separated by the thin walls of the vessels, blood from the fetus and from the mother exchange carbon dioxide and waste products for oxygen and nourishment. This makes blood circulation in the fetus vastly different from that in the adult.

This applies especially to the heart. The umbilical vein carries the blood of the fetus from the placenta to the right atrium. From there it passes to the left atrium through an opening in the central cardiac wall. Blood is subsequently pumped from the left ventricle into the aorta. The blood that enters the pulmonary artery is led directly into the aorta through a passage that is present only in the fetus. Both shortcut passageways close at birth. However, in some exceptional cases, the closure does not occur or is imperfect. The result may be that oxygenated blood mixes with blood deficient in oxygen and enters the aorta. The reduced oxygen content of such blood accounts for the name "blue baby" to describe the condition. The treatment for this condition is surgical.

When the fetus is only four weeks old, its vascular system is sufficiently developed for it to have a simple but beating heart. In early development, blood cells are formed in the walls of blood vessels and in the yolk sac. During the third month, however, blood-cell-producing bone marrow develops in the skeletal bones. Prior to delivery, blood cells are also produced in the liver.

The blood vessels in the legs and arms of a fetus about four months old. The photos give some idea of the development of the vascular system in fetal tissue. Growth is rapid and the need for blood-borne energy is great.

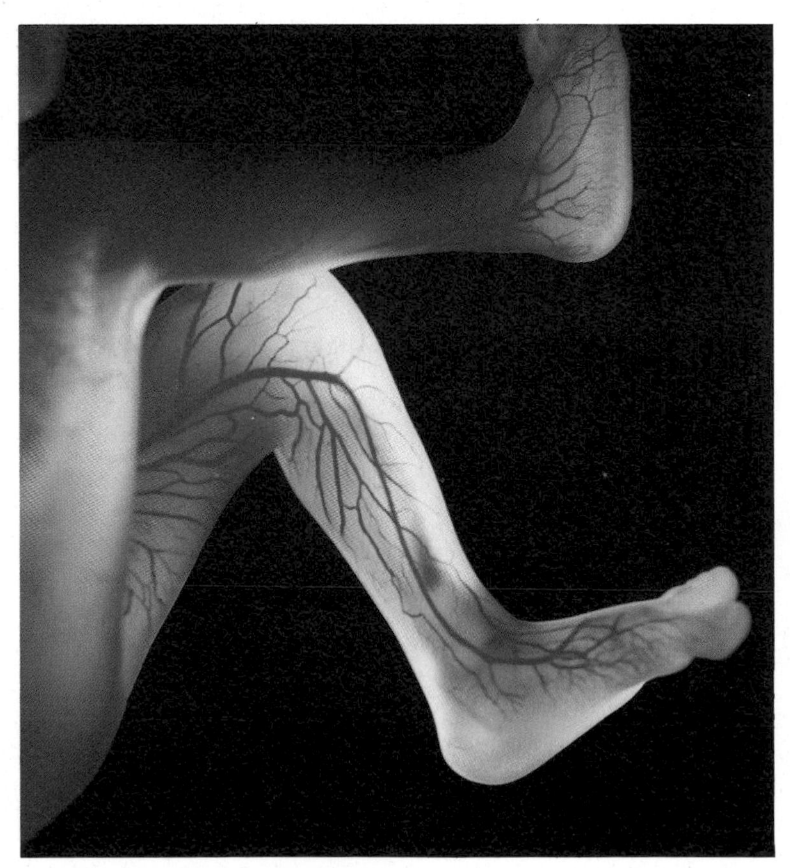

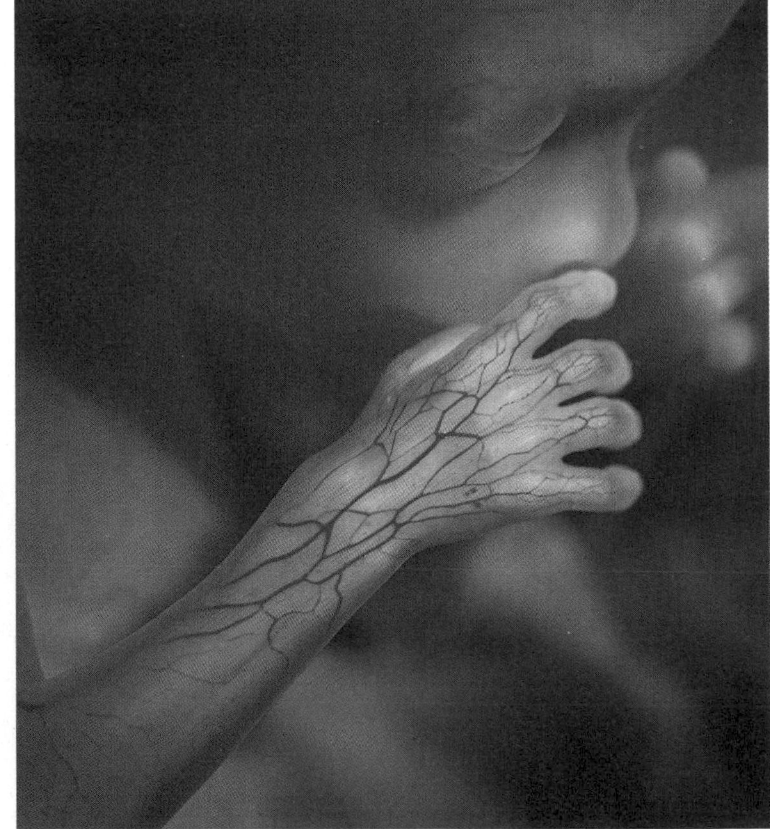

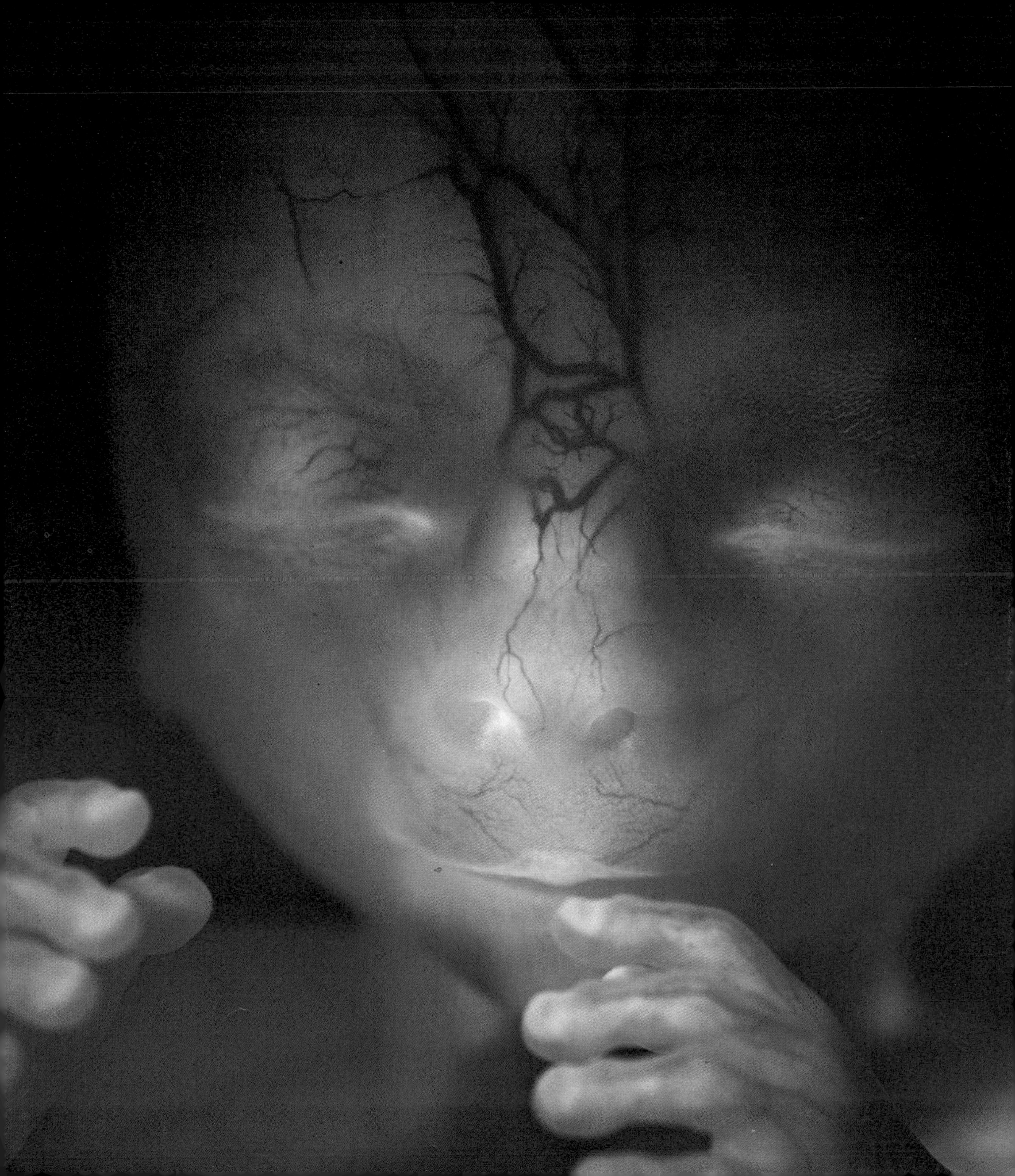

The Lymph Vessels

The Body's Drainage System

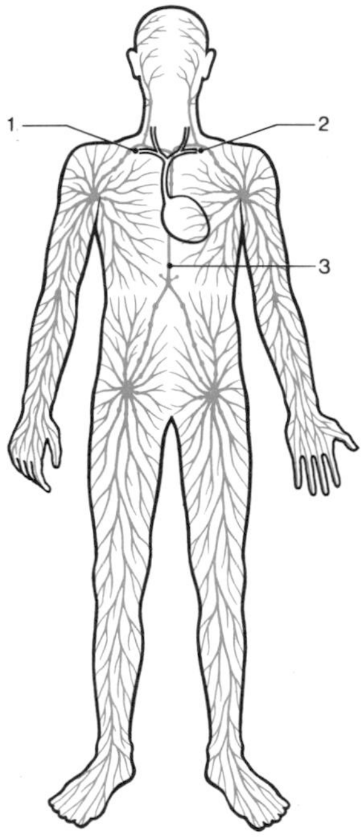

The lymphatic vessels

traverse the body and form a network among the lymph nodes. Here some of these nodes are indicated as blue spots.
From the thoracic duct, the lymph drains into the subclavian veins.

1 left subclavian vein
2 thoracic duct
3 right subclavian vein

All living cells in the body are bathed by the liquid lymph. Nutritious substances from the blood pass through the lymph into cells. The lymph itself contains no cells. It is slightly yellowish in color and composed of much the same elements as plasma, except that there are fewer proteins. Only about half the proteins in blood are able to pass from plasma directly into body cells. Nor can cell proteins that are the products of metabolism later enter the capillaries for removal, because the capillary pressure is greater than that in the surrounding tissue. Instead, waste proteins are returned through the lymphatic vessels, the walls of which are very permeable. The major lymphatic vessels resemble veins. They have valves that prevent backflow. Like venous blood, lymph is not pumped throughout the body by any heartlike organ, but is pressed forward by contractions of surrounding muscles. During a long period of standing, lymph pools in the legs and feet, which then swell. The largest lymphatic vessel is the thoracic duct, which empties into the left subclavian vein.

Lymph vessels carry fluid away from body tissue. They also play an important role in the body's defense against infection. The lymph nodes throughout the body are small "fortresses" where the circulating lymph deposits bacteria or foreign matter for disposal. There the infectious material is encapsulated and ingested by lymphocytes, a kind of white blood cell. Lymphocytes can divide, and they also produce antibodies. When the body is invaded by microorganisms, the lymphocyte count increases.

Cancer cells sometimes circulate through the lymphatic system. That is why, when a surgeon removes a cancerous growth, he usually also removes the lymph nodes in the drainage area where the tumor is located. If he does not, tumor cells may develop in lymphatic tissue and form secondary tumors, or metastases.

In the case of severe infection, lymph nodes swell and ache. Tonsilitis or a severe cold makes the tonsils and lymph nodes in the neck enlarge and ache when touched. An infection in a finger may result in swelling along the lymphatic vessels in the arm and lymph nodes in the armpits. An infection in the foot may affect nodes in the groin. If the infection is unchecked, it may affect nodes throughout the body, resulting in the condition commonly called blood poisoning (septicemia).

Lymph vessels in the intestines are responsible for the uptake of fat and its delivery to blood vessels. Fat droplets, unlike salts or other soluble substances, are too large to pass across the capillary wall directly, but can be taken up by lymph. After a fatty meal the lymph becomes milky in color.

The spleen is part of the lymphatic system. It is not only a reservoir for blood, but it produces antibodies and filters the lymph in the same way as the lymph nodes. Another organ that is considered part of the system is the thymus, which is located behind the upper part of the breastbone.

Lymphatic Vessels

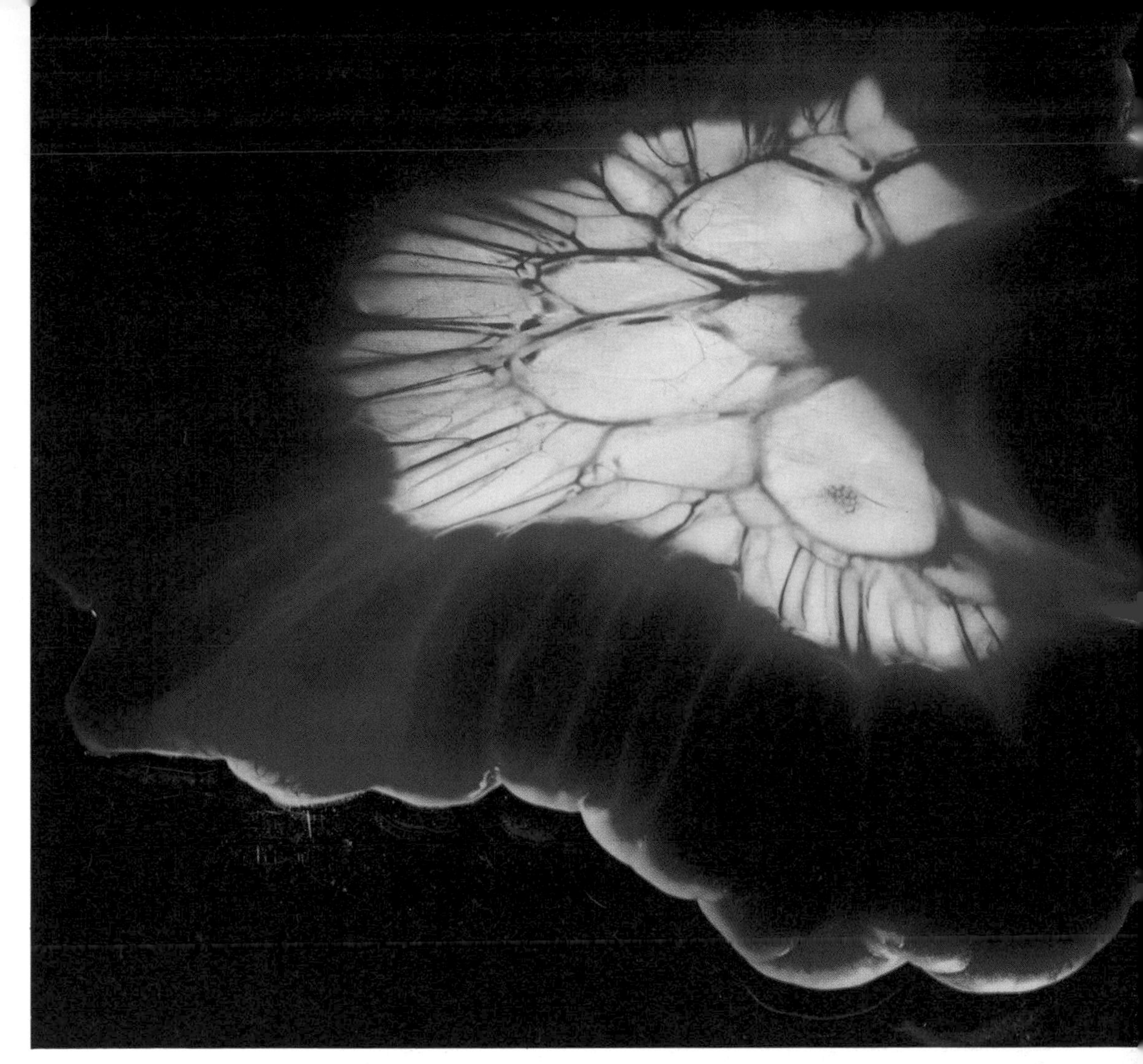

Lymph nodes in the mesentery
Lymph nodes are round structures. They are especially numerous in certain parts of the body, like the armpits, the groin, or, as shown here, in the mesentery, the fold of tissue attaching the intestine to the abdominal wall. In this magnification the lymph node appears as a group of small red rings with an extension to the right in the large yellow area at center right.

Lymph nodes function as filters for the lymph. Among other things, they trap bacteria and prevent them from entering the bloodstream. This could happen when the lymph drains into the subclavian veins through the major lymphatic trunk, the thoracic duct.

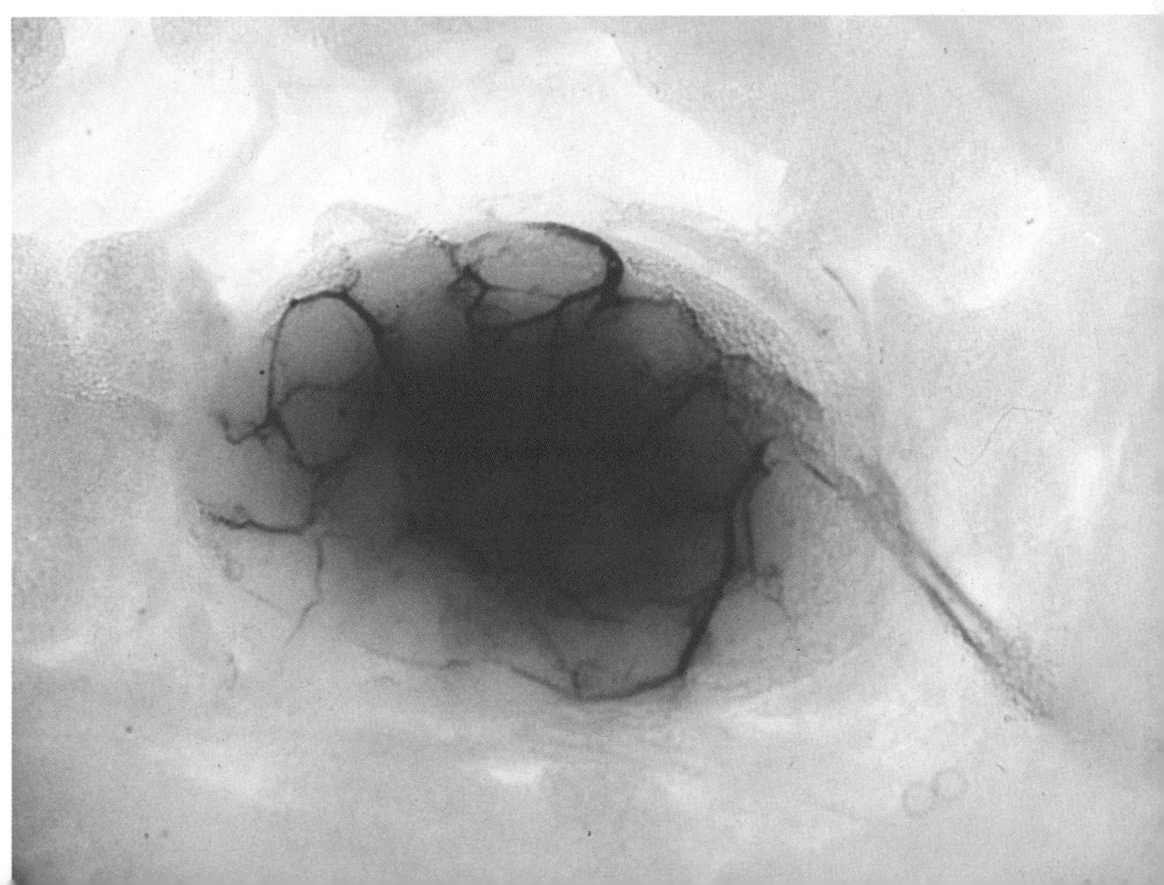

A lymph node
in the mesentery, enlarged from the photo above. Lymph passes through a number of such nodes as it collects from all parts of the body before emptying into the general circulation.

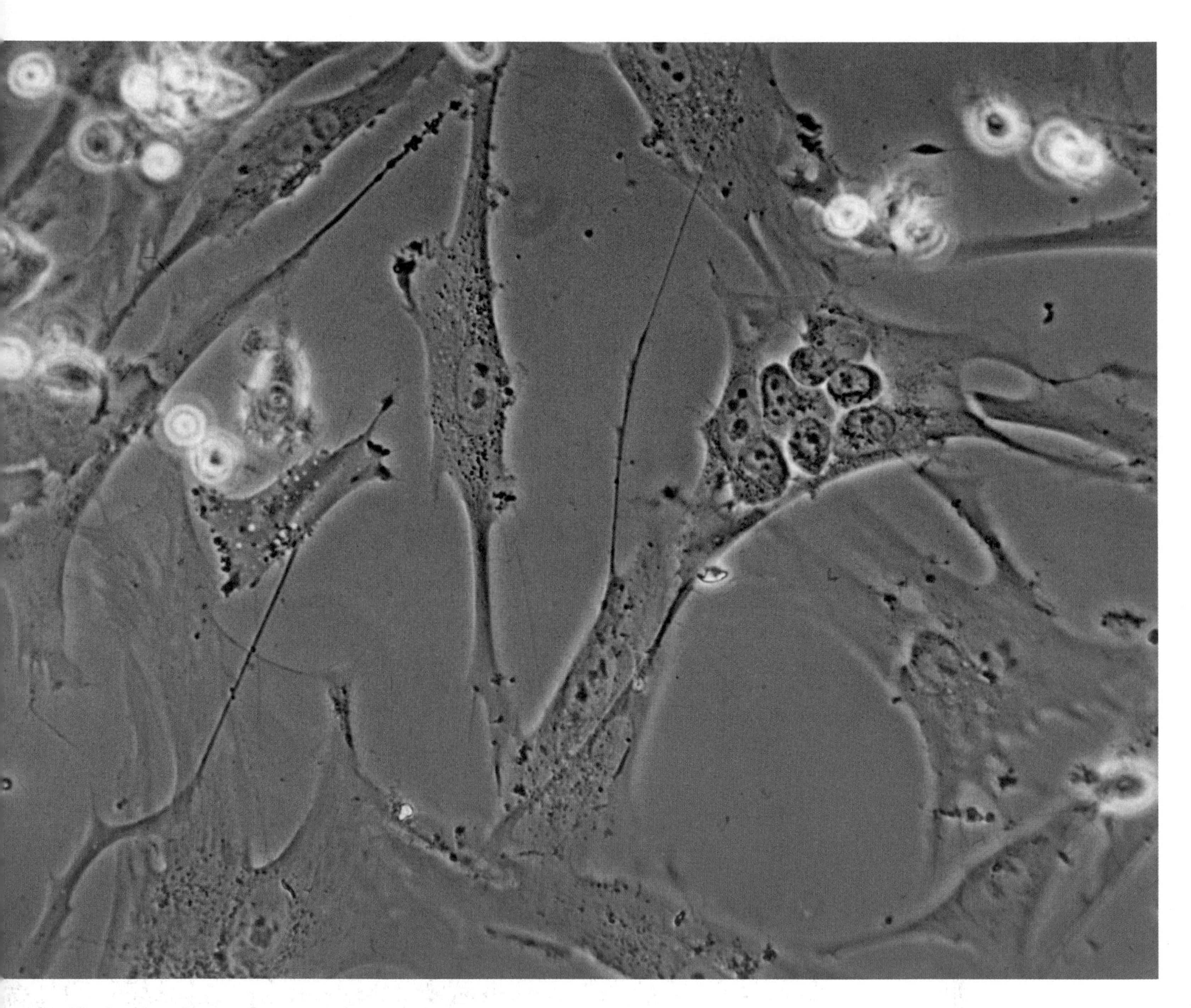

The function of macrophages is illustrated in this photo of a culture of live cells in a nutrient medium sandwiched between glass slides. The largest cells are fibroblasts, a type of connective tissue cell. Macrophages are relatively large cells with specific functions. They can be seen in several parts of the picture, but most clearly as the rounded granular structures at the upper right. Macrophages constitute the sanitary corps of the body. They are usually present in body fluid, as free cells. Some, however, are fixed. Macrophages ingest and destroy foreign bodies and microorganisms, such as bacteria, or, as in the photo, diseased and dead cells from body tissue itself. This is the normal way of disposal of such cells. Macrophages are always at work. The circular white spots in the photo are cells in planes outside the lens focus.

Respiration

Each cell in the body communicates with the atmosphere by way of the lungs. Blood, passing through the lungs, carries oxygen to cells to provide the necessary ingredient for the oxidation process that enables cells to work. One of the end-products of oxidation is carbon dioxide (in the presence of water, carbonic acid). Blood carries carbon dioxide away from tissues and back to the lungs, where it leaves the body in exhaled air.

Oxygen is a prerequisite for all higher life forms. Consequently, it is vital that all organ systems be supplied with it. Nerve cells in the brain are among the most demanding. At normal body temperature, they cannot endure a total failure of the oxygen supply for more than a few minutes without impairment. At lower body temperatures, the metabolic rate is reduced, and tissues can better withstand a lack of oxygen.

Almost all the oxygen in the air is biological in origin, produced by the process of photosynthesis in green plants. This process, which takes place in the presence of sunlight, enables plants to synthesize carbohydrates, such as sugar, from carbon dioxide in the air and water in the soil. The carbon dioxide in air that is used in photosynthesis is derived in part from the exhalations of air-breathing animals.

When life first began on earth, about three billion years ago, the oxygen content of air was not as high as it is today. As green plants spread over all the continents, the oxygen content of the atmosphere increased. With the exception of the anaerobes — microorganisms which can live in the absence of oxygen — all life today is dependent on the oxygen–carbon dioxide cycle. At present, the oxygen-producing and the oxygen-consuming forms of life are in balance. One expression of this stability is the relatively constant oxygen content in the atmosphere — about 20.96 percent.

In breathing, the lungs take up only about one-fifth of the

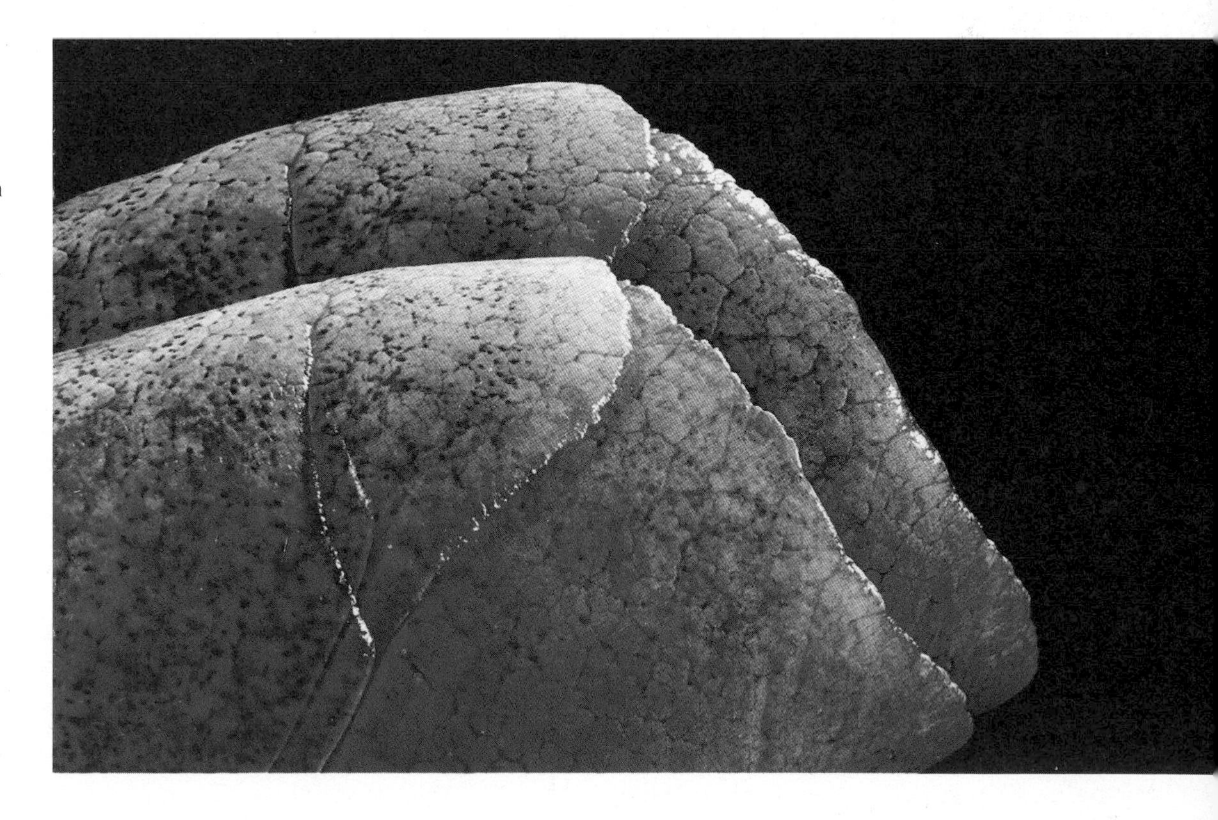

The same lung during inhalation and exhalation
The two photos, one behind the other, show the change in the lung volume — it amounts to several liters — from inhalation to exhalation. This lung is from an older man and is black with dust and smoke inhaled during a long life in a city. Some dirt particles enter the smallest branches of the bronchi, where they are neutralized by white blood cells and macrophages — the sanitary corps of the body.

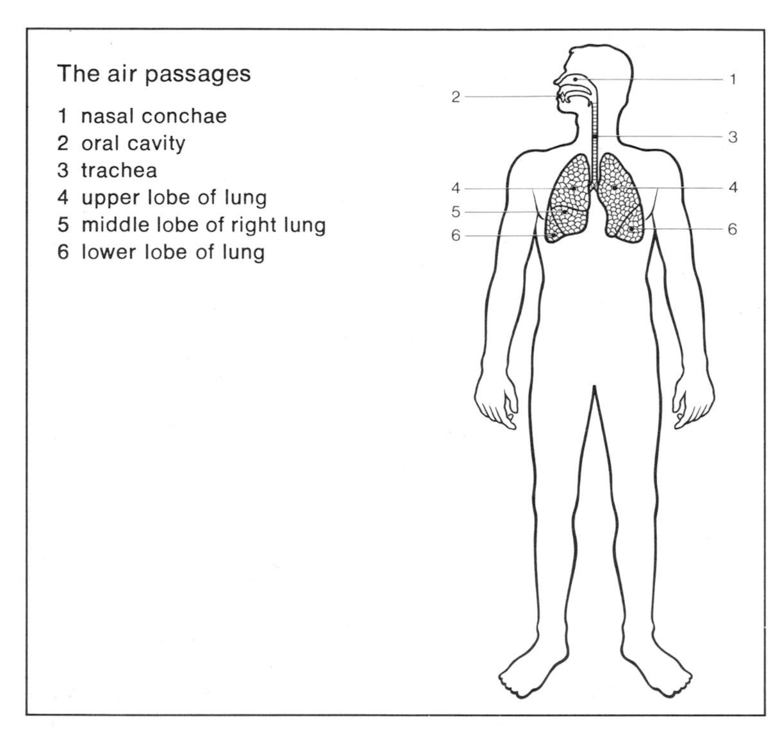

oxygen contained in the inhaled air. Exhaled air still contains about 16.6 percent oxygen, and this fact is made use of in the mouth-to-mouth method of resuscitation, when the rescuer's exhaled air fills the lungs of the unconscious patient. The carbon dioxide content of exhaled air — 4.4 percent — is, on the other hand, about a hundred times higher than that in the inhaled air.

The lungs' position in the chest, close to the heart, is important in circulation. The right ventricle of the heart pumps blood rich in carbon dioxide and low in oxygen to the lungs through the pulmonary arteries. There the carbon dioxide is excreted and oxygen taken up. The pulmonary veins carry the oxygenated blood from the lungs to the left atrium of the heart. This is the so-called lesser or pulmonary circulation. (A diagram of the blood circulation is given on page 64.)

The surrounding air is usually much cooler than the body. It also contains particles and microorganisms that may be directly injurious. In its travel through the air passages, the air we breathe is warmed, moistened, and filtered. These processes take place in the nose, mouth, pharynx, trachea (windpipe), and in the bronchi of the lungs.

The bronchi are covered by a ciliated and glandular mucosa. Dust particles and bacteria are trapped in the cilia and, through their movements, removed from the lungs. White blood cells in lung tissue are enlisted to cope with particles and microorganisms which may have slipped by the first line of defense.

Normally, the human body has a good chance of averting the menace of polluted air. Nowadays, however, air often contains a number of substances originating from chemical processes in factories and internal combustion engines. Such substances include mercury, cadmium, lead, asbestos, and other metals. The danger of breathing in such substances is that in many cases they are not excreted or broken down, but may be stored in tissue where they can poison individual cells.

Each lung is surrounded by a double-walled membrane, the pleural tissue, the outer layer of which is attached to the inside of the chest. Upon inhalation, the chest increases in volume and the lungs expand. This is due in part to the subatmospheric pressure in the intrapleural space. If the pleura is impaired so that air enters, equalizing pressure, respiration becomes much more difficult.

The gaseous exchange between blood and atmospheric air takes place in the alveoli, tiny air sacs clustered at the ends of the smallest air passageways, the bronchioles. In the adult, the pulmonary alveoli number in the hundreds of millions, providing a total surface area of between eighty and a hundred square yards. The walls of the alveoli are so thin, however, that, in spite of its large volume, pulmonary tissue is remarkably light. Drained of blood, the lungs weigh only a little over two pounds.

Fine capillaries run along the surface of the alveoles and in the spaces between them. Since the capillary diameter permits only a single-file passage of red blood cells, the cells do not

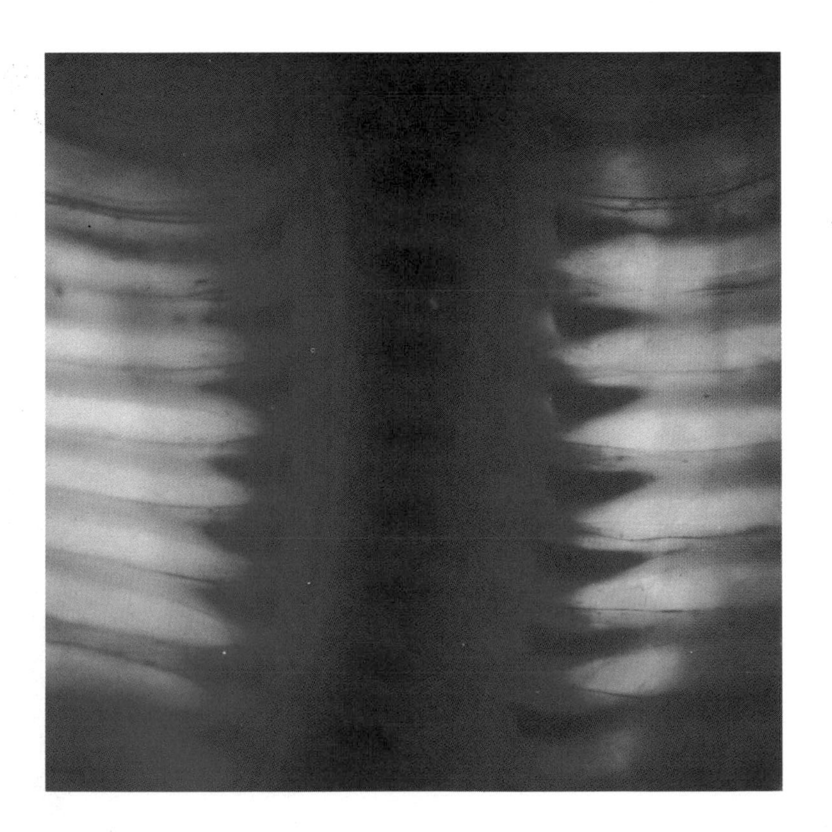

hinder each other in gaining access to air in the alveoli. With the interstitial fluid as a medium, contact takes place across two membranes, the capillary wall and the alveolar wall.

Oxygen and carbon dioxide are exchanged between red blood cells and the air in the alveoli by diffusion. This process occurs automatically, whenever there is a difference in concentration of gases in solution on either side of a permeable membrane. Relative to the blood, the air in the alveoli contains more oxygen and less carbon dioxide — consequently oxygen molecules diffuse into the blood and carbon dioxide molecules diffuse out of it.

At rest we breathe about 15 times a minute, each inhalation containing about half a liter of air, i.e., between 7 and 8 liters per minute. During strenuous effort we can increase the respiratory volume by about five times. The air in the lungs is exchanged successively. Even following a maximal expiration, more than a liter of air remains in the lungs. This volume of air is warm, and one of its functions is to act as a buffer against cold air. The air volume that can be expelled from full inhalation to full exhalation is called the vital capacity and amounts to about four liters. A high vital capacity is essential to singers, speakers, and athletes, among others.

Respiratory movements result from contractions of the muscles of the diaphragm and the chest. These muscles are governed by motor cells in the brain and in the spinal cord, which are in turn controlled by the respiratory center in the medulla at the base of the brain.

The regulation of respiration is very complex and involves chemical stimulation of the respiratory center by the carbon-dioxide content in the blood as well as stimulation by the degree of stretch in the respiratory muscles. Respiration is under the control of higher centers, too. We have all experienced an increase in respiration rate when nervous or excited, when we feel threatened, or in any stressful state. In such cases, impulses from higher centers in the cerebral cortex act on the respiratory center. Respiration increases — i.e., we breathe not only more frequently per unit time but take in a greater volume at each breath. During sleep, respiratory frequency and volume are correspondingly reduced.

Generally, regulation is related to a change in the level of sensitivity of the carbon-dioxide-sensitive cells in the respiratory center. When we are excited, threatened, or stimulated, their sensitivity to carbon dioxide goes up and the number of respiratory impulses increases in response to a given carbon-dioxide content of blood. When we relax, doze, or sleep, sensitivity to carbon dioxide goes down and, even though the carbon dioxide content in the blood may be the same, the number of respiratory impulses decreases.

◁ **The chest protects the lungs**
and consists of the breastbone (sternum), the ribs, and the thoracic vertebrae. Ossification in the newborn, as shown here, starts in that part of the ribs nearest the spinal column. The growth zone is darker. The light bands are muscles. The red streak below each rib is an intercostal artery.

The ribs in a newborn ▷
are reflected in the smooth surface of the lung. Each lung is surrounded by a double-walled membrane, the pleura, the outer wall of which is firmly attached to the chest. The pressure in the intrapleural space is subatmospheric. At inhalation, the chest is expanded and so are the lungs, with the result that air is sucked in.

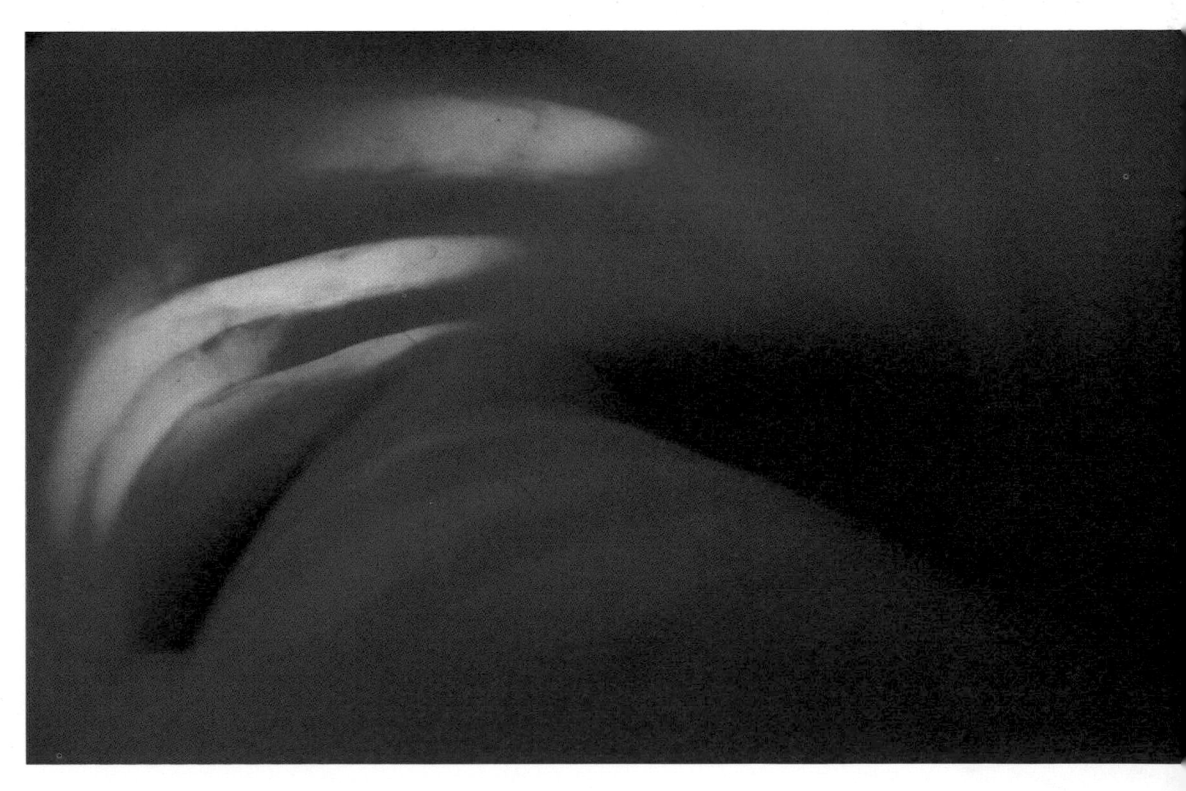

Gill slits in an embryo
In the first weeks of gestation, it is almost impossible to distinguish between human and other vertebrate embryos. However, the gill slits in the four- and five-week-old human embryo (upper and lower photos, respectively) do not develop into gills, but into the lower jaw, the hyoid bone (in the neck), and the larynx. The primitive heart and the primordial eyes are also clearly evident.

The photo on the right-hand page shows the gill slits in a six-week-old embryo. It was taken with a scanning electron microscope. Early fetal development provides a number of examples of how in the course of a few weeks the embryo passes through several stages of the evolution of man, a process that has taken millions of years. The fetus truly lives in a "sea," but it is never a fish and it never breathes through gills. The necessary oxygen is supplied to the fetus by the maternal blood through the placenta. The reason for an outward resemblance between embryos of different animal species is that the first stages of development of the principal organs take place in a similar way from similar types of tissue.

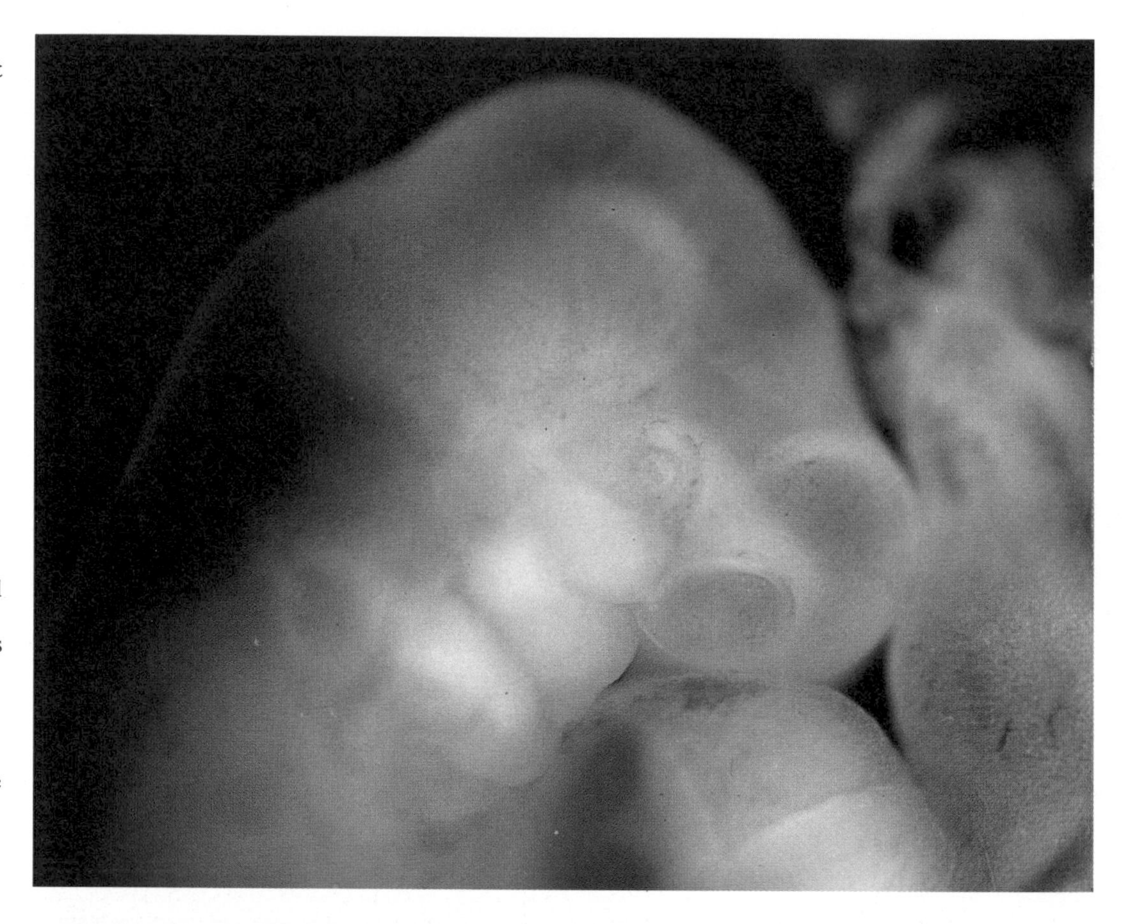

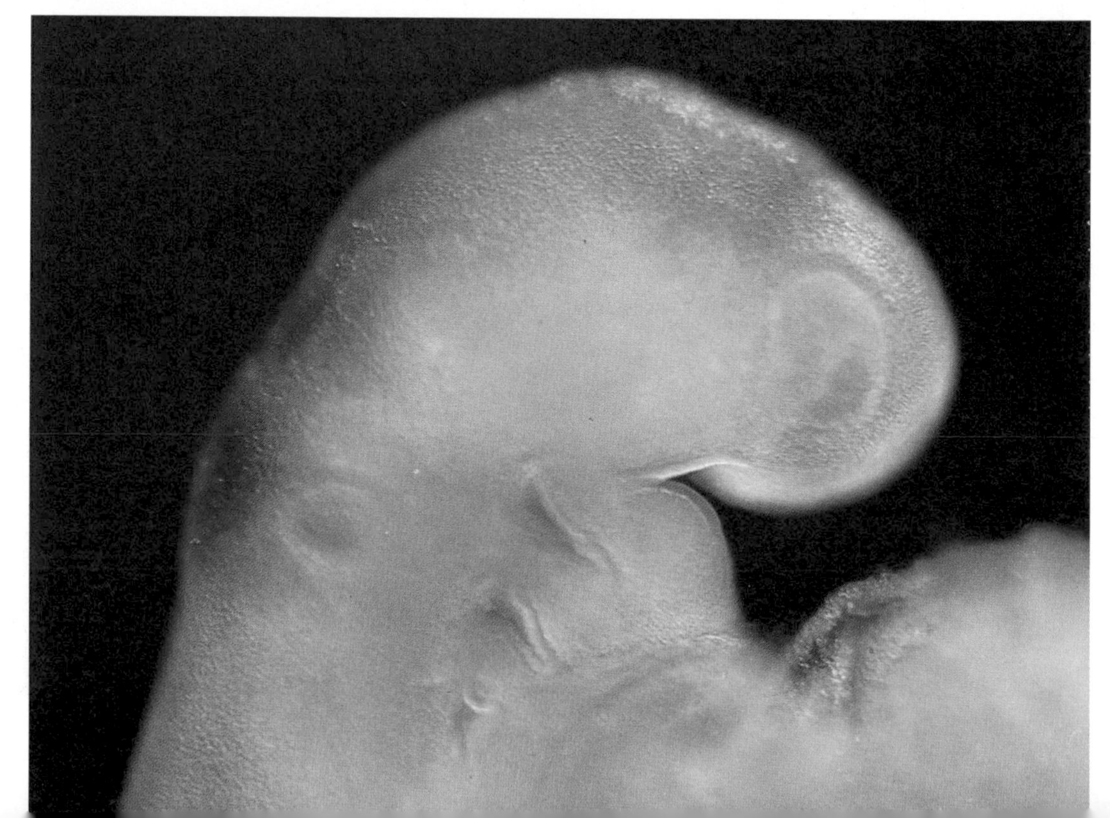

Respiration

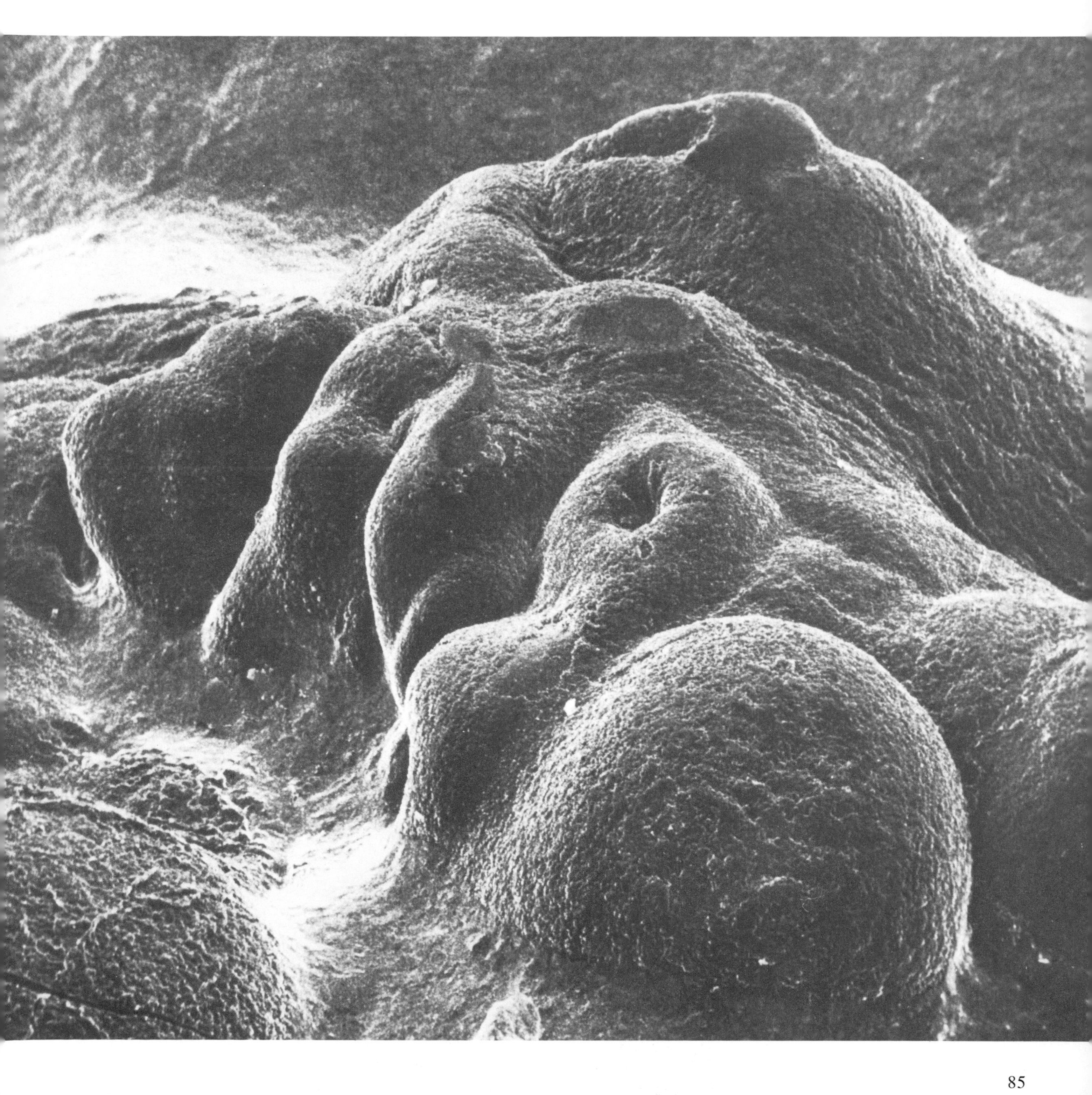

The Voice

One part of our speech apparatus is the larynx. It is formed by five cartilages, of which the thyroid cartilage is the largest and can be felt on the front side. Men sometimes have a protruding thyroid cartilage, the "Adam's apple." The cricoid cartilage, at the base, separates the larynx from the trachea, or windpipe. At the back, the vocal cords are attached to two small adjustable cartilages, the arytenoid cartilages. The epiglottis is located at the top. During swallowing, it is lowered and covers the opening into the trachea. The larynx is maneuvered by some ten muscles and ligaments. Human vocal sounds are produced when the vocal cords are set vibrating.

The opening between the vocal cords is called the glottis. When the vocal cords are stretched, they are set into vibration by air passing through the glottis. The resulting sound is modified by the arytenoid cartilages: the more the vocal cords are stretched, the higher the pitch of the sound. Loudness is controlled by the amount of air passing the glottis. The characteristic quality or timbre of the individual voice depends on features of the nasal and oral cavities and sinuses, the pharynx, and the thoracic cavity, all of which function as resonators.

In speaking, the vocal cords produce what is called a glide. In singing, the tone is fixed, sometimes with a vibrato added. The vibrato is a result of regular variations in the pitch, usually occurring 6 to 7 times per second. Our tone range is four, sometimes even five, octaves. In normal conversation, the frequency (perceived as pitch) of the male voice is about 80 cycles per second and that of the female about 400.

The process of speech is extremely complex. It is governed by speech centers in the cerebral cortex, which receive and modify impulses from the medulla. Most right-handed people have the speech center controlling the muscles of speech (the motor speech center) in the left hemisphere near the temporal lobe. Some left-handed people have theirs in the corresponding region in the right hemisphere. Other speech centers are located farther back in the brain. Whenever we express ourselves through articulated sounds, syllables, words, and sentences, our acts are governed by the speech centers, which also are influenced continuously by our conscious or unconscious monitoring of what we have just said. The human faculty for symbolic communication through speech, and its extension to reading and writing, is considered a major reason for the evolutionary success of the human race. Anthropological studies of our fossil forefathers and comparative anatomical studies of related species show that the human vocal apparatus has evolved in a direction that has favored speech, possibly at the expense of other faculties such as the sense of smell.

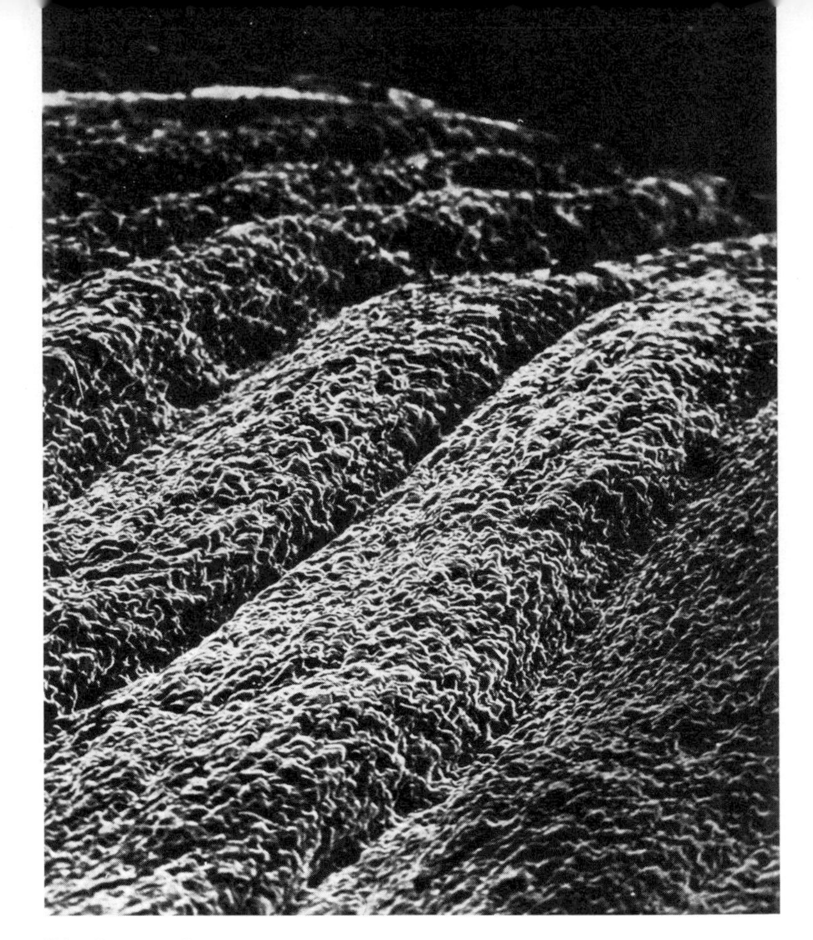

Diaphragmatic muscle
separates the thoracic and abdominal cavities. The diaphragm arches into the thoracic cavity and its contractions move it up and down. The photo illustrates how the diaphragm is composed of muscle fibers grouped into larger bundles. The tissue is magnified 20,000 times.

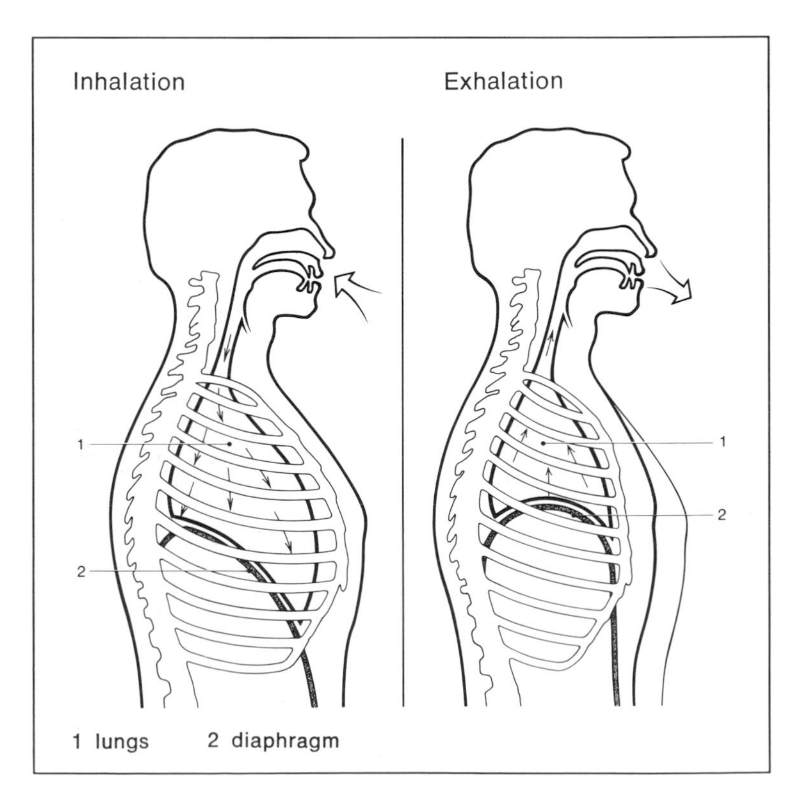

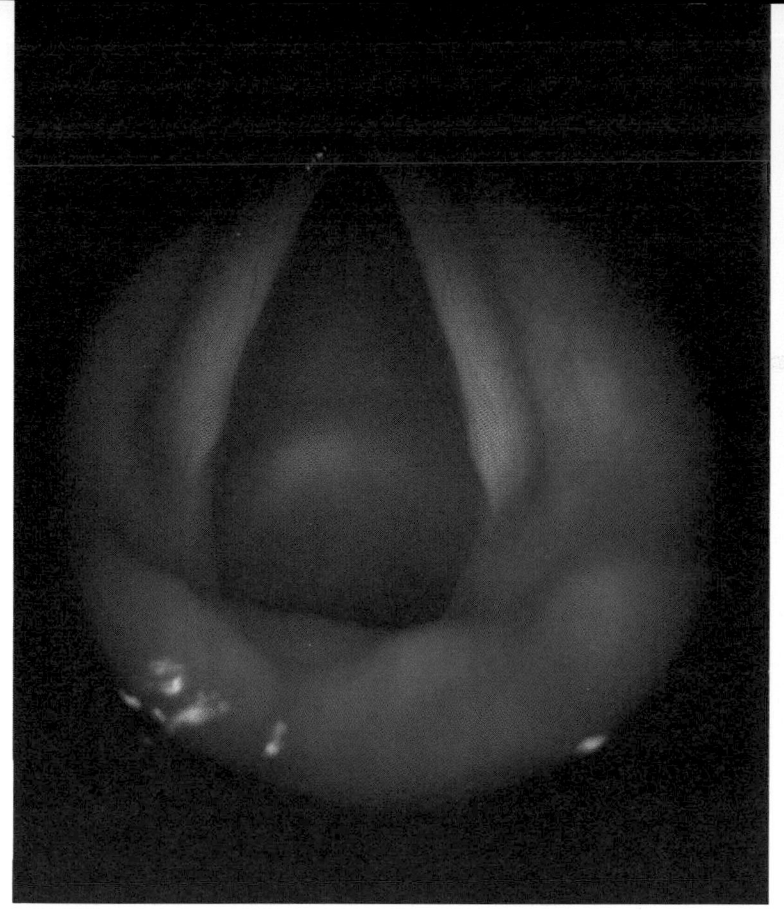

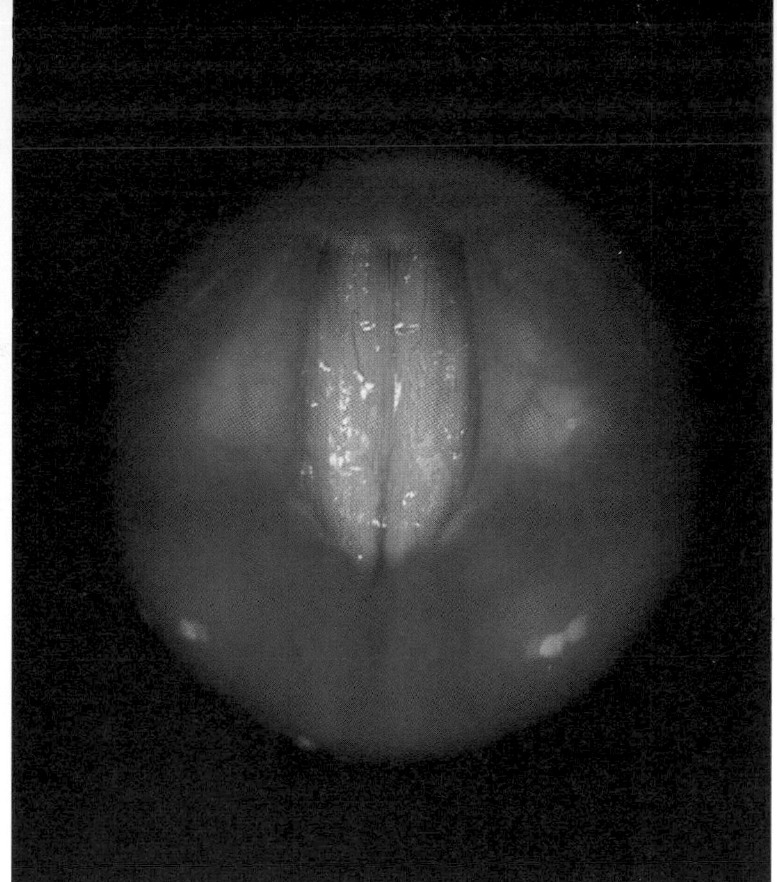

The vocal cords
appear white in those photos taken from the upper portion of the pharynx. The vocal cords are situated inside the larynx and consist of elastic bands covered by mucosa. By changes in the position of the laryngeal cartilages, the glottis can be opened or closed, and the vocal cords stretched or relaxed. The photo on the left shows the vocal cords at rest, i.e., when we breathe but do not speak, They are then relaxed and the glottis wide. The right-hand photo shows the vocal cords stretched and the glottis closed.

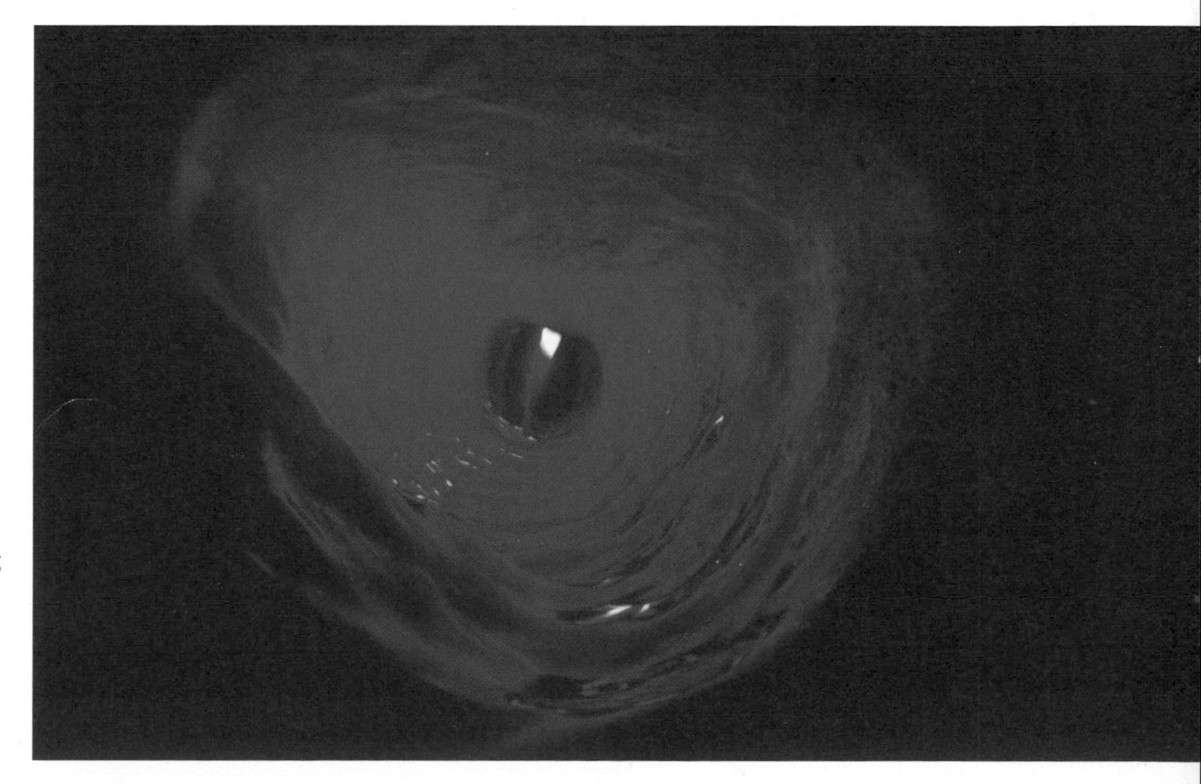

The glottis and the vocal cords viewed from below
with the larynx and its horseshoe-shaped cartilage rings in the foreground. During speech, the stretched vocal cords form a narrow opening through which air is forced. In passing, the air vibrates the vocal cords; the faster the vibrations, the higher the pitch of the sound emitted. Modifications in the quality of sound are added in the mouth and nasal cavity.

The Trachea

The trachea (windpipe)
viewed from the inside rear, showing some of the sixteen to twenty cartilage rings, which are open at the back. The trachea is a direct continuation of the larynx. It branches at its end into two primary bronchi, one for each lung. The points of ramification are strengthened by large, specially formed cartilage rings. The trachea is a tube approximately four inches long and a little less than an inch in diameter. It is very elastic and can be shortened or lengthened somewhat. The cartilage rings are not actually rings but are horseshoe-shaped, with the opening toward the back. The posterior wall of the trachea is a band of smooth musculature and connective tissue whose width is determined by the gap in the cartilage rings.

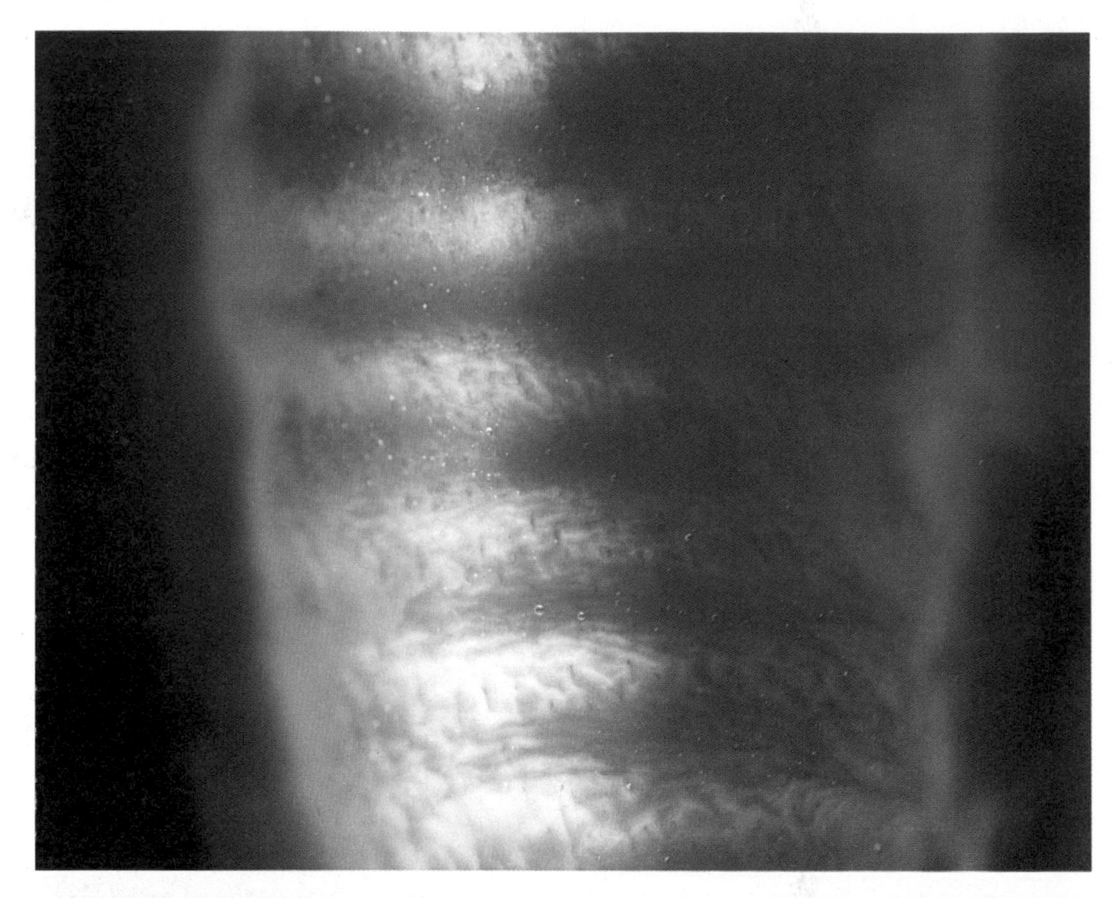

The bronchi viewed from above
The continuation of the trachea into the lungs looks like two dark holes. The primary bronchi are approximately one centimeter wide and ramify into the branches of the bronchial tree, the smallest subdivisions of which are the alveoli. The right primary bronchus is an inch or more long; the left is almost double this length. Like the trachea, the primary bronchi are kept open by cartilage rings. The front of the trachea points downward in this photo.

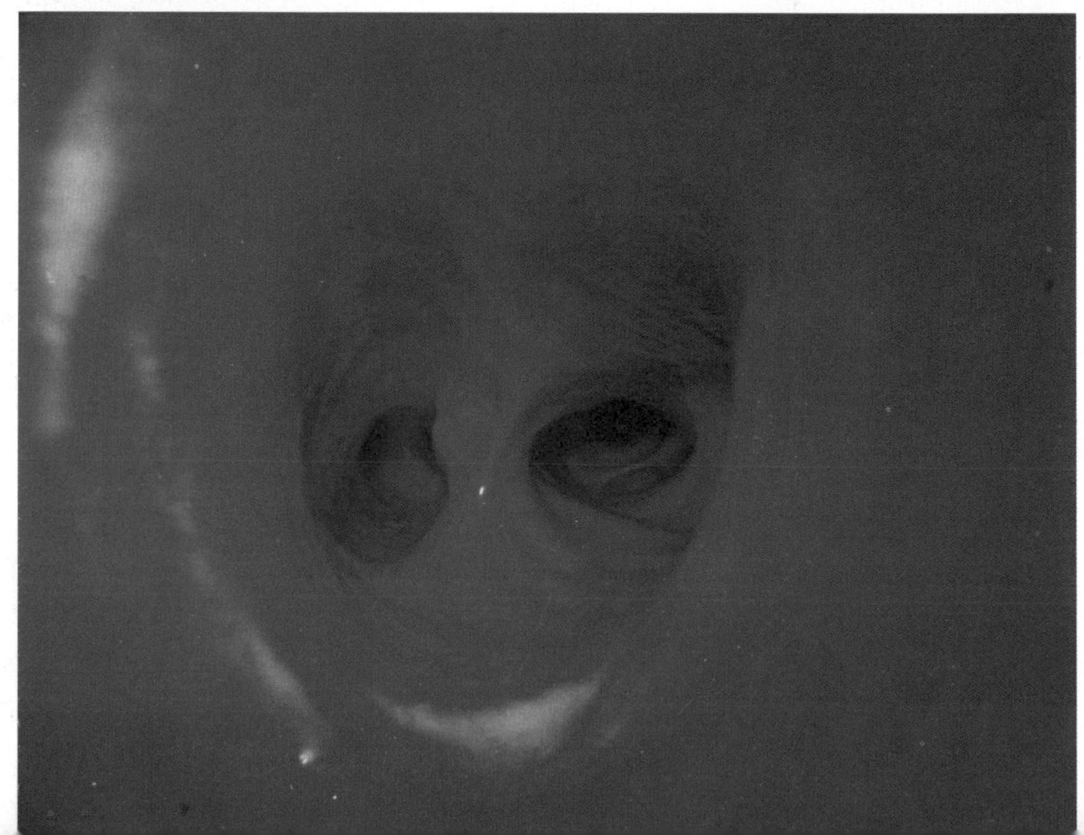

The pulmonary alveoli,
looking like bunches of grapes, form the ends of the smallest bronchi. Several hundred million alveoli, lining each lung, provide a total surface area of eighty to one hundred yards — about forty to fifty times the total skin area. It is this large surface area that permits the rapid exchange of carbon dioxide and oxygen in the lungs. The alveoli in the photo are magnified about 100 times. The capillaries surrounding the alveoli are so narrow that red blood cells are forced to move through them one at a time. Only two thin membranes separate the cells from air in the alveoli, facilitating the passage of gases to and from the blood. The process is an automatic one depending on the fact that there are differences between the partial pressures of gases present in the blood and in the air in the alveoli. The partial pressure of carbon dioxide in blood is greater than in the alveolar air, so carbon dioxide diffuses into the latter. Since the partial pressure of oxygen in the air is greater than that in venous blood, oxygen diffuses into the blood, where it combines with the hemoglobin in the red blood cells and is subsequently transported to body tissue.

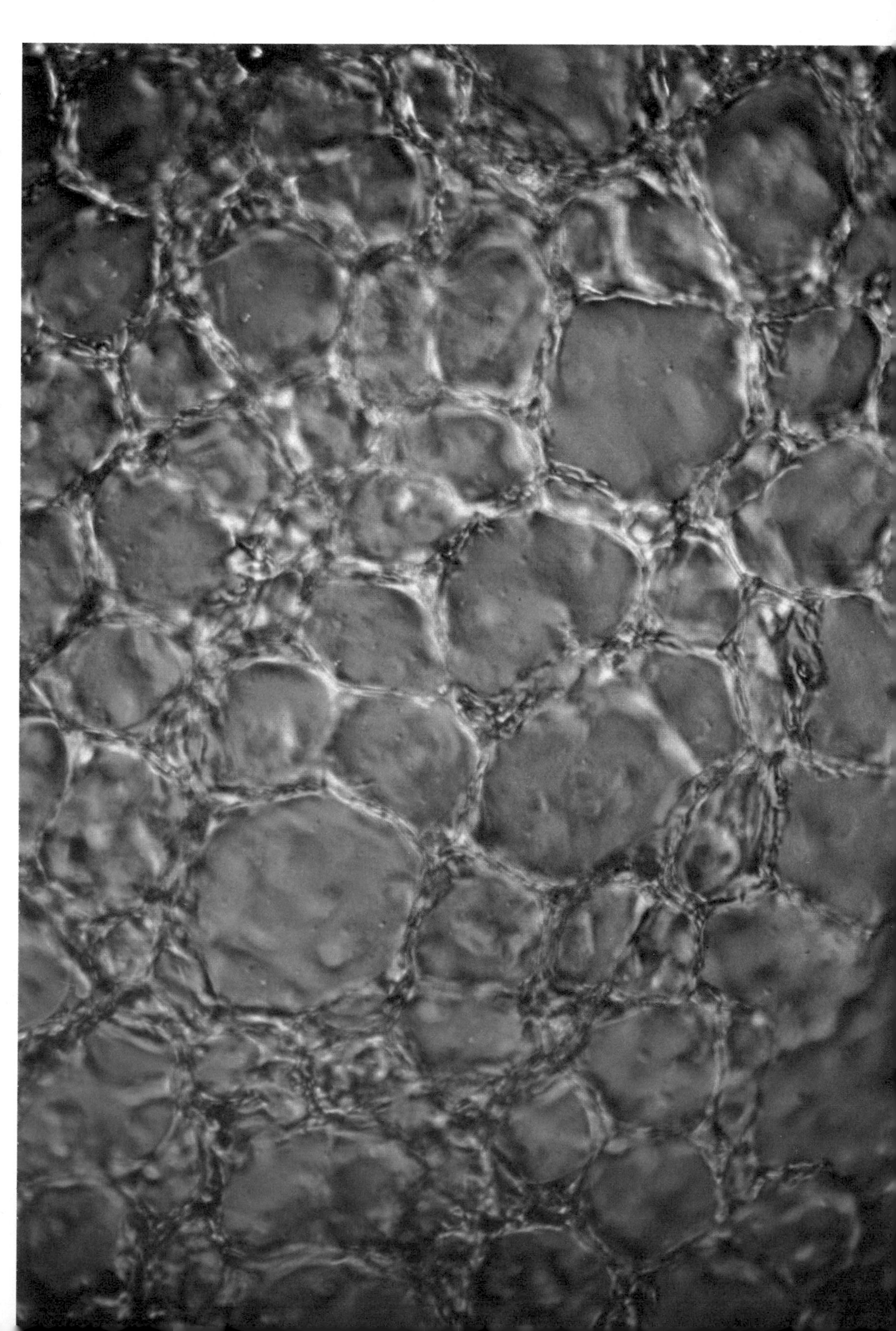

The Lungs

Lungs and the air we breathe
The lung of a three-year-old child is clean, as can be seen in the two photos on this page (1, 2). The lungs become dirtier with age, especially if exposed to city air, which is often polluted.

A variety of particles enter the lungs when we breathe. Many are eliminated by the lungs' own cleaning actions. However, if the air is heavily polluted, many particles will remain in pulmonary tissue, which then grows black as seen in the upper two photos on the right-hand page (3, 4).

Photos of the air surrounding us reveal how dirty it is. The two photos at the bottom of the opposite page show samples of country air and city air, each magnified 400 times. The country air on the left contains occasional dust particles, fine mineral granules, pollen, and spores carried by the wind (5).

City air looks completely different. A large amount of indistinguishable "rubbish" fills the space between such particles as can be identified. The large triangular particle is very likely a mineral chip (6).

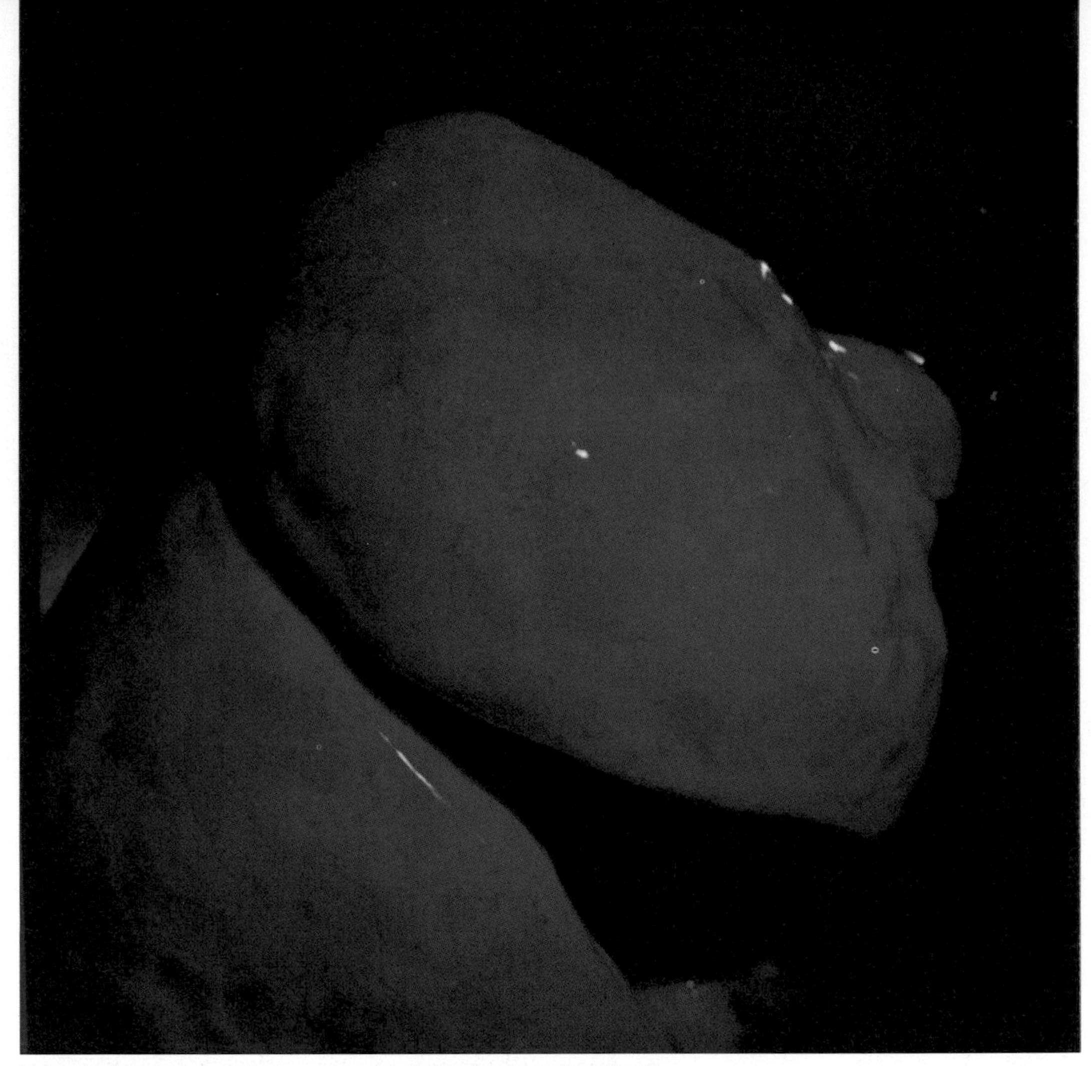

1 Part of a lung in a three-year-old child, about twice life size

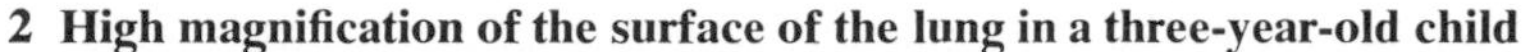

2 High magnification of the surface of the lung in a three-year-old child

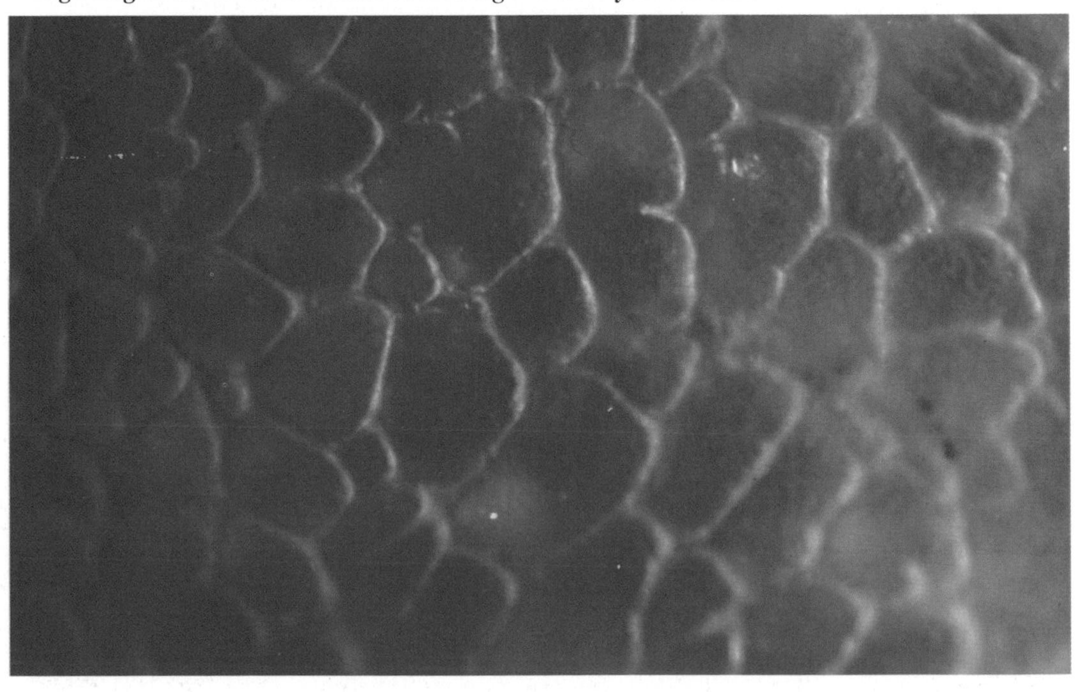

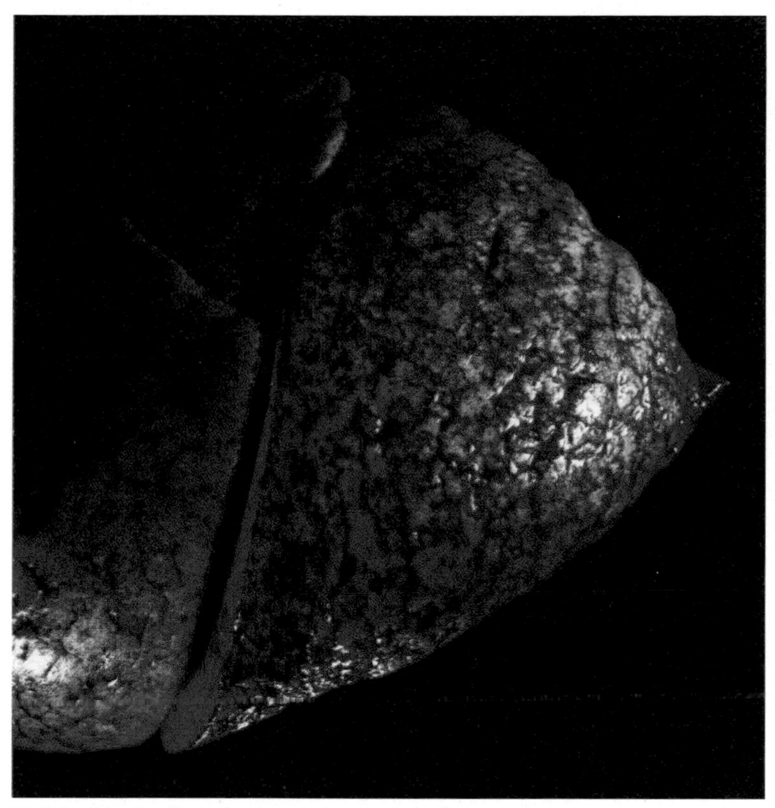

3 Dirt blackening the lung of a city dweller

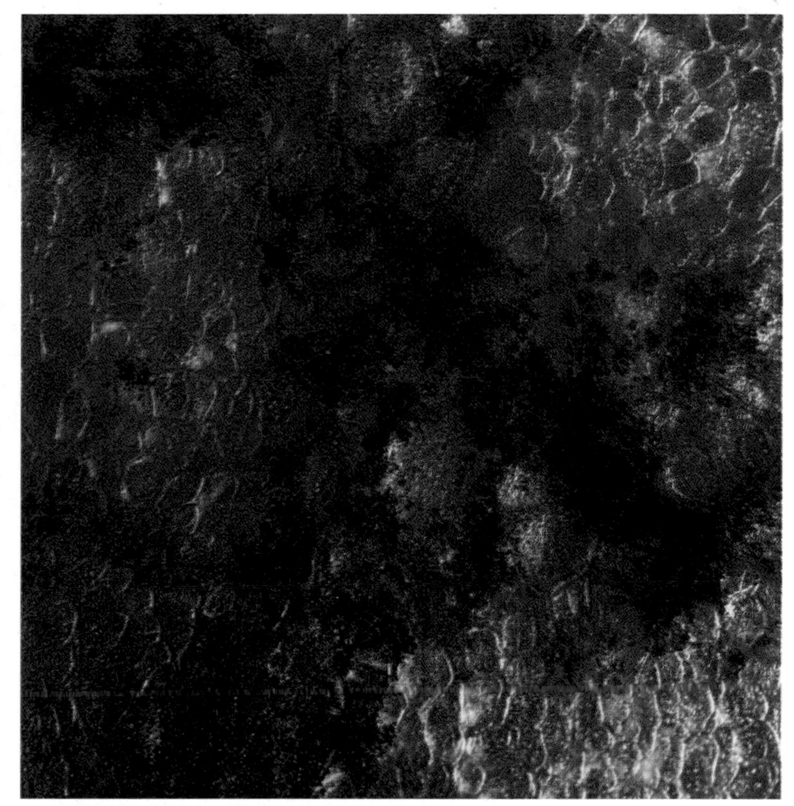

4 Close-up of the surface of the lung of a city dweller

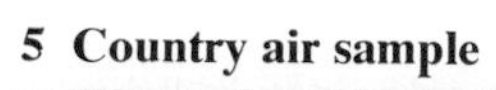

5 Country air sample

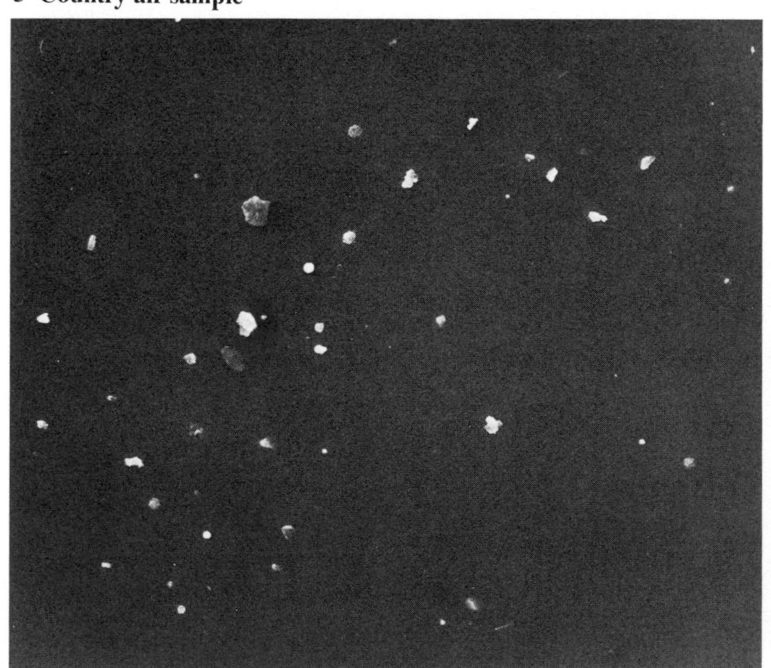

6 Urban air

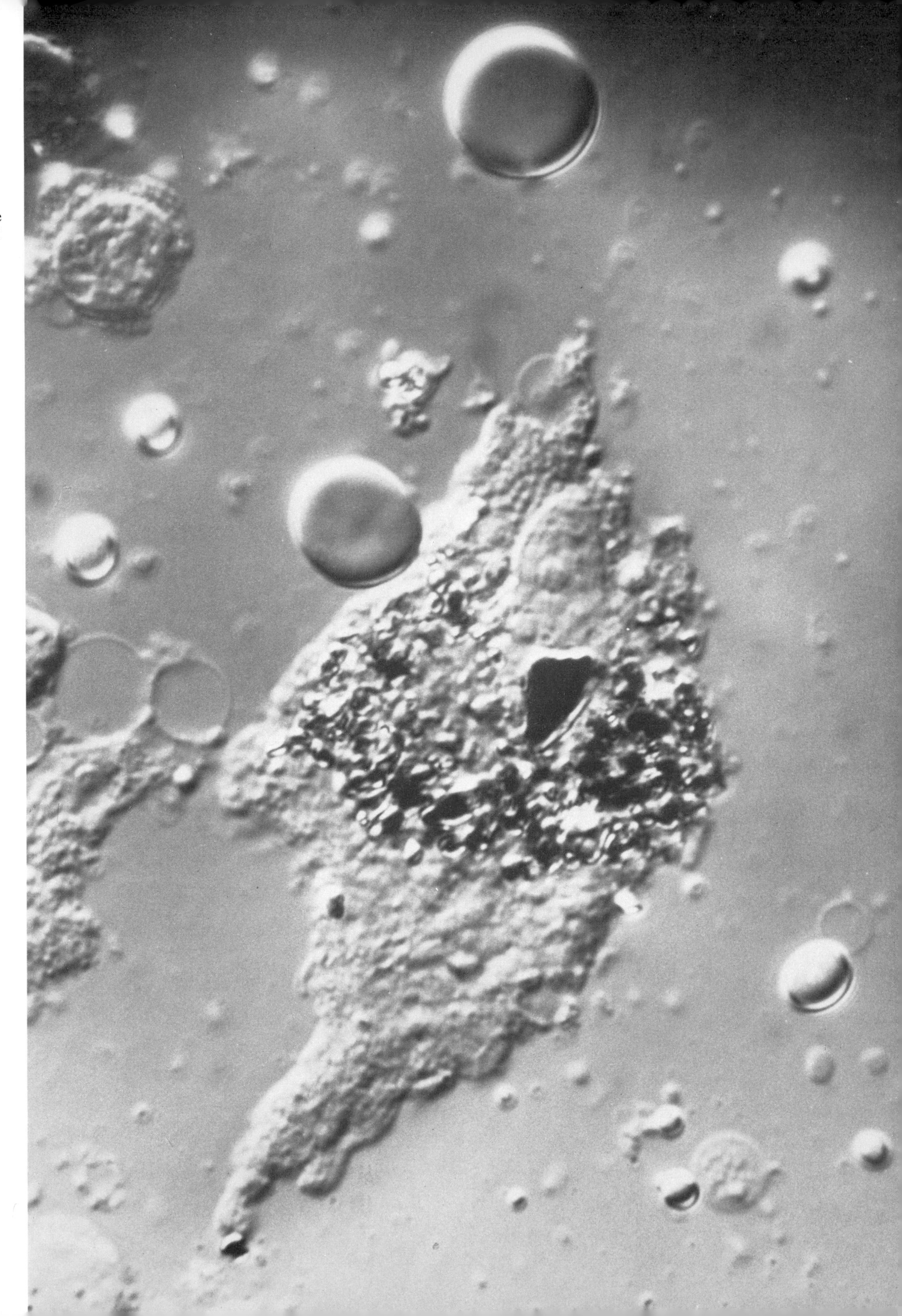

In a polluted lung
Macrophages and white blood cells try to engulf and render harmless the dirt particles and microorganisms that are dangerous. The macrophage in the center has ingested something angular, probably a stone chip. Many macrophages perish in the fight against invaders and impurities. When the photo was taken, the macrophages as well as the two red blood cells were alive.

Smoking

Tobacco smoke looks white and perhaps harmless, but it contains a variety of substances which may be injurious to tissues and organs. For one thing, it has a high carbon monoxide content. The concentration exceeds by far the amount we are exposed to from automobile exhaust on a heavily traveled street. Carbon monoxide inhibits the uptake of oxygen by red blood cells and consequently reduces the oxygen supply to tissues. In all probability, various tarry substances in the smoke can induce cancer in the air passages. Most scientists believe that excess smoking, especially of cigarettes, is dangerous. Tobacco smoke, moreover, inhibits the mobility of cilia and consequently their capacity to remove impurities and mucus from the lungs. Smokers often get bronchitis. The heart and the vascular system are also negatively influenced by smoking.

The photo below illustrates how tobacco smoke swirls down the windpipe, through the bronchi, and into the lungs.

Where the windpipe branches into the two primary bronchi, the air becomes turbulent and moves rapidly into the still finer bronchi.

Smoke sucked down into the lung

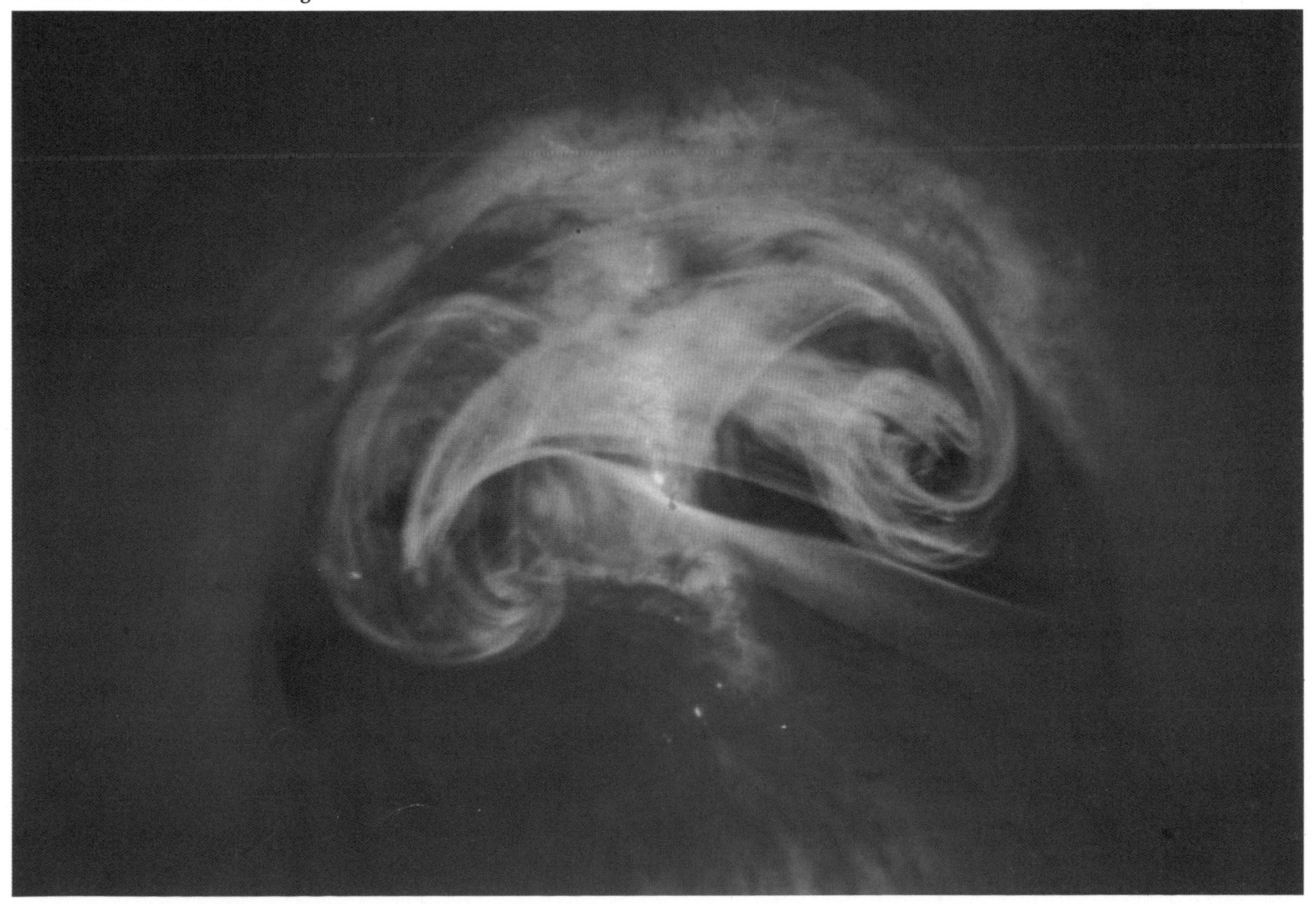

Body Temperature

In comparison to other warm-blooded animals, man has a relatively low body temperature. Our thermostat is set at approximately 98.6°F., whereas the temperature of the dog is about 102.8°F.; the pig, 103°F.; and the chicken, above 107°F. It is common knowledge that we do not experience such high temperatures except under unusual conditions; for instance, during long and strenuous work, in extremely warm environments, where heatstroke is possible, and during febrile (fever-producing) illness.

If the body becomes too warm or too cold, a variety of forces are set in motion to readjust its temperature to a normal level as quickly as possible. This level is determined by a kind of thermostatic regulatory center in the hypothalamus of the brain. The center consists of a group of nerve cells near other regulatory centers, concerned with thirst and hunger, for example. The "thermostat" notes the temperature of the blood passing the region. When the temperature is too high, blood vessels in the skin (cutaneous vessels) are dilated, permitting more heat to escape from the skin by radiation and, in part, by conduction.

If the temperature is not sufficiently lowered by this mechanism, the hypothalamus sends out impulses by way of the autonomic nervous system to stimulate sweating. We have about two million sweat glands, which are especially numerous in the forehead, the armpits, the palms of the hands, and the soles of the feet. The evaporation of sweat dissipates heat. During extreme muscular exertion, the sweat glands can excrete about 1 liter per hour. Much heat can also be removed through respiration, which increases during hard work. A deficiency in body heat produces the opposite reactions. The cutaneous blood vessels constrict so that the blood will not be cooled; we become pale. Large muscle groups are activated, and we generate heat by shivering. If this is not enough, we can resort to conscious measures. We wave our arms, put on more clothes, or move to a warmer place.

Why is a constant body temperature so important? The reason is that chemical reactions in the body are influenced by temperature. Many of these reactions are mediated by enzymes, which function optimally at 98.6°F. If body temperature is lowered by two or three degrees, the rate of all the reactions decreases. If body temperature is lowered to about 86°F., the brain is affected and unconsciousness occurs. Pulse rate as well as respiration become much slower. We find an environmental temperature of 77° to 86°F. warm, but a body temperature at this level is perilously cold. An elevation of the body temperature toward 102°F. is not pleasant either. Certain biochemical processes are influenced by high temperature. Oxidation in cells increases, requiring a faster supply of the raw materials necessary. A temperature above 104° or 105°F. is a great strain on the body; one above 108°F. is likely to be fatal.

Sweat glands open even at a fingertip
The openings appear as small dots on the ridges. Sweat glands are part of the body's temperature regulating system. Their density varies in different regions of the skin; they are most numerous in the palms of the hands, the soles of the feet, and the forehead, which has about 300 sweat glands per square centimeter.

A drop of sweat consists chiefly of water, but it also contains substances such as common salt, ammonia, and lactic acid. So, pronounced sweating leads not only to a loss of fluid, but also to a loss of salt. The drops of sweat convey heat to the surface of the skin, which is then cooled by evaporation, a process that removes still more heat.

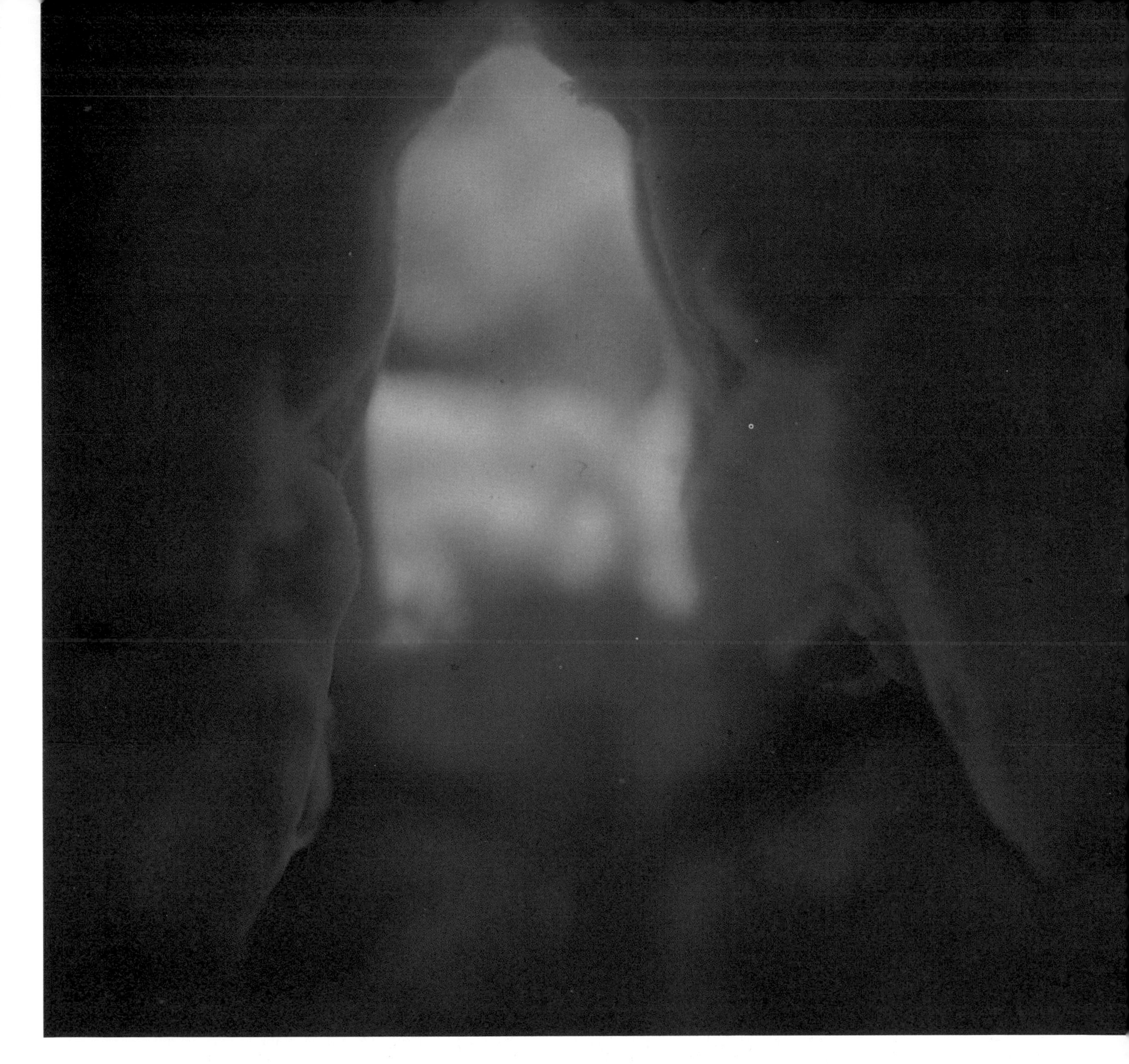

The hypothalamus
is the site of a number of important regulatory centers, including that for body temperature. The hypothalamus lies in the anterior portion of the brain stem and contains nerve cells that are stimulated by temperature variations in the circulating blood. Nerve impulses initiated here lead to adjustments of the heat balance in one direction or the other. Another center in the hypothalamus is concerned with thirst and fluid balance. The photo is taken from above, looking through the third ventricle of the brain. The various hypothalamic nerve cell groups (nuclei) lie inside the rough walls on either side of the bright midportion. They are seen here magnified 7 to 8 times.

The Kidneys

The Body's Cleansing System

The problem of taking in nourishment and excreting waste products is minimal in monocellular organisms. Food and waste both diffuse directly across the cell wall into and out of the cell, respectively. The organism will survive as long as the environment is supportive. If the environment changes, one-celled organisms may be impaired. In the case of multicellular organisms, the problems are not that simple. The cellular environment must be kept relatively constant. Special organs are developed to take in oxygen and nourishment, as well as organs to excrete the waste products and toxins produced by metabolism.

The waste product carbon dioxide is exhaled through the lungs. Some nitrogenous waste is removed with sweat. However, we excrete the main portion of organic and inorganic waste and toxins through the kidneys, which may be called the sanitation stations of the body. The two kidneys have a remarkable capacity, so large that if necessary we can manage with only one.

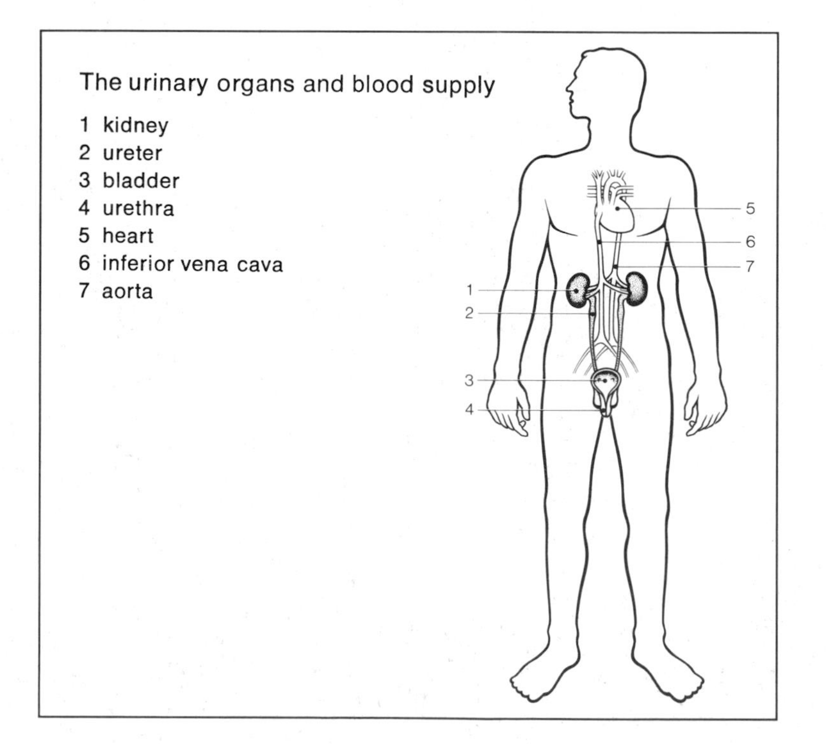

The kidneys lie in the posterior portion of the abdominal cavity at the level of the lowest ribs. An adult kidney is about 12 cm long, 6 cm wide, and 3 cm thick, and is protected by thick layers of fat. It weighs approximately 150 grams and is covered by a thin layer of connective tissue, the renal capsule. Nearest its smooth, dark red surface lies the renal cortex. Below it lies the medulla and the renal pelvis, a cavity in the center of the kidney which opens into the ureter. Each kidney contains approximately one million nephrons, the functional units of the kidney, which filter the blood. Each nephron consists of a glomerulus and a tubule. The glomerulus consists of a tuft of capillaries and a microscopically small, double-walled bulb, Bowman's capsule, which surrounds the tuft. Arterial blood passes through the tuft under considerable pressure so that a large quantity of fluid and solutes moves across the capillary walls. This fluid, the glomerular filtrate, collects in the space between the tuft and the walls of Bowman's capsule.

About 1,700 liters of blood pass through the kidneys every day, producing about 180 liters of glomerular filtrate in the glomeruli.

The glomerular filtrate contains a large number of substances that are vital to the body. Ninety-nine percent of the fluid and most of the solutes are reabsorbed as the glomerular filtrate passes down the tubules.

A tubule is an approximately 5 cm long coiled duct. It is surrounded by a network of capillaries. In the tubules, the glomerular filtrate becomes concentrated into urine, but in addition some excretion of waste products also takes place. Water makes up 95 percent of urine. The other components are normally substances the body wants to get rid of, such as urea, creatinine, uric acid, ammonia, and salts. The color of urine comes from transformed bile pigments. The kidneys excrete about 1.5 liters of urine a day, or approximately one-hundredth the amount of the glomerular filtrate.

Filtration in the kidneys begins with the straining of fluid from the blood. The strained portion includes everything except blood cells and larger protein molecules. This is followed by separation of the filtrate into the waste products to be excreted and the vital substances to be reabsorbed. In effect, when the kidney performs its cleaning operation, it first removes nearly everything from the room and later puts back only what is supposed to be there.

Urine in the renal tubules passes into collecting ducts, which drain into the renal papillae. The papillae open into the renal pelvis, where the urine from all the papillae collects and empties into the ureter. The ureter carries urine to the bladder, where it remains until the bladder empties through the urethra, generally two to five times a day.

In addition to filtration, the kidneys also are involved in maintaining the proper balance between fluid and electrolytes in the body. If the amount of body fluid decreases, the pituitary gland releases antidiuretic hormone. This hormone stimulates the renal tubules to absorb more fluid so that less is lost to the body in urine. The urine becomes more concentrated and the body saves water. If the body contains too much fluid, the kidneys produce a more dilute urine to get rid of the surplus.

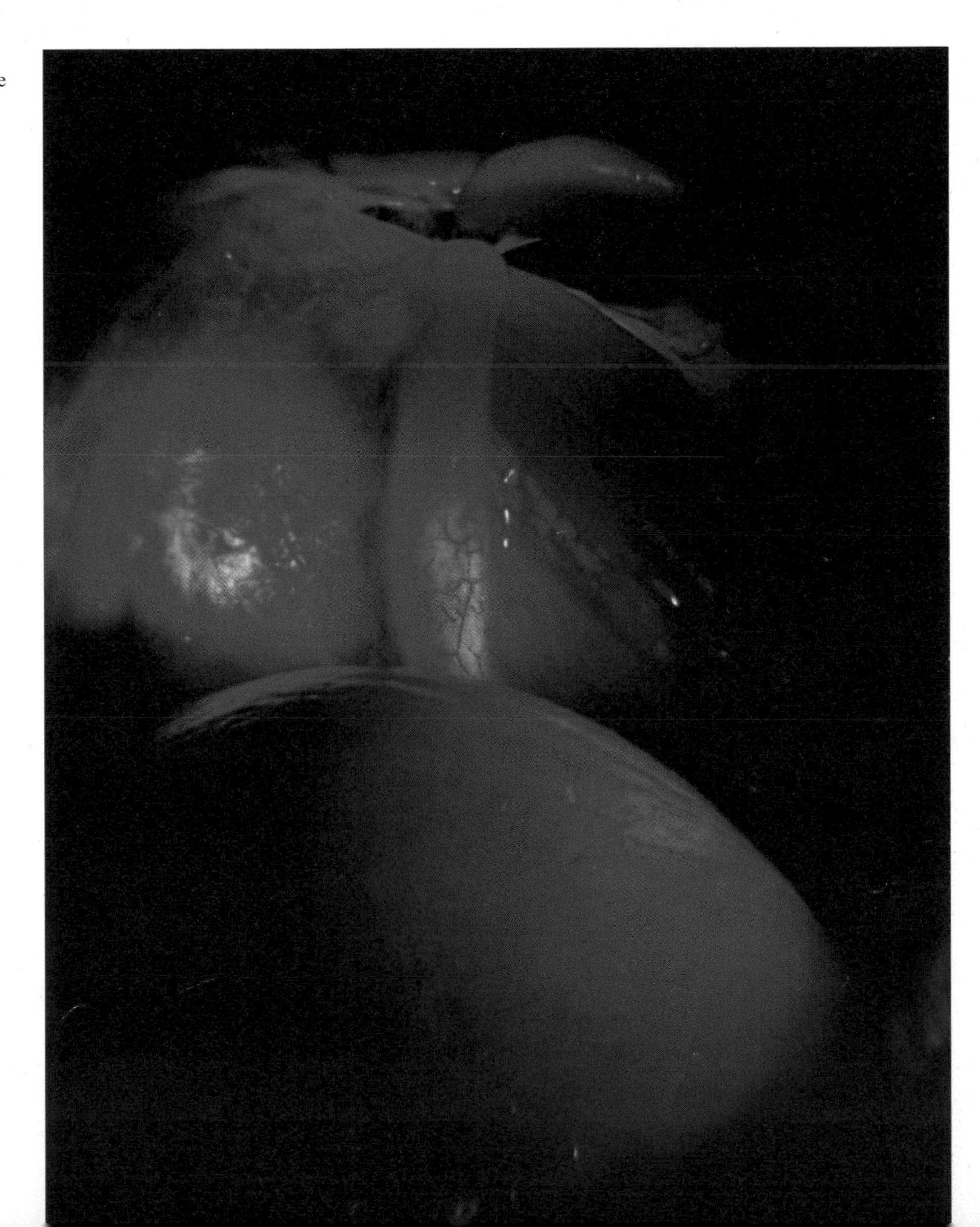

Kidney, ureter, and bladder
The bladder is in the foreground with the ureter leading up to the left kidney, visible at the upper right.

The Kidneys

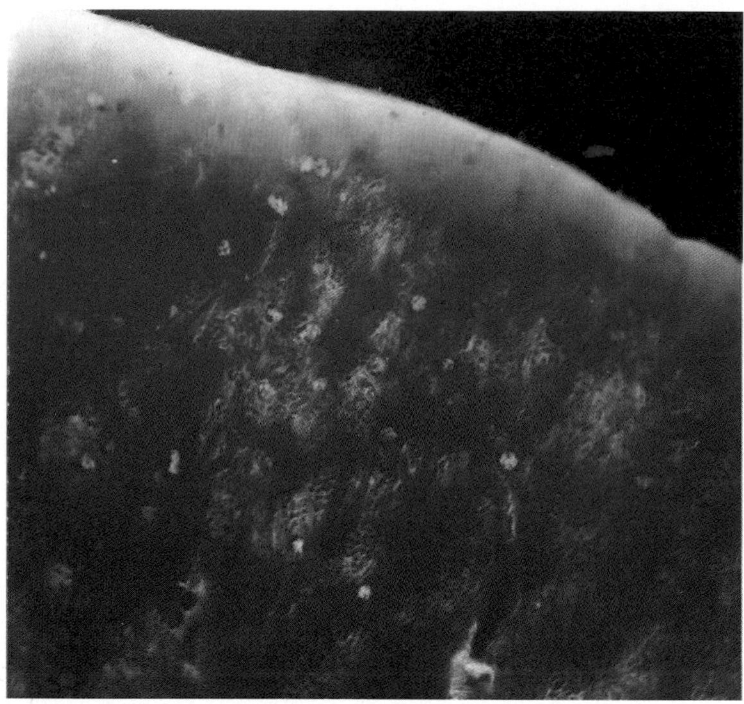

Section through the renal cortex. The fine capillary networks of the glomeruli are visible.

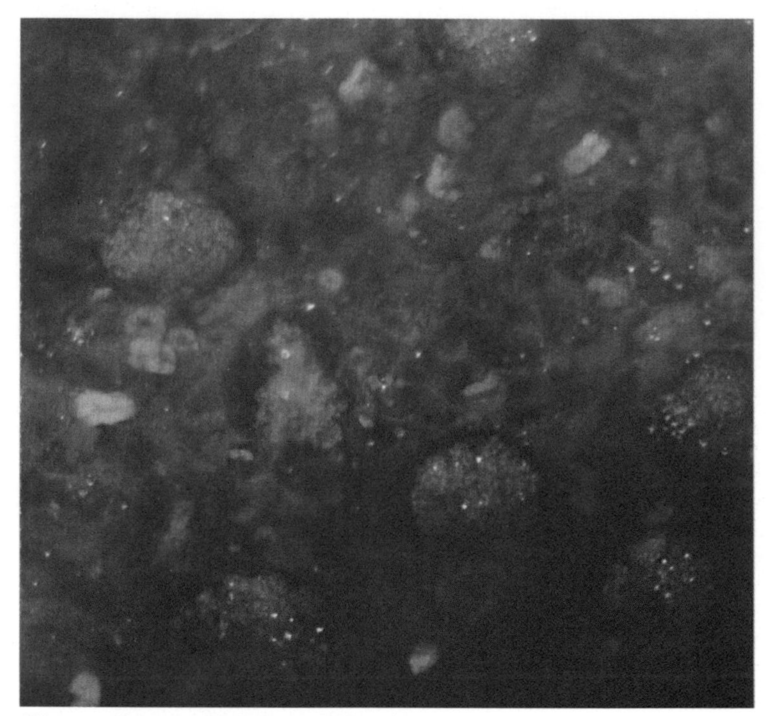

Glomeruli under higher magnification

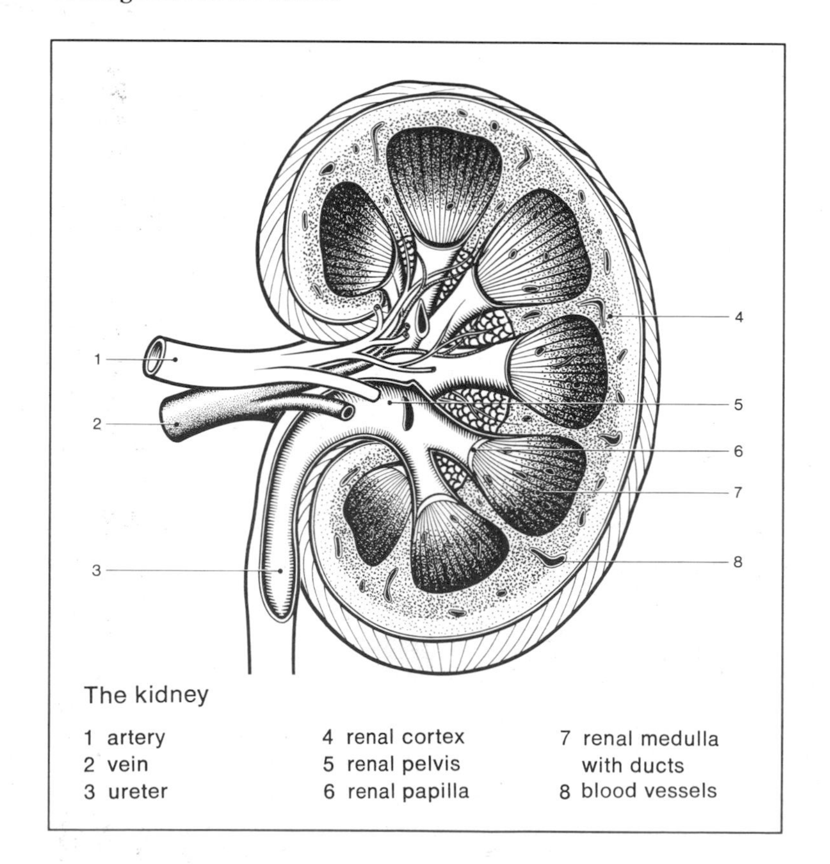

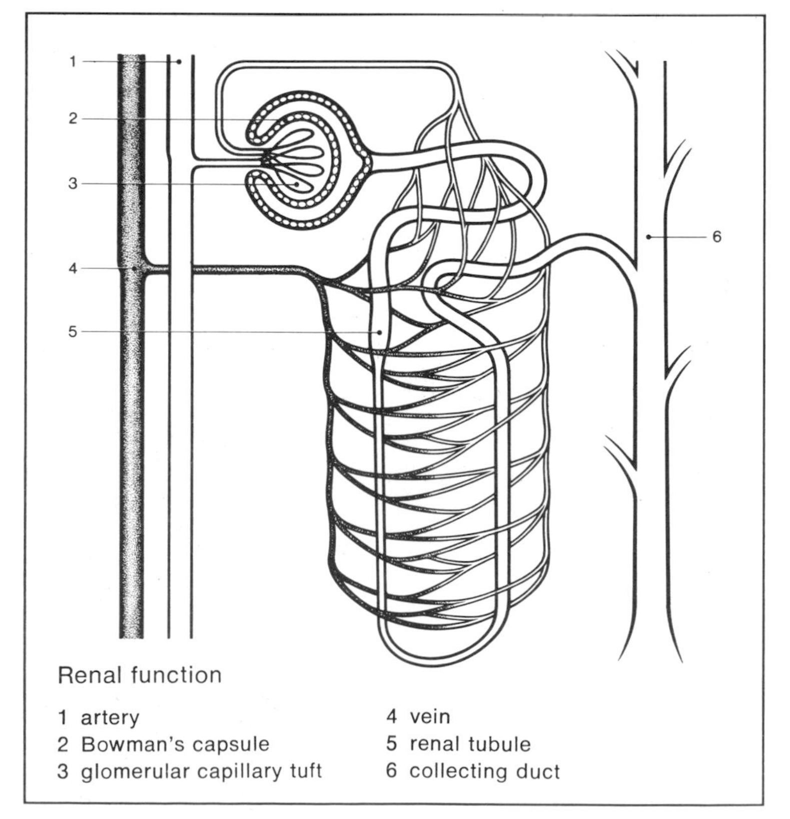

◁ **Section through the renal cortex**
The photo at far left shows numerous glomeruli in the cortex. The capillary tufts of the glomeruli filter the blood, producing the glomerular filtrate. The photo also shows the first convoluted parts of the tubules. In the tubules, 99 percent of the water in the glomerular filtrate is reabsorbed together with sugar and salts that can be reused. The tubules extend down into the renal medulla, seen at the bottom of the photo, and drain into collecting ducts. These converge in the renal papillae and empty into the renal pelvis. Together, a glomerulus and a tubule form a nephron, the blood-filtering and urine-producing functional unit of the kidney. Each kidney contains approximately one million nephrons. The pressure in the glomerular capillaries is considerably higher than in other capillaries, and permits rapid filtration. Plasma and solutes pass across the capillary walls, leaving only blood cells and proteins behind in the vessels. Reabsorption takes place in the tubules, which are surrounded by a dense capillary network. Vital substances and water are reabsorbed, while wastes and the amount of water required for keeping them dissolved are excreted.

The photo at near left shows glomeruli under a higher magnification. The capillary network lies inside Bowman's capsule, a bulb-shaped, double-walled container.

The capillary tuft of a glomerulus ▷
magnified about 400 times in this microscope preparation. The capillary network with blood cells is in the center, surrounded by Bowman's capsule. Blood is filtered in the tuft and the filtrate collects in the space between tuft and capsule. The glomerular filtrate totals about 180 liters a day. Renal tissue with cells and nuclei is seen outside the capsule.

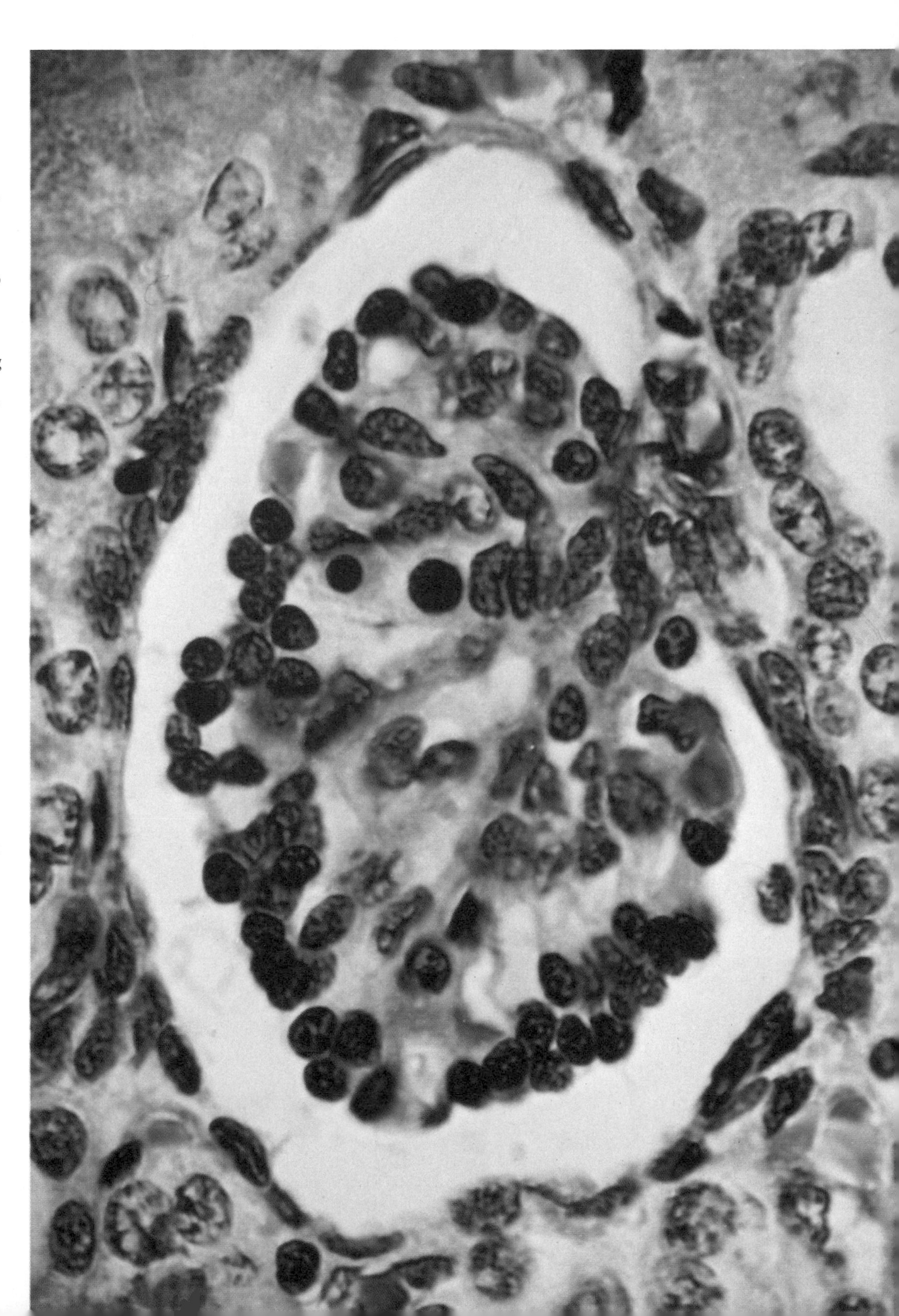

Renal cortex in a newborn, with glomeruli densely distributed and in some cases immature ▷

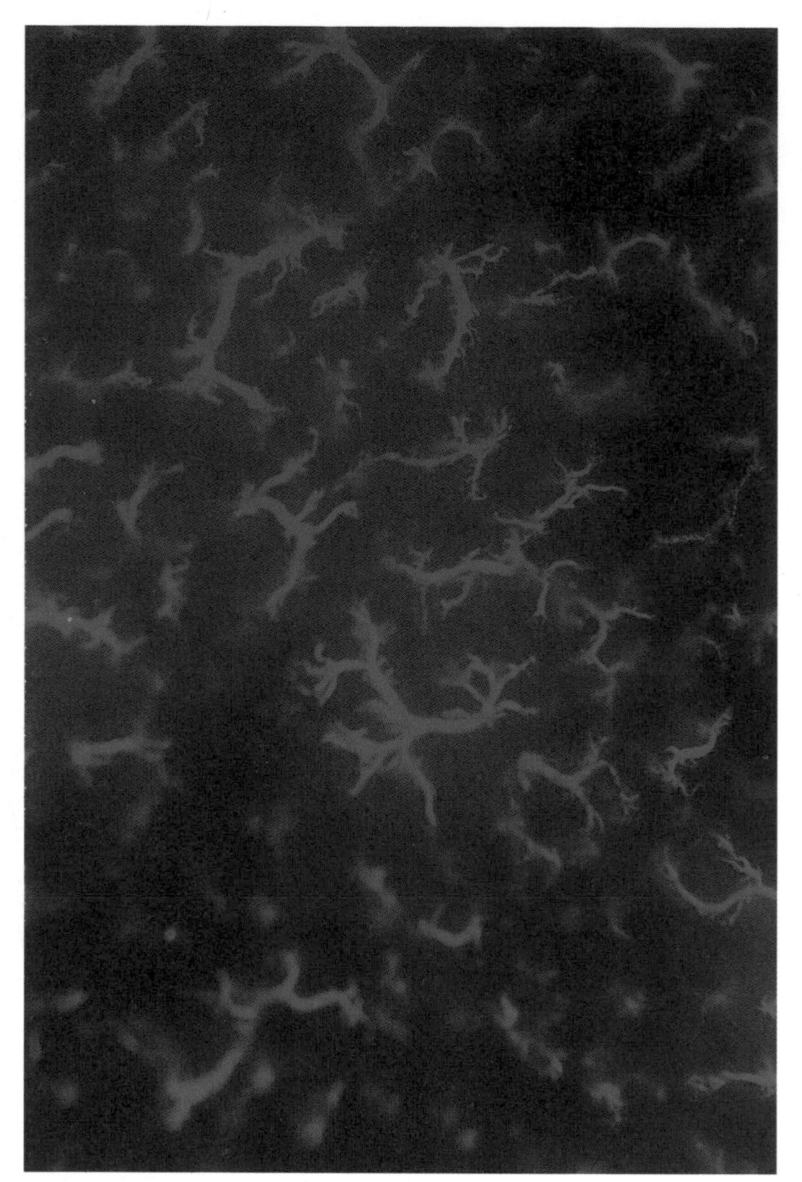

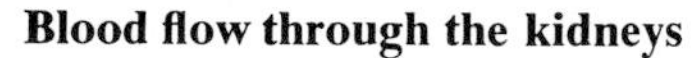

Blood flow through the kidneys
amounts to approximately 1,700 liters a day. The cleansing of the blood goes on continuously, which means that the kidneys must be well supplied with blood. The renal arteries, therefore, are relatively large. Renal tissue contains millions of tiny glomeruli where the blood cleansing process begins. Another important function of the kidneys is to regulate the water and salt turnover in the body. If we drink a lot of fluids, excretion of urine is increased and the urine is more dilute. Under conditions of dehydration, the pituitary gland releases an antidiuretic hormone. This increases the tubular reabsorption of water and the volume of urine is much reduced. The photo shows blood vessels on the surface of the kidney.

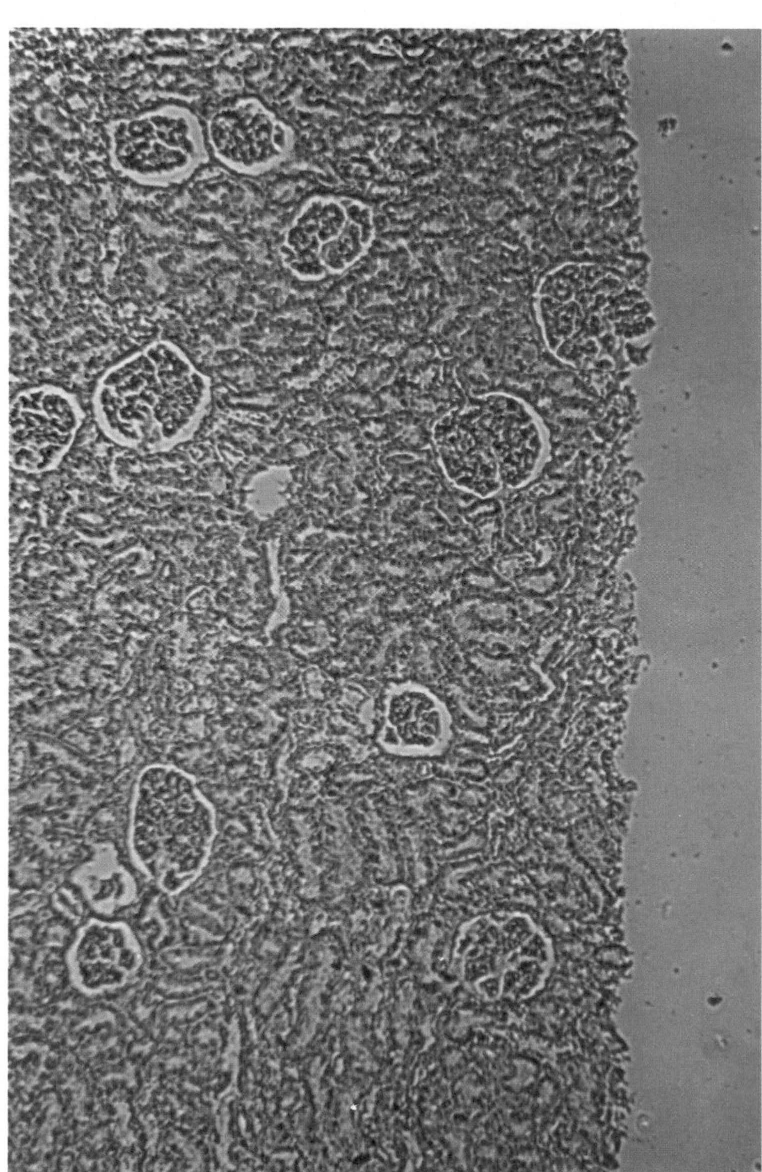

The renal cortex with glomeruli
appearing as globular structures in this magnified microscope preparation.

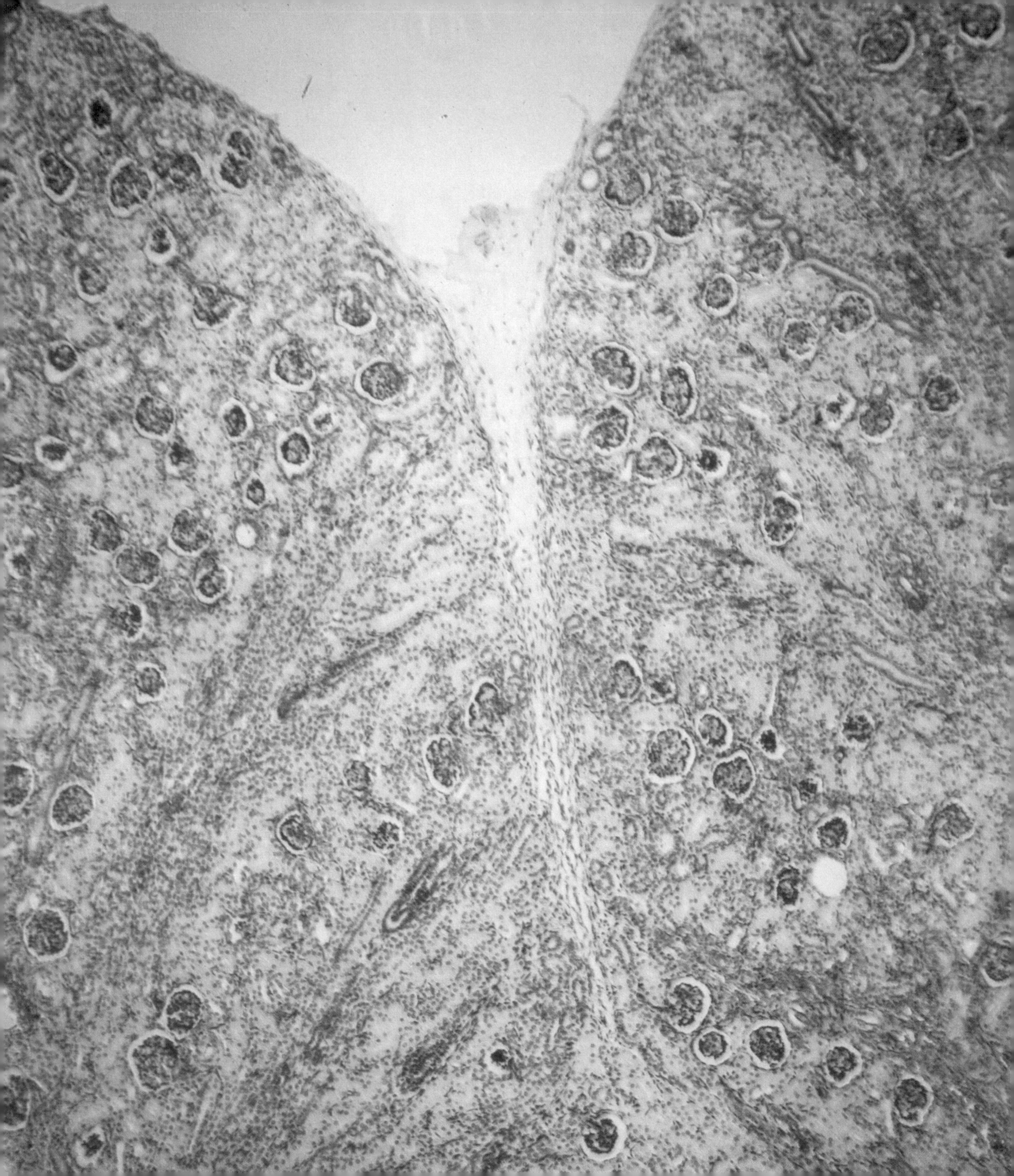

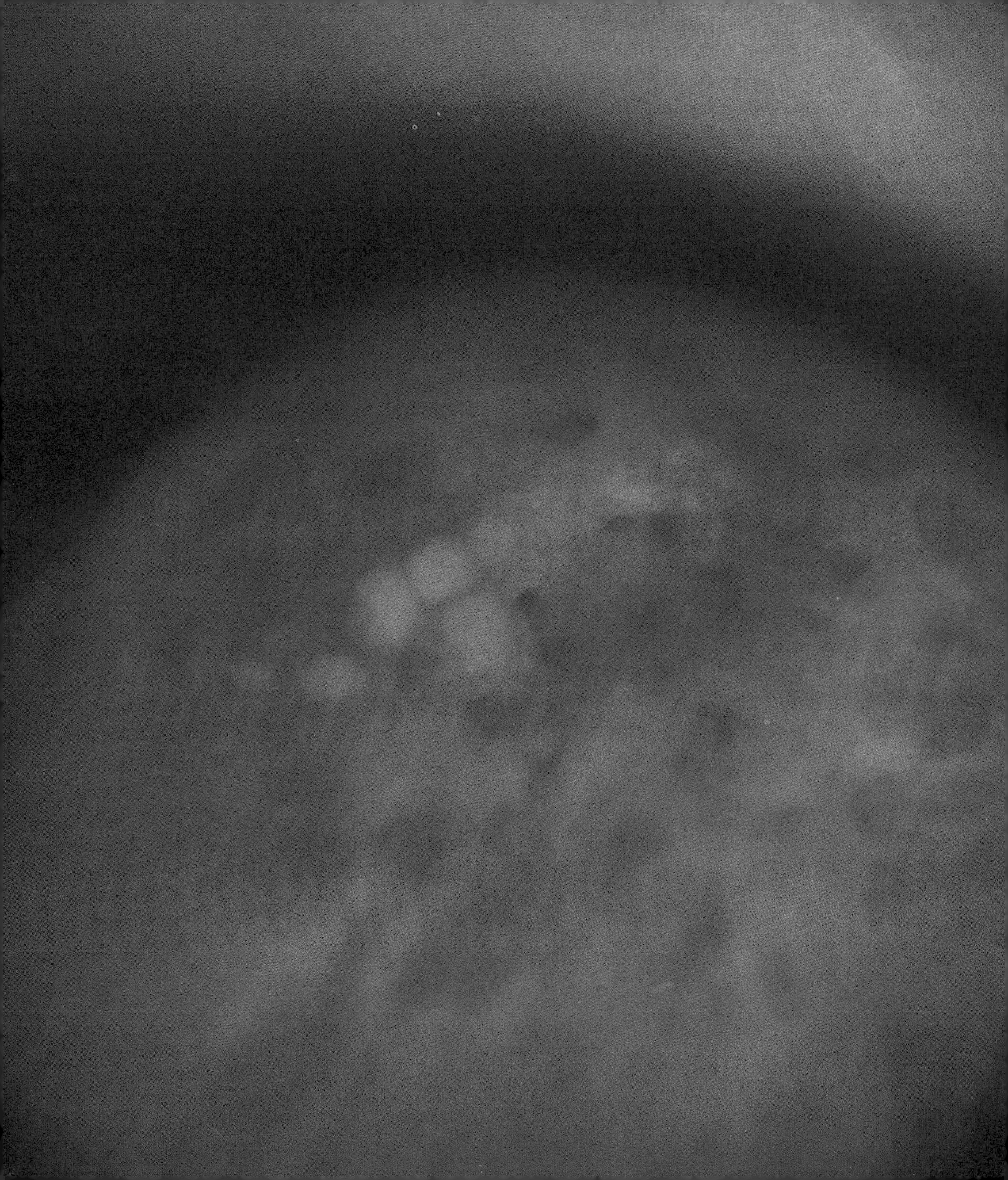

◁ **Drops of urine progress**
from the collecting ducts to the renal papillae. Urine is composed chiefly of water, in which nitrogenous waste products, salts, et cetera, are dissolved. Urea makes up the main part of the organic substances. Urine gets its yellow color from transformed bile pigments.

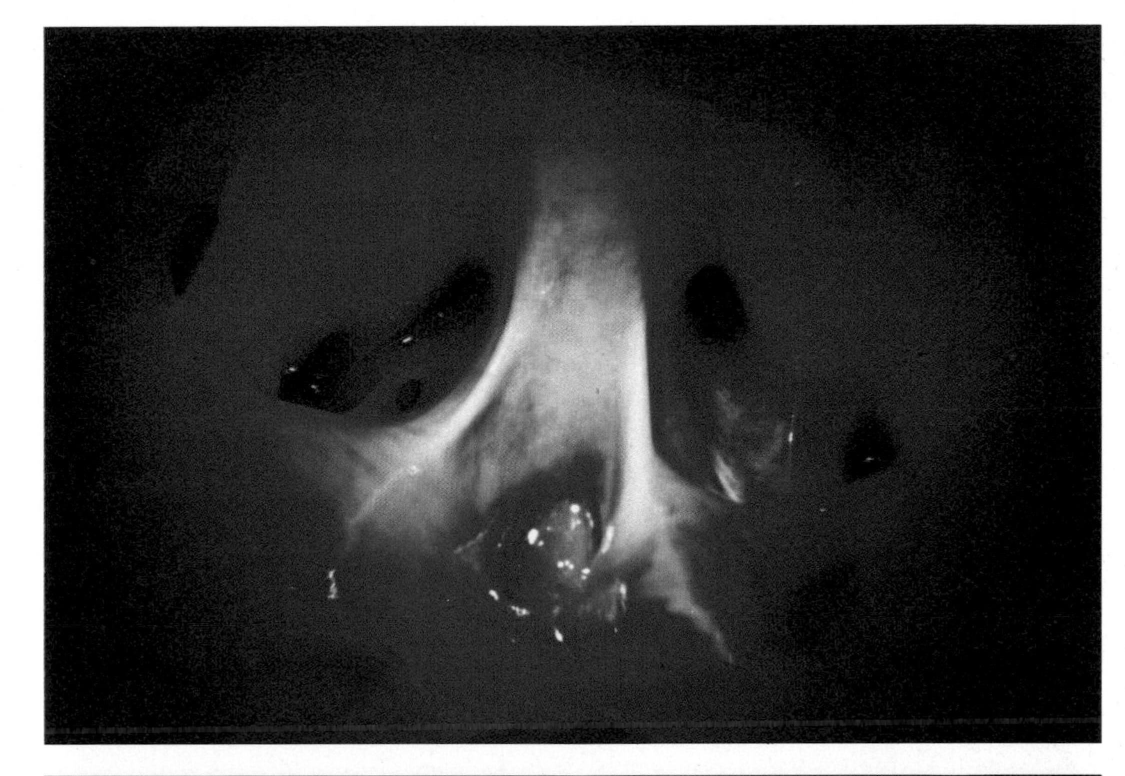

The openings of the papillae in the renal pelvis ▷
A number of collecting ducts converge in each papilla and drain into the pelvis.

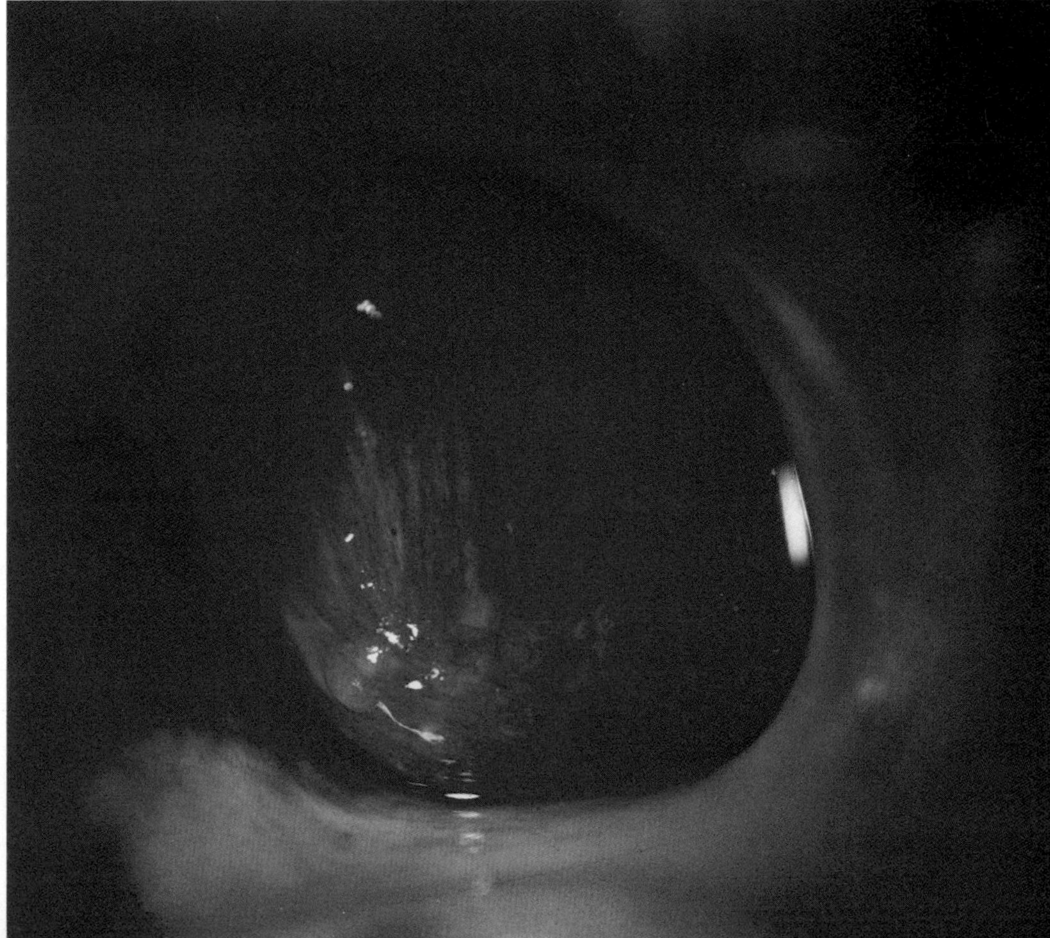

The openings of the renal papillae ▷
Through these openings, urine collected from the nephrons trickles into the renal pelvis. The urine is then conveyed to the ureter and the bladder. Normally, the two kidneys produce about 180 quarts of glomerular filtrate a day, but the concentrated urine leaving the body generally amounts to only about 1.5 quarts.

The opening of the ureter (1)
in the bladder. From each kidney, a ureter descends directly to the bladder, where the urine is stored. The ureter is a thin tube between ten and twelve inches long. A kidney stone obstructing the ureter may be extremely painful and lead to complications. If the pain stops suddenly, it usually means that the stone has passed into the bladder.

The closed opening of the urethra (2)
at the base of the bladder. The urine does not leak out of the bladder continuously, but is released by the action of a sphincter of smooth musculature encircling the opening of the urethra. The wall of the bladder consists of several layers of smooth muscles and can be considerably stretched. The bladder of an adult normally can contain about half a quart. In the healthy person, the voiding reflex occurs when the bladder contains .2 to .3 quart. In pathologic conditions, such as paralysis of the bladder (cystoparalysis) or prostatic diseases, it can hold considerably larger volumes.

Wide-angle view of the interior of the bladder (3)
The highlighted spots at the top are the openings of the ureters. The ureters connect the bladder with each kidney. At the bottom is the urethra. When the bladder needs emptying, a reflex is triggered which relaxes the sphincter muscles around the neck of the bladder. The bottom of the bladder sinks, and simultaneously the muscles in its wall press the urine into the urethra. The process can be voluntarily inhibited or facilitated. When the abdominal muscles are contracted, for example, the intra-abdominal pressure increases so that the bladder is emptied more rapidly.

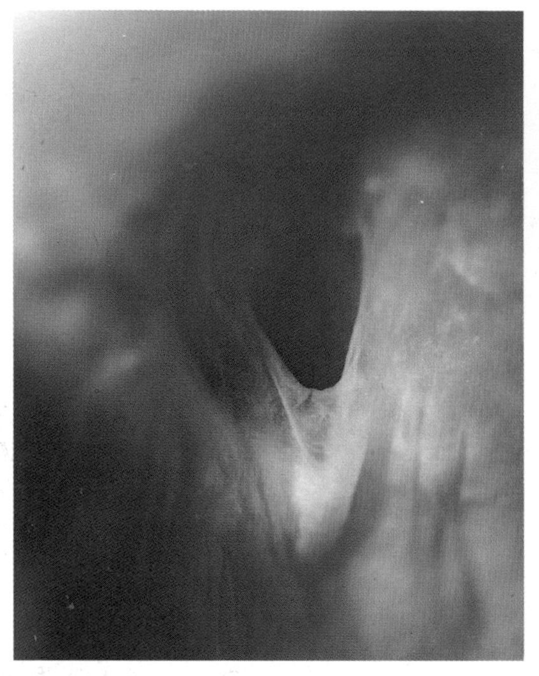

1 The opening of the ureter in the bladder

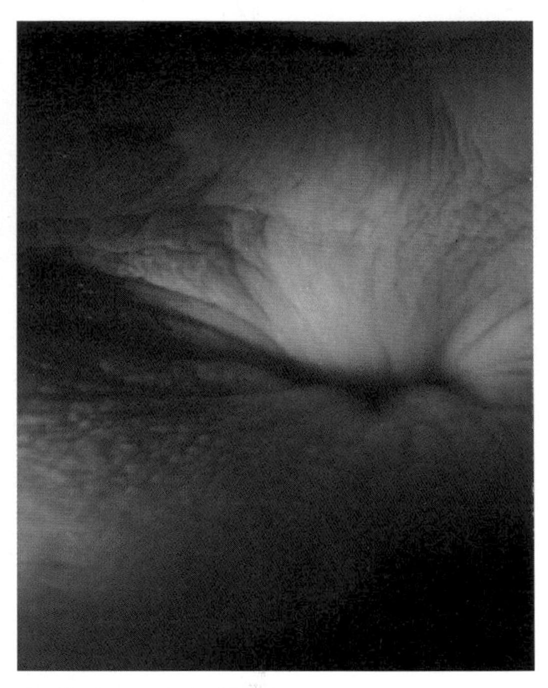

2 The base of the bladder with the opening of the urethra

3 Inside the bladder

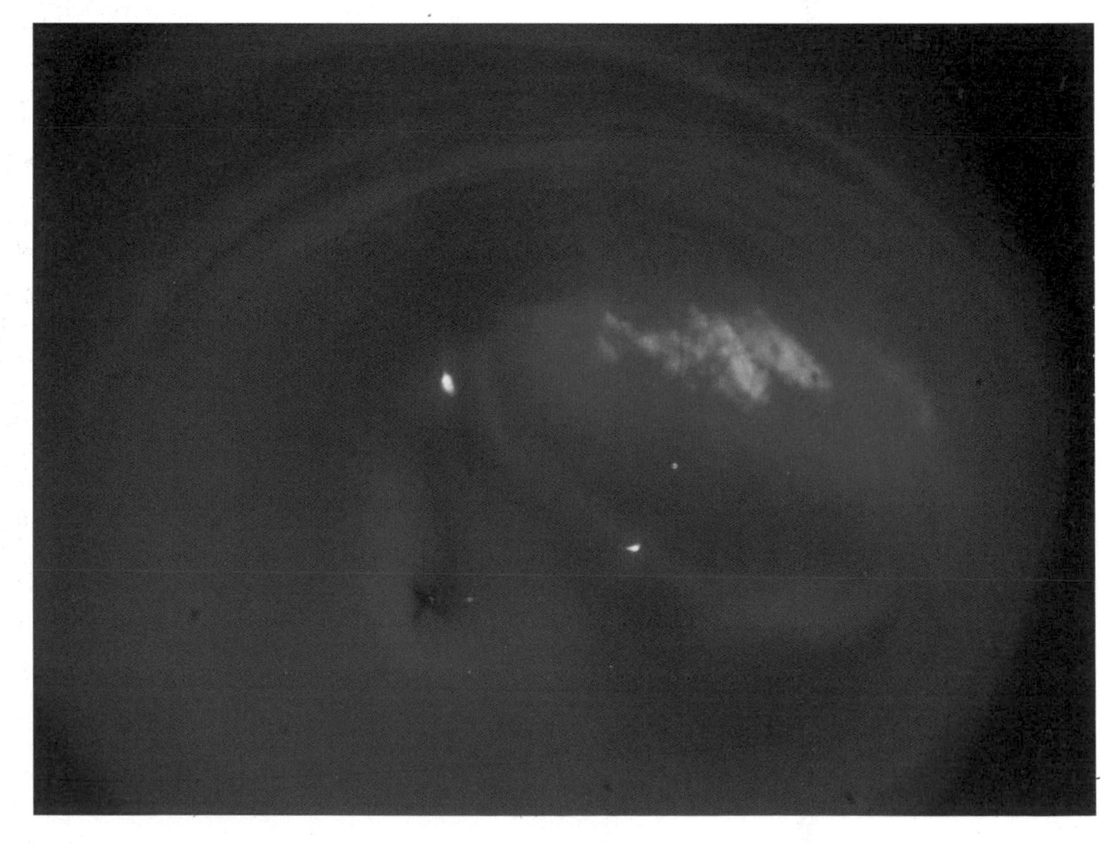

The Skin

The high degree of sensitivity of most organs in the body means that they require the best possible environment in order to function. Fluid balance must be properly regulated, temperature kept fairly constant, and the body protected against injury. A number of these functions are fulfilled by the skin. Often we do not realize the multiple services it renders.

Certain functions are obvious. The skin covers the surface of the body, an area which amounts to about two square yards in the adult (more in men than in women). The skin retains body fluids, it protects against mechanical injuries and against harmful radiation from the sun, it acts as a barrier to bacteria, and it insulates against heat and cold.

But the skin is also an active and vital organ in itself. Temperature regulation is mainly accomplished by the skin. When blood vessels there dilate, more blood passes through them. Since the skin is naturally cooler than other parts of the body, blood temperature is lowered. When an appropriate temperature has been reestablished, the capillaries in the skin constrict again. In cold, as little blood as possible should be exposed to cooling, so the blood vessels in the skin are constricted.

If there is a need for greater cooling, sweat glands are called into play. The sweat excreted spreads over the surface of the skin and evaporates. The process of evaporation requires heat, and so the surface of the skin becomes cooler. Sweat glands also aid in eliminating some of the waste products excreted by the kidneys.

The skin also provides the brain with continuous information about touch, pressure, cold, heat, and pain. A great many sense organs and nerve endings are present in the skin, and their activities provide us with the concrete sensations we have of our surroundings.

The appearance of the skin is of enormous importance in the impression we make on others. A whole industry thrives on making skin look attractive.

Skin is divided into three layers. From the outside in, these are the epidermis, the dermis, and subcutaneous tissue. The epidermis is a thin layer, normally no more than 0.2 mm deep. Its outer horny layer is composed of dead epithelial cells. These cells are continually being sloughed off and renewed from below. The epidermis becomes thicker with hard wear, typically on the fingers or the soles of the feet. Some of the deeper epithelial cells are pigmented. The pigment largely determines the color of the skin, but its primary function is to protect the body from ultraviolet radiation from the sun. As is true of other organs, the skin responds to variations in external conditions or to demands made on it. The tan people get when sunbathing serves as a protection from the sun. On the other hand, we need some exposure to sun, because it is through the action of ultraviolet radiation on skin that vitamin D is manufactured. Vitamin D is ultimately transported to the intestines, where it plays an essential role in calcium metabolism.

The dermis is a thin layer of connective tissue with elastic fibers arranged crosswise. They give the skin its elasticity. The arrangement of the fibers in segments creates natural cleavage lines in the skin. An incision along a cleavage line will heal faster than a crosscut, as all surgeons know.

The dermis has a rich blood supply. It is also the layer containing many fine, delicately branched nerve endings and muscles, mostly of the smooth type which is not under voluntary control. The shivering and "goose pimples" associated with cold are the result of the activities of these smooth muscles. These same muscles around hair cause it to stand erect. This is an excellent form of insulation in furry animals, since the coat becomes thicker and can hold more air, which provides further insulation. Other smooth muscles regulate the diameter of blood vessels.

The subcutaneous connective tissue is rich in fat cells and is thicker than either the epidermis or the dermis. Fatty tissue insulates against cold and acts as a shock absorber. In obese persons this layer can be very thick.

Sweat glands are present everywhere in skin. Sebaceous glands are found around the hair follicles. These glands excrete sebum, a fatty substance which when distributed on hair and epidermis makes them water-repellent, more elastic, and better able to endure cold.

Nails as well as hair are composed of the protein keratin, the same protein found in animal horns. A nail is built up of layers of dead epithelial cells closely packed together. The roots from which the nail grows are concealed in the skin. The growth rate of a nail is between 1.5 and 2 cm a year.

Sweat Glands

5 High magnification of an opening of a sweat gland. Bacteria (green spheres) are present here as everywhere else on the skin.

Section through the skin (1, 2)
The skin is composed of three layers — the epidermis, the dermis, and subcutaneous tissue. The photograph and the diagram are comparable and illustrate the structure. The epidermis consists of a thin layer of keratinized epithelial cells. Some of them are pigment-producing cells, melanocytes. The epidermis is kept soft and smooth by fatty secretions from sebaceous glands in the dermis. These open into hair follicles from which hairs grow.

Typical pattern of a fingertip (3)
Here, as in many other areas, the skin lies in folds which make it elastic. The pattern of folds on each fingertip is unique in each individual — it constitutes the fingerprint. As the surface skin wears off, the pattern is constantly renewed from the deeper epithelial cells. The small dots on the ridges between the folds are openings of sweat glands.

The specific character of the papillary pattern was discovered 150 years ago by Johannes Purkinje, a Bohemian anatomist. The fingerprint left on an object consists of fatty substances, salts, and dead epithelial cells from the surface of the finger, which stamp the object with the characteristic pattern.

A fingertip, magnified 150 times (4)
in the scanning electron microscope. The photo shows some dead epithelial cells and openings of the sweat pores.

Sweat emerging from a pore (5)
highly magnified in the scanning electron microscope. The green spheres on the surface of the skin are bacteria. Microorganisms are so abundant on the hands that no amount of washing can render them sterile. Folds in the skin, hair follicles, openings of glands, and minor cutaneous wounds are all ideal hiding places for bacteria. The tissue has been magnified approximately 3,500 times, and the photo colored to increase contrast.

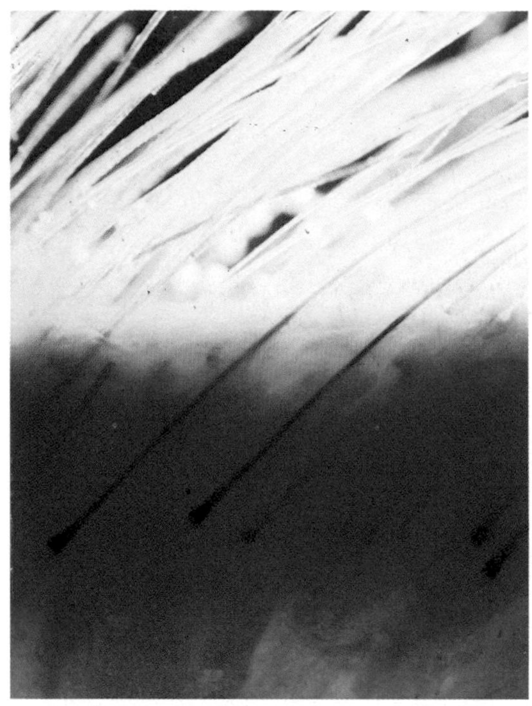

1 Section through hairy skin

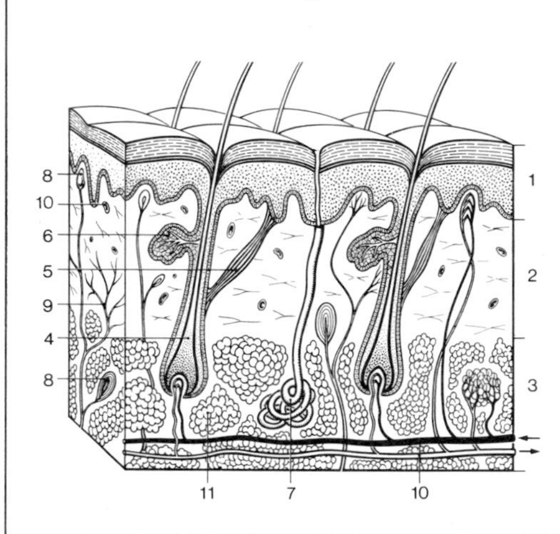

2

3 The fingertip has a typical pattern

4 The fingertip enlarged

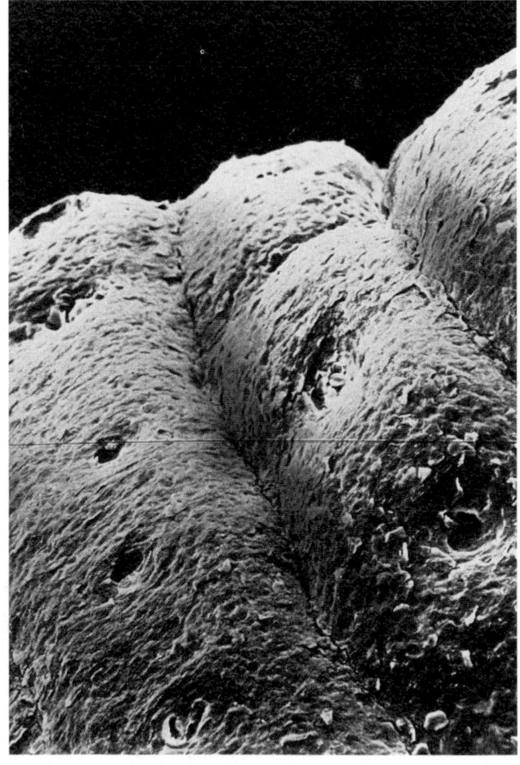

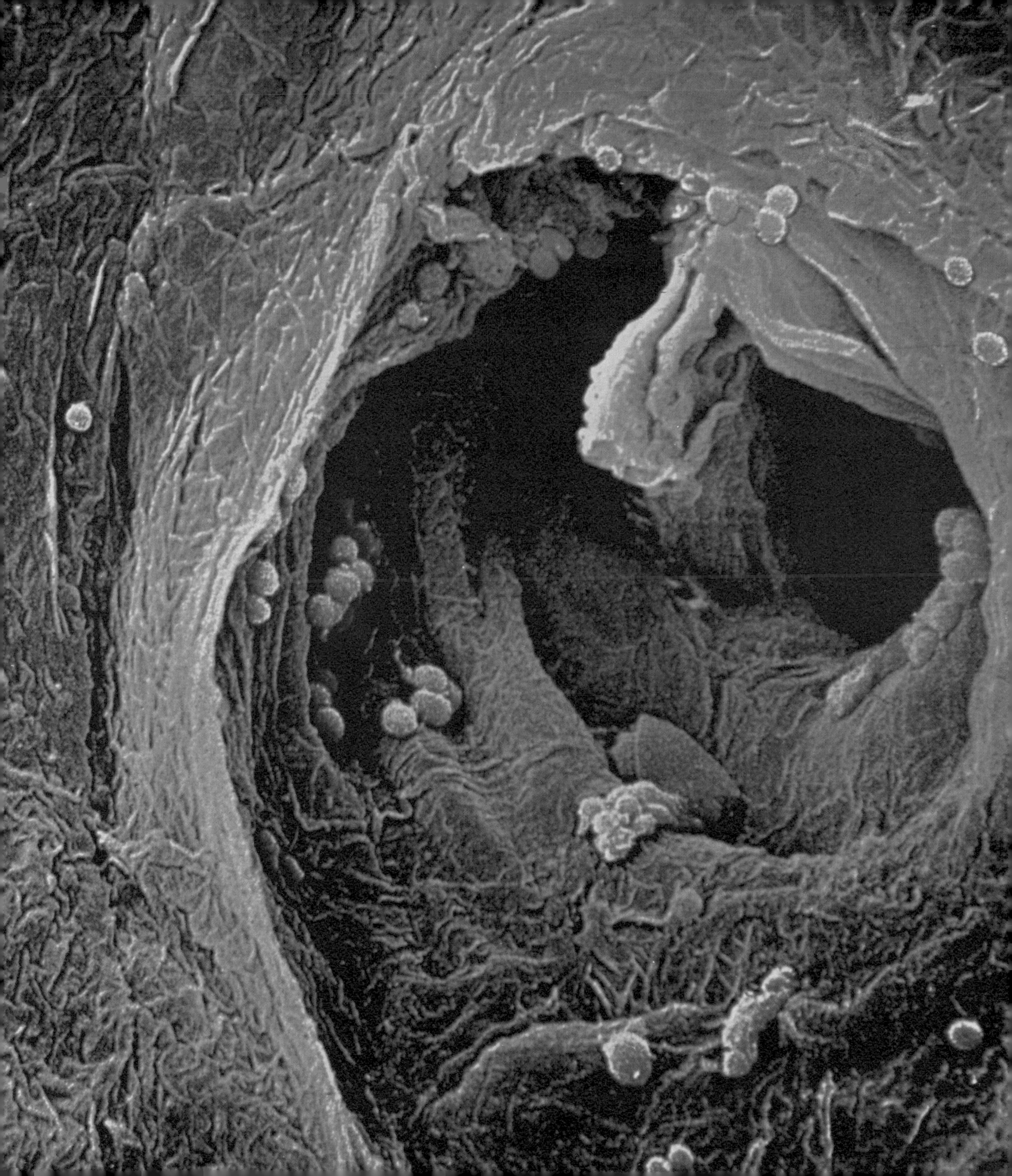

4 Epithelial cells of the hand in a seven-week-old fetus

Drops of sweat glisten on the palm of the hand (1)
where sweat glands are numerous — about 300 per square centimeter. The same is true of the soles of the feet. The number of sweat glands in an adult amounts to several million. Each day, they secrete about a pint of sweat, which consists mostly of water and small amounts of salts, lactic acid, ammonia, and other substances.

Minor injury to the skin (2)
Minor injuries occur easily. Those not affecting the dermis are hardly felt and heal without a scar. In deeper wounds, the often profuse bleeding has a cleansing effect. In small wounds, however, the blood soon clots, and a scab develops that protects the skin during healing.

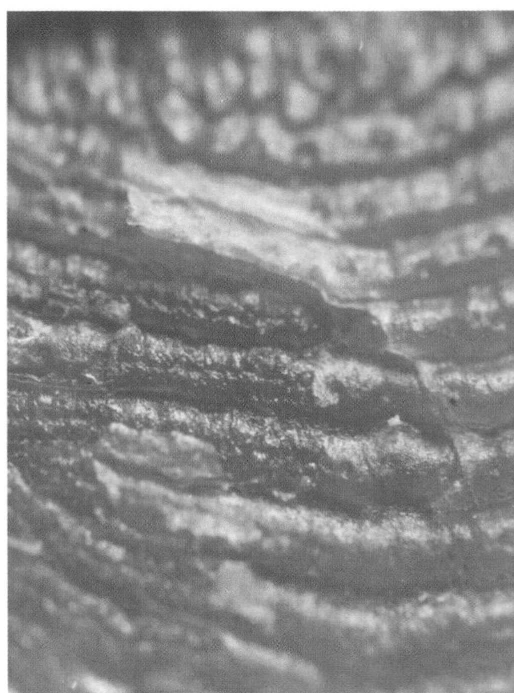

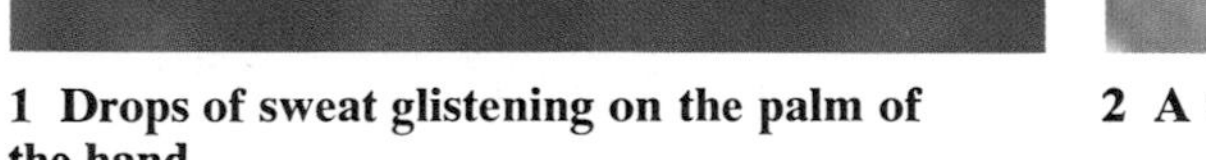

1 Drops of sweat glistening on the palm of the hand

2 A minor injury to the skin

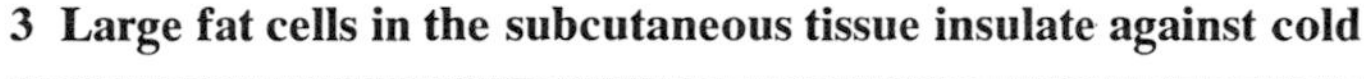

3 Large fat cells in the subcutaneous tissue insulate against cold

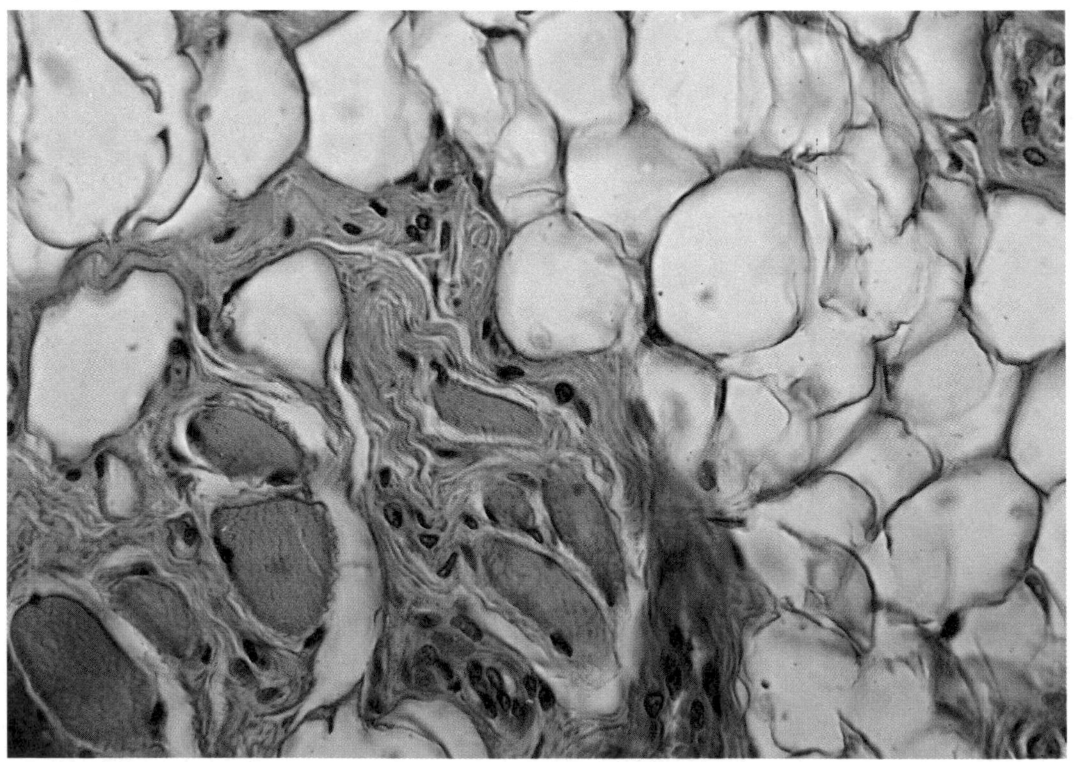

Large fat cells in subcutaneous tissue (3)
This stained section through skin shows fat cells lying among connective-tissue cells (the red streaks with the blue-black dots). Muscle fibers, which have been cut through, are also visible. In the upper right corner is part of an artery with muscle cells in its wall. Constrictions or dilations of blood vessels in the skin reduce or increase the transfer of heat, and consequently contribute to the temperature-regulating system of the body.

Epithelial cells of the hand are clearly seen (4)
in a seven-week-old fetus. The hand is not fully developed. The cells in the outer layer of the skin, the horny layer, are keratinized and dead. They are constantly being sloughed off and replaced from the lowest layer of the epidermis, just above the dermis. The fetus lies in fluid and its skin is protected by a fatty substance, the vernix caseosa, which also serves as a lubricant during delivery.

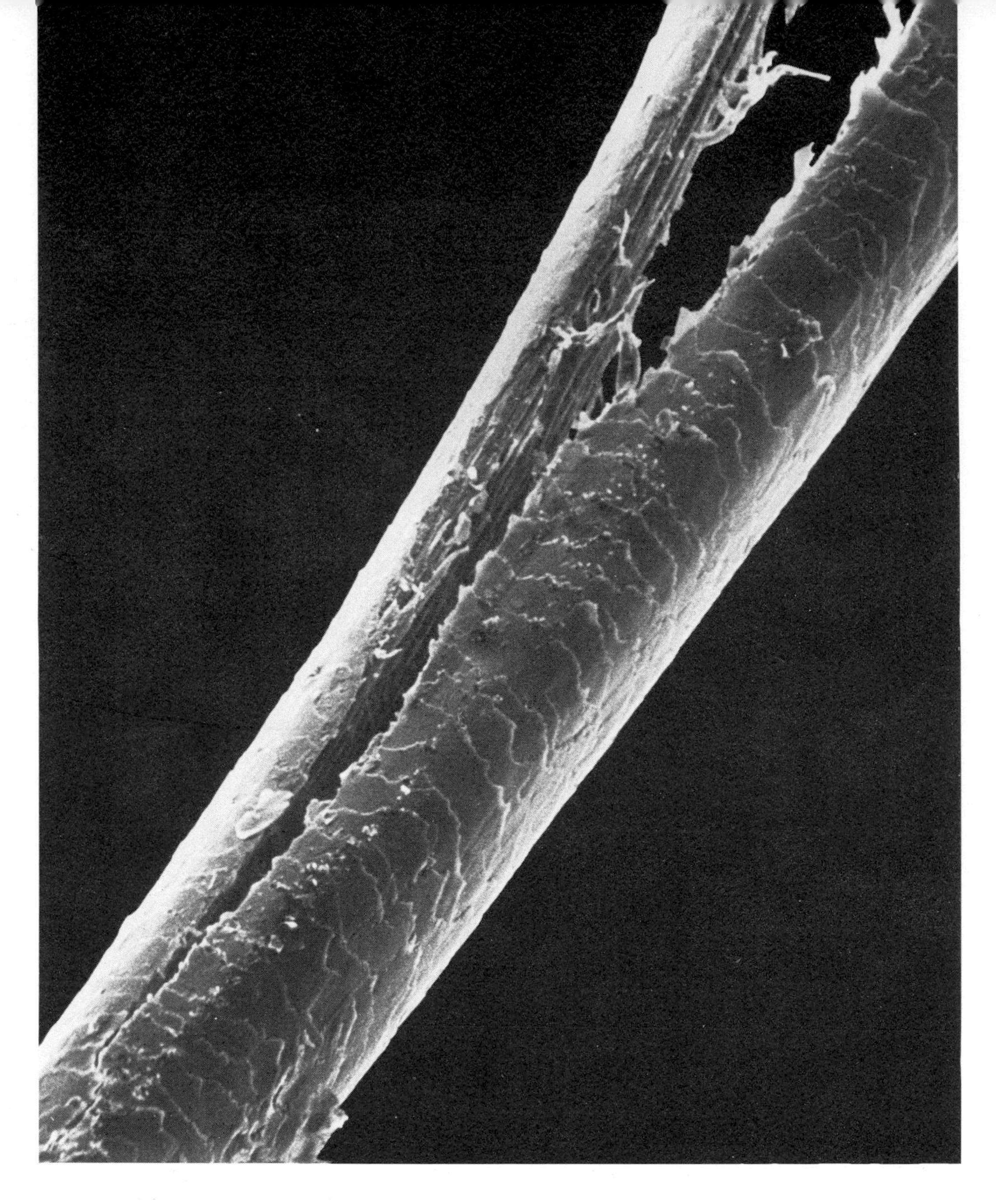

A split hair
When enlarged, "smooth" hair clearly has a scaly surface. Hair consists of fibers of horny tissue, which is also present in nails. Hairs grow from hair follicles in the dermal and subcutaneous layers. Hairs occasionally split at the ends for a variety of reasons.

Perspiration

Sweat consists of water and other substances such as salt, lactic acid, and ammonia. An adult excretes about a pint of sweat a day, but if one is very active or in warm surroundings, the sweat excreted can amount to as much as eight quarts. Sweat plays an important role in temperature regulation. It also helps to keep the skin soft and supple by providing a certain amount of water. The water in sweat comes from blood. If we perspire a great deal, we need more fluid and we get thirsty. One of the components of sweat is ordinary table salt, and if sweating is prolonged, it is essential for the salt lost to the body to be replaced.

Sweat glands lie like small balls of wool in the subcutaneous tissue. They open on the surface of the skin through small pores. The glands are controlled by the autonomic, or involuntary, nervous system. Sweat glands in the armpits and in some other parts of the body are large, and the sweat excreted contains some fat and proteins. Sweat is odorless and sterile when it first appears on the skin, but in air it starts to decompose, usually through the action of bacteria. This gives rise to its characteristic odor.

The Color of Skin and Hair

Skin color is determined by the amount of the pigment melanin stored in epithelial cells. Melanin is a copper-containing protein, and different proportions of copper in its structure result in different skin colors. Dark skin in some parts of the world is believed to be an adaptation to the intensity of sunlight. In strong sunshine, heavily pigmented skin protects against an overproduction of vitamin D, which would be harmful. In areas with less sunshine, people have lighter skins. Red-haired people have a special, high-iron-containing pigment in their hair and skin instead of melanin. This pigment does not offer the same degree of protection against ultraviolet radiation as melanin does. Therefore, redheads are usually more sensitive to sunlight than other people.

Man is a "naked ape," as the zoologist Desmond Morris has said. In contrast to the skin of our closest animal relatives, ours lacks the dense hair covering called fur. It is of interest in this connection that between the sixth and eighth months of gestation, the human fetus is covered by hair, called lanugo. Except in rare cases, this hair is shed before birth.

It has probably been an adaptive advantage for us not to have fur. Hairlessness makes rapid regulation of body temperature possible. As far as it is known, the first human beings lived in a warm climate. With no furry coat, the sweat glands can act quickly to cool a body hot from intensive and sudden movements — such as those associated with hunting or flight.

Even though we lack fur, the body is still covered by hair. The individual hairs are very fine and short except on the scalp, in the armpits, and around the sex organs. Head hair is important in protecting us against sun, cold, and, to some extent, physical injuries.

The razor blade cuts off a hair
The photo on the right shows the central core of the cut end and the surface scales.

A hair turns gray when air fills the canal in the hair and pigment no longer is formed.

Hairs on the head, highly magnified, seem ▷ like tree trunks amid mossy stones

A hair seen obliquely from above
The horny cells lie like tiles in rows above each other. They are pushed through the hair follicle by a papilla (a dome-shaped structure) on which the bulb of the hair rests.

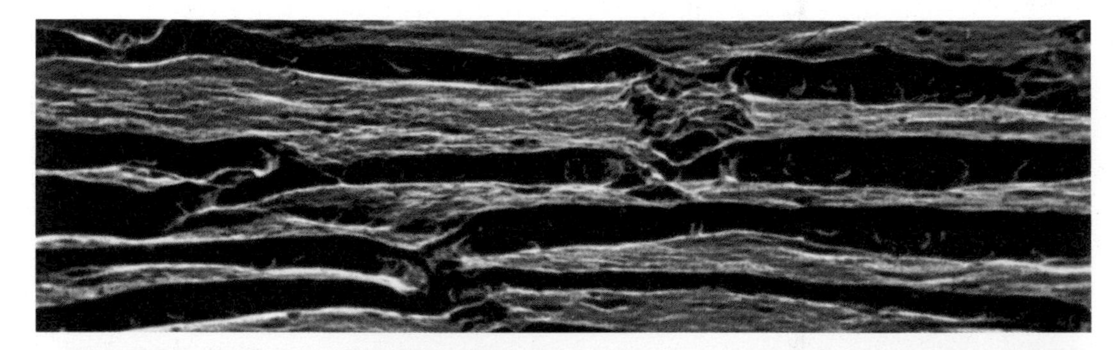

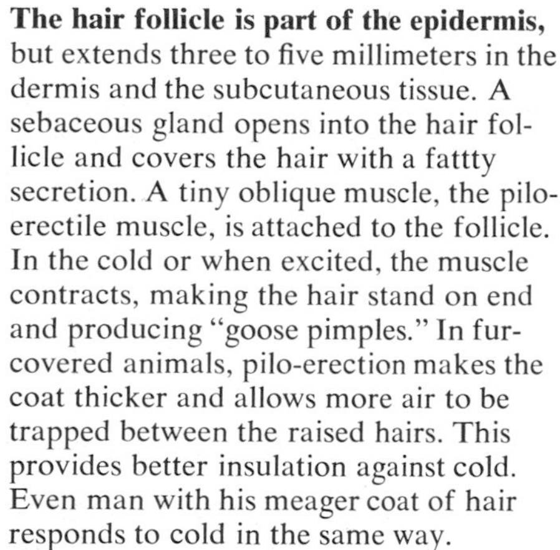

The hair follicle is part of the epidermis, but extends three to five millimeters in the dermis and the subcutaneous tissue. A sebaceous gland opens into the hair follicle and covers the hair with a fattty secretion. A tiny oblique muscle, the pilo-erectile muscle, is attached to the follicle. In the cold or when excited, the muscle contracts, making the hair stand on end and producing "goose pimples." In fur-covered animals, pilo-erection makes the coat thicker and allows more air to be trapped between the raised hairs. This provides better insulation against cold. Even man with his meager coat of hair responds to cold in the same way.

It is commonly believed that the shape of the hair follicle determines whether hair will be straight, curly, or frizzy. Cylindrical, straight follicles give rise to straight or slightly wavy hair. Curved, flattened or sword-shaped follicles are associated with frizzy hair. The cross-sectional shape of the individual hair also influences the type of hair. If oval, the hair is often wavy.

There are approximately 100,000 hair follicles on the scalp, and during a normal life span each produces about 26 or 27 feet of hair. Hormonal and hereditary factors both influence the production of the hair, slowing growth or leading to hair loss and eventual baldness.

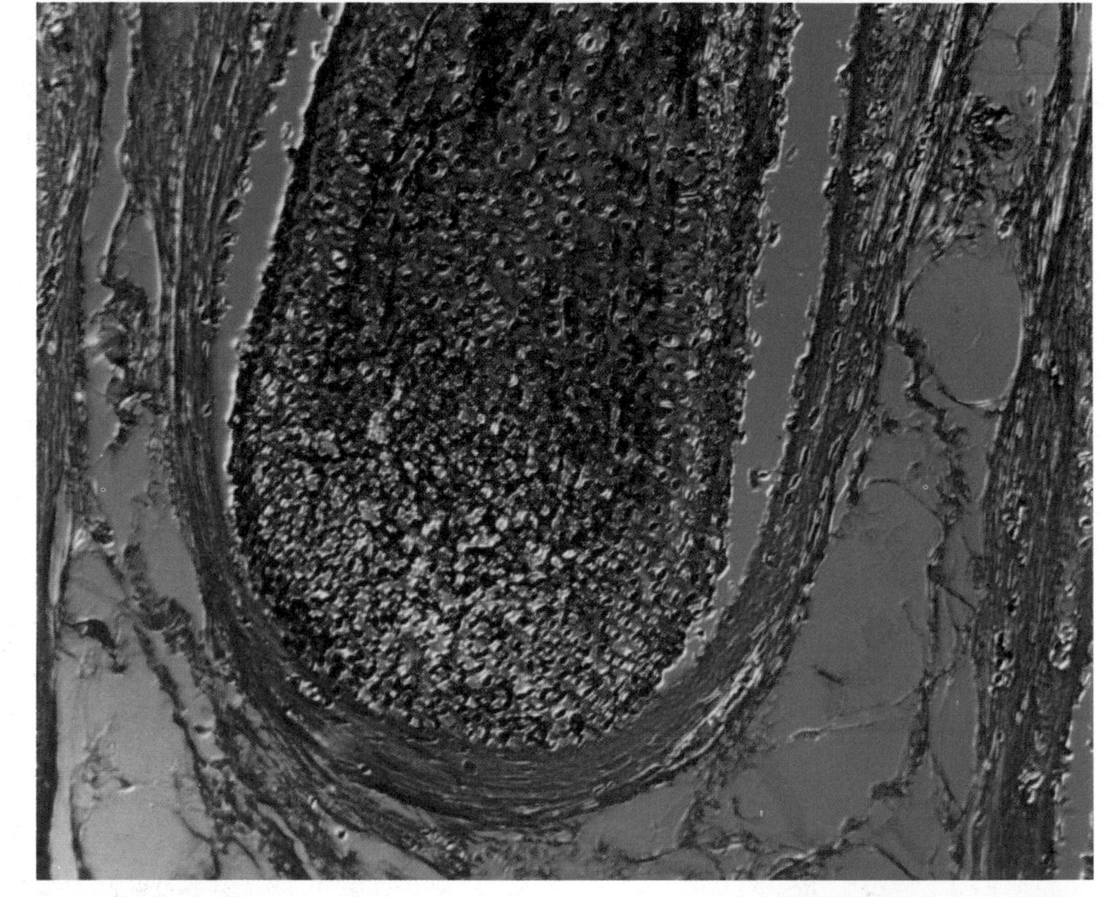

Cross section of hair follicles
The follicle consists of an outer and an inner hair sheath. The hairs form the dark rings in the photo. The core of each hair is lined by keratinized or horny cells. Hair gets its color from pigment cells in the hair sheath. In a white- or gray-haired person, the pigment cells have disappeared. Usually graying develops gradually and is determined by heredity.

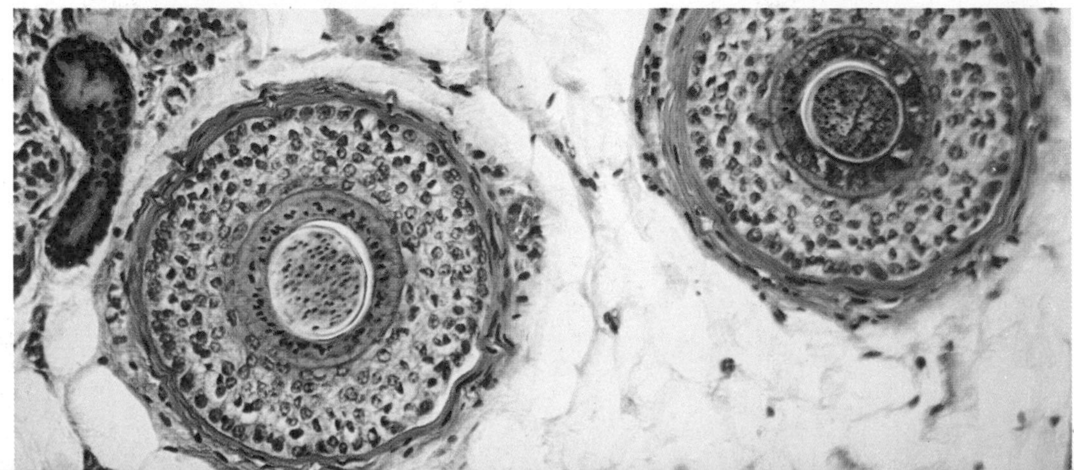

Hair

Fur-covered fetus (1)
During a short period of life we are covered with fur. Between the third and fourth month of gestation, short, unpigmented, downy hairs, called lanugo, start growing all over the body of the fetus. They are fully developed in the sixth month, but fall out before birth. The fine hair over most of the adult body (we are only completely hairless on the palms of the hands, the soles of the feet, the lips, and the external sex organs) is of a different type from the hair on the scalp.

Scanty beard in an old man (2)
The hairs of an old man's beard are far apart. With increasing years, the number of hair follicles decreases. On the other hand, hair may suddenly begin to appear in new places, such as on the external ear, in the auditory canal, in the nostrils of men, and on parts of the face in women.

Coarse hair and down hair (3)
At the left is the rough surface of a coarse hair. One of the three down hairs shown has been cut slantwise at the top. The scalelike arrangement of the horny cells appears clearly. The sebaceous gland in the hair follicle secretes an oily substance that keeps the horny layer of the epidermis soft and smooth, prevents drying, and increases the heat-insulating capacity of the skin.

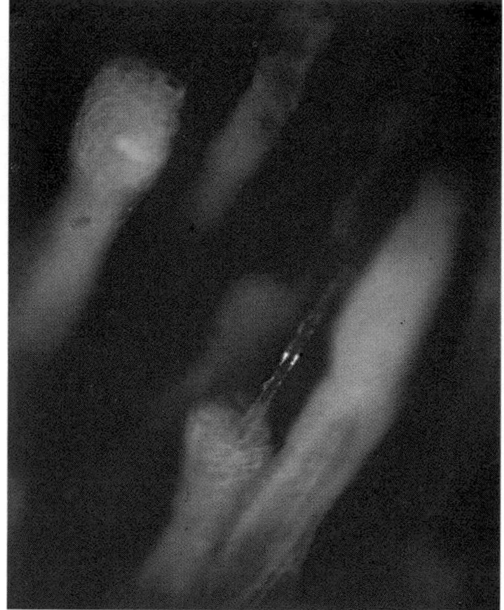

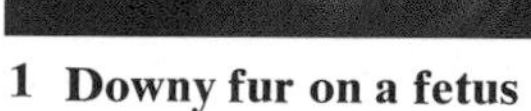

1 Downy fur on a fetus

2 Scanty beard of an old man

3 Coarse hair and three small downy hairs, greatly magnified

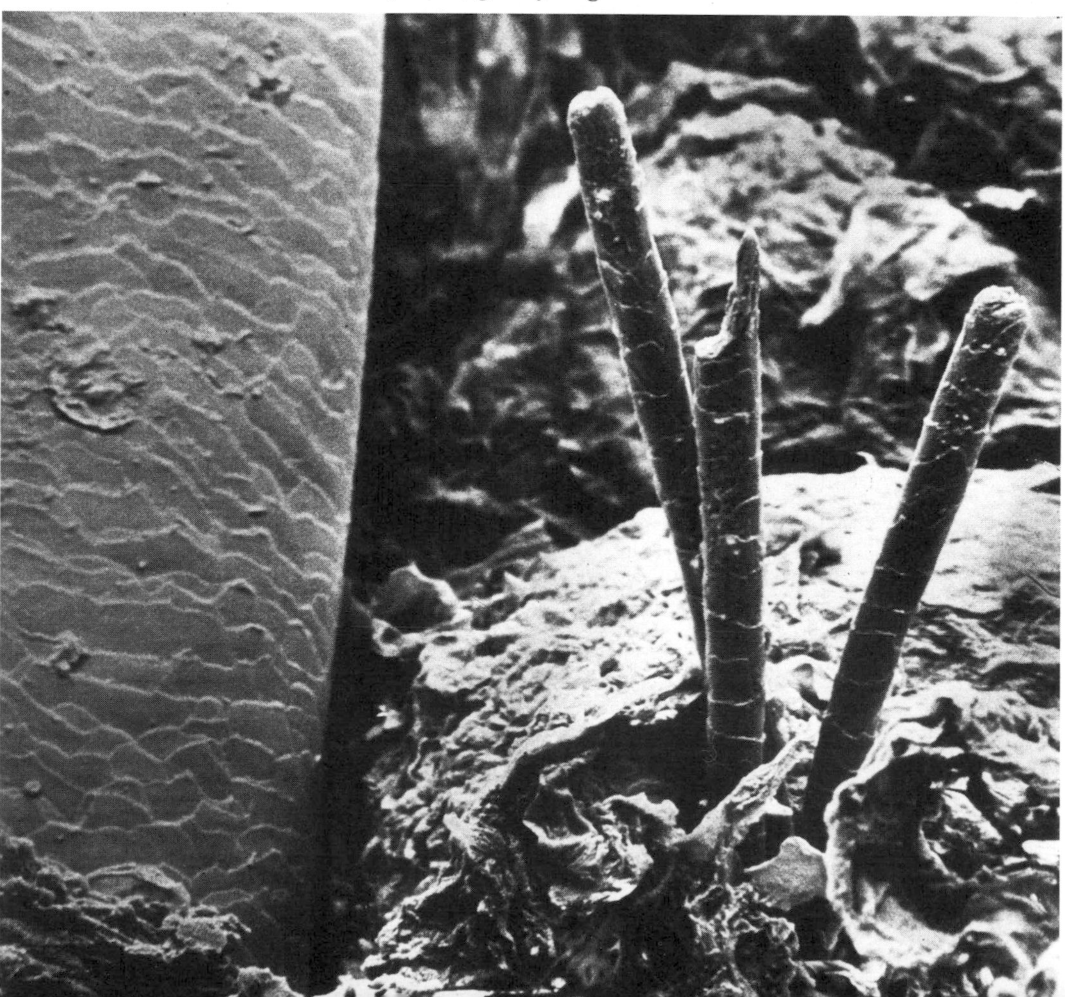

Muscles

Some important muscles

1 frontal muscle
2 orbicular muscle of the eye
3 orbicular muscle of the mouth
4 oblique cervical muscle
5 deltoid muscle
6 biceps muscle of the arm
7 radial muscle
8 greater pectoral muscle
9 straight muscle of the abdomen
10 external oblique muscle of the abdomen
11 quadriceps muscle of the thigh
12 gastrocnemius muscle

Every movement we make originates in the contractions of muscles. Some muscles — the skeletal or striated muscles — are under voluntary control. They are all directly or indirectly attached to the skeleton. Other muscles make up the smooth musculature, which has a different structure. Smooth muscles are not usually subject to voluntary control, and they perform actions which are often unnoticed or only partly perceived. Smooth muscles dilate or constrict blood vessels to maintain blood pressure; they move food through the intestines; they dilate and constrict the pupil of the eye. Heart muscle is composed of a third type of muscle tissue, cardiac musculature, which behaves like smooth muscle. It is governed by a nerve network located in the heart itself.

The working units in all muscle tissues are cells with the ability to contract. Fibers inside them, the myofilaments, which have a diameter of some millionths of a millimeter, can slide over each other to become shorter. The proteins myosin and actin are components of the myofilaments, and together they form a complex called actomyosin, which is organized into larger units.

The arrangement of myofilaments differs in smooth and striated musculature. In the small smooth-muscle cells, the filaments lie in the cell protoplasm and are difficult to see even in the microscope. In skeletal muscles, the myofilaments are arranged in long bundles of myofibrils, which are some thousandths of a millimeter thick. These muscles are called striated because they look so in the microscope: the myofibrils have banded portions which, when the fibrils are grouped in parallel bundles to make up muscle fibers, give the fibers a cross-striped appearance. The fibers themselves are surrounded by connective tissue.

The heart is our most important muscle. A complex system of muscles surrounds the cavities of the heart. In order to achieve the most efficient operation of the different parts of the cardiac musculature, the special cardiac nerve system controls the contraction wave. The heart contracts maximally at each beat. When it pumps less blood, it beats more slowly, but the extent of the contraction itself does not change.

Skeletal muscles, under voluntary control, behave quite differently. Contractions can be interrupted at any time, performed more or less rapidly, or with greater or lesser strength, as required. Striated muscle has a high rate of contraction compared to the smooth-muscle cells lining the intestinal walls or blood vessels.

Skeletal muscles are stimulated by nerve cells called motor cells in the spinal cord or brain, under the influence of other parts of the nervous system. Each nerve fiber to a muscle may innervate a bundle of one hundred to two hundred muscle fibers. A nerve impulse from the spinal cord ultimately reaches fine nerve branches in the muscles. The photo on pages 118 and 119 presents a remarkably good illustration of the innervation of a muscle bundle. Each nerve branch ends in a motor end-plate, which is the point of contact between the nerve and the muscle cell. When the nerve impulse arrives at the end-plate, a chemical agent is secreted which initiates the rapid contraction of the muscle fiber.

Frequent changes in body posture are an indication that many muscles are active more or less continuously. This is especially true of the muscles around the spine. Such a constant state of slight contraction of skeletal muscle is what is meant by muscle "tone."

Blood is the source of the energy muscles need in contracting. In the process, chemical energy is converted to mechanical energy, and heat is also generated. It is estimated that only about one-third of the energy used is transformed into movement. The heat produced by muscular contraction is not wasted, however. Conducted through the bloodstream, this heat provides still another means of temperature regulation.

If the efficiency of muscles is compared to a variety of man-made machines, it turns out that the effective output of muscle is of the same order of magnitude as that obtained in an atomic

A bundle of nerve fibers winding over muscle fibers to terminate in motor end-plates.

reactor. Here too, much energy is lost as heat. Man has constructed many machines with far greater efficiency.

The energy cycle in skeletal muscles is very complicated. In broad outline, it involves a stepwise oxidation of carbohydrates, some of which are stored in the muscles as glycogen. Oxygen is supplied by blood, and the process is accelerated by enzymes. Energy-rich compounds, such as adenosine triphosphate (ATP) and phosphocreatine, are involved in these chemical processes, ultimately enabling the proteins actin and myosin in the muscle cells to contract.

When the stores of ATP and phosphocreatine are depleted, muscle cells resort to glycolysis for energy, a way of breaking down sugar which does not require oxygen. This process leads to the formation of lactic acid in muscles. Large quantities of lactic acid reduce the efficiency of muscle cells. However, lactic acid also acidifies the blood, with the result that the heart starts beating faster and respiration becomes deeper. This means that more oxygenated blood is supplied to muscle cells. Glycogen is also mobilized from the liver and inactive muscles, and circulates in the blood to be used where needed. Such complex mechanisms make it possible for muscles to adapt to temporarily high demands.

Like other tissues in the body, muscles are also able to adapt to long-term demands. Continued exercise leads to an increase in muscle mass. Glycogen content also rises, and we become stronger. When muscles are little used, they diminish in size. This is especially apparent in the case of a broken leg kept in a plaster cast. Often the leg becomes thin in only a short while, and it may take a long time to restore it to its former size and strength.

The heart, the most important muscle in the body
The thick walls of the striated cardiac musculature enclose the four cavities of the heart.

With each beat, the heart pumps blood to the body, the lungs, and to its own musculature through the coronary arteries. Fluorescent dye makes the arteries show up clearly in the photo. Blood streaming through the fine branches of the coronary arteries supplies oxygen to the cells of the cardiac muscle and returns through veins to the right atrium. The heart shown is that of a six-month-old fetus. By the fourth week, when the embryo is only three millimeters long and the mother may not be aware she is pregnant, the primitive heart is already beating and pumping blood.

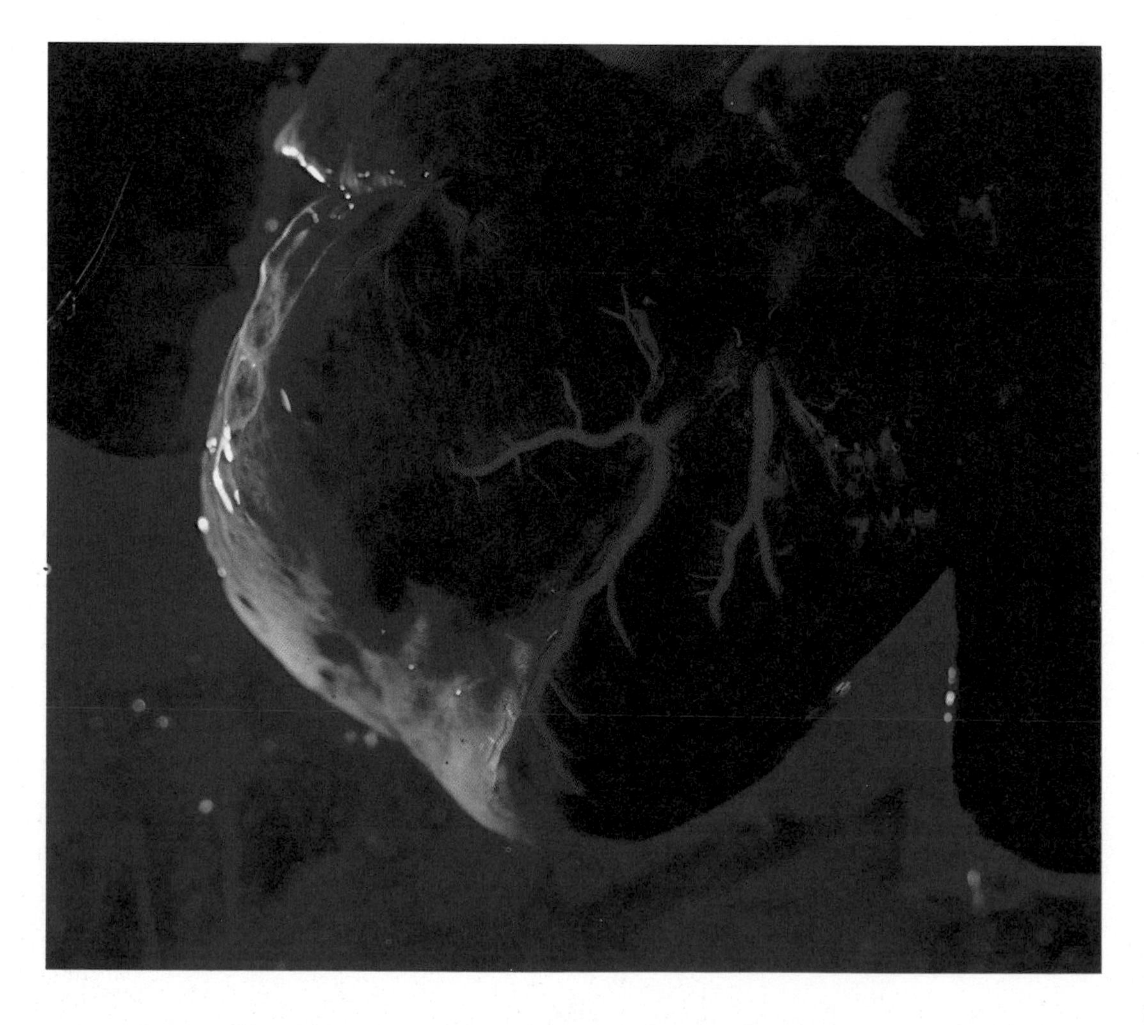

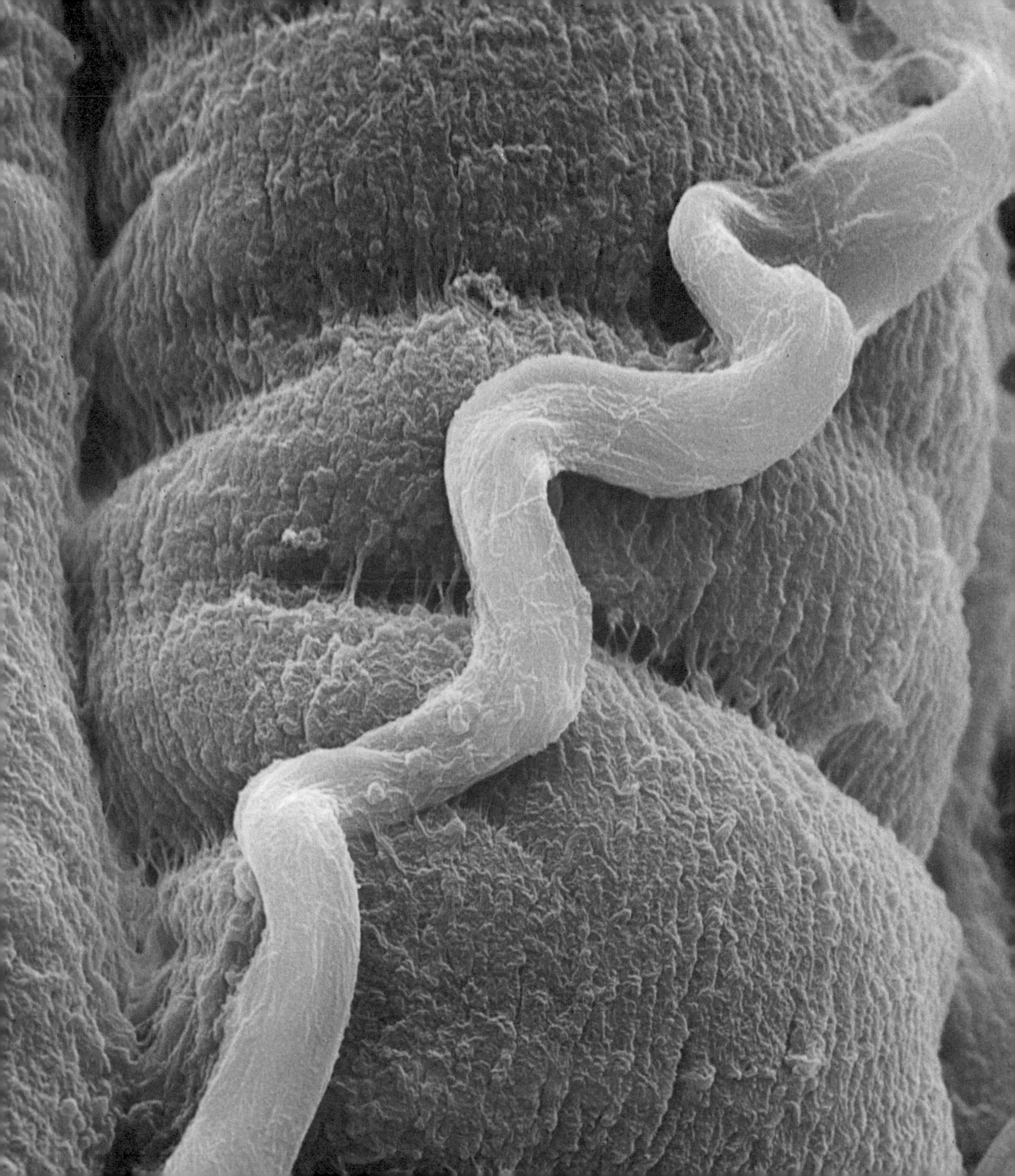

The motor end-plates
are the points of contact between nerve and muscle cells. The arrival of a nerve impulse at the end-plate initiates the release of acetylcholine, a substance which triggers the process of contraction in muscle cells. Acetylcholine is later broken down by the enzyme cholinesterase. The degree of contraction in the individual muscle and the balance between the activities of different muscle groups are regulated by the central nervous system. Not only does the nervous system signal muscular contractions through its efferent, or motor, nerves, but it also receives information about the degree of stretch in the individual muscle through special sense receptors in muscles called muscle spindles. Such reports provide information about the state of muscles in different parts of the body at all times.

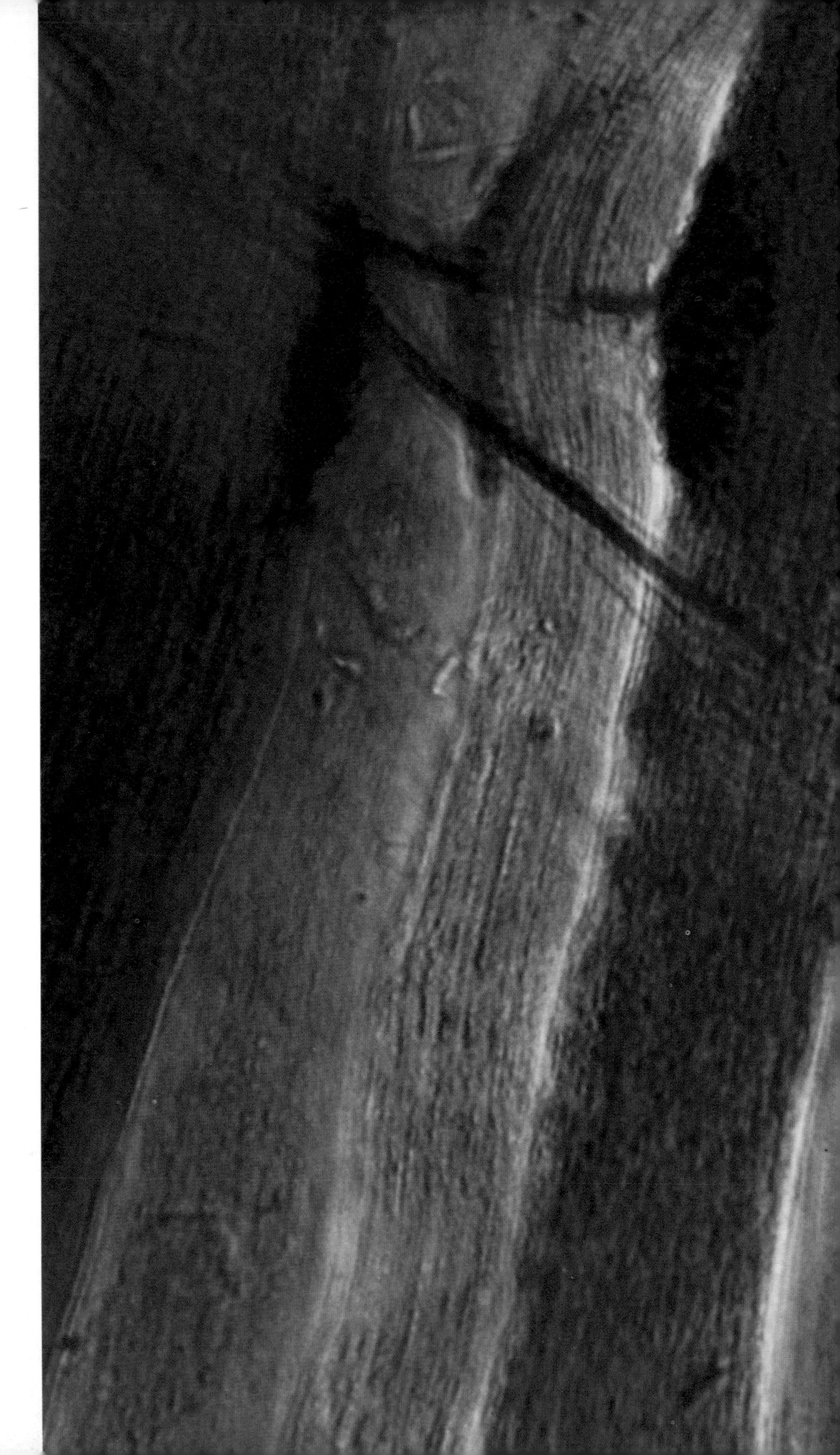

The Living Skeleton

The skeleton, especially the skull, has figured importantly in human thought and imagination. Death has been symbolized as a harvester, a skeleton carrying a scythe, who reaps all life. The skeleton remains, even after human flesh has decomposed, because the calcium and phosphorous compounds it contains do not easily decay.

But the skeleton is not dead tissue. It is alive, and is renewed at a rate of speed we seldom realize. We actually "change" skeletons many times in life: the period of complete renewal is less than two years. The skeleton consists of many types of bone as well as several types of cartilage and connective tissue. Like other tissue, the skeleton is rich in nerves and blood vessels. Skeletal bones also function as mineral depots. If the body is short of calcium, for example, bones can supply a certain amount. The composition of bones is approximately 21 percent water, 27 percent organic substances, and 52 percent inorganic salts, chiefly calcium phosphate and calcium carbonate. Skeletal bones are covered by the periosteum, a layer of dense connective tissue, well vascularized. The healing of fractures and other injuries to bone starts at the periosteum.

The beginnings of the spine can be seen when the embryo is only sixteen days old. About the fiftieth day, small cartilaginous vertebrae are present, and two weeks later ossification begins. The skeleton is still relatively soft and flexible at birth. This is necessary in order that the baby can pass through the birth canal. Bones that later join in the pelvis and in the skull are also separate.

The newborn has 350 separate bones. In an adult that number is reduced to 206, since many have fused.

The bones of the spinal column and the pelvis are strong and rigid. They carry most of the body weight, and permit upright standing and walking. However, the skeleton not only carries the body and holds interior organs in place, it serves a protective function as well. Important organs are well sheltered in the chest, the bones of the skull armor the brain, and the vertebrae form a bony canal surrounding the spinal cord.

The human skeleton is adapted to the multiple movements and actions that characterize human behavior, and individual bones show wide variations in size and shape. The mobility of

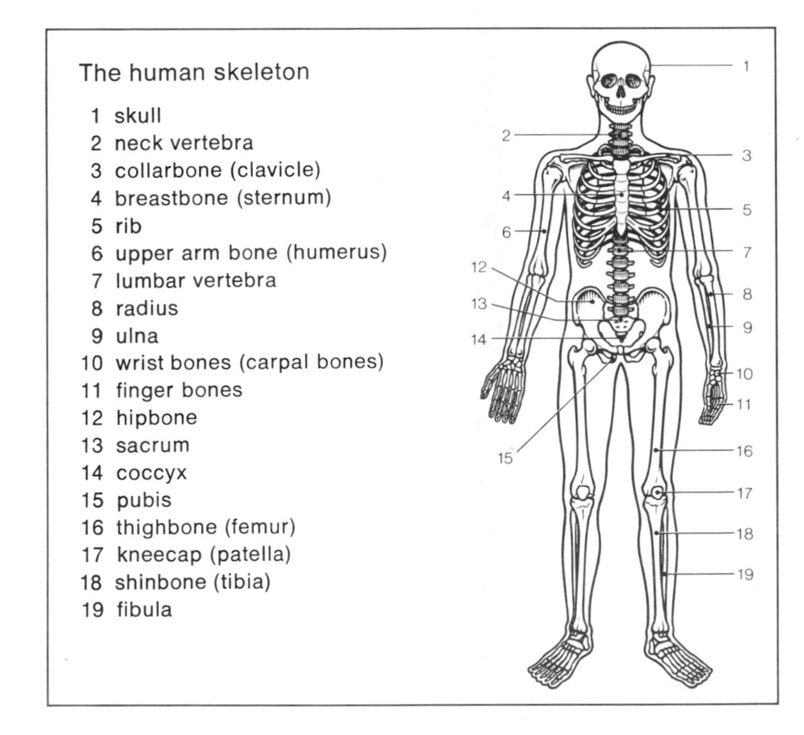

the arms is largely determined by the collarbones, the clavicles, which are placed so that the shoulders are forced out, away from the trunk. The existence of two separate bones in the forearm increases the mobility of the hand, since they make it possible to turn it more than halfway around. The significance of the position of the human thumb, which allows it to be placed opposite the other fingers, cannot be overestimated, since it is the key factor in making the human hand a skilled instrument. The hand of the ape does not have an opposable thumb. The coordination of the bones of the hand — twenty-seven in all — enables us to use our hands for precise movements or strong grasps allowing a great range of activities.

The skull is composed of twenty-nine bones; eight are cranial bones and fourteen, facial bones. The middle ear contains three very small bones which conduct sound from the eardrum to the inner ear. The inner ear is located in a cavity in the petrous part of the temporal bone. At the root of the tongue lies the hyoid

The Skeleton

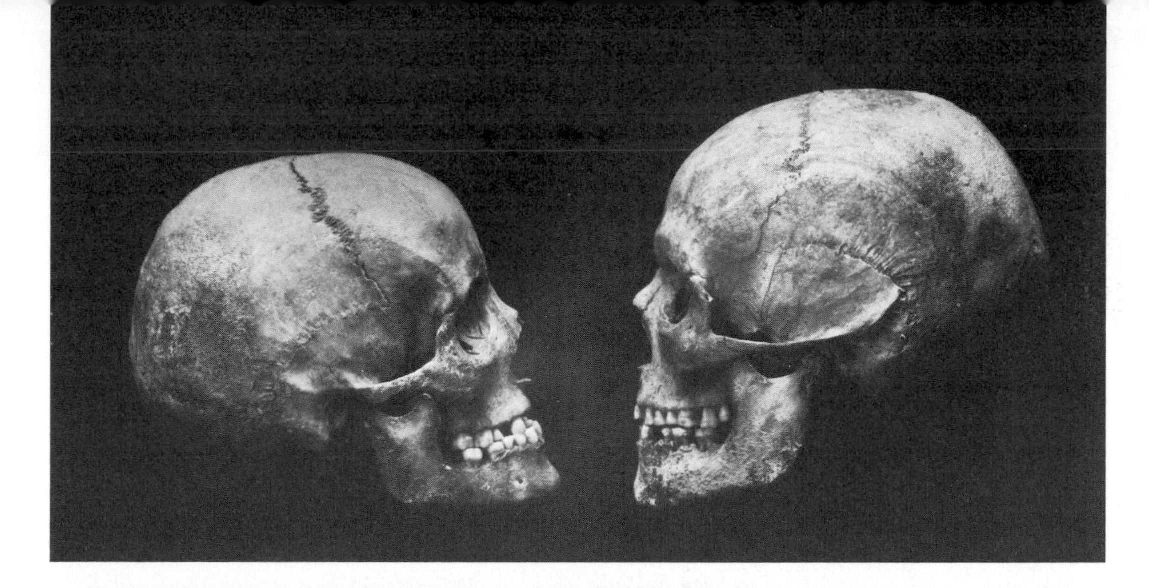

Skulls of an adult woman and man
The woman's skull is on the left. The two skulls were found in a medieval graveyard on Frösön near Östersund, Sweden, buried in earth for almost a thousand years.

Male and female skulls differ in various ways. The male skull is often larger, has heavier brow ridges, a recessed root of the nose, and stronger muscular attachments.

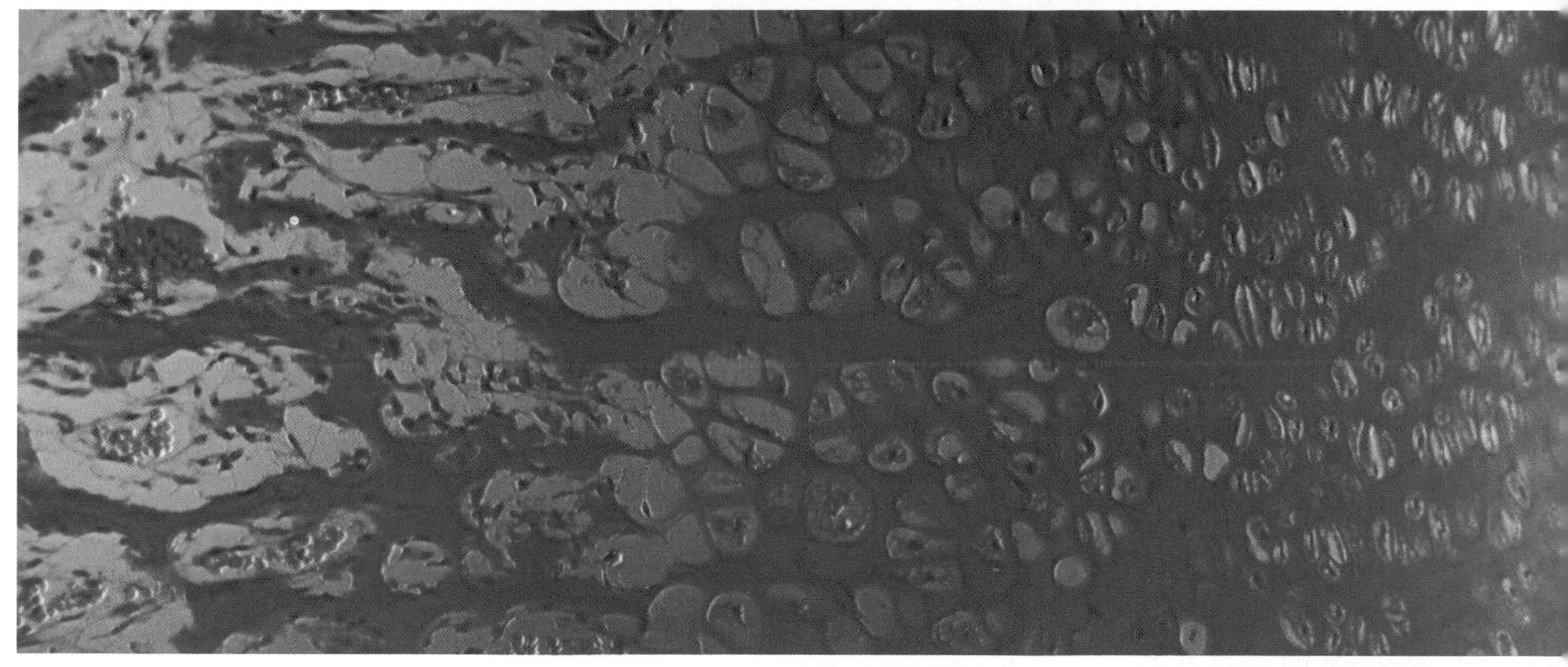

Growing bone
showing a distinct transition from round cartilage cells on the right to oblong yellow bone cells on the left. Bone formation takes place when minerals are deposited in cartilage cells — calcium salts "armor" the cartilage tissue.

Cross section of a tendon
viewed in an interference microscope. The oval structures are nuclei.

Tendons attach muscles to the skeleton. They consist of closely packed fibers of connective tissue. A tendon is most often formed as an extension of the sheath of connective tissue surrounding muscle cells. Where the tendon reaches the bone, the fibers fuse with the periosteum.

3 Bone tissue in a spinal vertebra forms an elegant latticework attached to cartilage

bone, which does not articulate with the rest of the skeleton. The eyes lie protected by the bony sockets of the skull.

There are fifty-one bones in the body trunk, of which half form the spinal column (the backbone) and the pelvis, the others the chest. The trunk is strong and rigid, but also flexible. At rest, the spinal column is a graceful S-shape. The vertebrae are held together by strong ligaments composed of connective tissue. Between successive vertebrae are shock-absorbing disks of cartilage, and the ligaments and muscles attached to the vertebral arches and processes enable a variety of trunk movements. At the back of the pelvis, between the two shovel-shaped hipbones, the five lowest vertebrae are fused into the sacrum. Below the sacrum is the coccyx, which is also composed of fused vertebrae.

The uppermost vertebra is called the atlas; it supports the head, just as, in Greek mythology, Atlas supported the heavens. The atlas is ring-shaped and revolves on a pivot on the second vertebra. The neck vertebrae are held in place by strong ligaments. If a violent jerk displaces the pivot on the second neck vertebra so that it slips into the spinal cord, death may be

Cross section of compact (solid) bone (1) The section of solid or dense bone shown here is magnified 100 times. It consists of concentric layers (lamellae) of bone tissue with deposits of calcium salts in between. The lamellae have a central canal for blood vessels and nerves. The dark spots are nuclei of bone cells. These cells, osteocytes, have long, thin extensions which connect with blood vessels to supply nourishment to the tissue. Other types of bone cells, osteoblasts and osteoclasts, participate in the formation and resorption of bone, respectively.

The long bones of the skeleton have a shaft of solid bone that encloses a marrow cavity. In the enlarged ends, the solid layer is thinner and surrounds porous, spongy bone tissue. This tissue is composed of many spikes, trabeculae, which give great strength to the bone.

1

Central canal in solid bone tissue (2), magnified approximately 200 times. This diagonal section shows the individual lamellae as well as the nuclei of the bone cells (black spots).

2

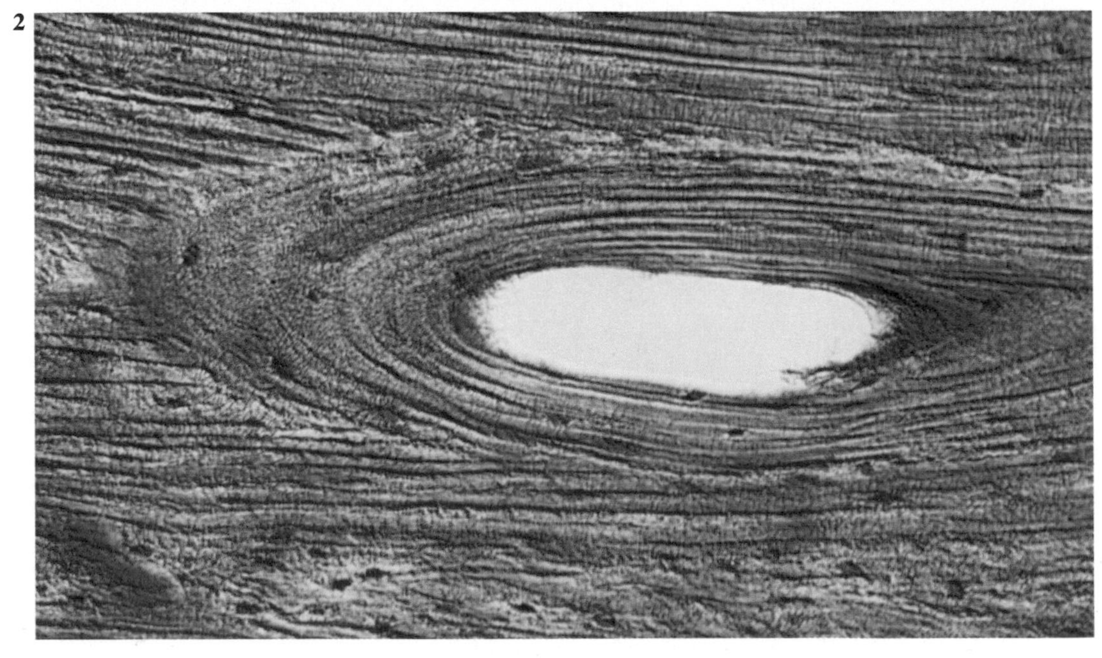

Spongy bone tissue in a vertebra (3) The white material on the left is cartilage from an intervertebral disk. The trabeculae (spikelike forms) describe a latticelike pattern which absorbs external pressure and strain. The vertical spikes are especially strong. (They appear horizontal, since the photo is reproduced with the intervertebral disk at the side rather than at the bottom.) The same trabecular construction principle is applied in building an airplane to provide strength and lightness. Normally, spongy bone contains bone marrow, the cell-rich tissue where red blood cells are manufactured. In this preparation the bone marrow was removed. Magnification is approximately 50 times.

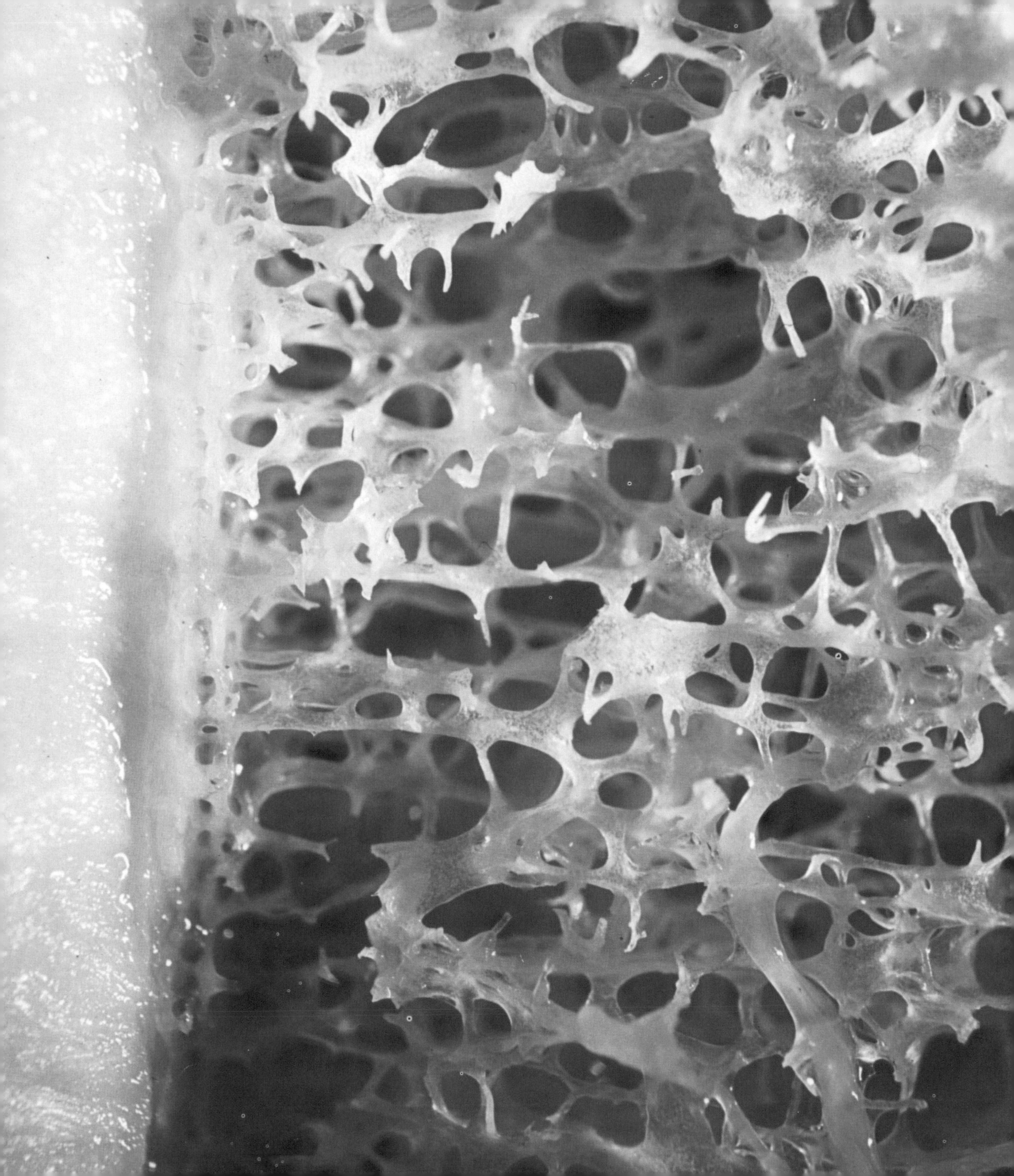

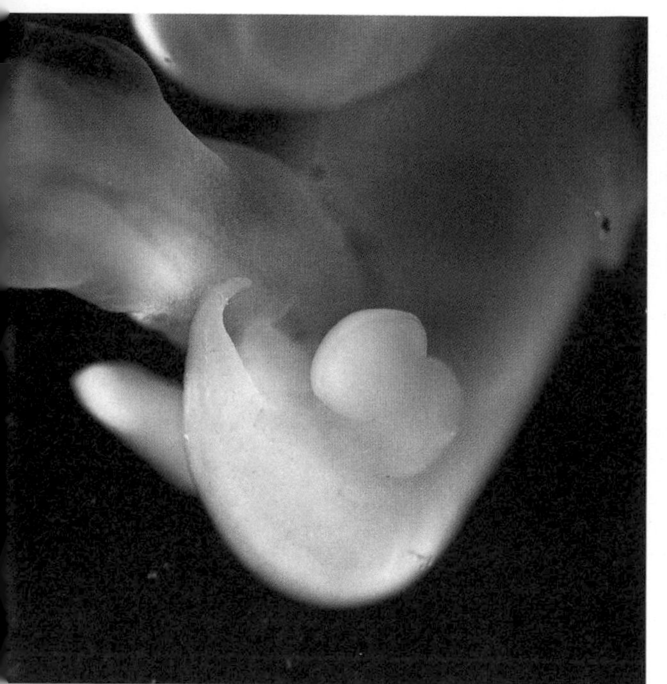

1 The human embryo has a tail in the fifth week of gestation

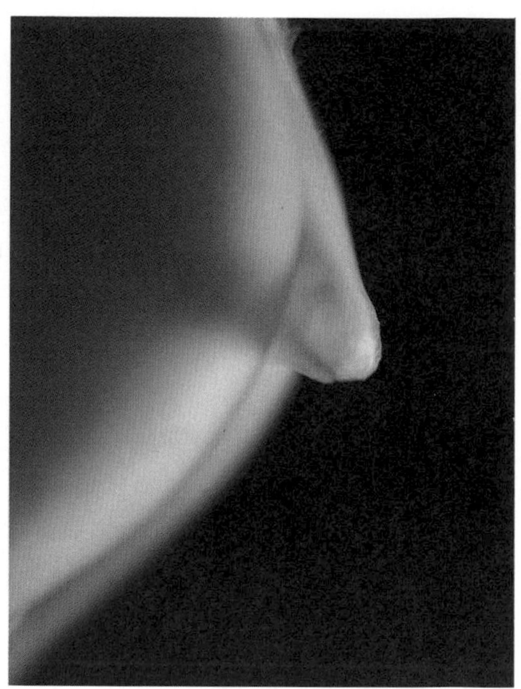

2 The tail is still there in the third month

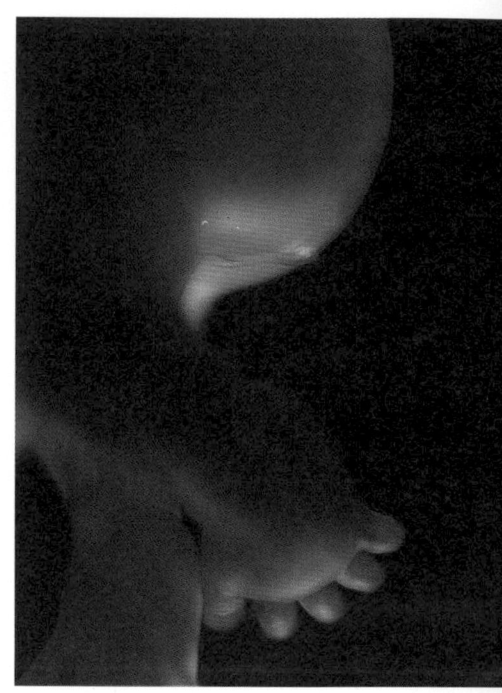

3 In the fourth month the tail is almost gone

instantaneous. This can occur in certain kinds of automobile accidents.

The hip joint, which holds the femur or the thighbone, is strong and firm, as it carries the body weight. The foot, whose structure is comparable to the hand, is an example of how part of the body has changed in response to demands. Some of the twenty-six bones in the foot form the arch, which makes the walk springy.

Joints define the direction and extent of motion of the bones of the skeleton. In the knees, the elbows, and the fingers are hinge joints. The ribs articulate with the vertebrae in gliding joints. The hip and shoulder joints are ball-and-socket types. A saddle joint, which gives a high degree of mobility, joins the thumb and the wrist. At a joint, the periosteum is replaced by a thin layer of cartilage, and the surfaces which come in contact are lubricated by synovial fluid to keep them in smooth working condition. All joints are surrounded by joint capsules and ligaments which hold the joint in place.

Bone formation starts during gestation. Calcification takes place either directly from connective tissue and cartilage, or in the border zone between bone tissue and cartilage. There, special cells called osteoblasts take up calcium salts from the blood and deposit them between the fibrocartilage as apatite crystals. Milk and cheese are important sources of these calcium salts. Calcium uptake in the digestive tract depends on vitamin D, which is formed in the skin in the presence of ultraviolet radiation from the sun. Food also contains some vitamin D. The deposition of apatite crystals is regulated by hormones from the parathyroid glands.

The combination of mineral salts and elastic fibrocartilage makes skeletal bones hard, but not so brittle that they break easily under pressure. The tensile strength of bones is amazingly high. When a full-grown man steps on his foot, the bones are exposed to a pressure of nearly 2,000 pounds per square inch. The pressure increases many times in the case of more rapid movements, such as in lifting or jumping.

Changes in demand influence bone tissue. Under stress, bone becomes stronger; during inactivity, it loses calcium. The changes in circulating hormones following menopause result in a reduction of calcium in the female skeleton. For this reason fractures are more apt to occur in older women than in older men.

The skeleton makes up approximately 18 percent of the body weight. Its design combines strength with lightness. The outer portion of a bone is solid, the inner is more porous and spongy. The spongy tissue is composed of trabeculae, supporting strands, a means of construction which provides optimal strength with the minimum amount of material. The cell-rich bone marrow fills the cavities between the trabeculae. Here red blood cells and a majority of white cells are formed.

Joints

◁ **The human fetus has a tail (1–3)**
for several months. It is comparatively long in the fifth week (1). The embryo is then only a few millimeters long and can hardly be distinguished from a fish or a bird embryo. In the third month (2) a small tail can still be seen. Some weeks later (3) it has almost disappeared. The caudal vertebrae become inverted and fuse into the small coccyx.

When the fetus is less than two months old (4–5) ▷
and approximately 3 cm long, the hands and feet begin to form. The individual bones are clearly seen in the transparent hand, which also has primordial nails. The right photo shows the feet. The individual bones as well as the joints are visible, even though the hands and the feet are only about 2 or 3 mm long.

During the fifth week of gestation, the limbs appear as small knoblike buds.

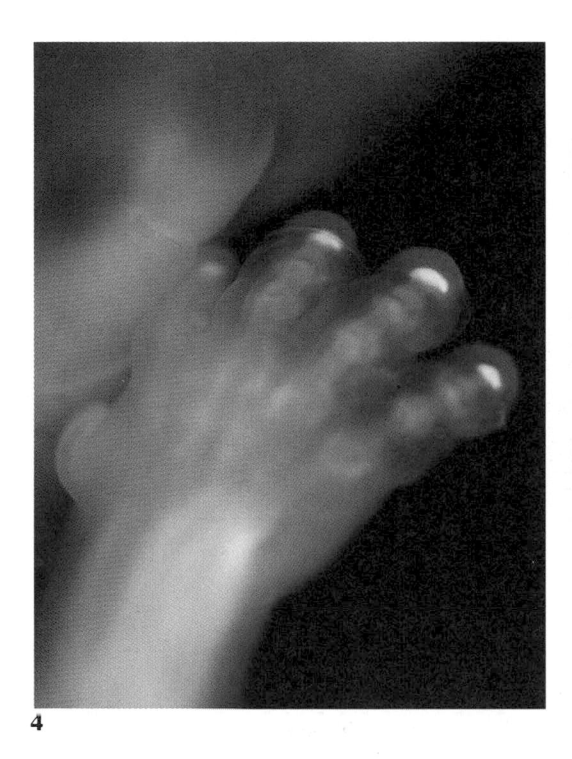

4

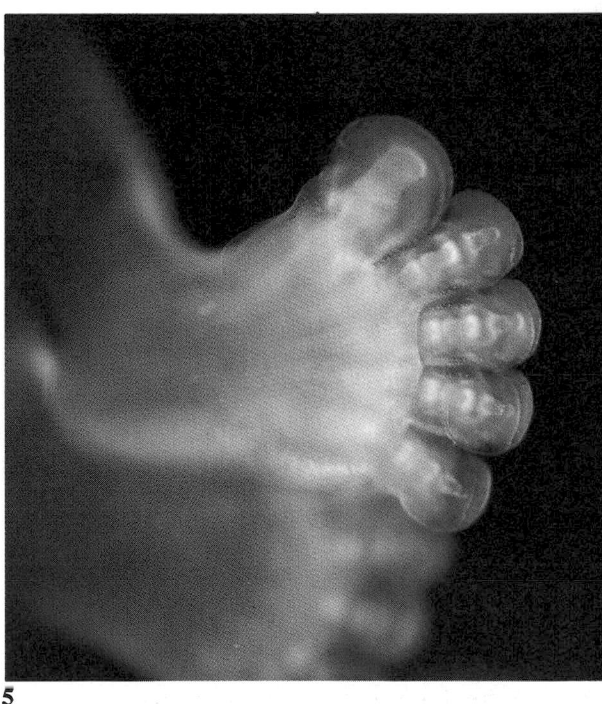

5

The strong joints in the body ▷
The knee joint at the upper left is the largest joint in the body. Here the thighbone — the femur — and the shinbone — the tibia — articulate in a hinge joint, secured by ligaments and a strong joint capsule. The rounded ends of the thighbone can be seen under the opened kneecap, or patella. Inside the knee joint are ligaments that run crosswise between the thighbone and the tibia. The cartilage surfaces are thin, indicating that the subject is a young person.

The shoulder joint (upper right) is a ball-and-socket joint. It has a great range of movement. One reason for this is that the shoulder blade does not articulate with the chest. In the photo, the head of the upper-arm bone has been lifted out of its socket in the shoulder blade.

The hip joint in the bottom photo illustrates how the head of the thighbone fits into the socket of the joint. A tendon from the head of the thighbone aids in keeping it in its proper position in the joint. Often the whole weight of the body rests on the hip joint; this is especially true when something heavy is being carried.

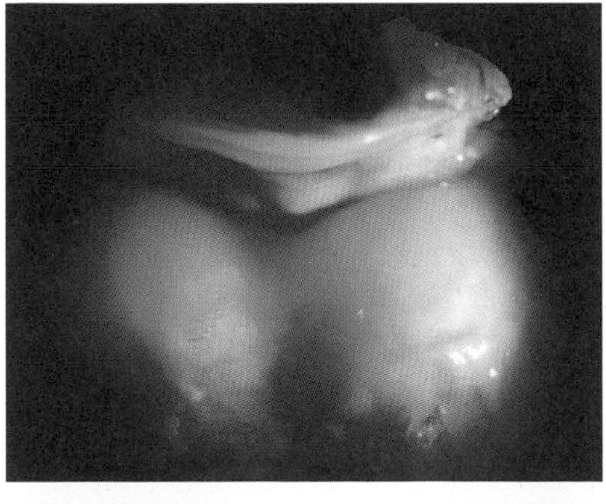

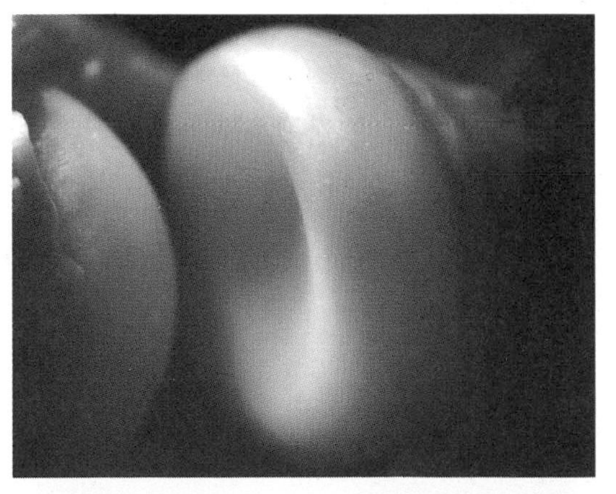

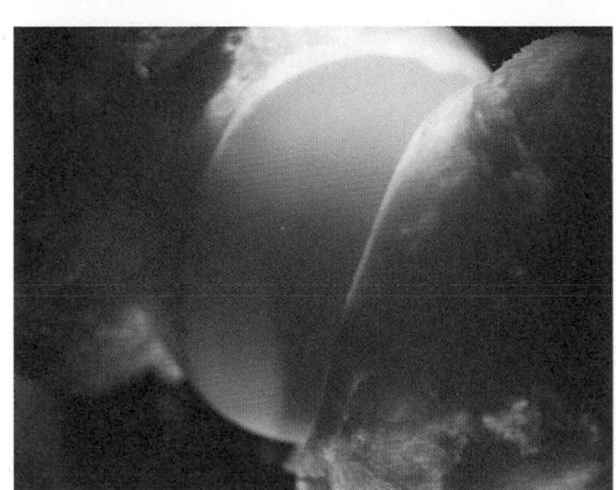

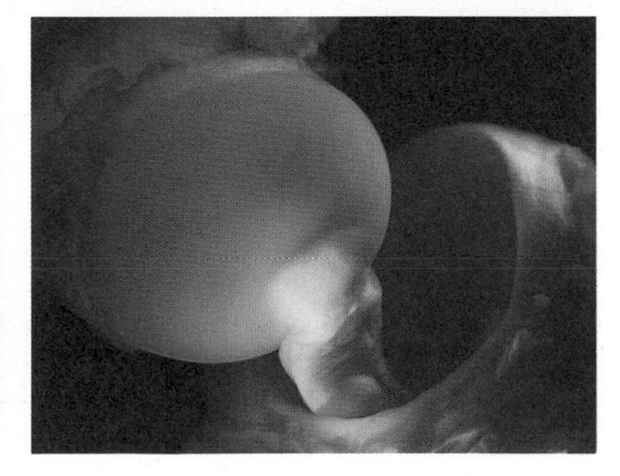

Teeth

Teeth have other functions besides grinding food. They are of considerable importance in how we look — well-shaped, white teeth make a favorable impression. Without teeth, speech articulation becomes blurred and indistinct. Sharp *s* sounds can hardly be produced without teeth, and try to produce a *v*, *f*, or *t* without them!

The grinding of food by teeth allows it to be thoroughly mixed with enzymes in the saliva. This is the first step in digestion.

Teeth develop from the same primitive tissue as hair, skin, and mucous membranes. The primordia of the permanent teeth are formed early in gestation, but these teeth do not begin to erupt until the age of six. For the first five or six years, the child has a primary set of about twenty teeth, the milk teeth. The jaws are far too small to hold the large teeth of the adult. As the jaws enlarge, the permanent teeth start appearing. The second molar is normally cut at the age of twelve. The third, the wisdom tooth, does not appear until eighteen or twenty. In about one person in five, it may not appear at all.

A tooth is composed largely of dentin, also called toothbone or ivory, rich in calcium and phosphorus. The crown is protected by a thin layer of enamel. This is the hardest tissue in the body, nearly as hard as glass. Enamel is built up of five- or six-sided calcium crystals closely packed together. They stand at right angles to the dentin, which supplies the enamel with nourishment. Thus tooth enamel, although largely inorganic in makeup, is still living tissue that grows and can be renewed. Injuries to it can heal if they are not too large. Unlike dentin, tooth enamel has no sensory nerve endings.

Inside the dentin is a cavity filled by the pulp, a tissue with blood vessels, nerves, and soft connective tissue. The dentin as well as the pulp is very sensitive to pain. The pulp canal, which opens at the tip of the root, is the passage to the jawbone for blood vessels and nerves.

The roots of the teeth are embedded in cavities in the jawbone. Numerous fibers of connective tissue anchor the teeth firmly there. The root is covered by the cementum. Like the skeleton, teeth have a greater resistance to decomposition than other tissues. Teeth may be among the remains after a fire and can be used to identify victims. Teeth are also often preserved in fossils, enabling anthropologists to trace the development of teeth during the course of evolution.

Section through a tooth

1 enamel
2 toothbone (dentin)
3 gum
4 pulp
5 cementum
6 root canal with blood vessels and nerves
7 jawbone

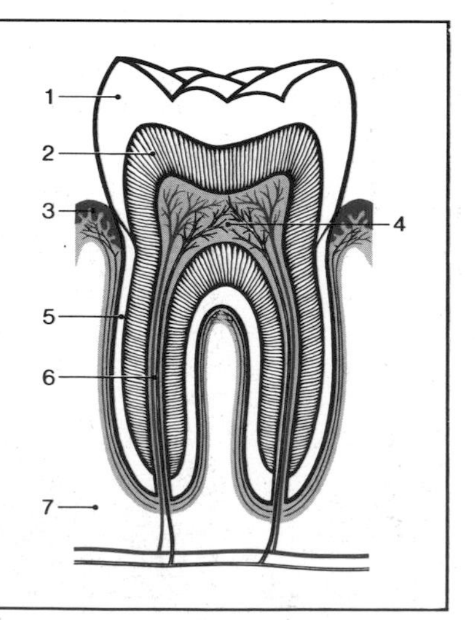

Section through a tooth viewed in an interference microscope
This molar tooth has a crown of hard enamel covering the toothbone. The black region in the center is the pulp. Nerves and blood vessels pass through canals in the tips of the roots to supply the pulp.

Teeth

Enamel and toothbone magnified 200 times
Few people have flawless teeth. Dental caries, or decay, is a very common disorder. It is not known how caries develops, but sugar and lactic acid bacteria probably play an important role. When the destruction process penetrates the enamel, bacteria can reach the dentin and advance to the pulp, causing toothache. If the pulp becomes infected, it becomes necessary to remove it and replace it with a filling.

A carious tooth
The cavity in the enamel is seen at the left in this section through a tooth, photographed with an interference microscope.

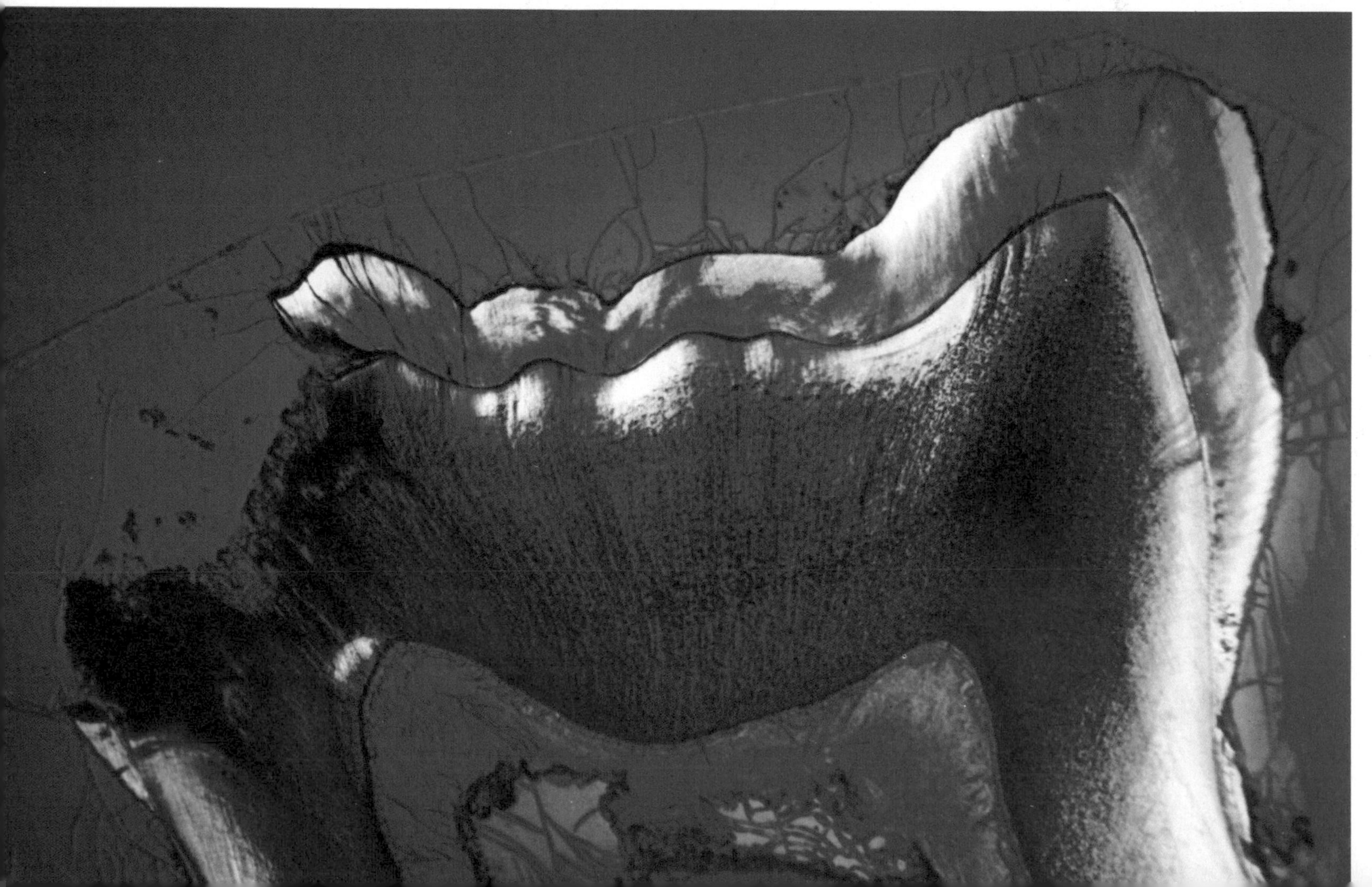

What Happens to Food

Each cell — and there are hundreds of millions in the body — is a miniature factory. It produces goods and services of various kinds and needs a continuous supply of raw material. This material comes from food. Carbohydrates and fats are used chiefly as fuel, proteins as building materials. Vitamins, minerals, and salts are necessary components at various stages in the process. Metabolism, the general term to describe the building up and tearing down process, depends largely on hydrolysis — the splitting of substances into simpler units by the addition of the components of water.

Digestion, or the breakdown of food, is an example of a destructive process — a catabolism — which occurs by oxidation: oxygen combines with food matter and energy; water and carbon dioxide are produced. The oxygen used is provided by air in the lungs. The digestive tract, or pathway of food in the body, is in essence a long tube which starts at the mouth and ends at the anus. Fully extended, its length in an adult would be twenty-seven to thirty feet. It can be divided functionally into six segments. Its general purpose is to reduce food to simple components which can be taken up by blood or lymph and delivered to the cells of the body. Some twenty different chemical substances in salivary, gastric, and intestinal juices participate in the chemical decomposition. Of these, enzymes are the most important. These substances act as catalysts, chemicals which even in small concentrations can initiate and accelerate biochemical processes without themselves being changed. Without the aid of the enzymes, virtually no nutritive substance could be absorbed in the intestines.

The mouth is the first station on the route of food in the body. Here food is chewed, subject to the approval of gustatory and olfactory senses. As it is chewed, food is mixed with saliva, which helps to liquify it and thus facilitate its passage through the alimentary tract. Salivary enzymes also begin the process of decomposition of carbohydrates. The secretion of saliva takes place all the time, but it increases when food is put into the mouth. Secretion is also heightened by the sight or aroma of attractive food, and sometimes even just by thinking of something good to eat.

The next station is the pharynx. After food has been initially prepared in the mouth, the tongue moves it to the rear of the oral cavity. This triggers a complex series of muscular contractions which result in swallowing: the soft palate closes the nasal cavity, the larynx is raised, the epiglottis is lowered over the trachea, and the food is propelled into the esophagus.

The esophagus is the third food station. It is a muscular tube approximately a foot long. This is only a brief stop for food, however, because in a few seconds the muscles in the esophageal wall push food into the stomach. The muscular contractions occur in progressive waves of contraction and relaxation. Such movements are called peristalsis, and they describe the transport mechanism in the esophagus, stomach, and intestines. Peristalsis is stimulated partly by the vagus nerve in the brain, partly by the autonomic nervous system, but ultimately by a network of nerves in the walls of the hollow organs themselves. Impulses in nerve fibers are triggered by food stretching the walls of the tract. Sense organs respond to the tension and signal nerve cells to release acetylcholine. This causes the smooth musculature to contract. As food progresses, a wave of contraction moves down the tract.

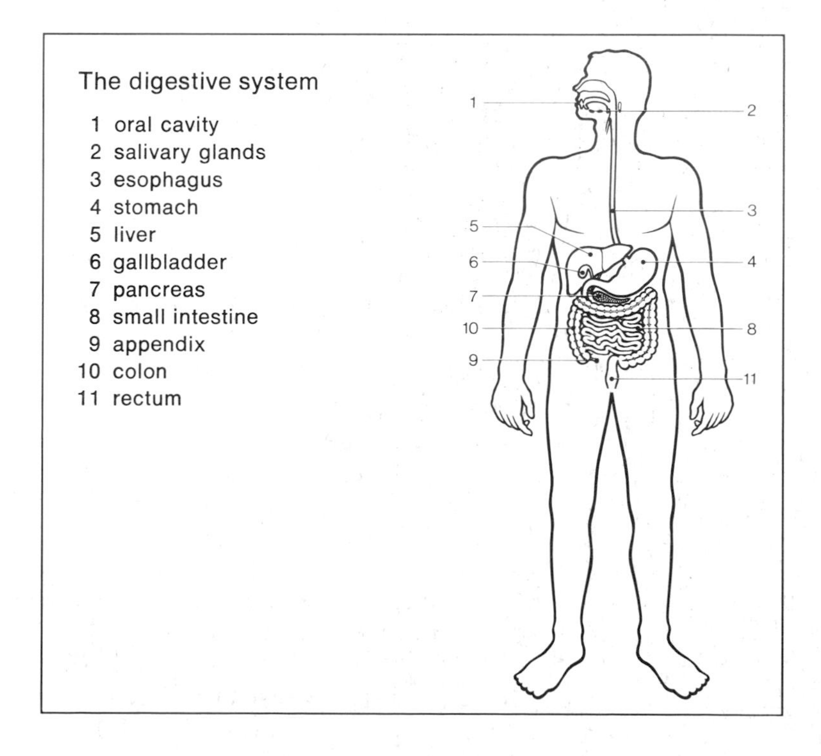

The Tongue and Pharynx

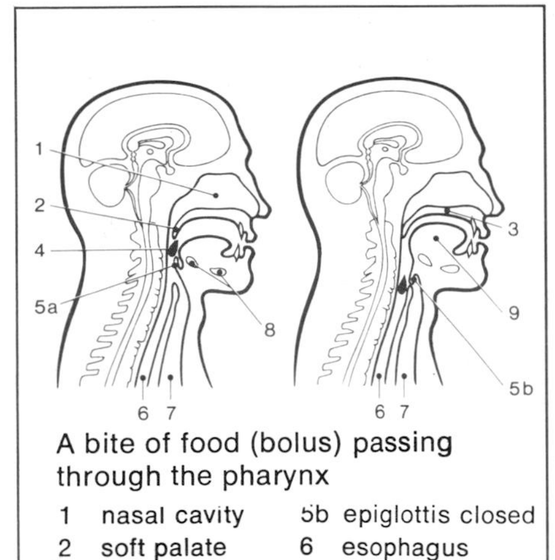

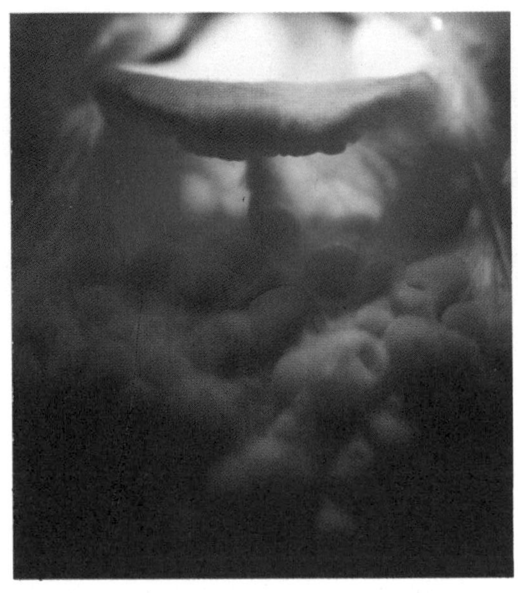

1 The epiglottis is open during respiration

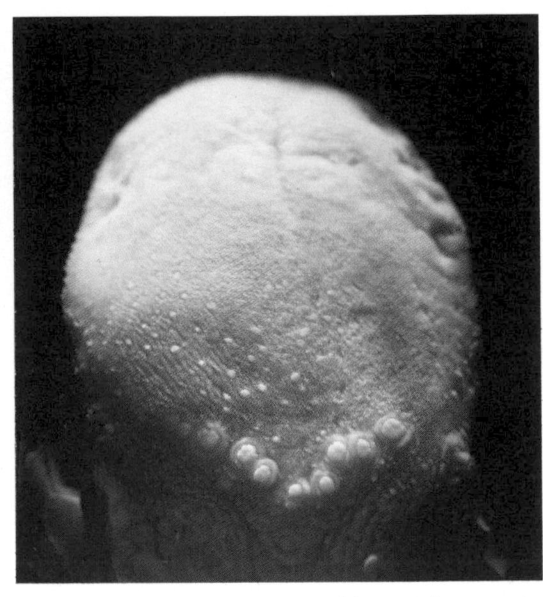

2 The tongue, our most movable muscle

The epiglottis opens during respiration (1)
The back of the tongue is in the foreground. An accumulation of lymph nodes and follicles here, together with the tonsils, helps guard against infection that might be introduced by bacteria in food.

The tongue is a highly flexible muscle (2)
with three main functions: it is the organ of taste; it mixes saliva with food and transports it toward the pharynx; and it is a most important organ of speech.

The striated musculature of the tongue is arranged in bands oriented in three directions. This gives the tongue its high range of movement. The taste buds are sensitive to salt, sweet, sour, and bitter flavors. They are situated in the vallate, the fungiform, and the filiform papillae — small rounded swellings on the surface of the tongue. The tongue has several thousand taste buds, most of them concentrated in the large vallate papillae, which are arranged in the form of a V near the root of the tongue. The photo shows the different types of papillae with the large vallate papillae at the bottom. At the root of the tongue is the hyoid bone, a bone that does not articulate with any other skeletal bone.

Saliva (3)
secreted from a salivary gland under the tongue. One of the components of saliva is the enzyme amylase, which breaks up starch to a more simple sugar, maltose.

3 Saliva secreted from under the tongue

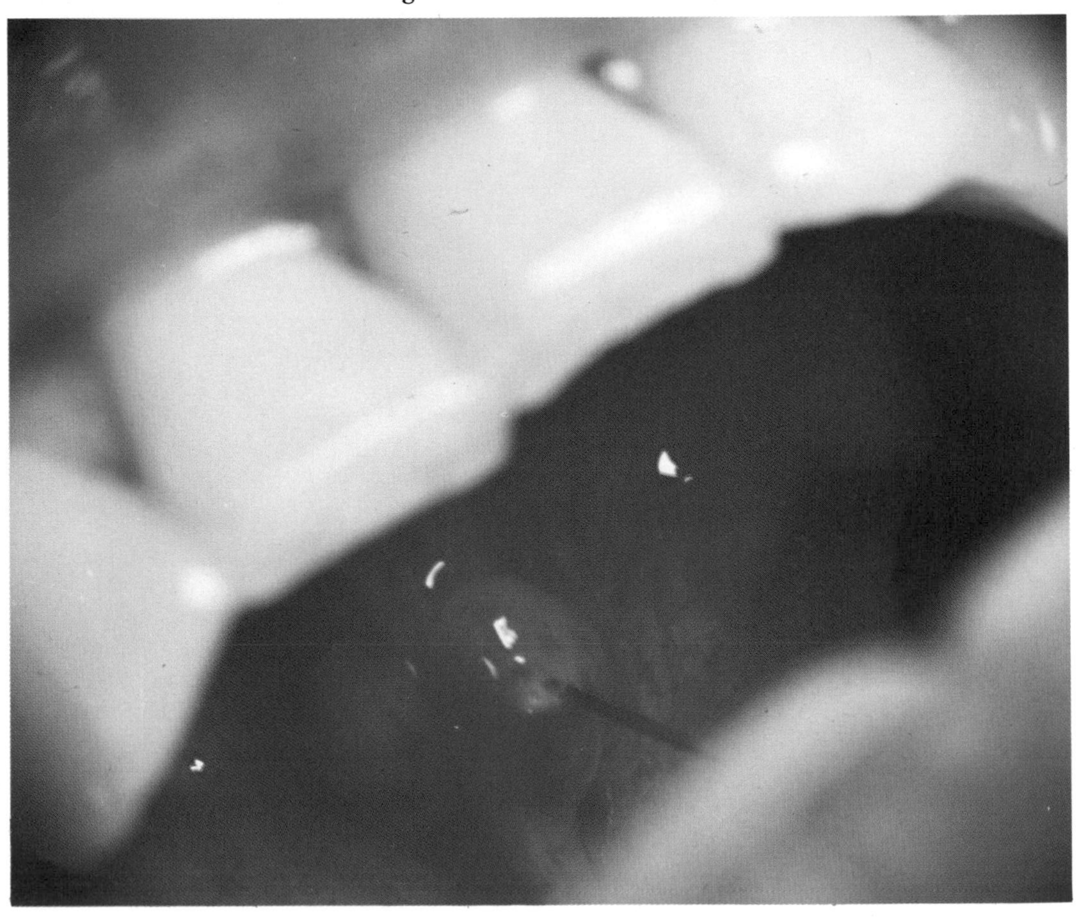

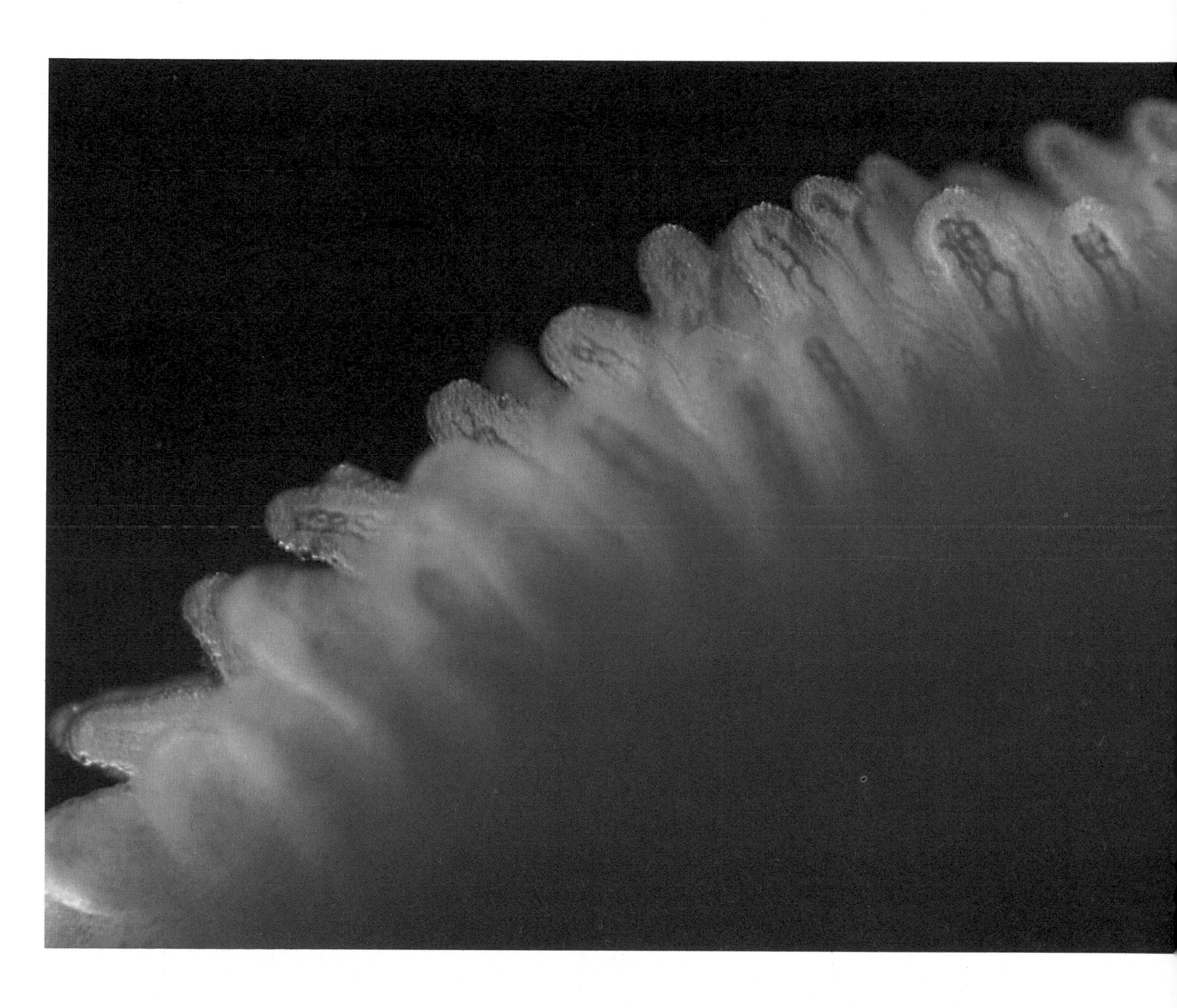

Filiform papillae at the tip of the tongue in a newborn, highly magnified. The taste buds are situated in the papillae. It was long believed that each papilla was sensitive to only one taste. Recent findings, however, suggest that individual taste buds may respond to more than one taste.

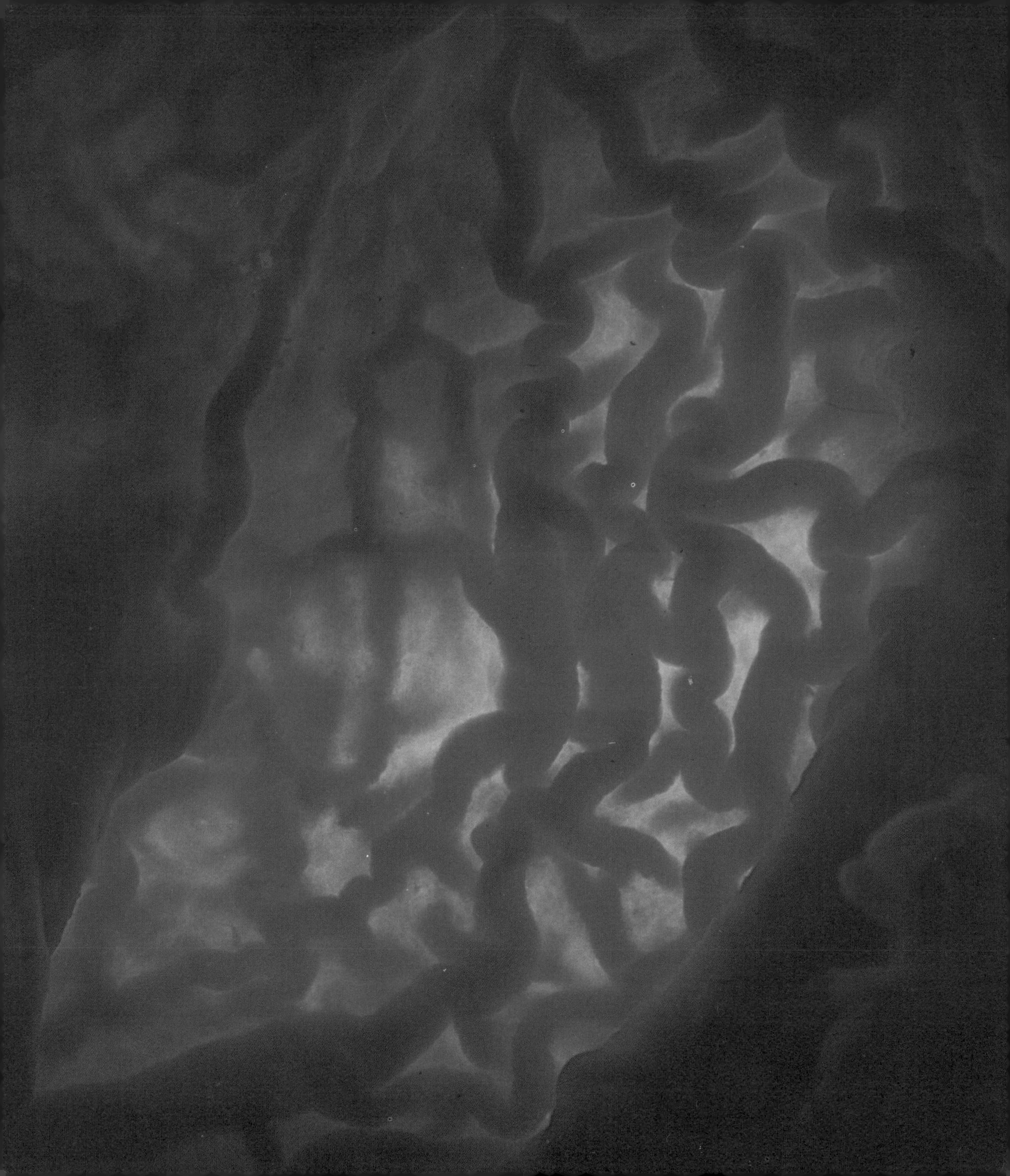

The Stomach

The stomach is the fourth station. Normally it can hold about one and a half quarts. The stomach wall consists of three muscular layers. Sphincters encircling the opening of the esophagus into the stomach and of the stomach into the duodenum, the first part of the small intestine, function as "gatekeepers." The two openings are called the cardia and the pylorus, respectively. The gastric mucosa contains glands that secrete hydrochloric acid, mucus (to protect the stomach from the acid), and the protein-splitting enzyme pepsin. When food enters the stomach, the hormone gastrin is released in the blood to stimulate the secretion of gastric acids. During its two- or three-hour stay in the stomach, food is processed to a thin mash. The hydrochloric acid makes it highly acid and this increases the efficiency of pepsin. With the exception of substances such as water and alcohol, no absorption of food takes place here.

Gastric contractions propel the stomach contents bit by bit past the pylorus and into the small intestine, the fifth station.

◁ **The interior of the stomach wall is highly convoluted,**
a construction which allows it to be distended. In the folds of the mucosa are the glands that secrete gastric juices. To protect the stomach lining against this very acid substance, the mucosa also secretes a thick mucus.

In some diseases, the secretion of gastric juice increases, so that the mucosa becomes irritated. Gastric ulcer may then occur. Stress over a long period of time may also increase secretion and thus be a contributing factor in gastric ulcer.

The stomach and small intestine in a newborn ▷
The liver, which is dark red, is visible at the left. The stomach opens into the small intestine at the pyloric orifice, where one of the several sphincters in the body is located. It opens at intervals of a few minutes and releases small portions of stomach contents into the duodenum. The capacity of the stomach in adults is about one and a half quarts.

The stomach continues the process of digestion, which will be completed by intestinal juices. Contractions of the very muscular wall of the stomach mix the gastric juice with food. The stomach can contract even when empty, producing hunger pains or pangs.

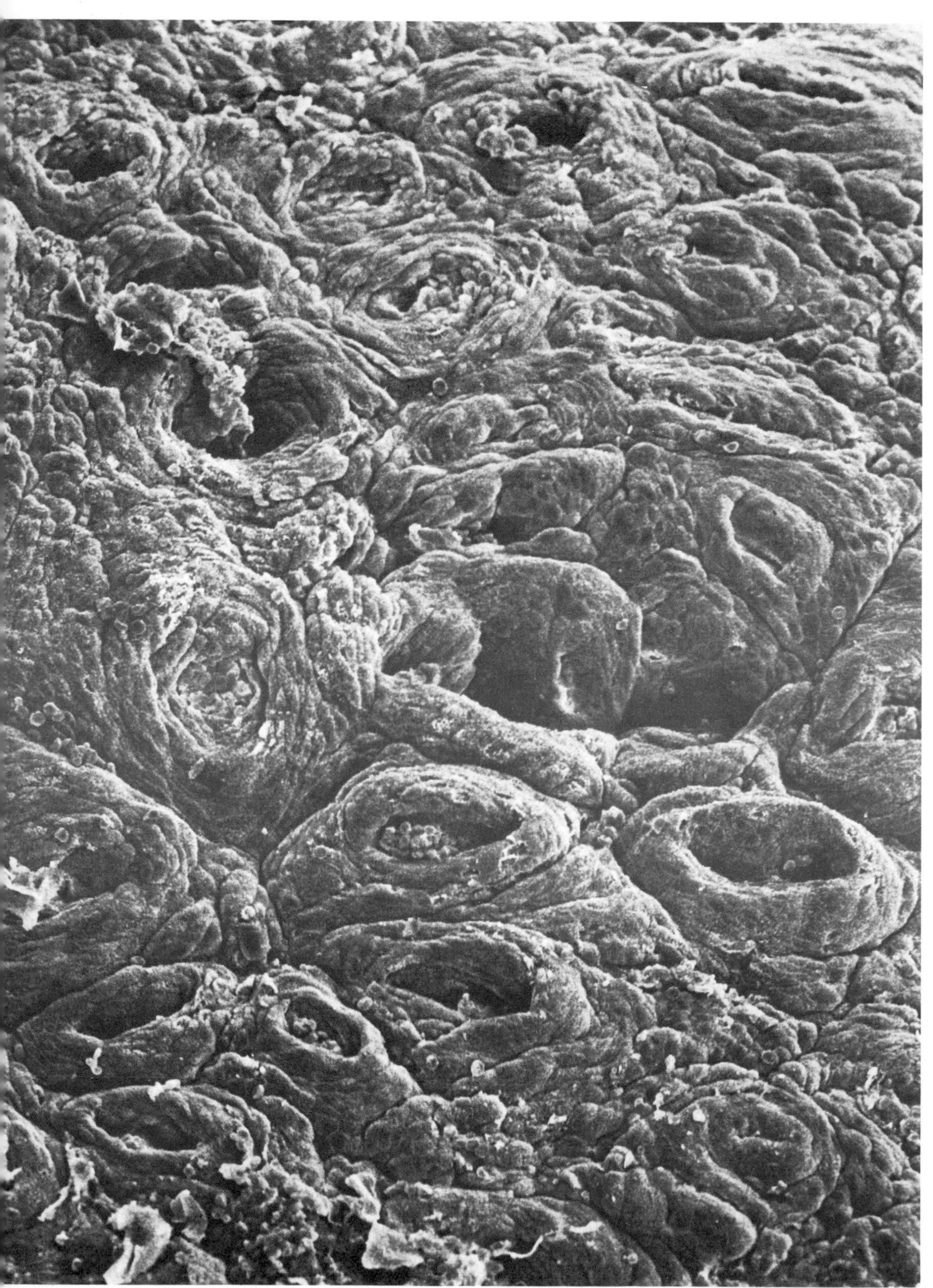

◁

The stomach
contains three different types of glands. The photo shows the gastric mucosa in the lower portion of the stomach, magnified 500 times. The area is near the pylorus, where the pyloric glands are situated. These are simple mucus-secreting glands. Mucous glands are also present near the upper end of the stomach. Throughout most of the stomach are the "fundus" glands, which secrete enzymes and hydrochloric acid.

The opening of a mucus-secreting gland ▷
near the pylorus, magnified approximately 3,000 times in a scanning electron microscope.

The bottom photo shows mucus emerging from the gland opening.

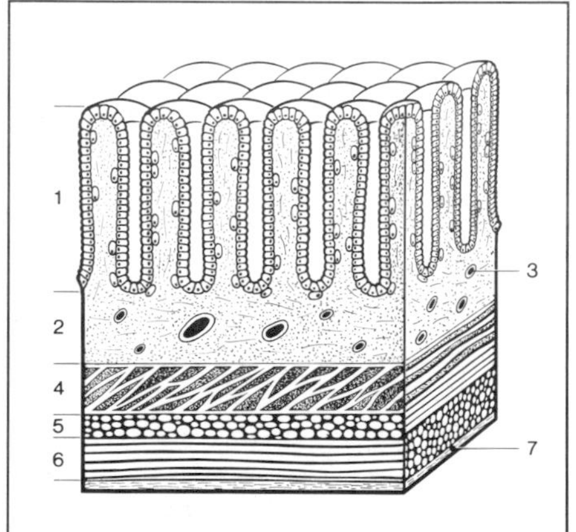

The layers of the gastric mucosa

1 glands producing gastric juice
2 connective tissue
3 blood vessels
4 oblique muscles
5 circular muscles
6 longitudinal muscles
7 peritoneum

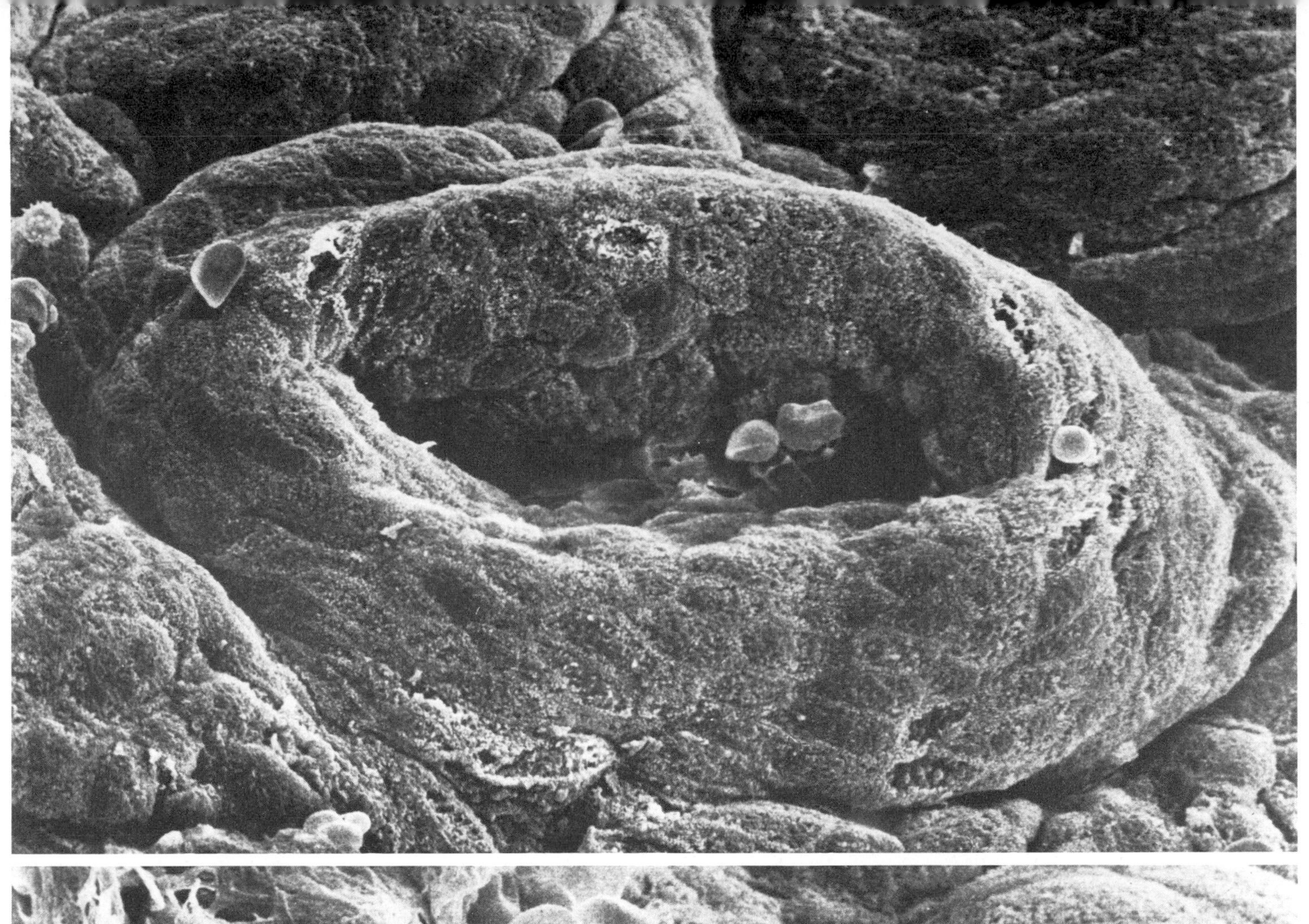

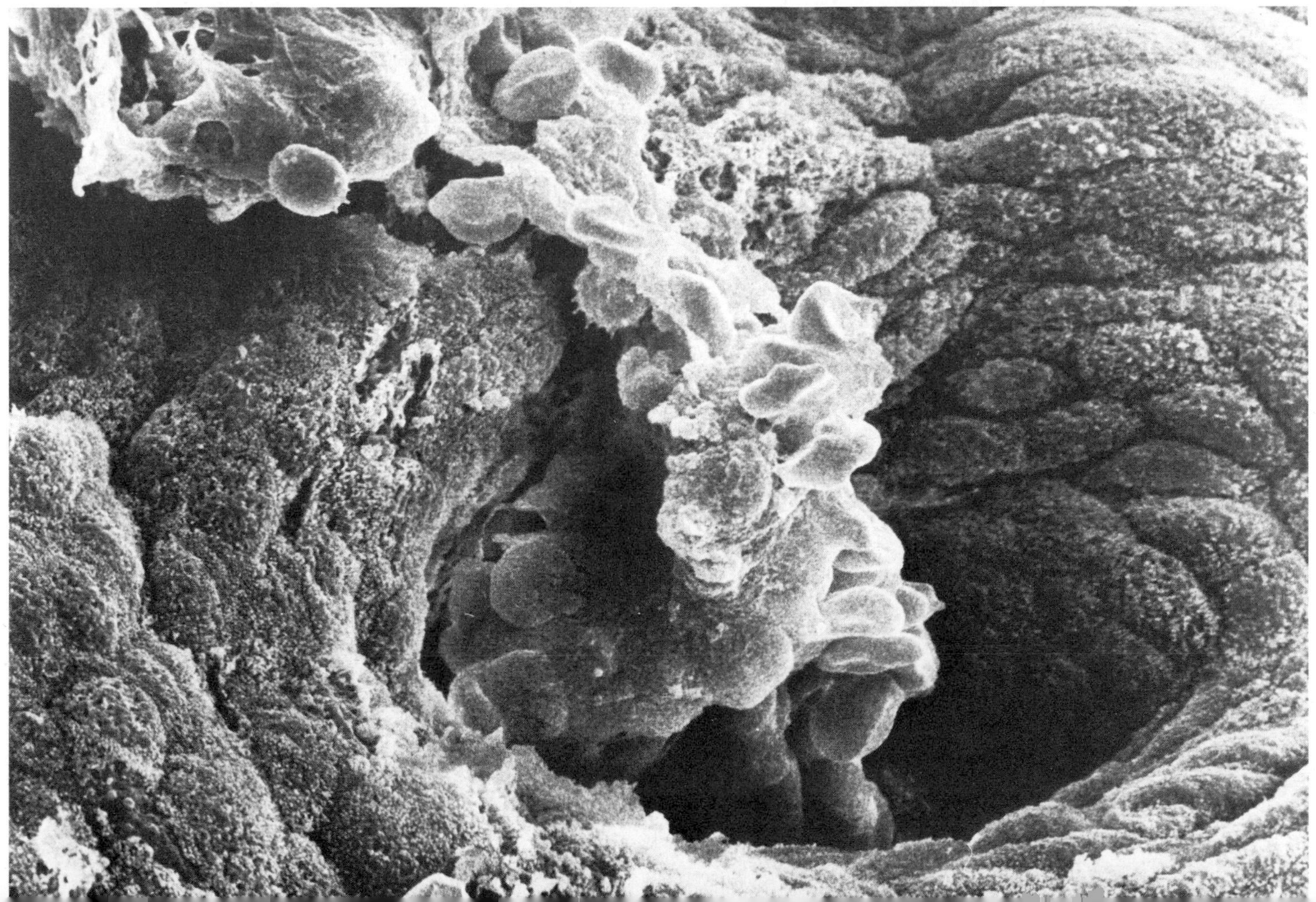

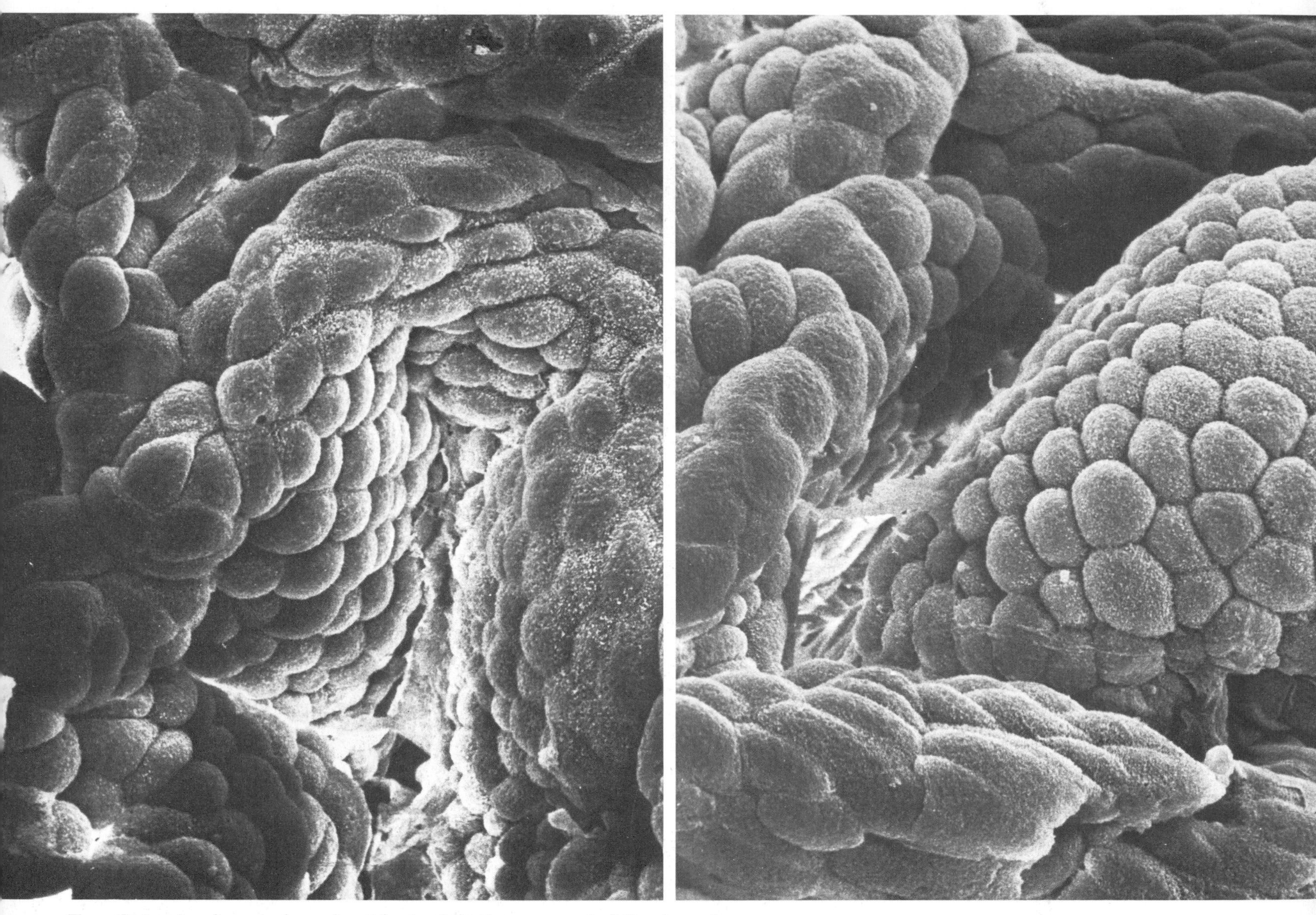

Two photos showing openings of gastric glands in the upper part of the stomach

The Small Intestine

Food from the stomach passes into the intestines in small amounts. The intestines are designed for maximal absorption of useful substances. The major divisions of the intestines from the stomach down are the small intestine, the large intestine or colon, and the rectum. The first part of the small intestine is the duodenum. Its name derives from the fact that its length corresponds approximately to twelve fingerbreadths — about ten inches. (The Latin word for twelve is *duodecim.*) The next portion of the small intestine is called the jejunum, meaning the "empty" intestine. It is about eight feet long and it passes into a somewhat longer segment, the ileum, which ends in the colon. The structure of the colon differs from that of the small intestine, and it is much larger in diameter.

The digestion of food is brought near completion in the duodenum. Peristaltic movements propel the food forward here at the rate of 2 to 3 cm a minute. The intestinal wall is composed of four layers: an outer lining of connective tissue, two layers of smooth muscle, and, innermost, a mucosa. Since the movements of the intestines are produced by smooth musculature only, they are not under voluntary or conscious control.

The arrival of the stomach contents in the duodenum liberates

◁ **The gastric mucosa,**
showing openings of the gastric glands, magnified 3,000 times. The gastric glands in the stomach wall number in the millions. In digesting a normal meal, the glands secrete between a pint and a quart of gastric juice. The secretions begin before food enters the stomach — at the sight or smell or even the thought of food when hungry. "Watering" then occurs not only in the mouth but also in the stomach. The gastric glands are controlled partly by nerve impulses and partly by the hormone gastrin. When the stomach contents are released into the duodenum, another hormone, enterogastrone, is liberated from the duodenal mucosa. When this hormone reaches the gastric glands it inhibits further secretions of gastric juice.

Intestinal organs in a newborn ▷
The liver (red) and the gallbladder (green) are seen at the top. Just below on the right is the stomach, and in the foreground, the coiled distended small intestine. The intestine has ample leeway to move in the course of its activities. It is attached to the back of the abdomen by the mesentery, which contains blood and lymphatic vessels. The intestines, the mesentery, and the abdominal wall are covered by a membranous lining, the peritoneum.

The intestines in a newborn

hormones in the duodenal wall. One hormone stimulates the gallbladder to release bile, others stimulate the production in the pancreas of pancreatic juice. The pancreatic juice is alkaline, and to some extent neutralizes the hydrochloric acid added with the gastric juice. The pH measurement, which indicates the degree of acidity or alkalinity, increases toward alkalinity, and this provides proper working conditions for the many enzymes present in the pancreatic and the intestinal juices. Bile emulsifies the fat in the intestinal contents, called the chyme, so that it can be broken down by enzymes. Pancreatic juice contains protein-splitting enzymes, as well as enzymes which can convert starch and other carbohydrates into substances absorbable by the intestinal wall.

When food reaches the jejunum, it has by and large been broken down into its basic components. Digestion continues through the action of enzymes in the secretions of glands in the wall. The most important function of this intestinal segment, however, is to absorb the nutritive substances and transport them to the blood and lymph so that food becomes available to all parts of the body.

Intestinal Villi

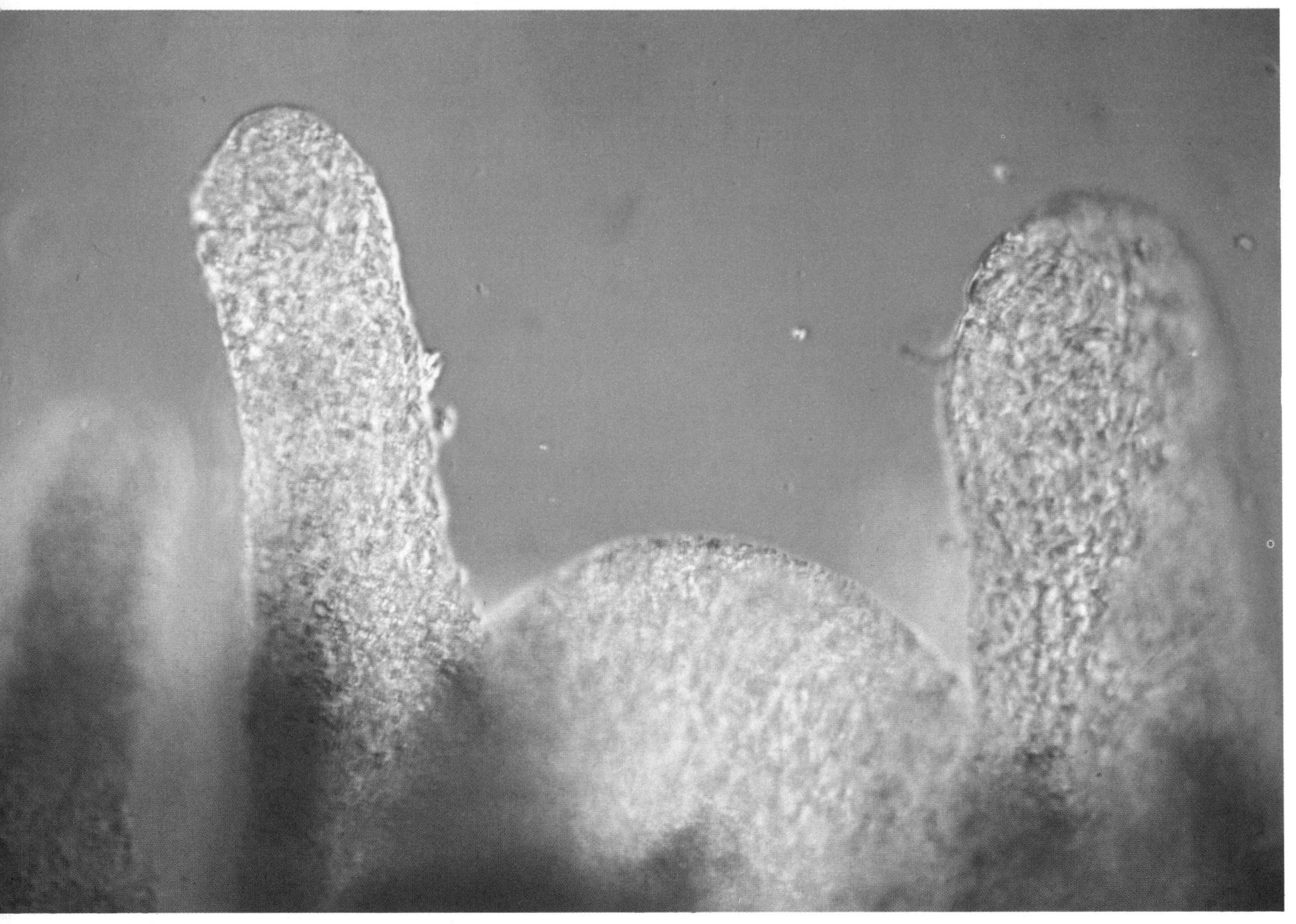

Intestinal villi

The intestinal wall is covered by a vast number of small fingerlike projections — villi — through which the chyme passes. These projections increase the surface areas of intestinal wall to a total of about ten square yards — several times greater than the total skin area. The large surface facilitates the absorption of food. Each villus contains a rich network of capillaries and lymph vessels. The individual villus can be shortened and lengthened by smooth muscles. This means that the nutritive substances are, so to speak, pumped out to the blood and lymph vessels leading from the villi. Blood vessels transport the basic ingredients of food, in particular glucose and amino acids, while the lymph vessels transport fat. The nutrient-rich blood vessels drain into the portal vein, which leads to the liver and then into the general circulation. Lymph vessels, rich in fats (which make them look milky), drain into larger vessels and ultimately into the main lymphatic trunk, the thoracic duct. This duct ascends through the abdominal cavity and empties into the left subclavian vein, which via the superior vena cava carries blood to the right atrium of the heart. Thus the greater part of the nourishment supplied by food is able to reach all parts of the body.

Intestinal Villi

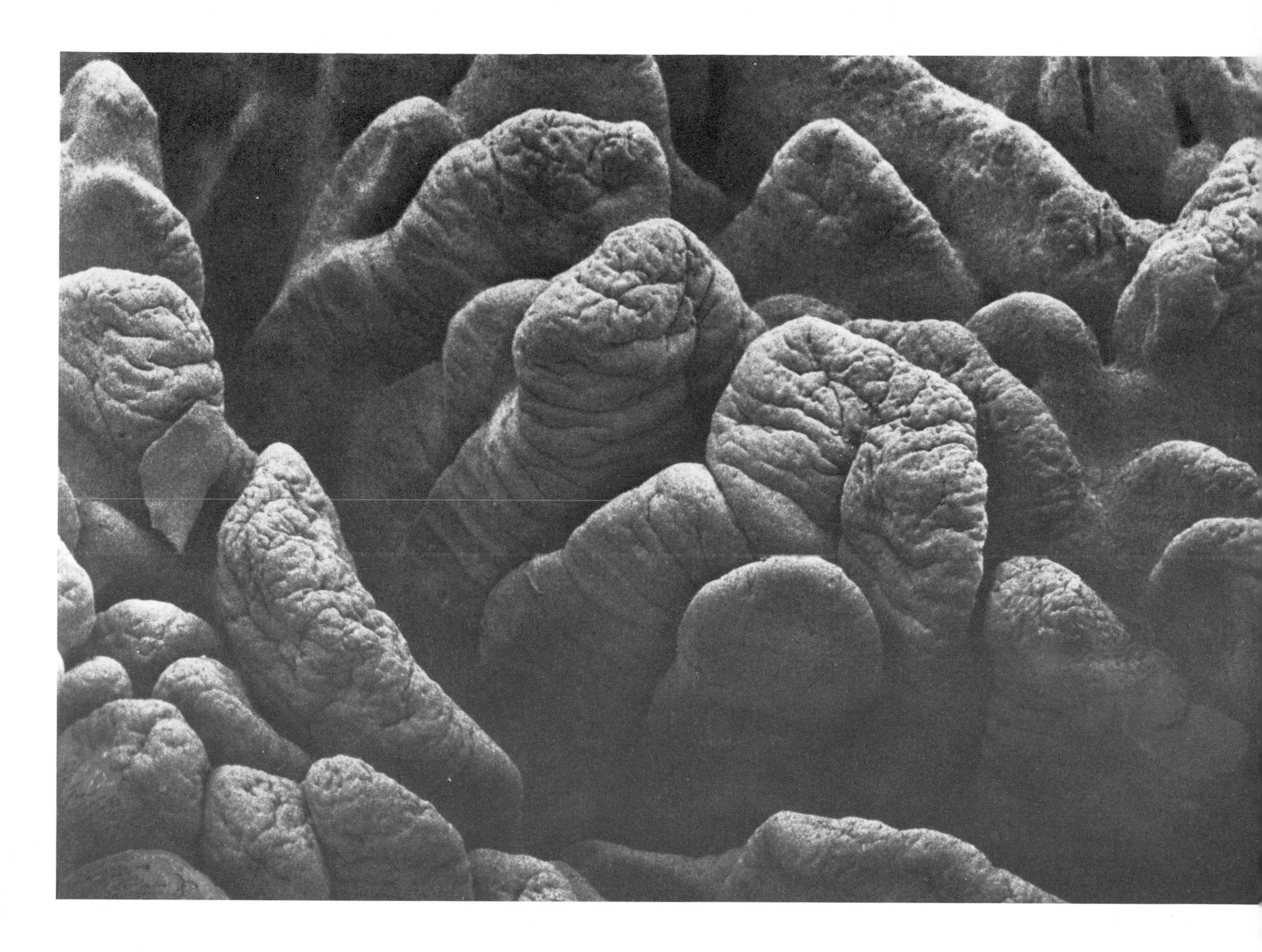

◁ **Villi in the small intestine, magnified 1,000 times**
Most of the end-products of digestion are absorbed by the villi of the small intestine. Each villus contains a network of capillaries and lymph vessels. Proteins, sugar, and some fat are absorbed by the blood vessels, while most of the fat passes into the lymphatic system.

Intestinal villi △
seen in the scanning electron microscope. These fine projections, 0.1 to 1.5 mm long, fringe the total inner surface of the small intestine.

The Intestines

Cross section of the small intestine magnified about 10 times
Note the villi lining the wall. About 23,000 pounds of solid food and 45,000 quarts of fluid pass through this tunnel during a normal life span. In an adult, food travels at a rate of about an inch a minute, more slowly further down in the intestines. Most of the water is absorbed in the large intestine, so that the contents become concentrated. Peristaltic movements likewise become slower. It normally takes about five hours for food to pass from the pylorus to the colon; about twelve for the total route from mouth to rectum.

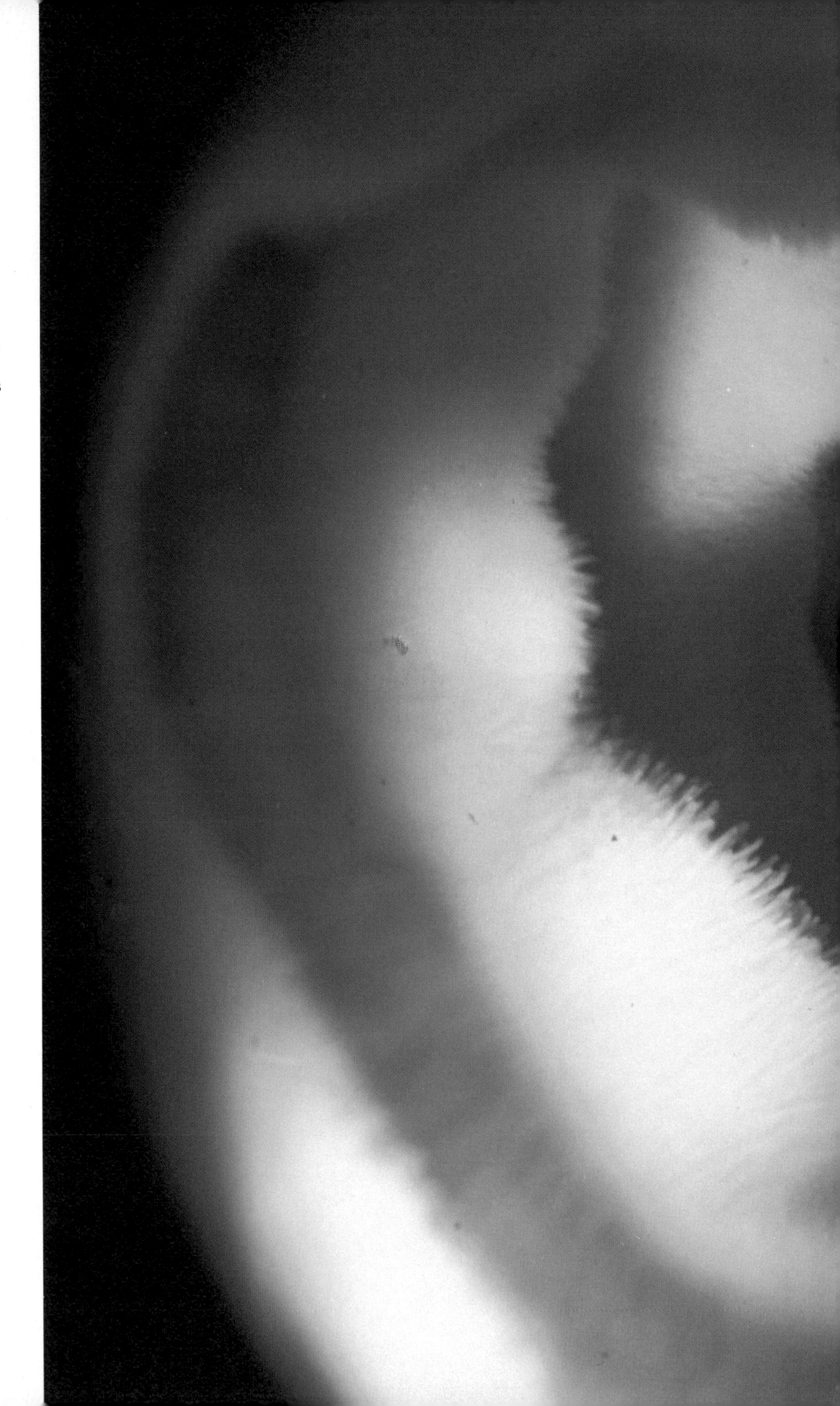

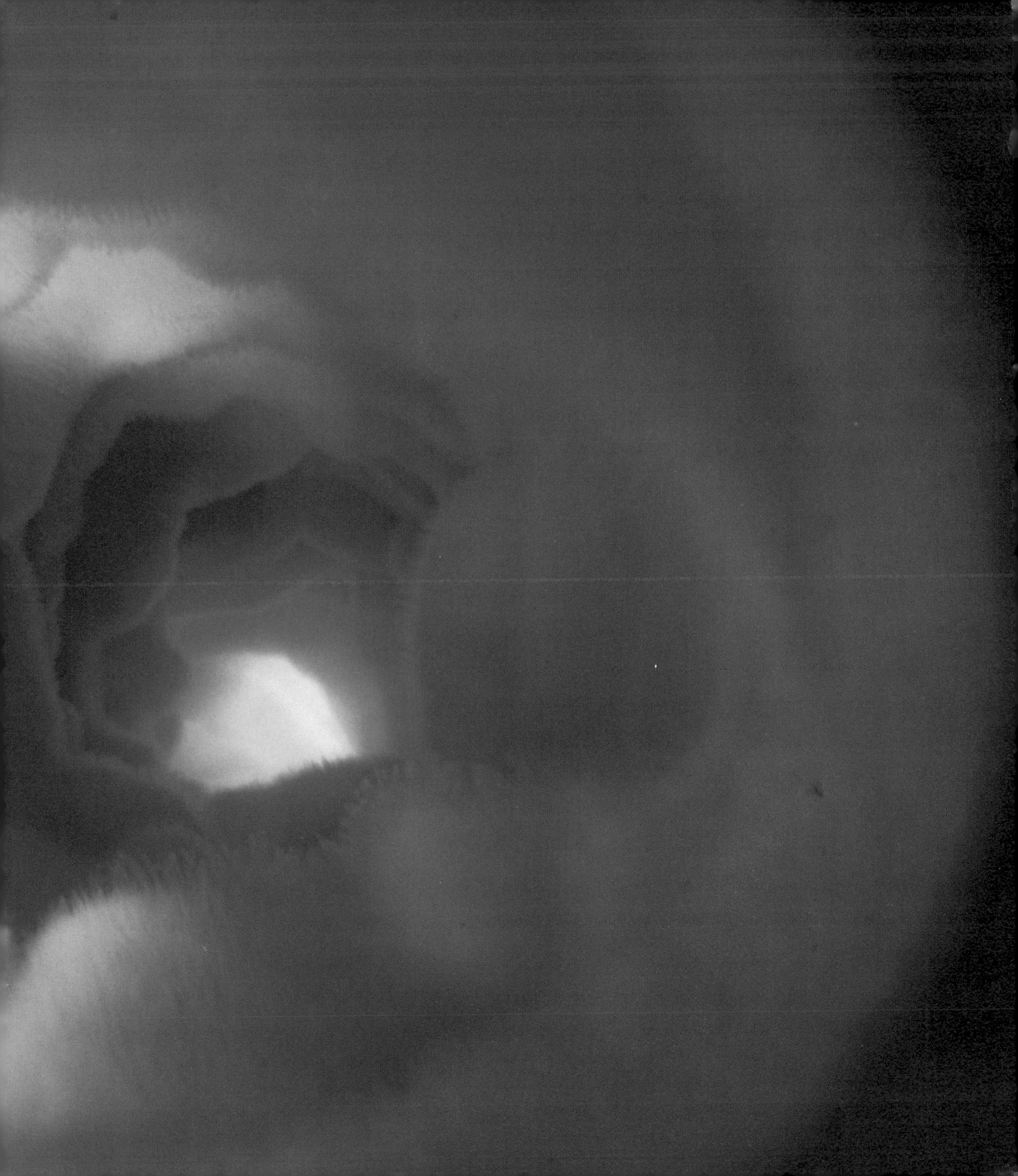

The Pancreas

Next to the liver, the pancreas is the largest gland in the body, with a length of about six inches. In structure it is somewhat like the salivary glands. It secretes between a pint and a quart of pancreatic juice a day, several times its own weight. This means that the pancreas is one of the hardest working organs in the body. Pancreatic juice is necessary for the final breakdown of food. It contains inorganic salts such as bicarbonate which make it alkaline and help to neutralize the acidity of the stomach contents. It also contains a number of enzymes. Pancreatic juice flows into the duodenum through the pancreatic duct, which joins the last part of the bile duct. The enzyme amylase breaks starch into a sugar, maltose. Lipase breaks down fat into glycerides and fatty acids. Trypsin and chymotrypsin split the proteins into simple components. Together with the intestinal juice, the pancreatic juice supplies the enzymes necessary for the final breakdown of the food — the process that began with the action of saliva in the mouth.

The production of pancreatic juice is regulated by the stomach contents in a simple way. When the very acid stomach contents pass through the pyloric orifice into the duodenum, the hormone secretin is liberated in the duodenal wall. Secretin is then carried in the blood to the pancreas, where it stimulates secretion. The contractions of the gallbladder producing bile are regulated in a similar manner by another hormone.

If, for some reason, pancreatic juice is not present in sufficient quantity, food cannot be broken down with a maximum of efficiency. Part of it will then leave the body unutilized.

Throughout the pancreas are scattered the islets of Langerhans, which are collections of cells responsible for the production of insulin, another vital function of this gland. The hormone insulin, which enters the general circulation, is necessary in order that sugar present in the blood can be taken up by body cells. Insulin deficiency is a characteristic of diabetes mellitus. One of the symptoms of this disorder is excretion of sugar in the urine. The disease, if untreated, may lead to disturbances of metabolism and secondary complications that may be fatal.

Part of the pancreas with the liver in the foreground

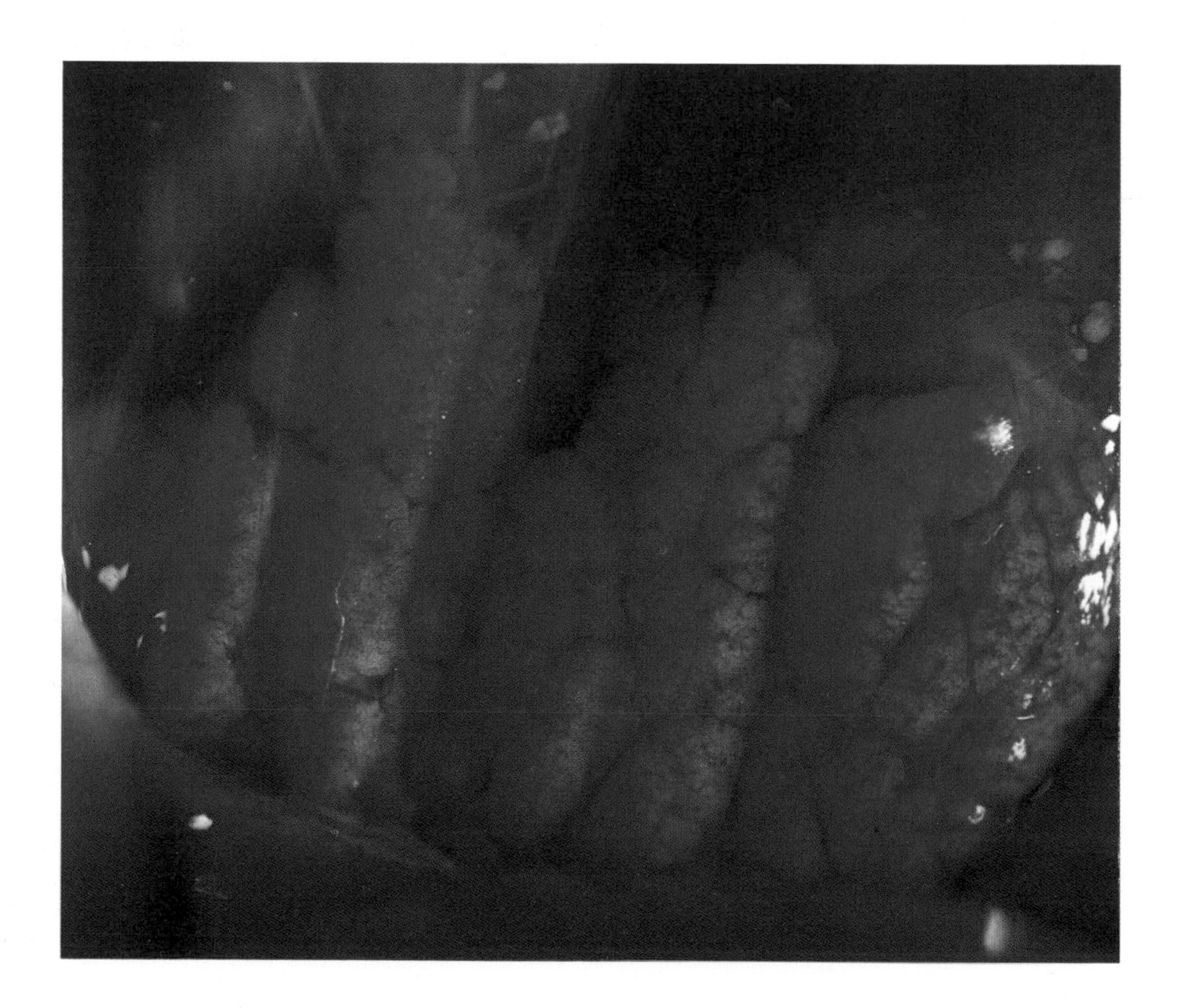

The liver, seen from below, and the bladder (greenish yellow)

The Liver

The liver is the largest gland in the body and has a variety of functions. It can best be compared with an extremely advanced industrial chemical plant. The portal vein carries the nutrient-rich blood from the intestine to the liver. The blood contains finished products as well as many semicomplete ones. Vitamins, iron, and hormones are among the substances deposited in the liver. Sugar from the intestine is stored as glycogen. When cells need more fuel, the liver converts glycogen to glucose, which is circulated in the blood to supply the necessary energy. The liver also stores proteins for later use, and it plays an important role in the metabolism of fat. The liver is the major detoxification center for poisonous wastes. It converts ammonia, a nitrogenous waste product of cells, into urea. It also breaks down such toxins as alcohol, drugs, and other substances that we voluntarily or involuntarily consume. A by-product of the many biochemical processes carried out by the liver is heat, so the liver is also important in temperature regulation. Another major function of the liver is the production of bile.

The liver is located below the diaphragm, toward the rear on the right side. Liver cells are arranged in millions of cylindrical columns with small veins down their centers. These veins drain into the hepatic vein, which carries blood toward the heart. It should be noted that the liver has a double blood supply: nutrient-rich blood comes from the intestine via the portal vein, and oxygenated blood from the aorta via the hepatic artery.

The wall of the gallbladder with cholesterol deposits ▷

Bile flows into the gallbladder, where it is stored and concentrated. The production of bile in the liver goes on continuously. It is the liver's contribution to the digestion of food, especially the breakdown of fat. Bile contains sodium bicarbonate, so it is alkaline, like pancreatic juice. It also contains bile pigments, which are derived from the breakdown of red blood cells, as well as bile salts and cholesterol. In the intestine, bile acts as an emulsifying agent on fat; that is, it breaks fat up into fine droplets which can be handled by fat-splitting enzymes.

Bile capillaries drain into larger ducts which transport bile to the gallbladder. The passage of food from the stomach to the duodenum is the signal for smooth muscles in the wall of the gallbladder to contract, and thus release bile.

The interior of the gallbladder with the bile duct in the center
The green color of the bladder wall comes from bile pigments; the yellow represents cholesterol deposits.

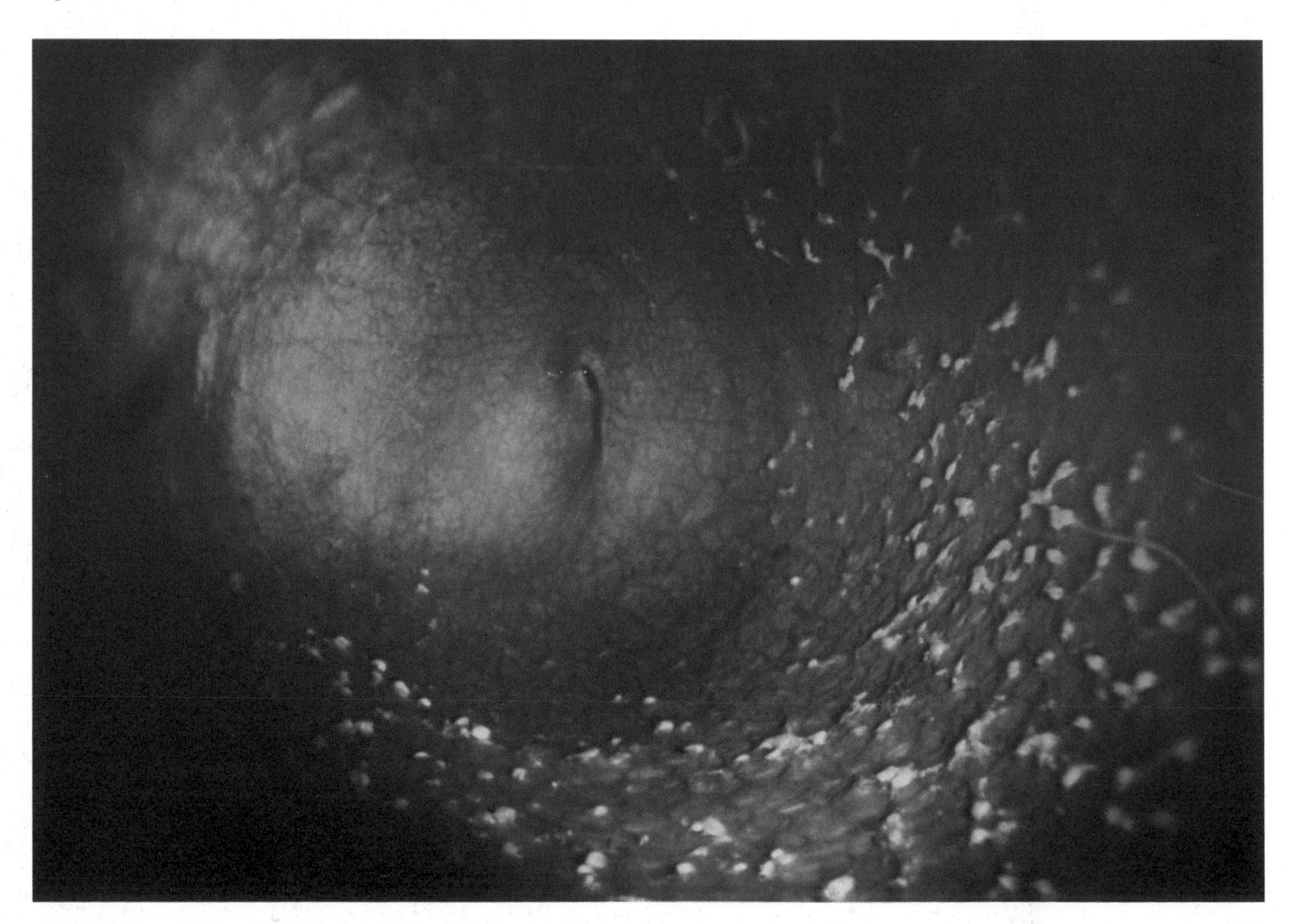

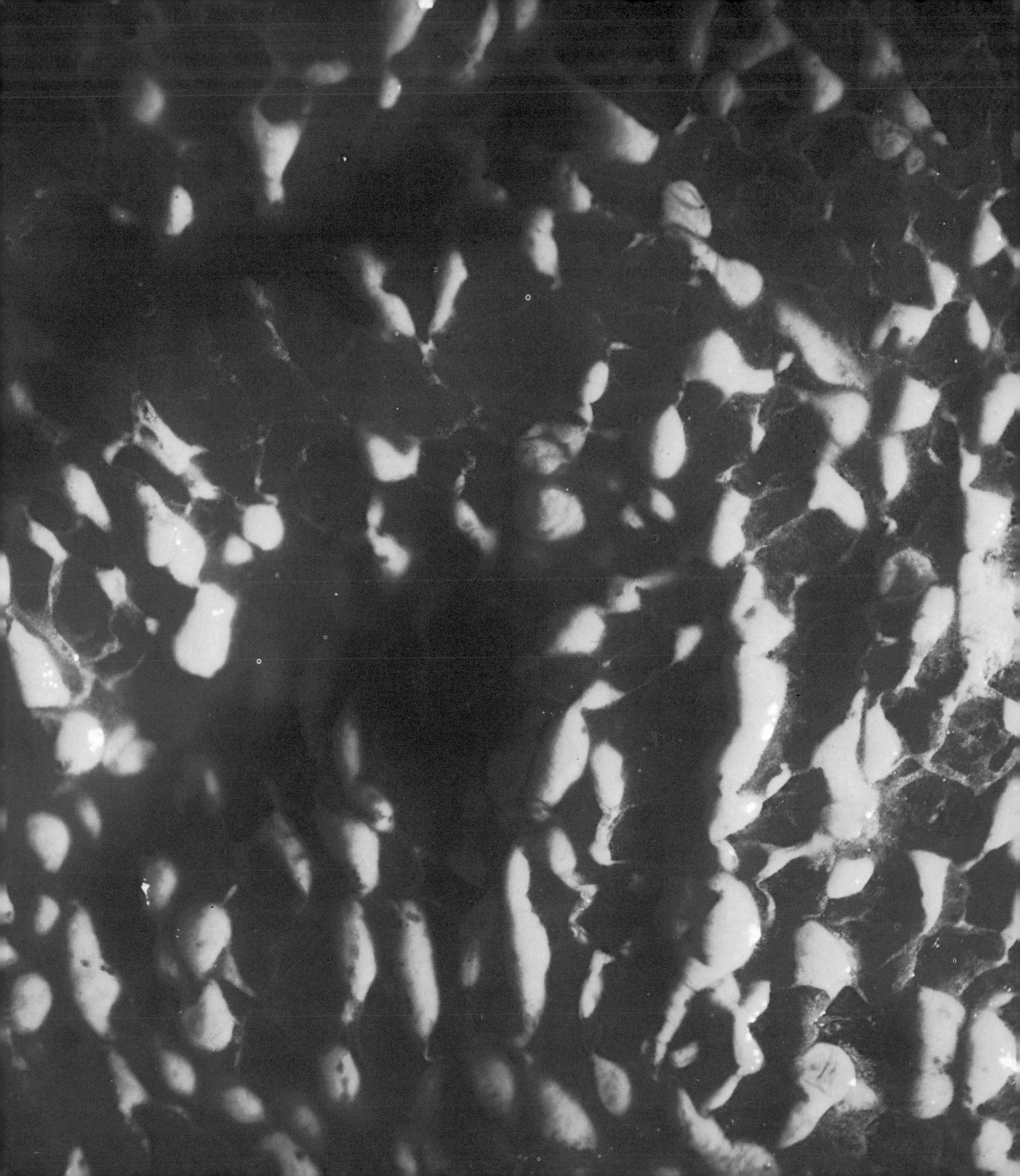

Small intestine suspended by the mesentery
Note the fine vascular network in the intestinal wall. Through these vessels oxygenated blood reaches the intestine, and nutrient-rich blood passes from the intestine into the portal vein to the liver, from which it will go to the heart and out into the body.

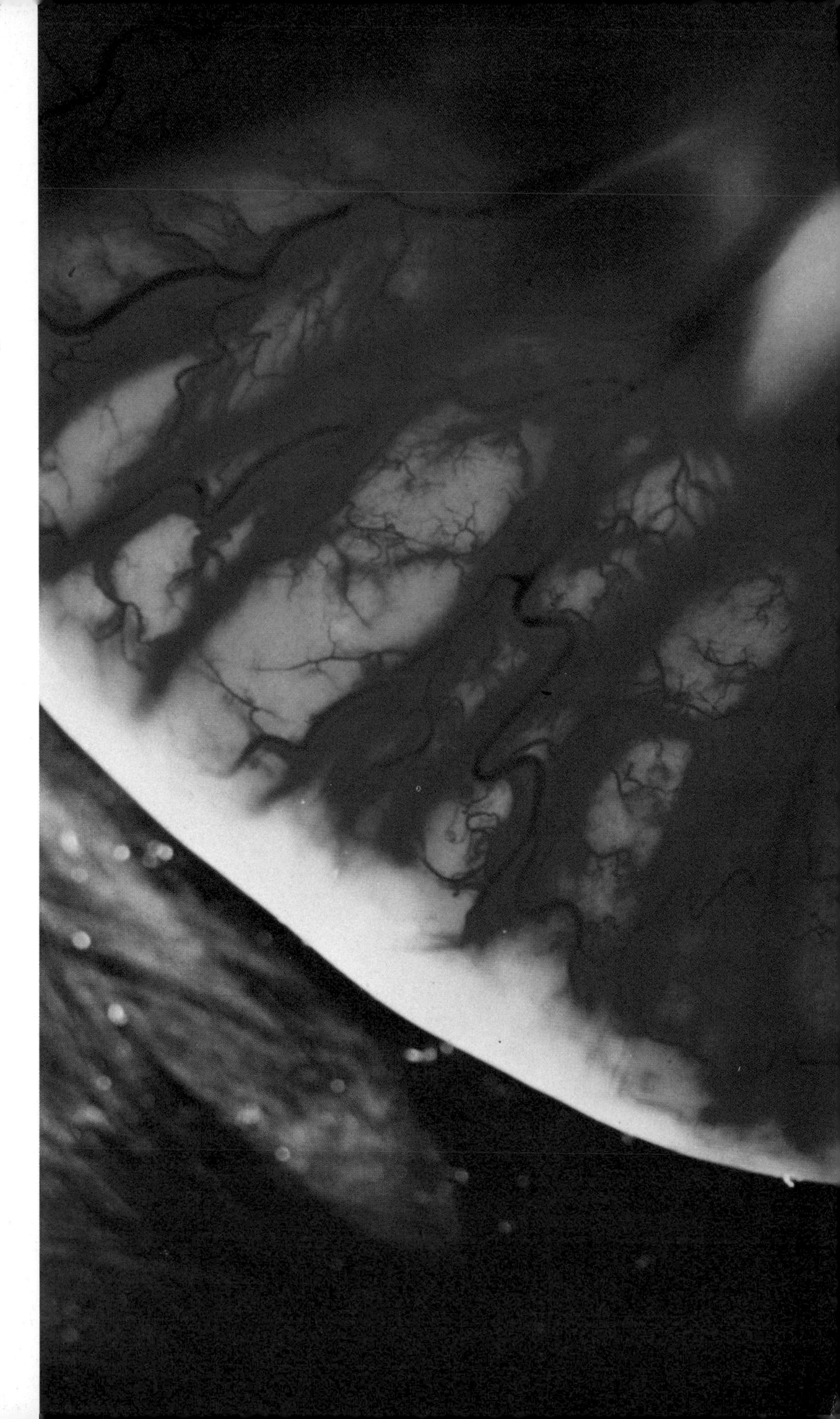

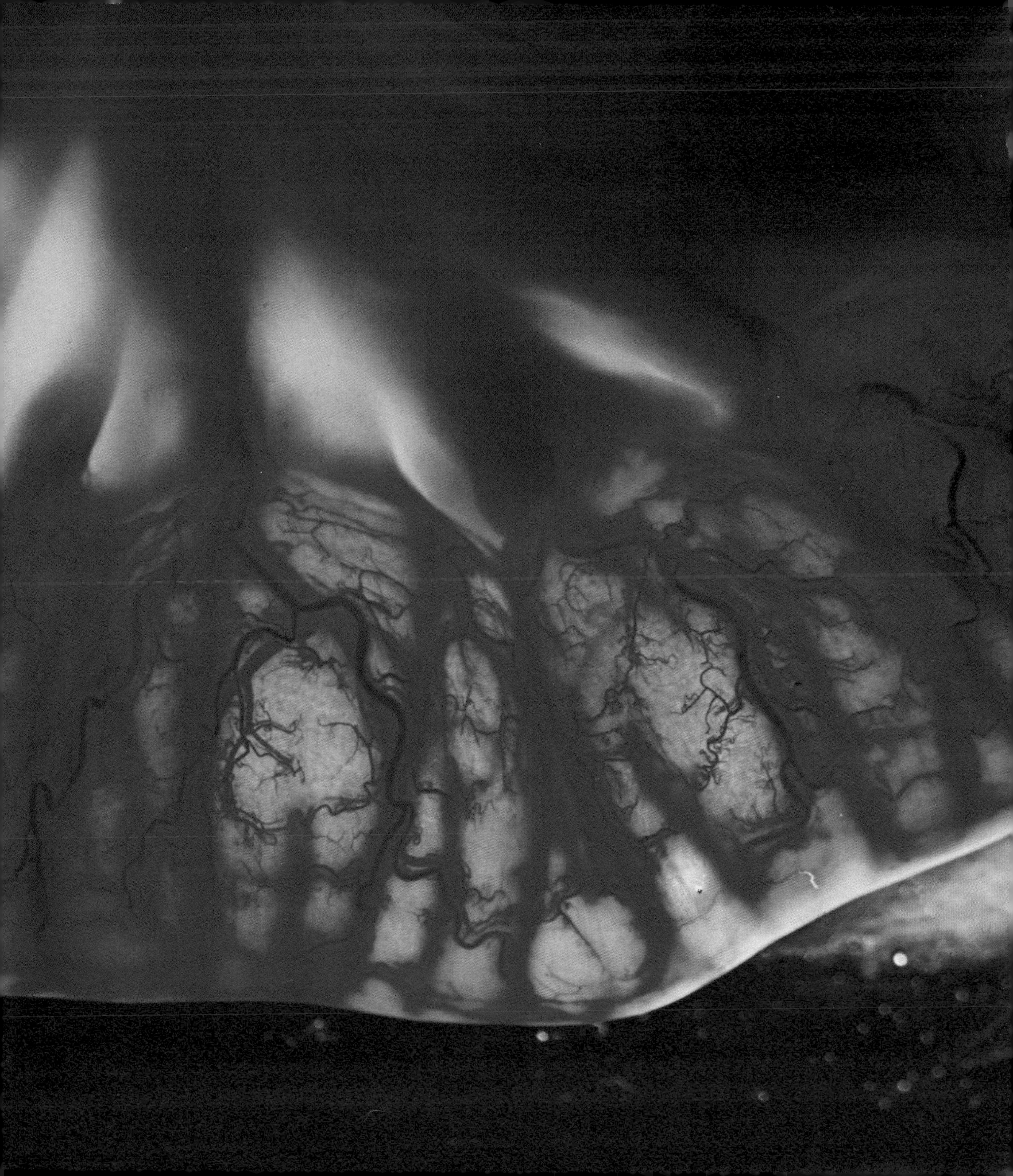

The Colon

The sixth and last station on the route of food is the colon, or large intestine. Its main function is the absorption of water from the chyme, which, as a result of the efficient operations of the small intestine, is now low in nutritious substances.

At the point where the small intestine enters the colon is the cecum, a small pouch, with a wormlike attachment, the appendix. The appendix is easily inflamed. The condition, appendicitis, can be highly dangerous, especially if the appendix bursts and infectious matter spreads to the abdomen. Appendectomy is the surgical removal of the appendix.

A valve regulates the passage of intestinal contents into the colon. At this stage the material is 80 percent water, swallowed as such or derived from saliva and bile, or gastric, pancreatic, and intestinal juices. Progress through the colon is slow, and as the water is absorbed the contents change into a semisolid mass. If the body were not able to reabsorb most of this water, we would have to drink far more often.

The colon contains a number of useful bacteria, essential in the production of important vitamins such as vitamin K and nicotinic acid. Small but metabolically vital quantities of certain salts, such as calcium, magnesium, and iron, are also absorbed in the colon.

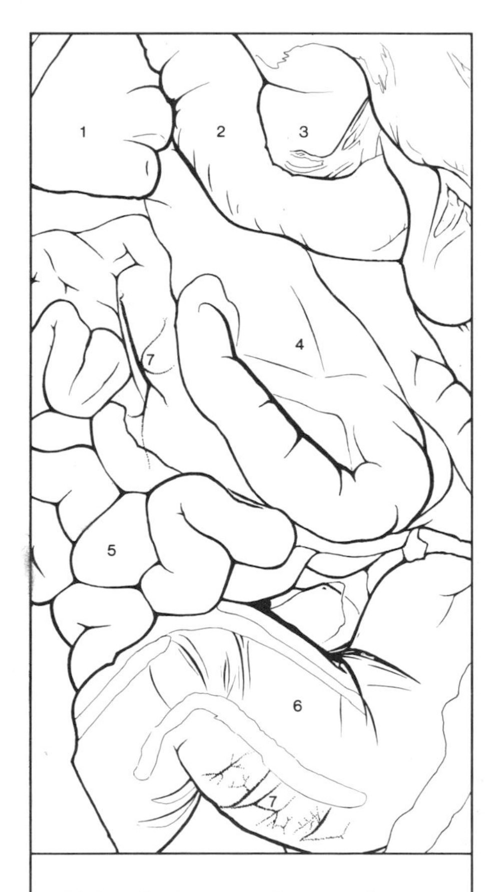

Abdominal organs in a newborn

1 liver
2 stomach
3 spleen
4 mesentery
5 small intestine
6 colon
7 blood vessels

Interior, convoluted wall of the colon ▷
The small intestine does not end abruptly at the colon, but extends partway into it. The first sacklike portion of the colon is called the blind gut.

The colon first ascends along the right side of the body, then runs transversely across the upper part of the abdomen, and finally descends along the left side, ending in the rectum. The colonic mucosa secretes mucus, but not enzymes.

◁

The principal abdominal organs in the newborn
The spleen is at the upper right and the liver at the left. The curved white organ at the center top is the stomach. Parts of the small intestine and the thicker colon are also seen.

The wall of the colon
seen in a scanning electron microscope. The smaller photo shows a number of openings in the mucosa, one of which is enlarged in the photo on the right. These are channels for mucus produced by cells lining the cavities.

The main function of the colon is to absorb water from the colonic contents, and to store them temporarily.

In the rectum, waste products from the digestive tract accumulate. Nearly everything of value in the food consumed has now been absorbed.

The emptying of a full rectum is voluntary. Defecation involves a series of peristaltic contractions in the colon and the simultaneous relaxation of the sphincter.

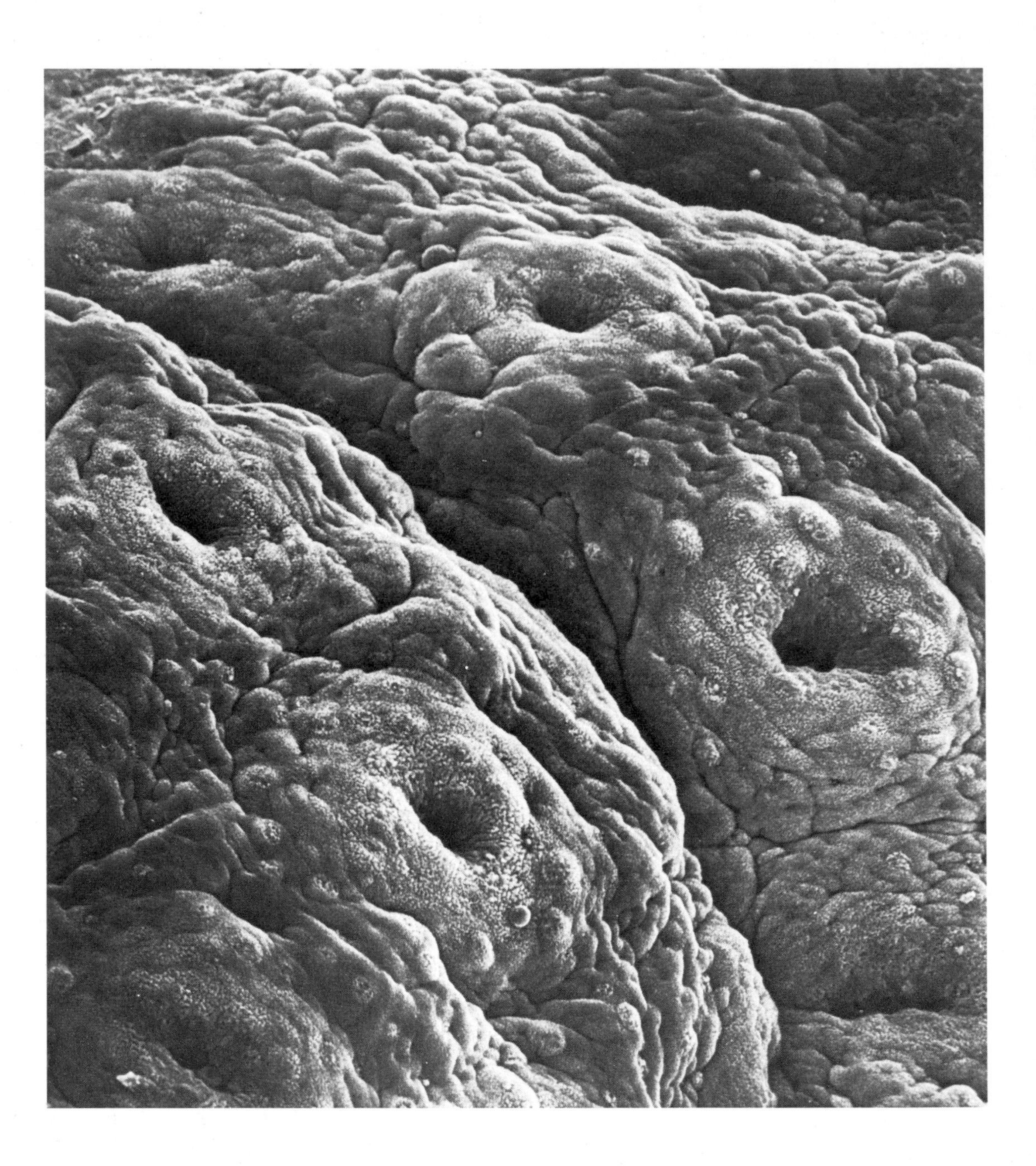

The Rectum

The colon is about a yard long and ends at the rectum. When the contents of the colon are slowly pressed into the rectum, they have a semisolid consistency. The brown color of the stool is due to bile pigments. The stool contains mainly dead mucosal cells, bacteria, and indigestible residual material. A sphincter encircles the rectum at the anus, to control the ultimate elimination from the body of the solids it cannot use.

Metabolism

The origin of the proteins, fats, carbohydrates, minerals, and vitamins required by the body is of no consequence to the digestive system. By the time food reaches the pharynx, the difference between filet mignon and sausages becomes unimportant. From now on it is only a question of nutritional value. The energy required to maintain the body at rest, not digesting food — the so-called basal metabolism requirement — is on the order

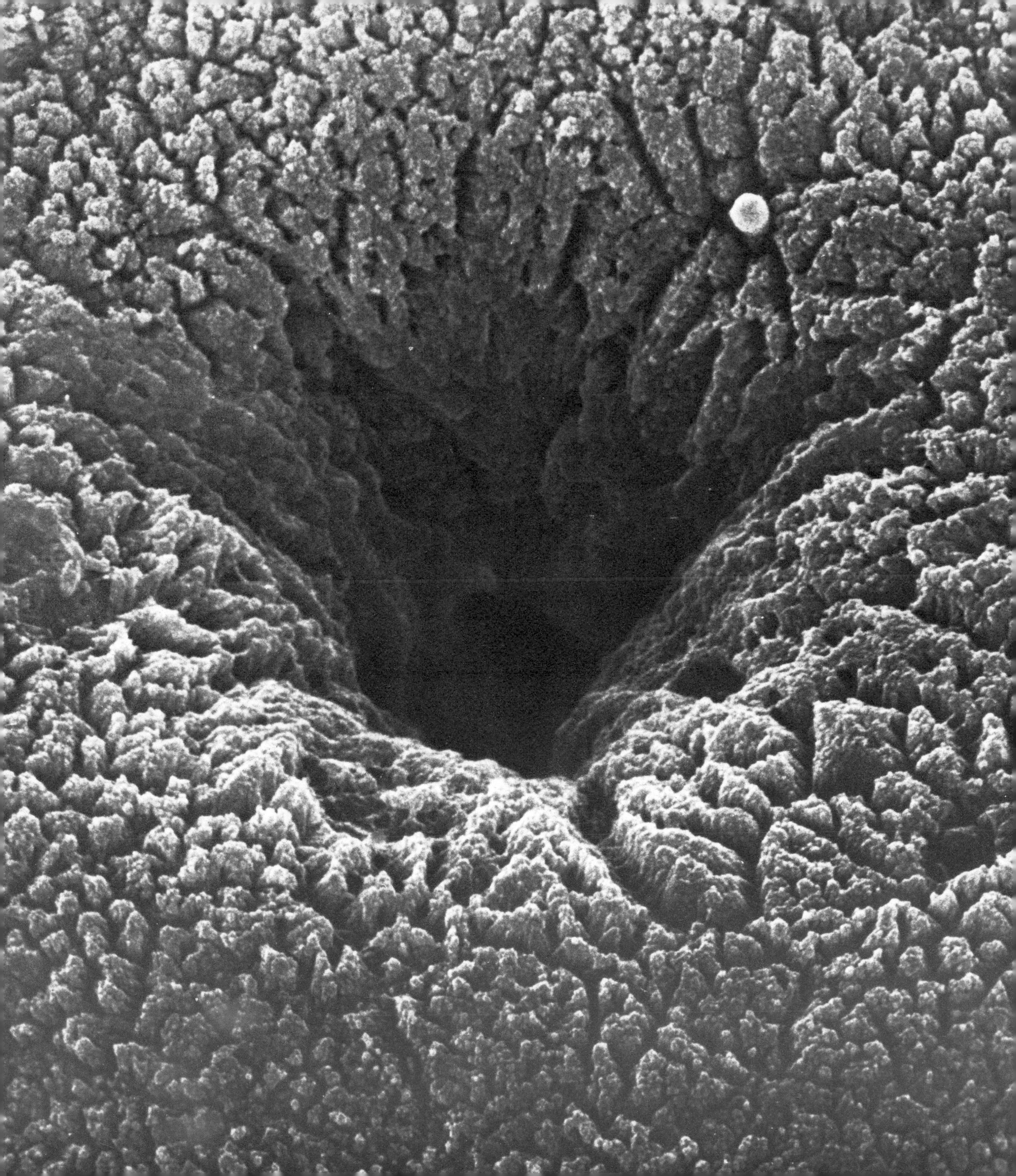

of 1,500 calories a day. This is the energy necessary to maintain circulation, body temperature, and other vital processes without performing external work. All forms of muscular activity naturally increase the energy requirement. A lumberjack working hard may use four times that amount of energy a day — 6,000 calories. From an energy point of view, sedentary work is obviously less demanding. A total of 2,000 to 2,500 calories may be sufficient, and could be supplied by a diet consisting of 100 grams of fat, 100 grams of protein, and 400 grams of carbohydrates a day. If we eat more than we need, the excess is stored; if less, we must draw upon our energy stores.

Cell metabolism involves a series of complex energy-consuming events. Energy not used directly is bound in phosphate compounds and stored in the cell. The most important of these compounds is adenosine triphosphate, ATP, which liberates energy on conversion to adenosine diphosphate, ADP. This process is reversible. Thus, with a high energy level in the cell, ATP can be resynthesized from ADP.

The citric acid cycle plays an important role in these pathways. It is a sequence of biochemical reactions which liberates energy. Citric acid is a chemical combination of acetic acid and oxalacetic acid which, by the action of enzymes, is converted into seven other compounds. At each step, energy is placed at the disposal of the cells, and the efficiency of the process is impressive. While two molecules of ATP are generated when glucose is broken down by glycolysis, the citric acid cycle generates thirty-six! The energy-producing processes run parallel to those energy-consuming processes used in building up the cell. Cells grow, reproduce, and function according to the code laid down by the chromosomes in the nucleus. The raw materials are supplied by the intestines with oxygen coming from the lungs. The vital processes of the body are regulated by a variety of hormones and enzymes which, in the intact organism, assure a high degree of coordination and cooperation among the different parts.

In addition to major dietary needs, food must contain small amounts or minute traces of other substances. These include vitamins and elements such as cobalt, copper, and molybdenum.

Vitamins are components of enzymes and thus have an indirect influence on metabolism as a whole. Vitamin deficiencies cause pathological changes. A few facts about the most important vitamins are of general interest.

- Vitamin A is necessary for the formation of skin and mucosa, and is a component of rhodopsin, the light-sensitive pigment in the visual receptor cells. It is found mainly in milk, cheese, butter, egg yolk, and liver. Carrots contain carotene, which is converted to vitamin A in the intestine. Symptoms of vitamin A deficiency are poor vision at night and reduced resistance to infections.
- The B vitamins are a group of vitamins important in carbohydrate and protein metabolism, as well as in the formation of blood. B vitamins are found in beans, yeast, meat, liver, and vegetables. A deficiency of vitamin B_2 may cause sores in the corners of the mouth and gastric and intestinal disturbances; a deficiency of vitamin B_{12} may lead to pernicious anemia and give rise to metabolic disturbances based on the body's inability to utilize fats in food.
- Vitamin C is assumed to regulate the exchange of oxygen in cells and is of importance to connective tissue, bones, teeth, and blood vessels. It is found in citrus fruits, potatoes, parsley, black currants, tomatoes, and cabbage. Deficiency symptoms are easy bruising of the skin and bleeding from mucosae and interior organs, as well as reduced resistance to infection.
- Vitamin D is important in the regulation of calcium and phosphate balances in the body. It is formed in the skin by the action of the sun but is also found in milk, egg yolk, and liver. One symptom of deficiency is the bone disease rickets.
- Vitamin K is necessary in the normal process of blood clotting. It is formed by the action of bacteria in the colon, but it is also present in cabbage and spinach.

The anus in a newborn
The digestive tract begins and ends with a sphincter.

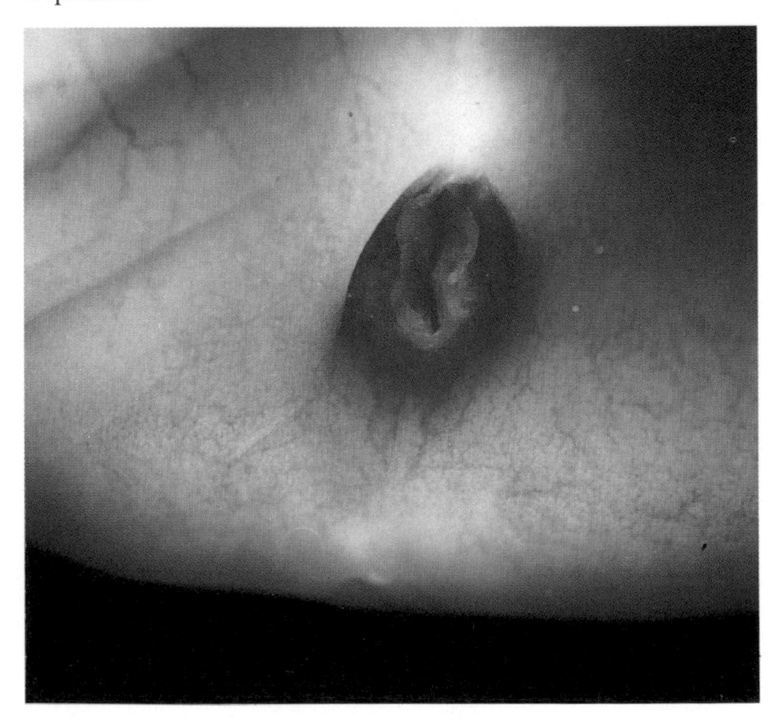

The Nervous System

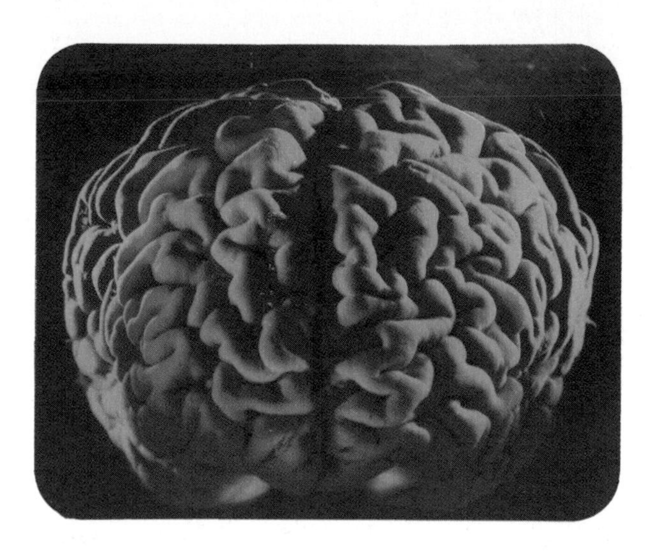

An exploration of man should properly begin with an exploration of the nervous system, for that great mass of cells and fibers contains the way stations and pathways which determine what is uniquely human in our nature. Our capacity to reason abstractly — to use symbols — distinguishes man from all other species. This nervous-system talent forms the basis for human speech and writing and for the accumulation of techniques and inventions which have flowered into the arts and sciences of human culture. It is through our brain that we perceive the world, and, willy-nilly, it is through our brain that we can and do change that world in ways not possible for other species.

The nervous system can be regarded as a complex computer. Its essential components are the nerve cells, or neurons. Slender extensions (processes) from some cells carry information from the sense organs in toward the central parts. This information is processed en route and may give rise to other impulses conveying orders to muscles and glands. In this way nerve impulses become translated into behavior — either external and visible or internal and hidden from view.

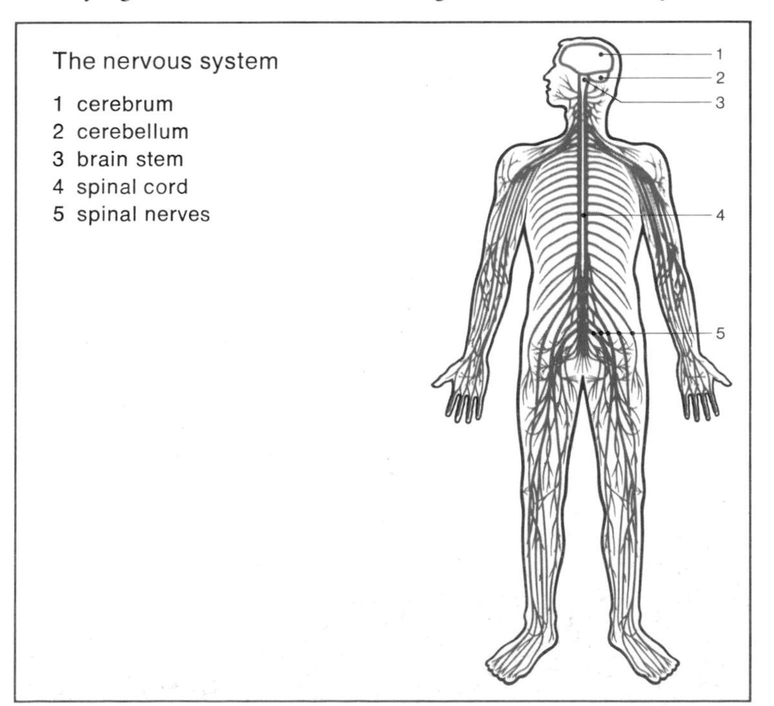

An enormous number of neurons participate in this process. The figure generally cited is ten billion, but that is only an estimate. The number of functioning neurons increases during development and may peak at maturity to ten times ten billion. But neurons themselves grow old and die, and when they do they are not replaced. So in the course of human aging the population of nerve cells irrevocably declines.

The nervous system is anatomically divided into central and peripheral parts. The peripheral nervous system is composed of spinal and cranial nerves distributed throughout the body and the head. Generally the peripheral nerves contain both sensory (afferent) fibers, coming into the central nervous system from peripheral sense organs, and motor (efferent) fibers, going out from the central nervous system to muscles or glands. For this reason the peripheral nerves are called "mixed" nerve trunks or simply "mixed nerves." Many also contain fibers leading to and from the visceral organs or glands which function in digestion, respiration, and other automatic or involuntary processes. These nerve fibers and their cell bodies constitute the autonomic nervous system, which has parts in both the central and peripheral divisions.

The central nervous system is composed of three major interconnected elements: the cerebrum, the brain stem (including the cerebellum), and the spinal cord.

The translation of sensory messages takes place at many levels in the nervous system. Relatively simple chains of neurons within the spinal cord control a number of involuntary acts called reflexes. When your knee jerks up in response to a physician's tapping the patellar tendon with a hammer, or when you quickly withdraw your finger from a pinprick, you are experiencing a sequence of neural events — a reflex — which was programmed into the nervous system at birth. Such protective responses occur when a sensory ending — a touch fiber in the fingertip or a "stretch" receptor in the knee joint — reacts to stimulation and fires impulses in toward the spinal cord. There the impulses are relayed to other cells and give rise to the appropriate muscle contractions. The motor signals activated

The spinal cord
is lodged in the spinal column and fills the tunnel formed by the arches of the vertebrae. The white bands flanking the cord at each segment are nerve roots. The motor fibers which innervate muscles exit from the ventral side of the cord. Nerve fibers from sense organs in the skin muscles enter the dorsal portion of the cord.

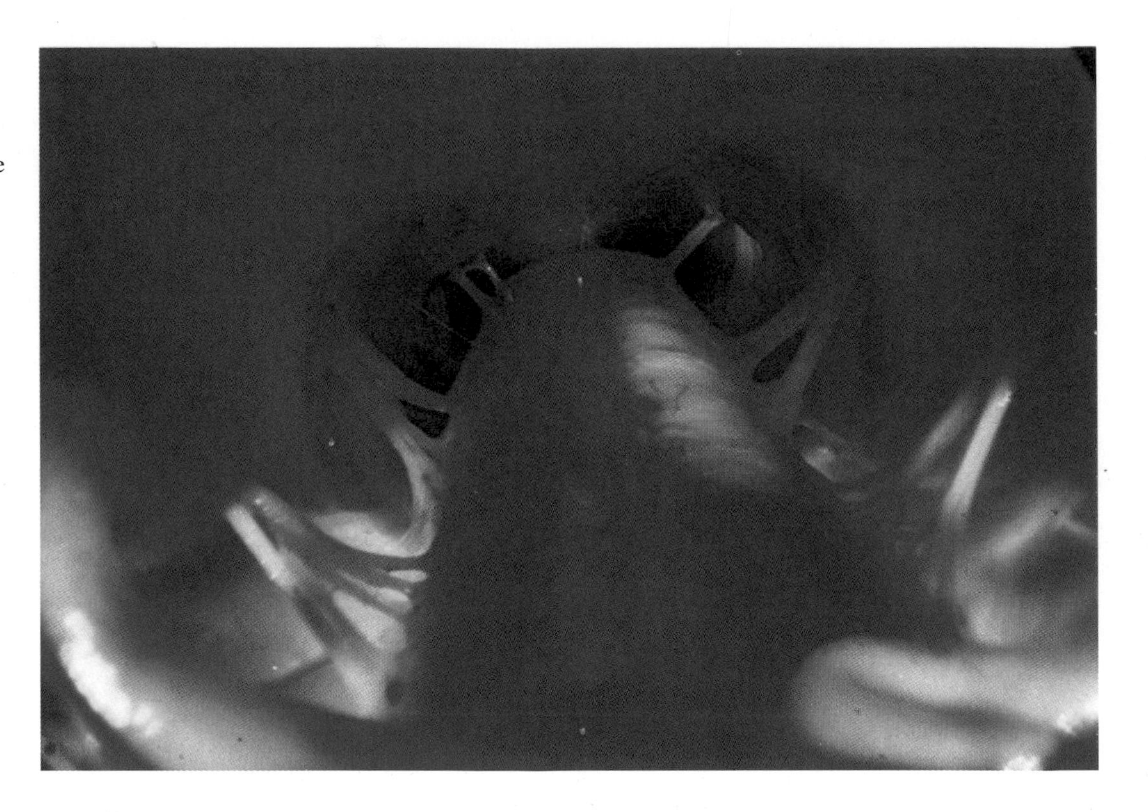

in spinal reflexes are among the fastest acting in the nervous system. Their speed of conduction down the nerve fiber may reach velocities of 100 meters a second (about 200 miles an hour).

The spinal nerves emerge in pairs from either side of the spinal cord. The motor fibers destined for skeletal muscles arise from motor "roots" coming off the front, or ventral, parts of the cord. Fibers conveying sensory information from the skin or other parts of the body are grouped in roots which enter the back, or dorsal, parts of the cord. Each composite (mixed) nerve contains individual fibers of varying thickness. The thinnest have diameters of only some thousandths of a millimeter, while the thickest, encased in sheaths of a fatty substance called myelin, may be ten times that size. Experiments have shown that there is a direct relationship between diameter and conduction velocity. The thickest nerve fibers conduct at speeds as high as 150 meters a second, while the thinnest meander along at about 1 meter a second.

The spinal cord in the adult is approximately eighteen inches (45 cm) long. It extends from the large opening at the base of the skull, the foramen magnum, to the second lumbar vertebra, falling short of the length of the spinal column. The spinal nerves are named and numbered according to the vertebral divisions of the spinal column into cervical, thoracic, lumbar, and sacral segments. The cervical nerves control the neck and arms. The thoracic portion innervates the chest. The lumbar nerves are distributed to the lower extremities, the legs and feet, and the sacral nerves mainly supply the organs of the pelvis and the buttock muscles.

In cross section the spinal cord appears as a slightly flattened tube with a small hollow portion — the central canal — running down the middle. This shape reflects the embryological origin of the nervous system in the form of the neural tube. Surrounding the central canal is an H- or butterfly-shaped area of gray matter consisting of nerve cell bodies. Generally, sensory neurons are grouped in dorsal parts of the H; motor cells, in the ventral regions. The central gray area is in turn surrounded by white matter. The color reflects the myelin sheaths covering nerve fibers, for this is the area representing pathways of fibers coursing their way up and down the cord.

Immediately above the cord is the brain stem, the "shaft" of the brain. The nerve cells and pathways are more complex at this level, both in structure and function, and represent a more advanced stage of development in evolution. At this level, sensory information from all parts of the body and most of the sense organs in the head reaches integrating centers which may trigger reflexes or more complex sensations or reactions. Among the automatic responses controlled at this level are ones involved in the regulation of breathing, digestion, heart action, or blood pressure. A respiratory center in the lower brain stem,

The Spinal Cord

The spinal cord (1)
viewed from the rear. The roots entering the cord here consist mostly of afferent fibers from sense organs in the skin and muscles.

Cauda equina (2),
or horse's tail, is the term applied to the collection of mixed spinal nerves that emerge at the lower end of the cord. They contain efferent as well as afferent fibers, which are distributed to the lower extremities, the bladder, buttocks, et cetera.

The spinal nerves (3)
consist of hundreds of thousands of individual nerve fibers. These include motor fibers from motor nerve cells in the cord destined for the skeletal musculature, as well as sensory nerve fibers from peripheral sense organs. Note the small blood vessels in the thin layer of connective tissue on the surface of the nerves.

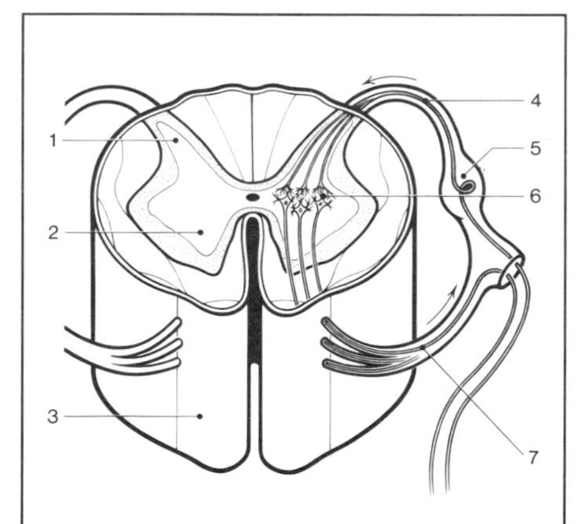

Cross section of the spinal cord

1 posterior horn
2 anterior horn
3 spinal meninges
4 afferent nerve (dorsal root)
5 ganglion
6 synapse between nerve cells in gray matter
7 efferent nerve (ventral root)

1

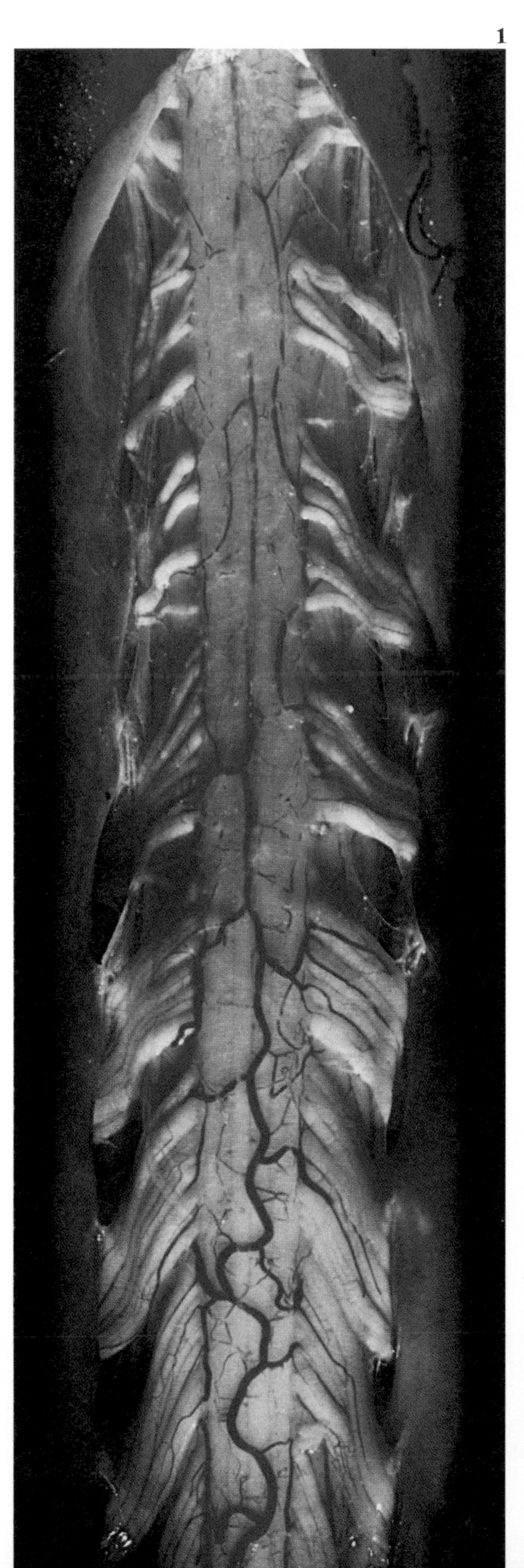

2

3

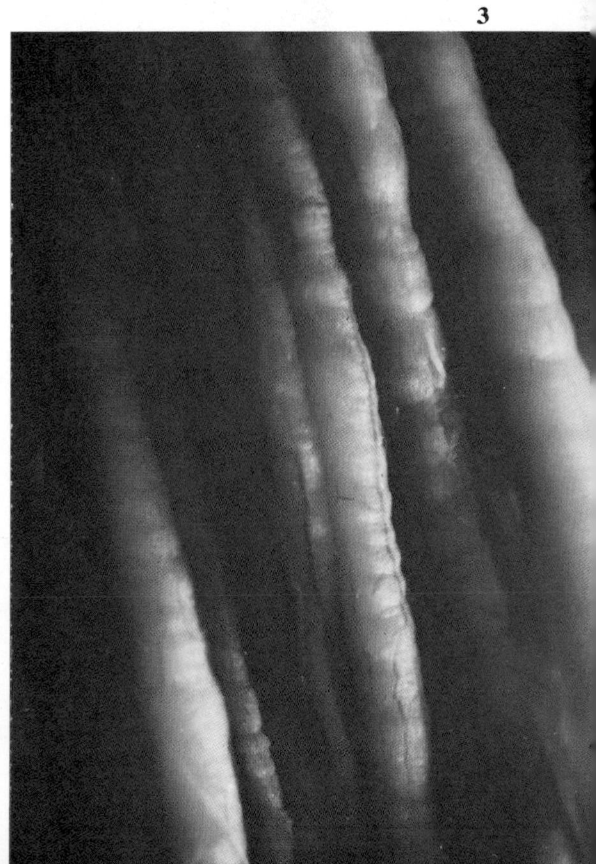

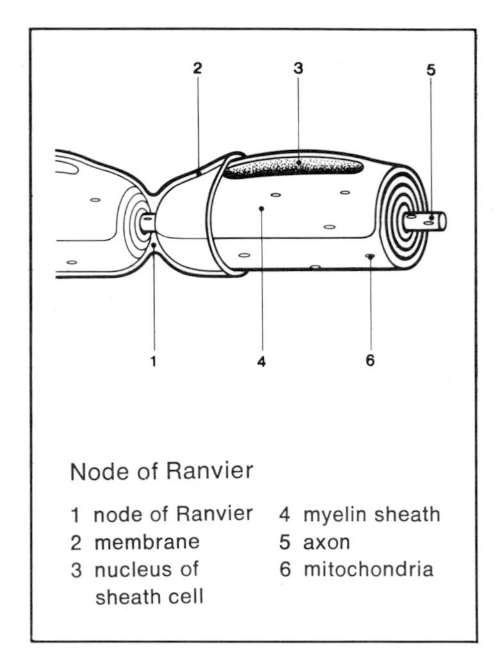

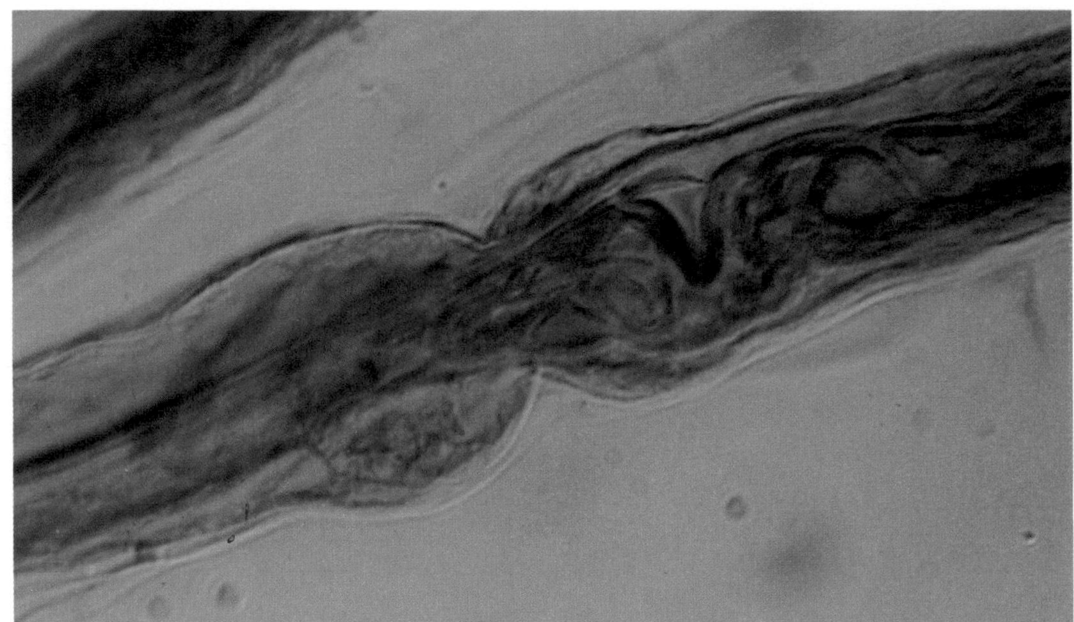

Myelinated nerve fibers are interrupted periodically at the nodes of Ranvier

for example, can alter the depth or frequency of breathing in response to chemical receptors sensitive to the carbon dioxide level in the blood. Some of the nerve cells are themselves sensitive to the carbon dioxide content in the blood, but the respiratory center also responds to sense organs located in the walls of arteries in the neck. When the CO_2 level increases, respiratory volume increases and carbon dioxide is exhaled.

The centers for control of respiration, blood pressure, heart action, digestion, elimination, and other vital processes are generally located in lower parts of the brain stem. Interacting with them but involved in more complex responses to changes in the environment are centers in the higher parts of the brain stem. Here are a variety of cell centers which trigger protective responses to rising or falling body temperature, or mediate feelings of hunger, thirst, sexual arousal, aggression, or other states of emotional or motivated behavior. In this regard, a small body of gray matter, the hypothalamus, and the surrounding gray areas are of particular importance.

Above the brain stem bulges the pair of cerebral hemispheres of the cerebrum, the most complex part of the nervous system and the highest integrating center. Impulses at this level are thought to mediate the higher mental processes of thinking and memory, perception, skilled behavior, and consciousness. It is as though the cerebrum contained parts equipped to analyze and act on information arriving at other parts. A good deal is known about certain of these areas. For example, sensory information from all parts of the body ultimately is projected to a thin strip of cortex (the outer layer of gray matter covering the hemispheres) behind one of the principal furrows of the brain. Just to the rear of this central sulcus, or groove in either hemisphere, is a map of the body — a rather distorted map, because more space is devoted to those parts rich in sensory endings. Thus, the face and the hands take up a much larger part of the map than do the areas representing the legs and trunk. On the anterior side of the central sulcus lies a similarly distorted map corresponding to the distribution of motor nerves to skeletal muscles. Here the areas devoted to the thumbs and fingers or to the mouth and face are huge in comparison to the trunk or lower limbs. Thus the amount of brain space taken up by sensory or motor projections of the body surface is a direct reflection of our capacities. We can make the most subtle distinctions in touch with our fingertips, and we use our thumbs and index fingers to perform the most delicate movements of all. In contrast, our abilities to discriminate sensations on the back or to make subtle movements with our toes are considerably less.

In addition to these general sensory and motor projection areas, the cerebral cortex also contains a visual map receiving impulses from the eyes. The visual cortex lies in the occipital lobes at the rear of the brain. Again, the area where vision is most acute, a small central part in each retina, takes up as much room in the projection area as does the entire periphery

◁ **A spinal nerve**
is composed of many nerve fibers of varying thickness. The thickest ones have an insulating sheath consisting of a fatty compound (myelin). The sheath is interrupted at intervals, called nodes of Ranvier. The nerve impulse moving down a myelinated fiber is able to jump from node to node and so can be conducted at high speed. The velocity of impulses in myelinated nerve fibers is up to 100 times higher than in unmyelinated fibers.

Cross section of a nerve ▷
The small dark rings are myelinated nerve fibers. They are grouped in bundles of varying size and encased in connective tissue. The nerve itself consists of bundles of these bundles, also wrapped in connective tissue.

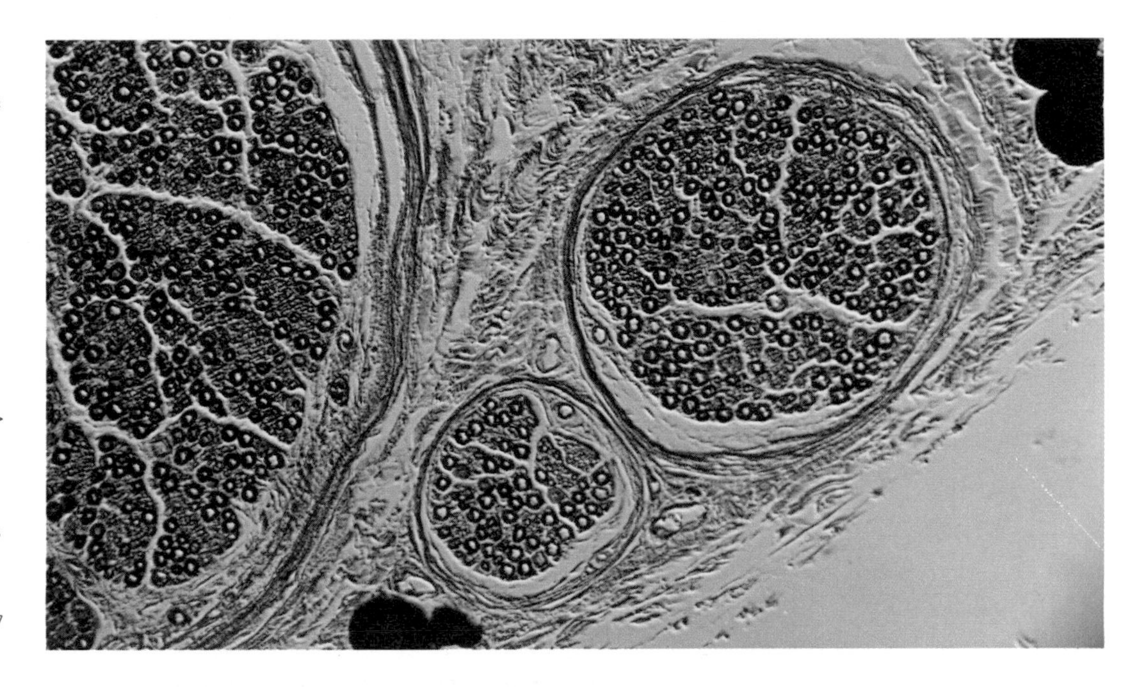

Motor nerves ▽
leading to skeletal muscle. Their terminal portions are richly branched.

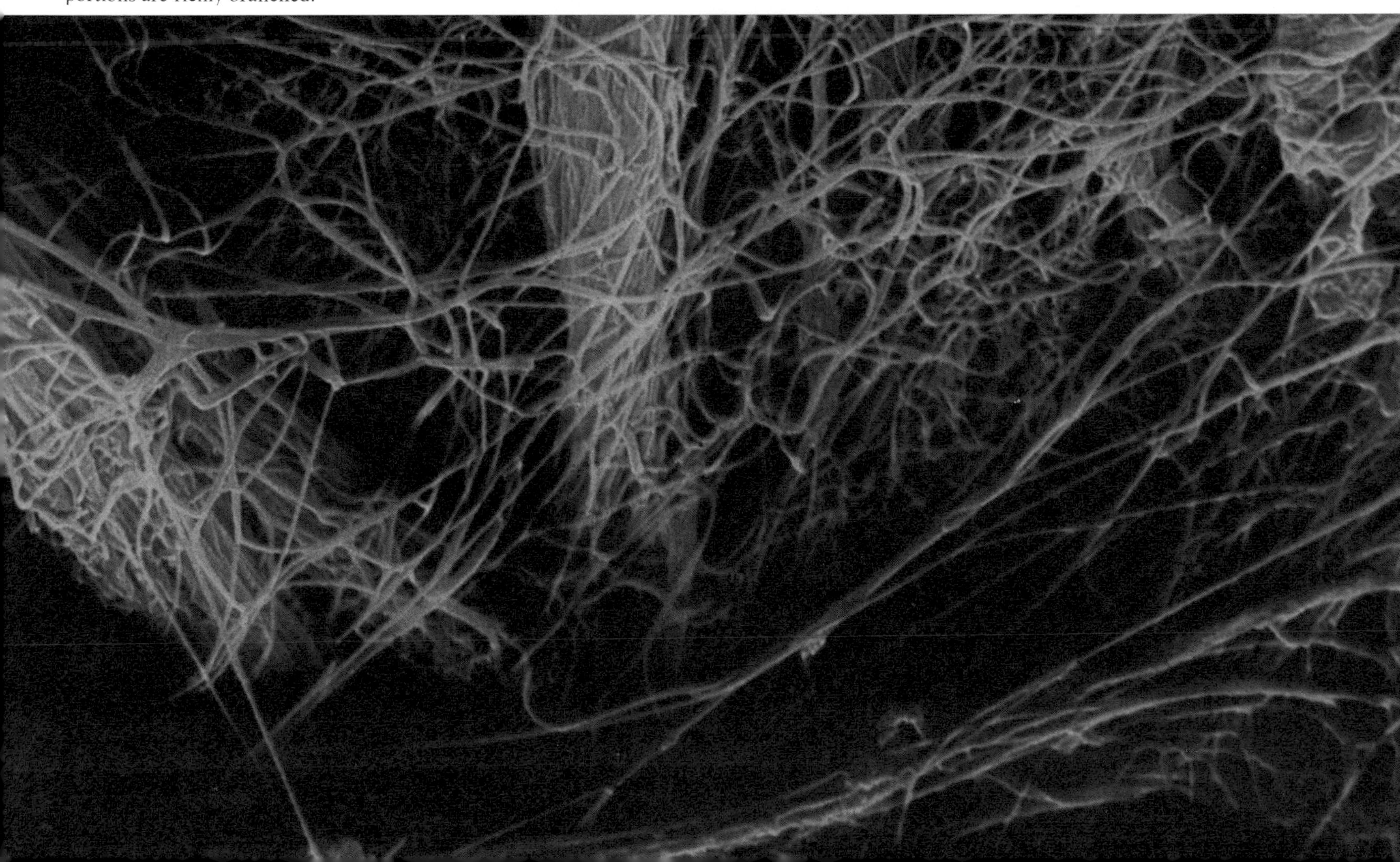

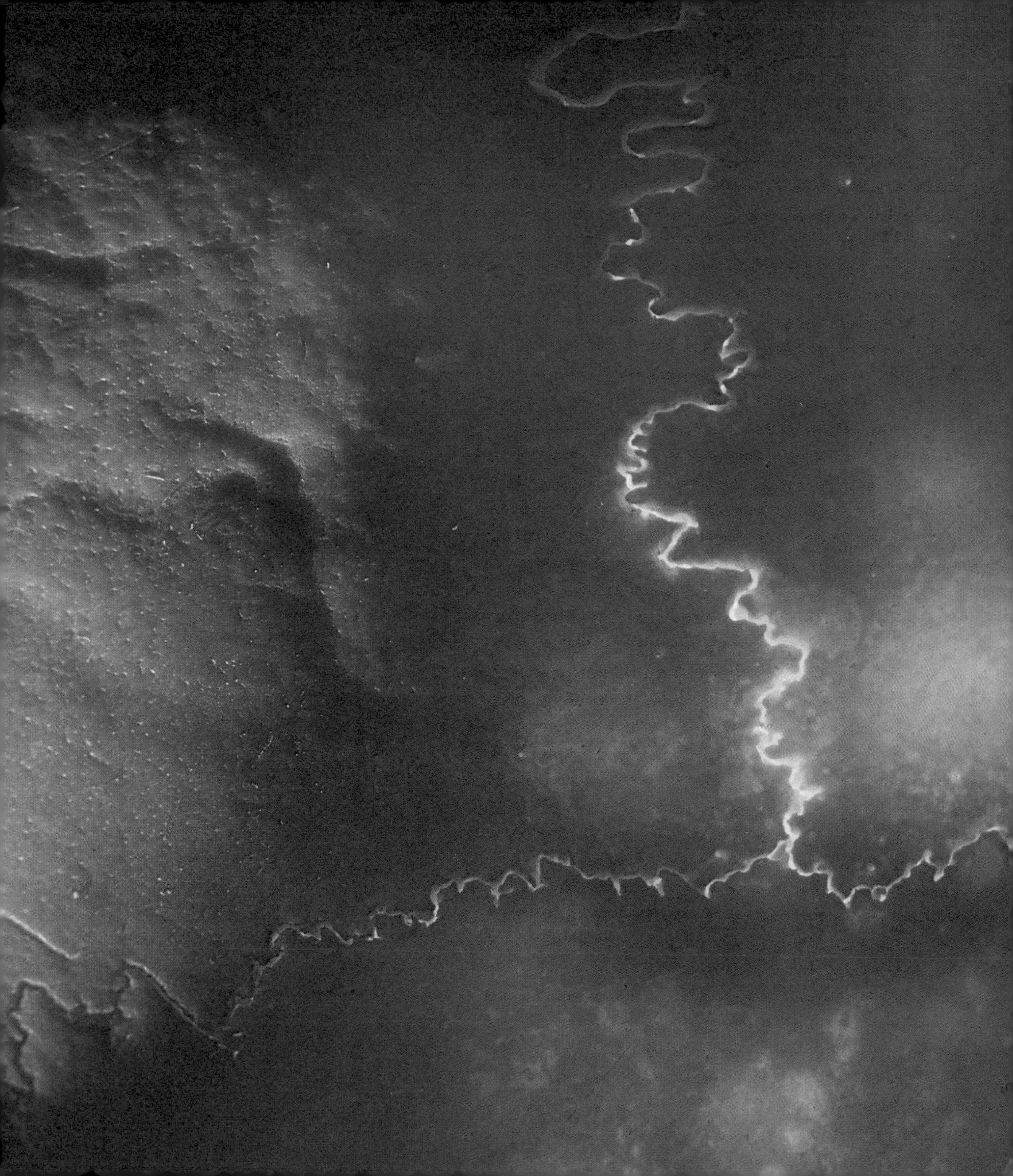

◁ **The lines of fusion (sutures) of the cranial bones**
These bones fuse during early childhood. This is how the sutures look when a light is shone through the skull. The irregularity of the sutures increases their strength. The vertical suture unites the parietal bones on each side of the skull and extends to the occipital bone at the rear. The transverse suture joins the frontal bone with the two parietal bones.

The base of the cranium ▷
This "face" appears when light is shone through the base of the skull. The "eyes" are openings for vessels and nerves. In the middle of the bridge of the "nose" is the sella turcica, a saddle-shaped depression where the pituitary gland is located. Above it are two oblong openings (small black hollows in the center) through which the optic nerves pass to either eye. The "nose" is the foramen magnum, a large oval opening in the occipital bone at the base of the skull through which the brain and spinal cord are connected. A small bony process projects into the center of the foramen magnum. It is a part of the second cervical vertebra which acts as a pivot, permitting head movements from right to left. Fine grooves for blood vessels are visible on the sides of the skull.

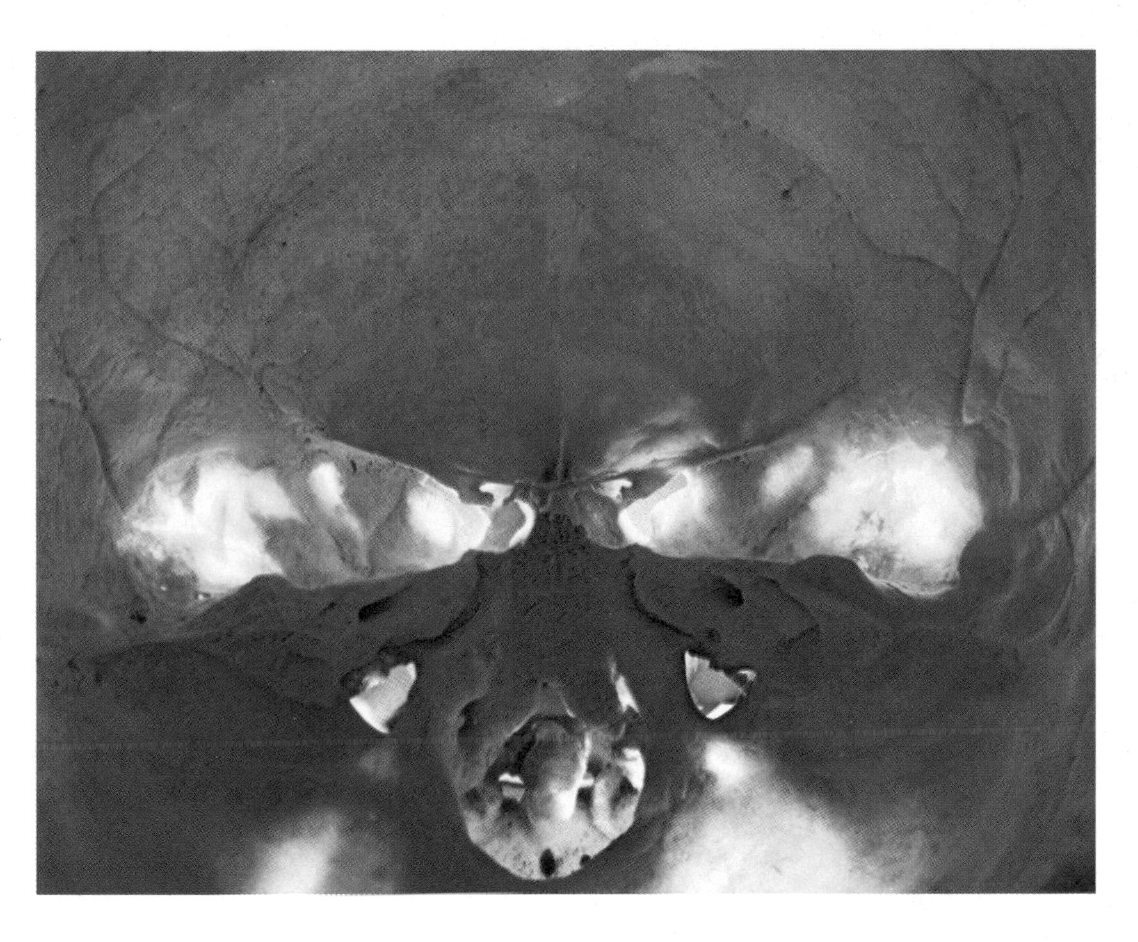

of the visual field. Thus, objects we fixate on sharply are usually seen with a higher degree of sharpness than objects in the background (assuming good lighting conditions). Hearing too has its map in the cortex, which registers impulses from the ears for identification and recall.

All these maps are present in duplicate, one in each cerebral hemisphere. The arrangement is such that the right half of the body is mapped on the left cerebral hemisphere and vice versa. The left portion of the visual field is projected to the right occipital lobe and vice versa. However, the maps on both sides are connected, impulses are sometimes conveyed to both hemispheres and continuous comparisons are possible.

Because of the position of our eyes, the retinas reflect two slightly different images of an object in the visual field — one seen somewhat from the right, the other somewhat from the left. But we don't see double. Normally, the two images are fused into a single one, much as the two slides in a stereoscope form one three-dimensional view. Stereoscopic vision is a major reason we are able to perceive depth, a phenomenon that is still not fully understood.

Interestingly, the maps concerned with the external environment — with touch, vision, hearing, and other senses — occupy only a small portion of the cerebral cortex. The rest is made up of "association" areas, assumed to be responsible for further processing or integrating of the input. It is here, presumably, that audible sensations can be combined with sights or smells, or we can compare something we see with how it feels. It is also here that present sensations are compared with past experiences, and decisions are made as to whether the present situation requires some action or can be ignored. Some association areas of the brain may also be involved in the storage or recall of complex memories, a phenomenon of much interest today.

Higher forms of learning also occur at the cerebral level, but simple kinds of learning are possible at lower levels in the nervous system.

Nervous tissue, like all other living tissue in the body, is dependent on a blood-borne supply of nourishment. Biochemical mechanisms in nerve cells require a source of energy in order to maintain a difference in potential across the cell membrane and the intercellular fluid. One way in which the difference is maintained is by the active transport of sodium ions out of

The formation of the nervous system
The central nervous system develops in the embryo from the ectoderm, the outer germ layer. A portion of ectoderm at the back of the embryo folds in to form a tube. The hollow center of the tube is preserved in the adult as the central canal of the spinal cord. The head portion of the canal widens to form the cavities of the brain, the ventricles. The photos show the stage of development in a four-week-old embryo. Note the segmented arrangement of tissue around the neural tube at the back of the embryo (the series of scallops) in the photo on the left. The two small finlike protuberances are the primordial arms. The photo on the right shows the head of the embryo with the parts of the developing brain — the brain "vesicles" — and even a primordial eye. The umbilical cord extends from the abdominal region, just below the heart, which appears as a red ball. The bottom photo is a front view of the embryo, with the brain vesicles below and the infolded neural tube at the top.

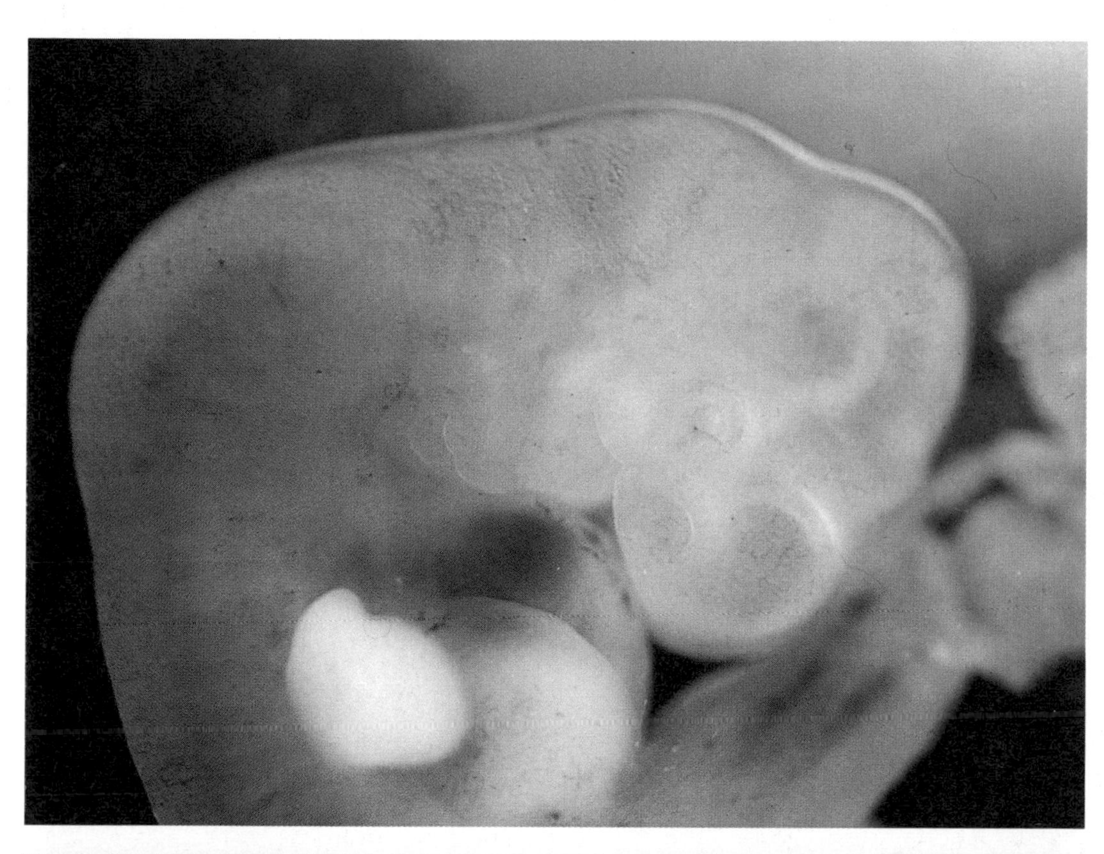

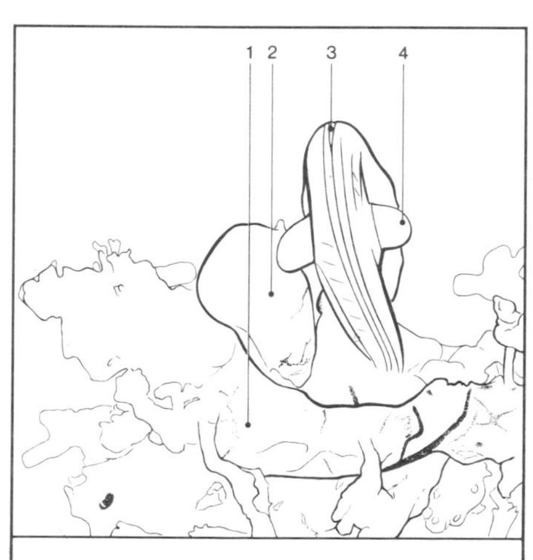

The nervous system
in a four-week-old embryo

1 placenta
2 yolk sac
3 neural groove and neural tube
4 primordial arm

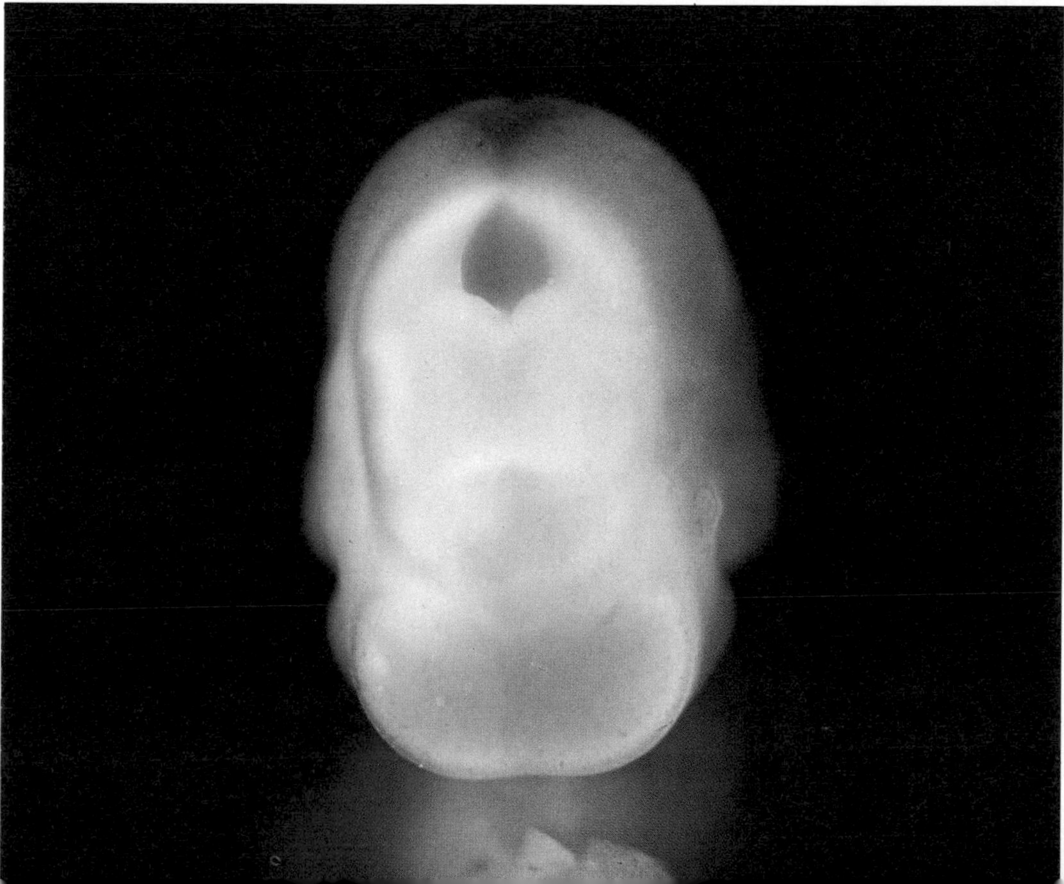

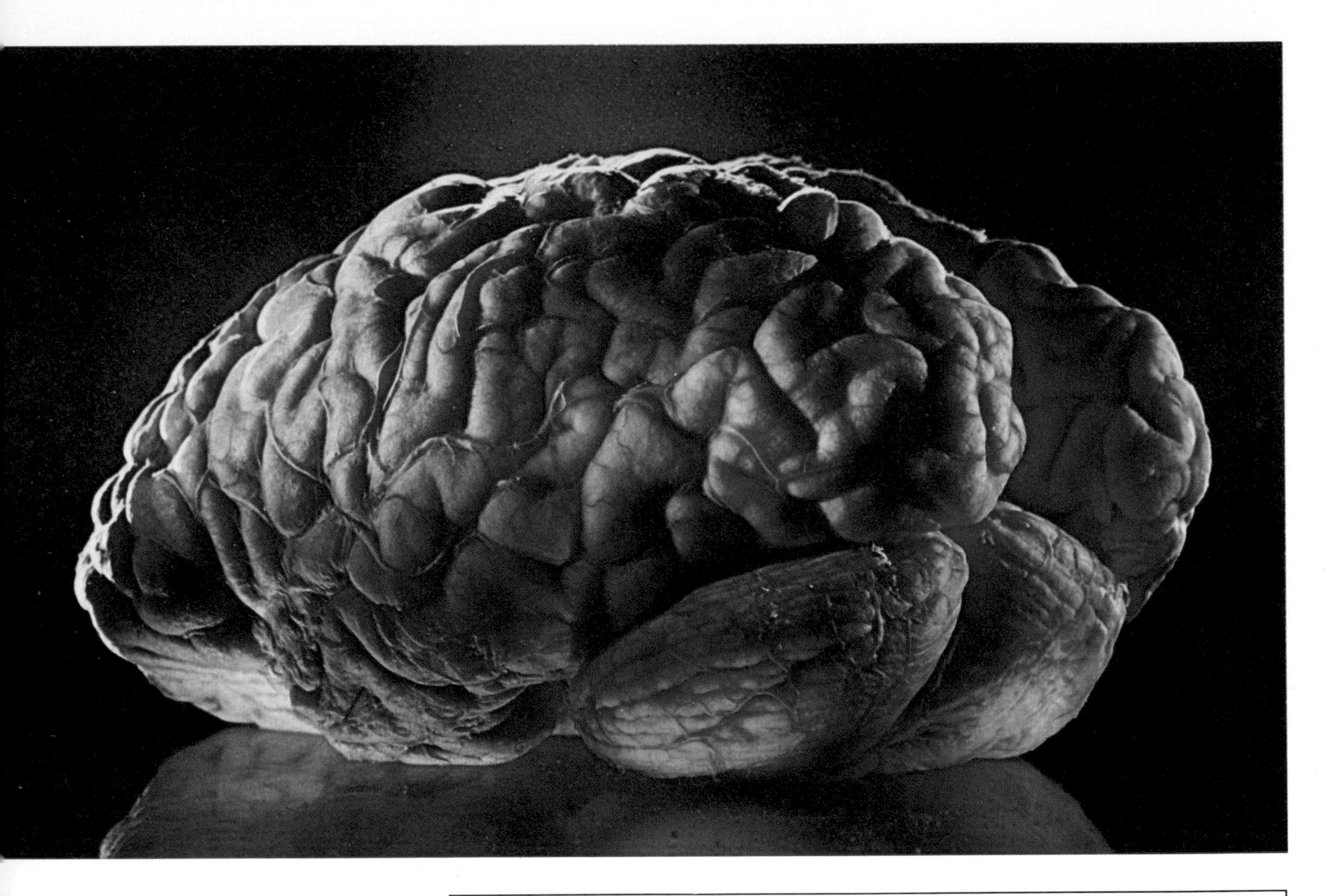

The human brain
seen from the left. At the front of the brain (far left) is the frontal lobe; at bottom, the temporal lobe; at the top, the parietal lobe. At the back (far right) the occipital lobes of both hemispheres can be seen. The cerebellum is visible below the occipital lobes.

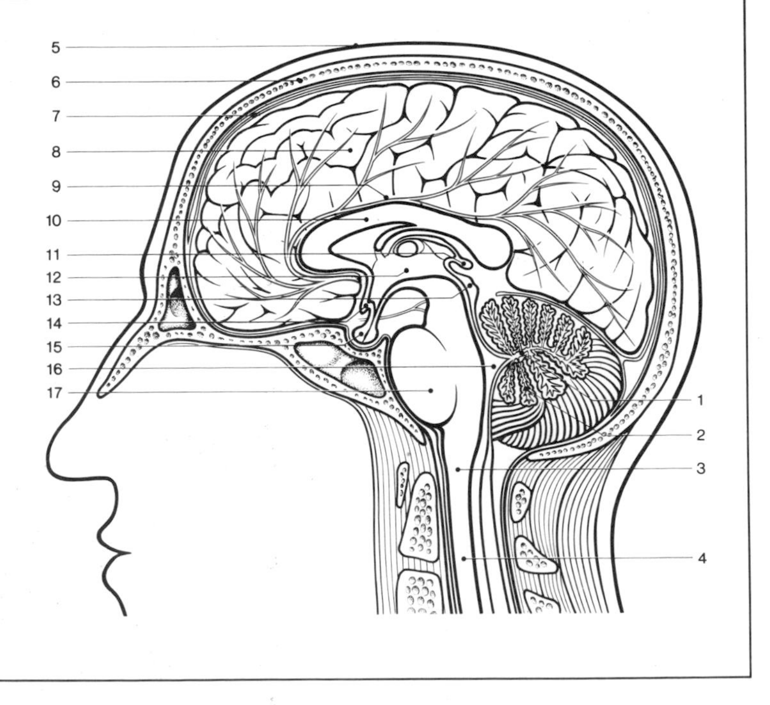

The convolutions of the brain
In man and other higher species, the surface of the brain is much convoluted in order to accommodate the myriad cells of the cerebral cortex. In lower species the surface of the brain is much smoother. In this photo, the three meninges covering the brain have been removed, and the naked surface of the cerebral cortex is seen, together with some small openings for vessels. The major convolutions and furrows in the human brain have special names in the anatomic nomenclature.

A cortical map of the body

The human body drawn at the right illustrates the comparative importance of different parts of the body in terms of cortical representation. This is a section through the motor cortex of the right hemisphere, seen from the front. Nerve cells in the numbered areas control muscles in the part of the body shown above them. This part of the cortex is a thin strip immediately in front of the central sulcus, a deep fissure in the brain. The lips (4) take up more room than the neck (8); the fingers (9–13) more than the toes (22).

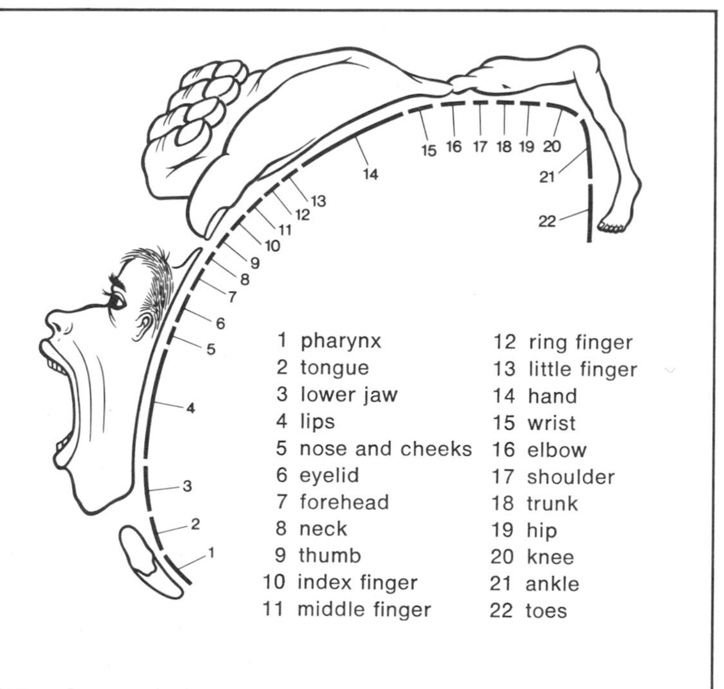

the cell. The energy for this pumping comes chiefly from the blood sugar, glucose, which is oxidized into carbon dioxide and water.

Some energy is also used for the maintenance or repair of neurons and the large population of supporting cells around them, the glial cells. Proteins, fats, and carbohydrates are the essential building blocks, while the instructions for the particular products to be produced are coded in the nucleic acids contained in the cell nucleus. The synthesis of proteins in nerve cells is of some interest in the study of memory mechanisms. Some specialists believe that particular proteins are synthesized or that alterations in the structure of ribonucleic acid occur in the process of laying down memories and thus are the biochemical means by which memories are preserved or stored.

Neural tissue makes especially high demands on fuel and oxygen throughout life. That is the reason for the rich blood supply to the brain, carried by four large arteries. While the brain makes up only 2 percent of the total body weight, it consumes 25 percent of the oxygen quota of the body at rest. Numerous mechanisms protect the brain and its supply of nourishment — the skull bones are only one of them. It could be said that the chief function of the body is to protect its principal organ, the brain, and to ensure that it is properly nourished. If the total blood supply to the brain is cut off, depriving it of oxygen, it is only a matter of minutes before the brain dies — and with it, the individual.

The total oxygen consumption in the nervous system, in both nerve and glial cells, can be measured. Every day, year in and year out, the human brain consumes about 3.3 ml of oxygen per 100 grams of tissue per minute. Considering the brain as a whole, the rate of oxygen consumption is remarkably constant. It has been found, however, that within its energy budget the brain can vary its consumption considerably. The process of concentrating, of thinking hard, may mean that certain areas in the cerebral cortex double their oxygen consumption relative to the resting level, while it is diminished in other areas.

According to present knowledge, the total cerebral oxygen consumption does not decrease during sleep, but a change in its distribution may take place. Brain injuries which cause unconsciousness, on the other hand, result in a reduced oxygen consumption. So does general anesthesia in surgery. A higher-than-normal consumption has been measured during intense

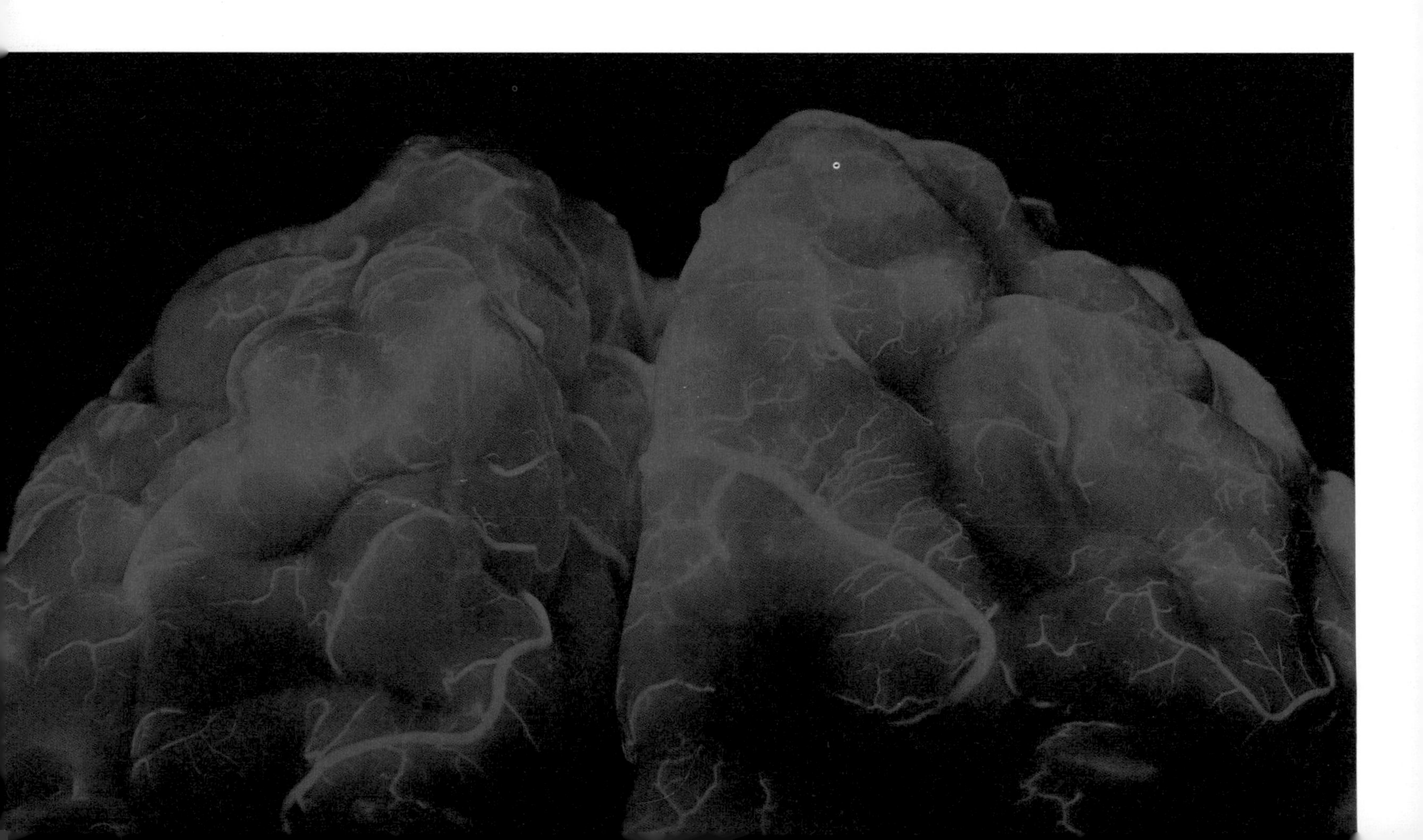

fright and during epileptic seizures.

The electrical activity of the millions of cells in the cerebral cortex can be measured externally through the skull. An electroencephalogram (EEG) is a recording of the variations in electrical potential picked up by electrodes fastened to the scalp. The variations have a dominant frequency of 8 to 12 cycles per second and an amplitude of 10 to 100 microvolts. This mass activity, recorded as an EEG curve, occurs continuously throughout life, just as does cerebral metabolism, and normally there are typical differences in the pattern of EEG activity in an alert state and during sleep. It is still not known what determines the very regular rhythms of the EEG. Certain cells in the cerebral cortex and other groups in the thalamus, in the brain stem, appear to be involved.

The waking state is associated with consciousness. We perceive our existence as an independent individual with an ego distinguishable from that of other individuals. In terms of electrical activity, such self-consciousness requires that the nervous system have an energy input of at least 20 watts and produce an EEG pattern of 8 to 10 cycles per second. Above and below certain energy limits, consciousness disappears. Modern neurophysiology can thus outline some necessary conditions for the maintenance of a "mental life." However, an explanation of exactly how this takes place is not yet known. The question of how the body's principal organ, the brain, can produce a psyche is still moot. Obviously, many fundamental questions about the nervous system remain unanswered. Since we do not have a clear idea of how mental life is produced, we have difficulty in understanding thought processes or the unique human ability to use numbers, letters, or other abstract symbols. But we have gained some insights. The Greeks believed that mental life was embodied in the fluid-filled cavities of the brain — the ventricles. According to Hippocrates, the most famous of the Greek physicians, this fluid gave the spark of life to the various parts of the body. Today we know that cerebrospinal fluid protects the nervous system both mechanically and chemically, but it can be completely drained from the system without impairing mental life. Instead we believe that mental life is embodied in the neural tissue itself and that it is by the action of nerve cells in the brain that we are able to think and to possess that degree of self-consciousness that makes us human beings.

◁ **Cerebral blood vessels**
Energy metabolism is higher in the brain than in most other organs of the body. Nerve cells are irreversibly damaged or die if they are cut off from oxygenated blood for only a few minutes. To meet this high demand, brain tissue is richly supplied with blood through an extensive vascular network.

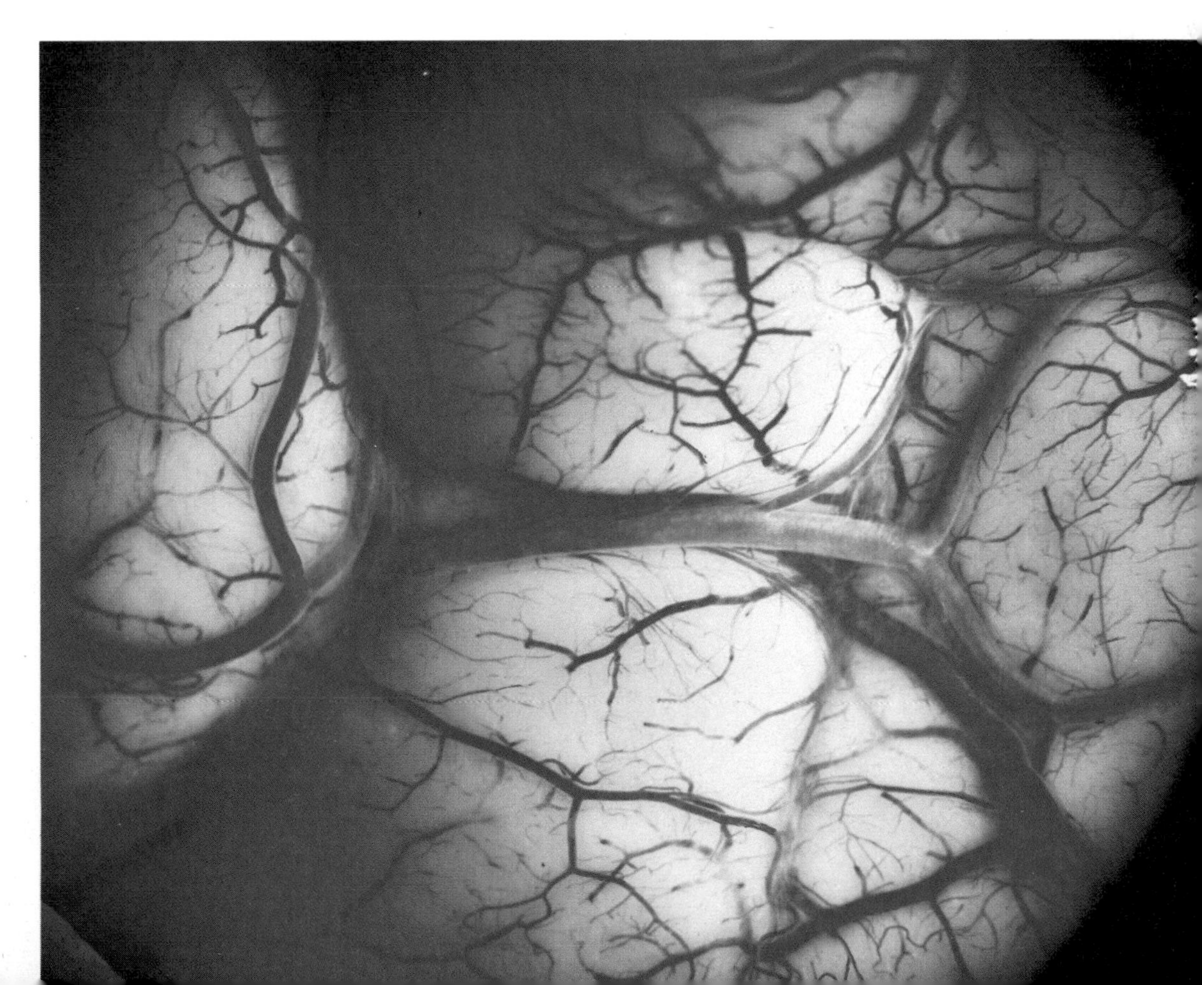

Close-up of the brain surface ▷
The photo shows veins and arteries on the surface of the cortex. Some vessels appear to end abruptly in rounded structures. These are the points where arteries enter the cortex or veins make their exit.

A nerve cell
There are many types of neurons. The largest, with the longest extensions, are the motor cells of the cerebral cortex that control skeletal muscles. Their cell bodies are approximately .05 to .075 mm in diameter. Most nerve cells have smaller cell bodies, about .005 to .015 mm in diameter. Together with its extensions (processes), the nerve cell constitutes the functional unit of the nervous system. Certain extensions, dendrites, convey electrical impulses toward the cell body, which contains the nucleus. The axon is the slender process which conveys impulses away from the cell to other neurons or muscles or glands. The photo shows a nerve cell as seen in the scanning electron microscope.

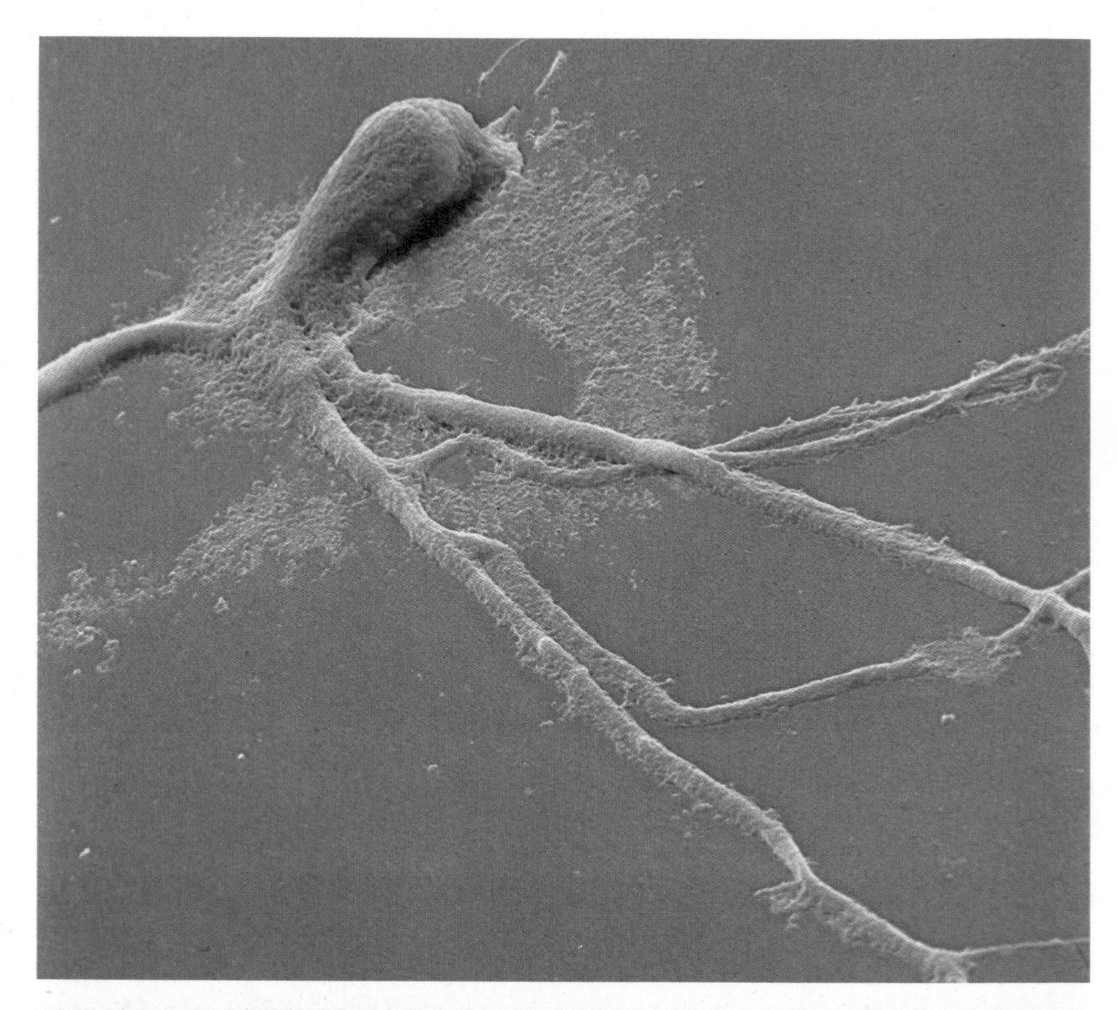

A large nerve cell from the cerebellum
The two main extensions from the cell body later ramify. The nucleus can be seen inside the cell body. The long diagonal bands are branches from other nerve cells.

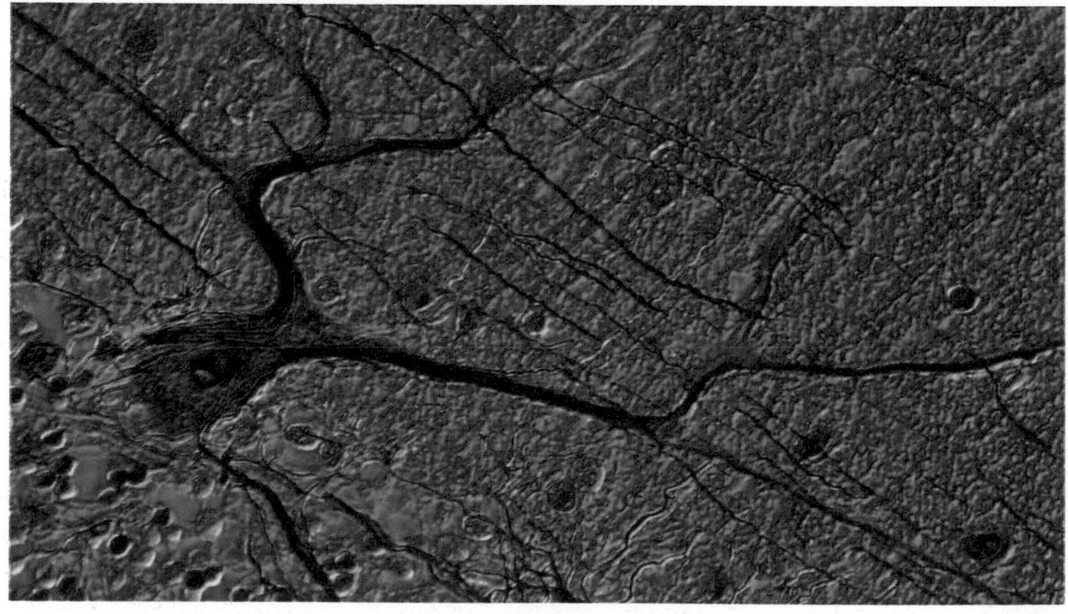

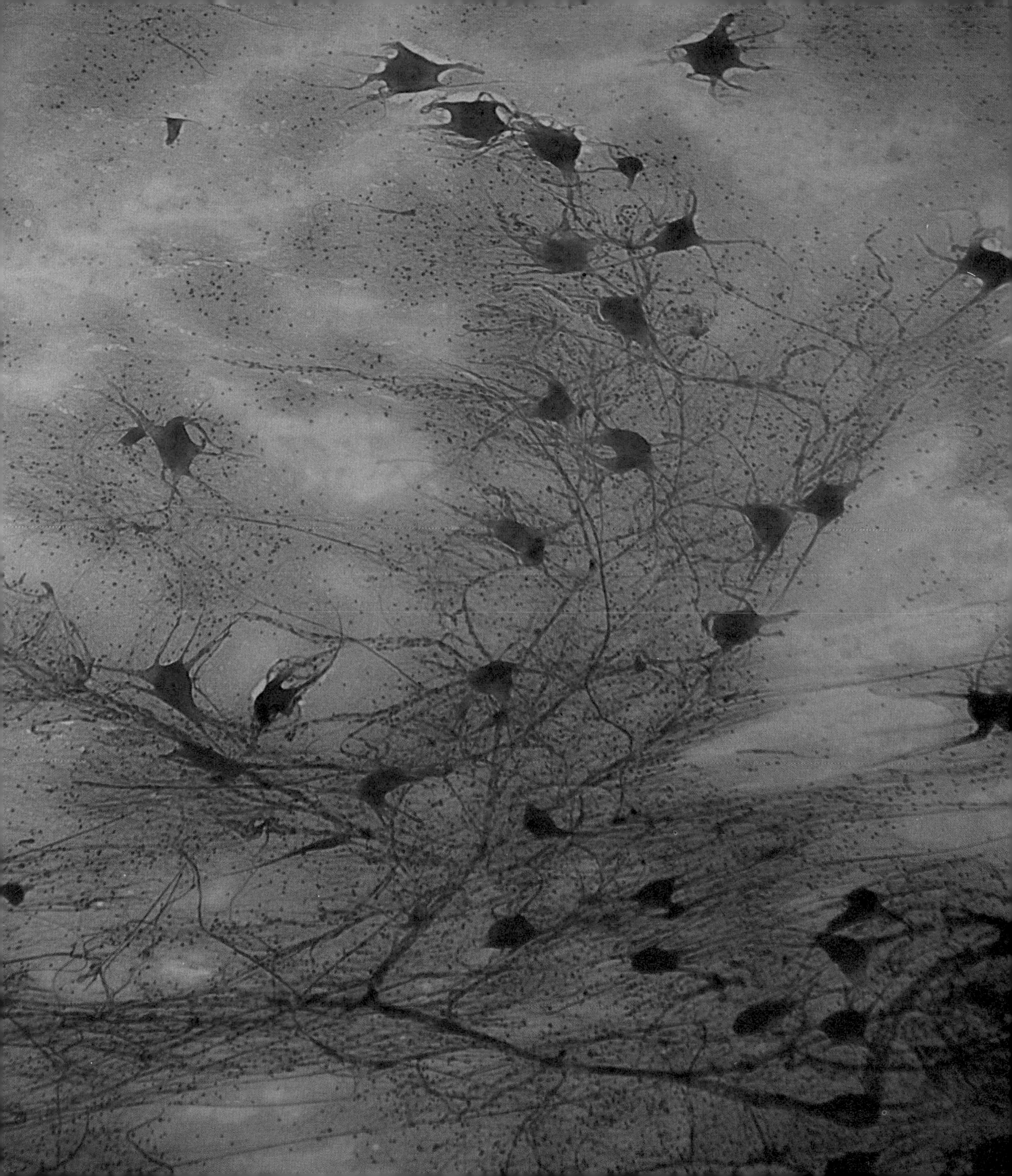

Nerve Cells

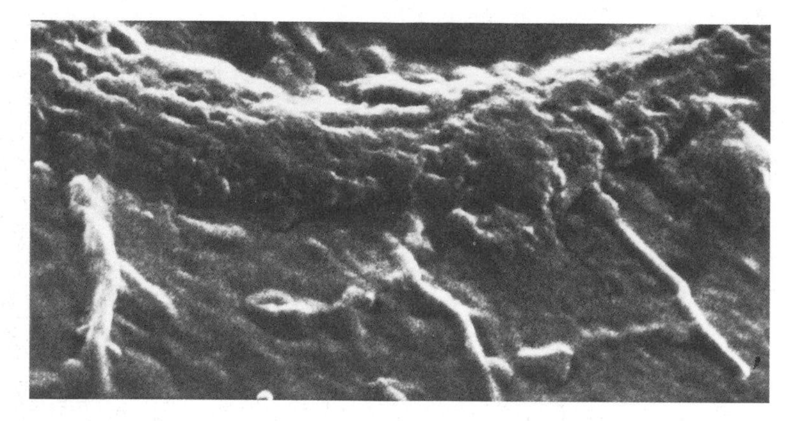

Synapses
are regions of contact between nerve cells. Here nerve impulses cross the gap separating one cell from another. The photo shows fine nerve fibers in contact with a larger one at small terminal bulblike expansions called end-feet or *boutons*.

An isolated nerve cell,
magnified 20,000 times. The diameter of the nerve cell body is less than .01 mm thick. The cell body is in contact with other cells through numerous extensions. The photo opposite is a detail of the smaller one (indicated by box). It illustrates how nerve fibers cross each other to form an elaborate network.

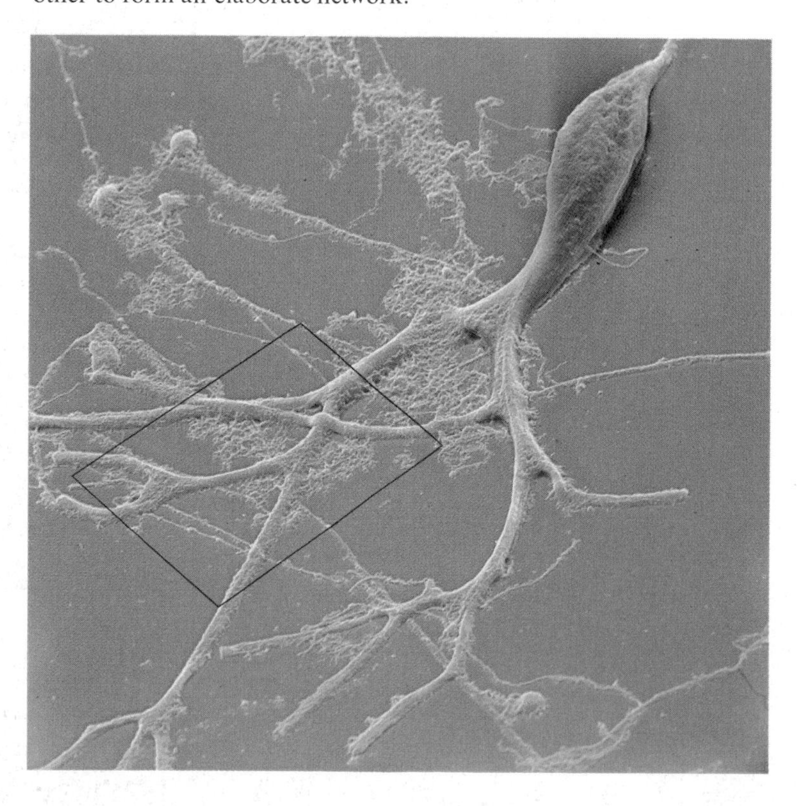

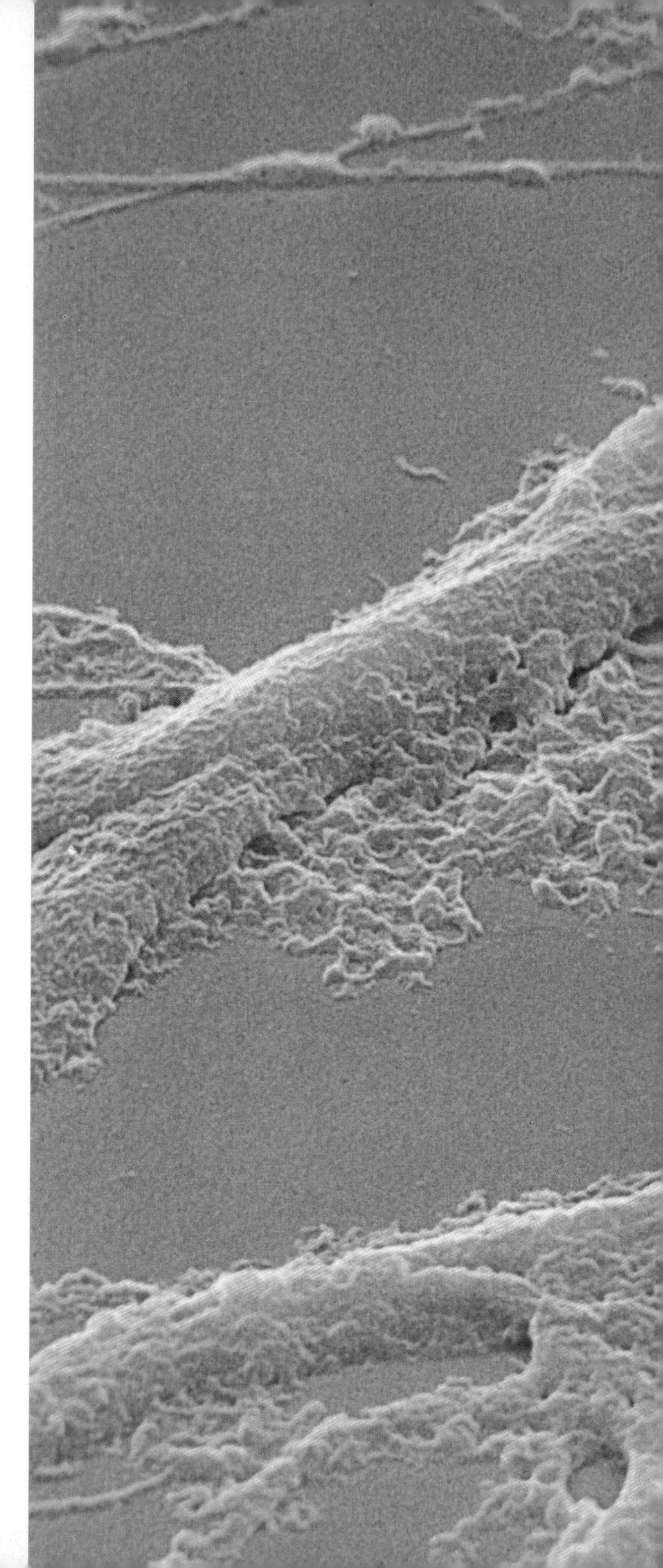

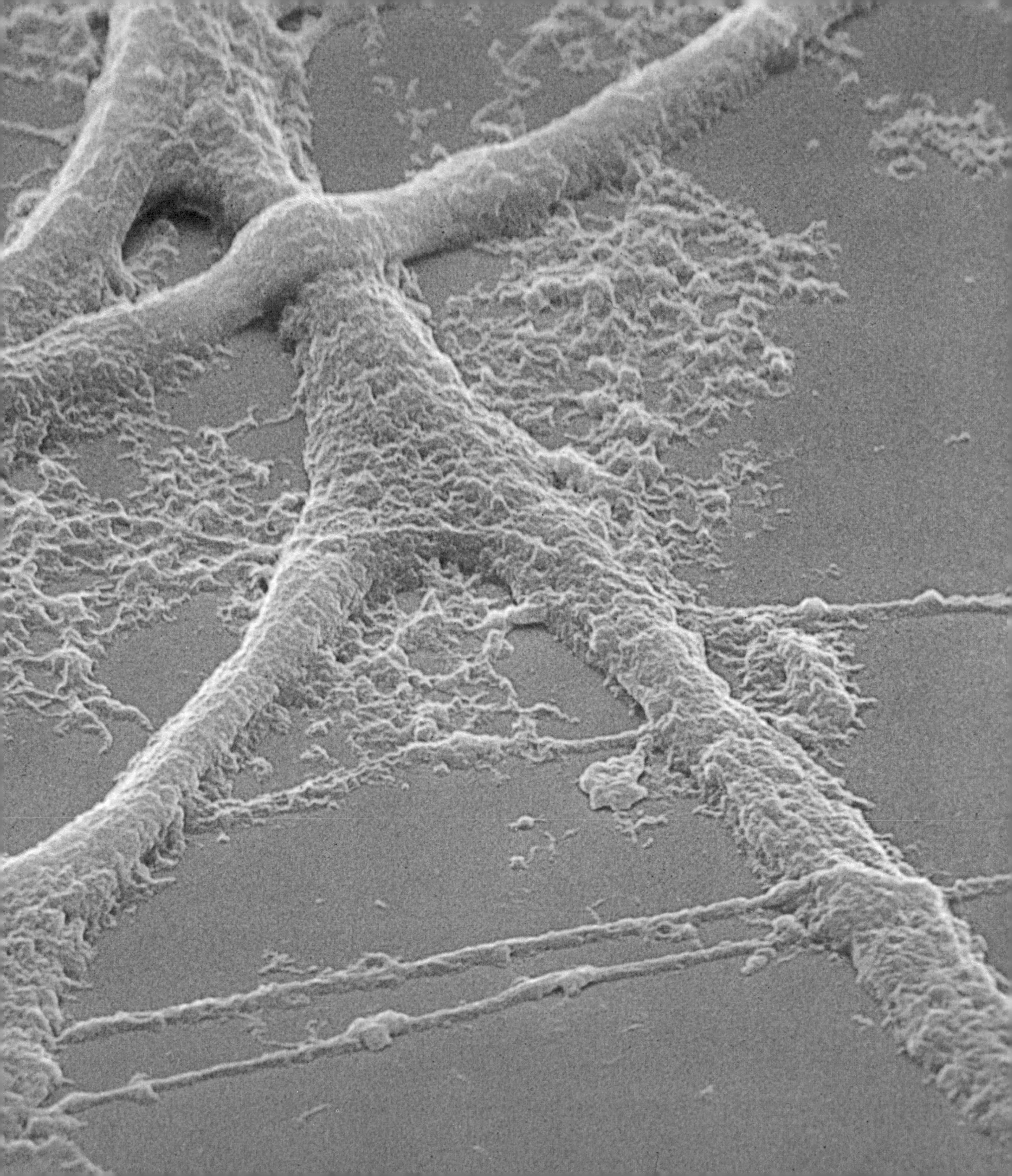

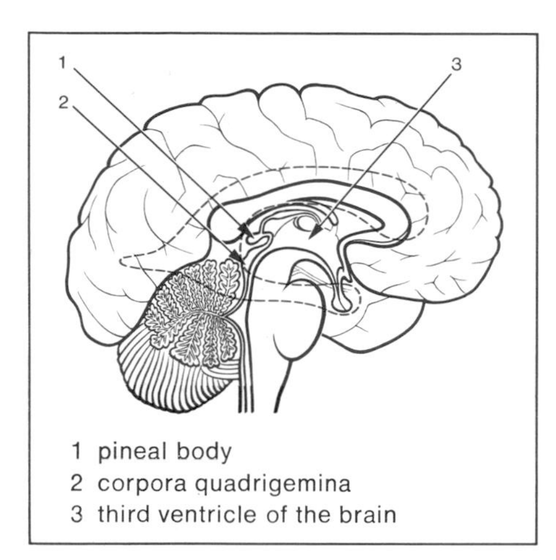

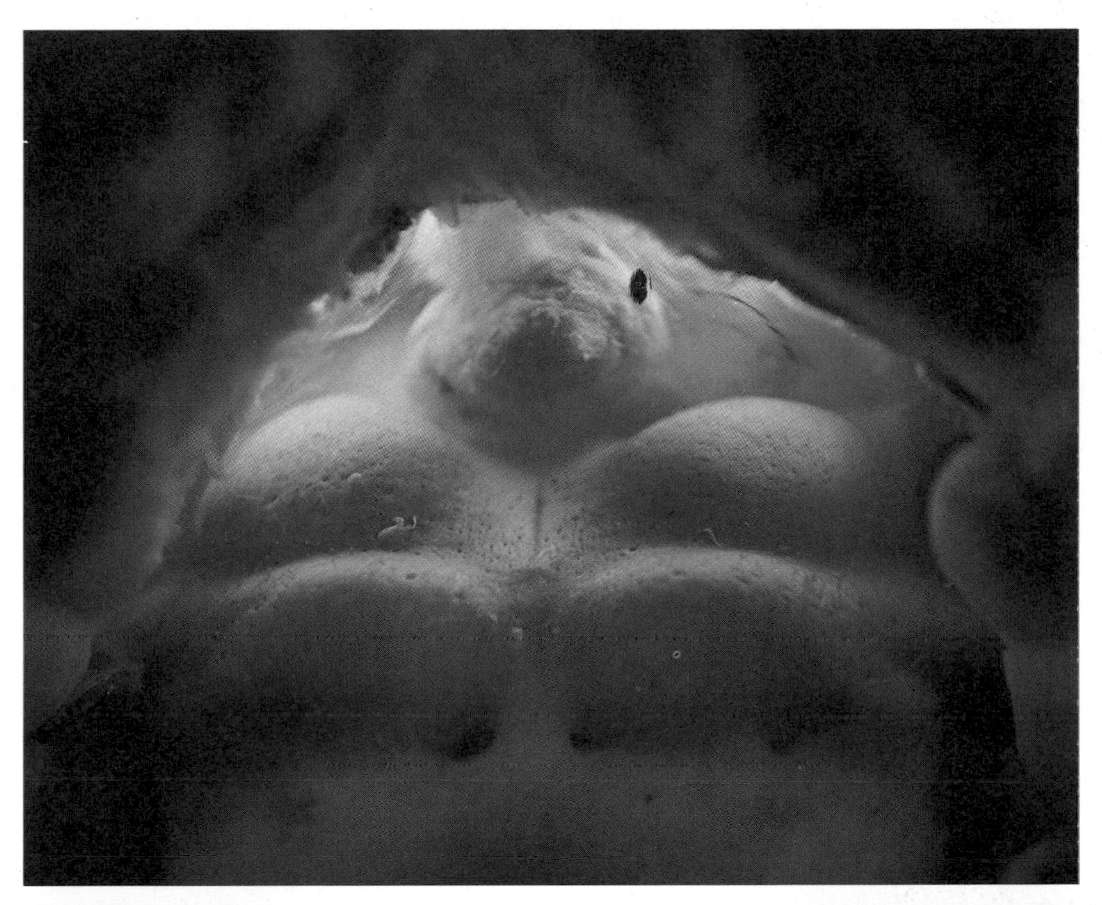

◁ **The pineal body — the "seat of the soul"**
Between the hemispheres, below the posterior portion of the corpus callosum and above the corpora quadrigemina (see diagram), is a small protuberance called the pineal body, or epiphysis. Its function is still not clear. In former times it was considered the seat of the soul.

The corpora quadrigemina ▷
are four small eminences on the back (dorsal) side of the brain stem, arranged in two pairs. They are relay stations for auditory and visual impulses. Centered above them is the pineal body. Part of the corpus callosum, a major pathway connecting right and left hemispheres, is also visible.

The cerebral ventricles ▷
The brain is not a solid mass of tissue. In its interior are interconnected cavities filled with a watery fluid, the cerebrospinal fluid. The fluid is produced by a tissue rich in blood vessels, the choroid plexus, mainly in the lateral ventricle of each hemisphere. The plexus is the light red formation at bottom left. The lateral ventricles connect with the third ventricle, a cavity lying between the two hemispheres. A narrow canal, the cerebral aqueduct, running under the corpora quadrigemina, connects the third ventricle with the fourth. Openings in the fourth ventricle, located under the cerebellum, enable cerebrospinal fluid to leave the ventricular system to bathe the outer surface of the brain and spinal cord.

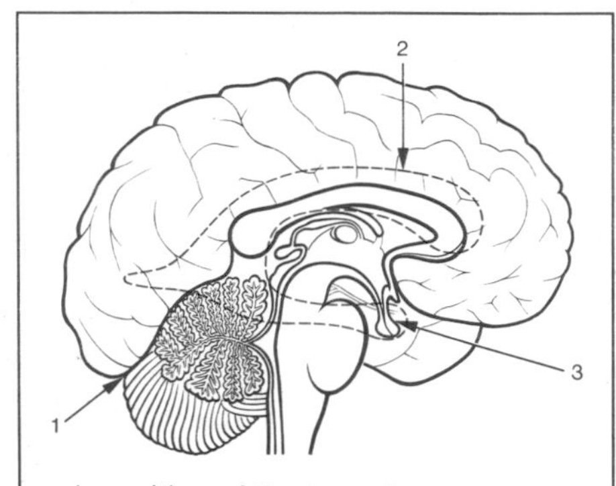

1 position of the tentorium
2 outline of the lateral ventricle
3 pituitary gland

The tentorium of the cerebellum ▷
is the tentlike portion of the dura mater, the outermost of the three membranous connective tissue coverings of the brain, which forms a protective partition between the cerebellum and the occipital lobes of the cerebrum. It also lines the inside of the skull. The falx cerebri is a sickle-shaped fold of dura that extends down the middle of the brain and separates the two hemispheres. The membranes below the dura mater are the arachnoid and the pia mater. The pia consists of fibrous tissue rich in blood vessels and fine nerve fibers. The arachnoid, its outer portion, is separated from it at numerous places, called subarachnoid spaces, which contain cerebrospinal fluid. The photo is a view into the rear of the skull, where the two hemispheres of the cerebellum are located in their protective tent. The tip of the tentorium reaches up into the falx cerebri. Its side walls are beside the occipital lobes.

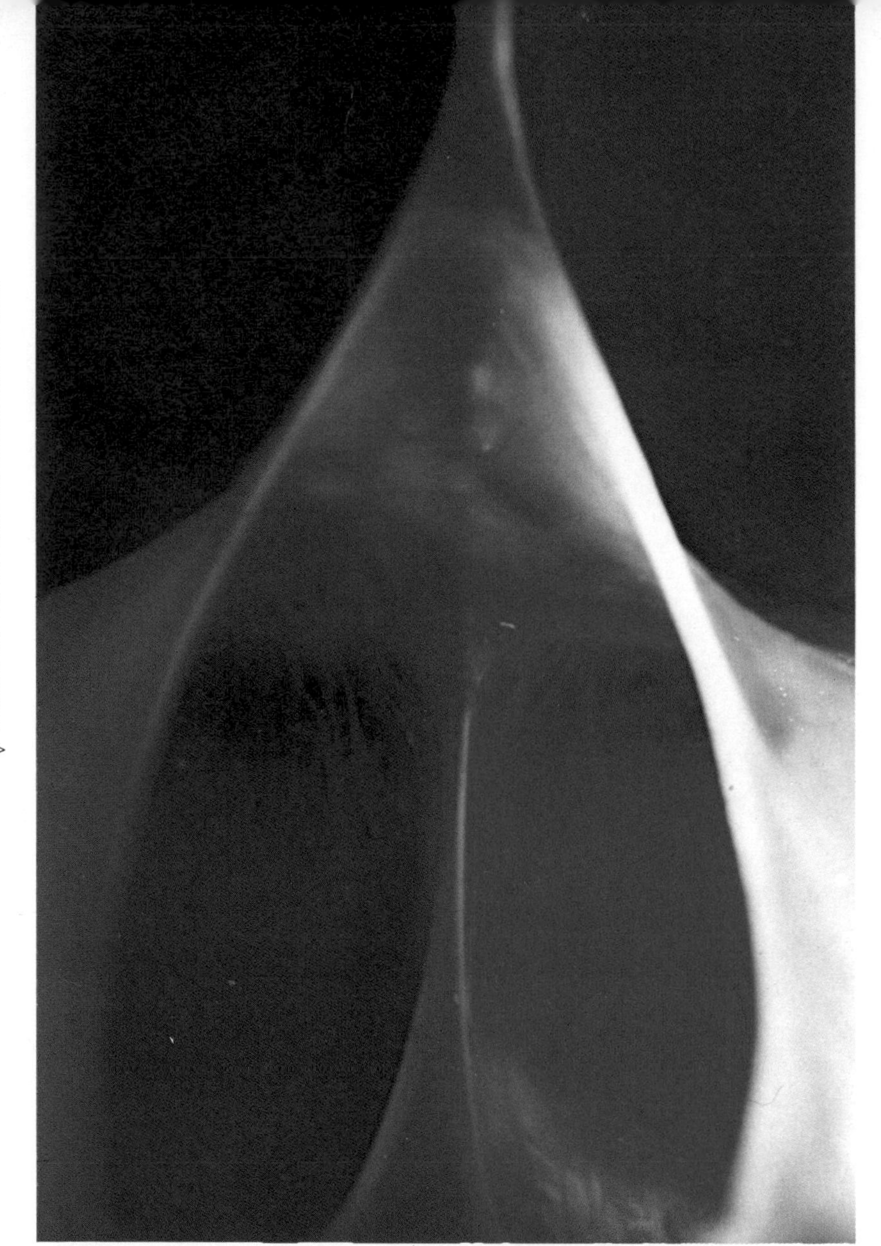

The choroid plexuses ▷
of the cerebral ventricles contain cells which produce cerebrospinal fluid. It fills the ventricles and spreads over the brain surface in the subarachnoid spaces and down around the spinal cord. The choroid tissue resembles a bunch of grapes.

The pituitary gland in the sella turcica
Under the brain is a saddle-shaped depression in the bone forming the base of the skull called the sella turcica. Here lies the pituitary gland, perhaps the most important endocrine organ in the body. In the large photo is the gland, a dark, rounded lump in the center. It is suspended by a stalk attached to the hypothalamus at the base of the brain.

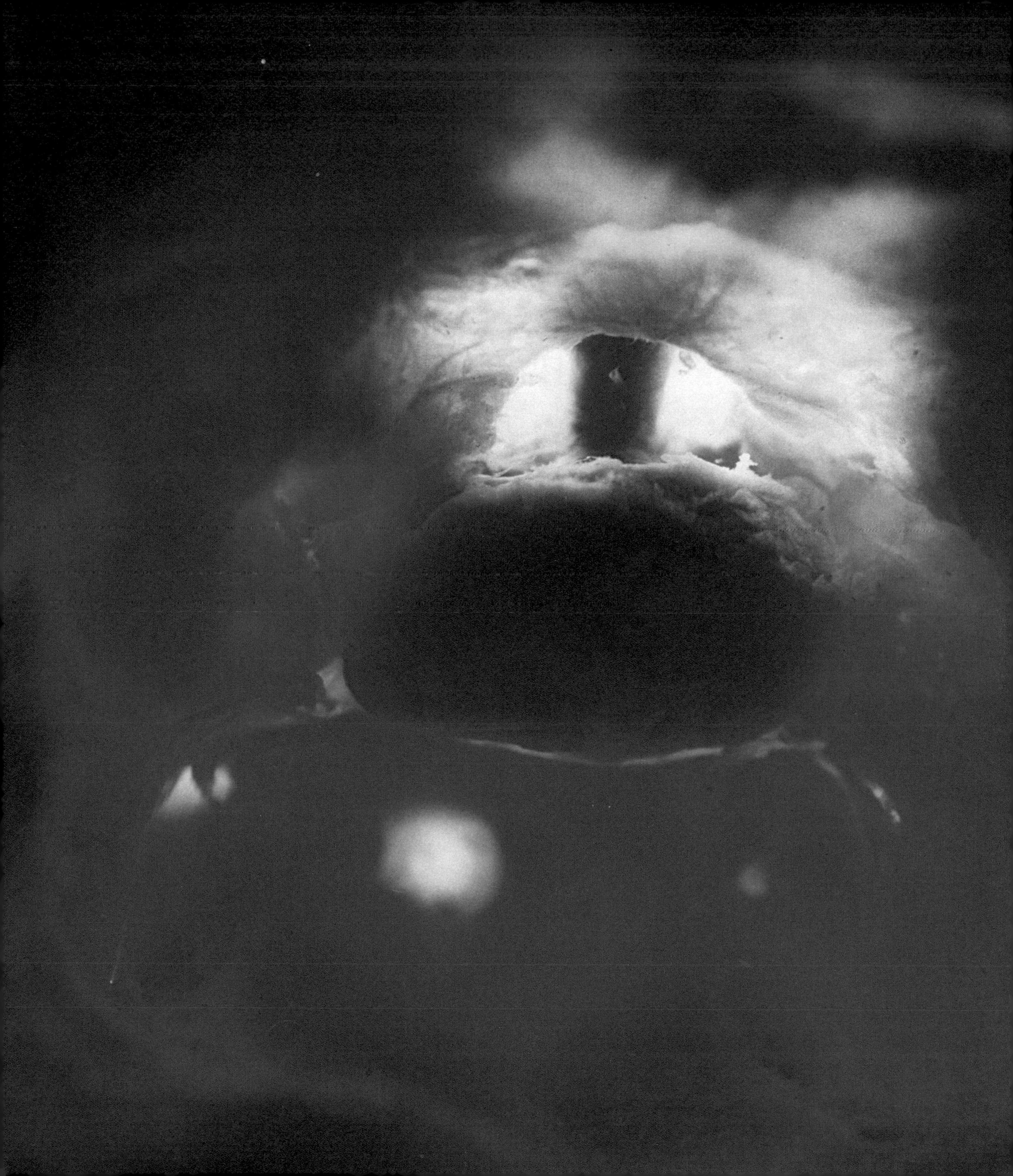

Arbor vitae, ▷
or the "tree of life," is the term anatomists have applied to the beautiful structure that appears when a section is made through the midline of the cerebellum. The white branches of the tree are myelinated fiber pathways that conduct nerve impulses. The grayish brown cortex (see detail photo) contains nerve cells important in the automatic regulation and coordination of body movements. They help to make the acts of walking, standing, sitting, or running smooth and well balanced.

Detail of cerebellum, ▷
under higher magnification

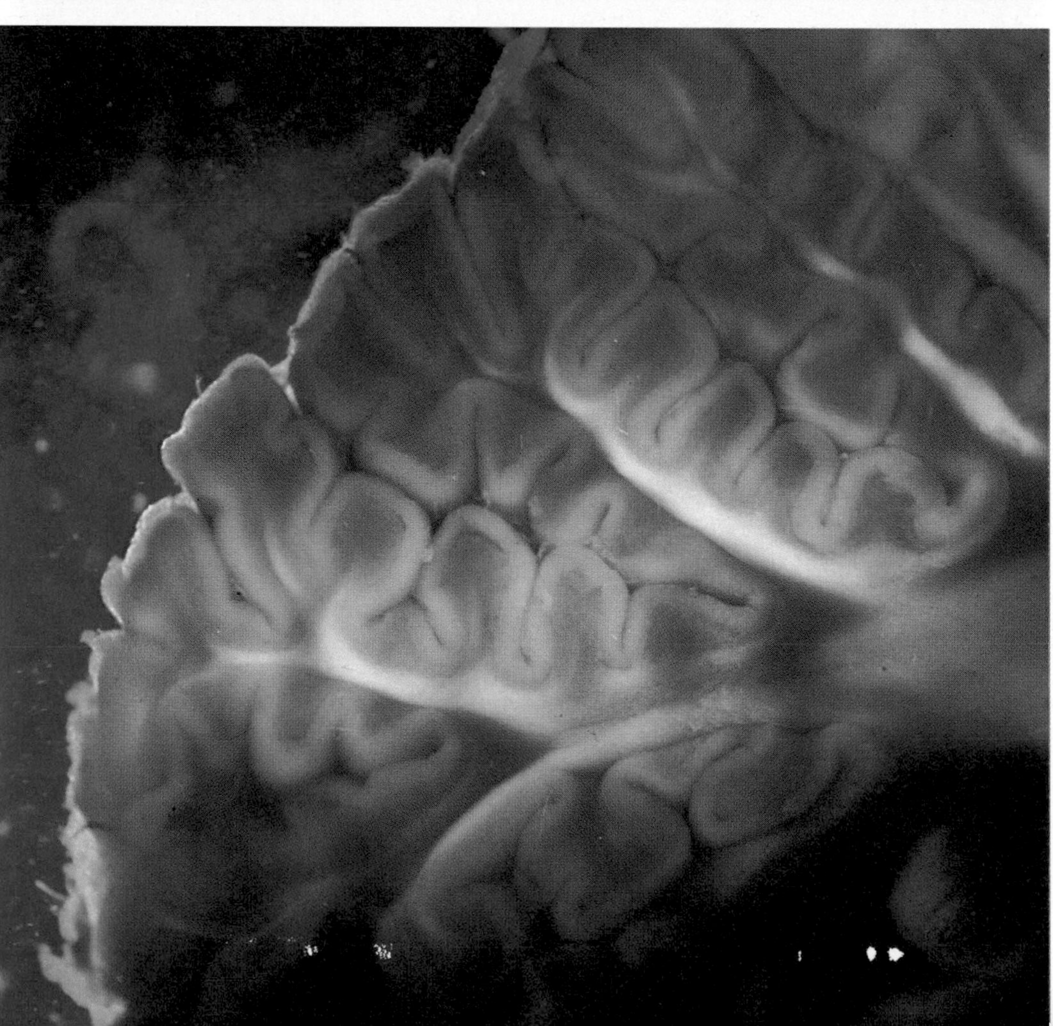

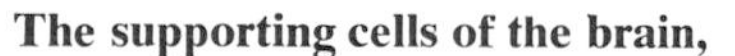

The supporting cells of the brain,
the glia, greatly outnumber nerve cells. There are several kinds of glial cell. In the photo on the right-hand page, the cells appear as rounded or oval structures with extensions, or processes. In addition to their supportive and protective roles, glial cells are presumed to serve the metabolic needs of nerve cells, transporting nutrients or removing waste products.

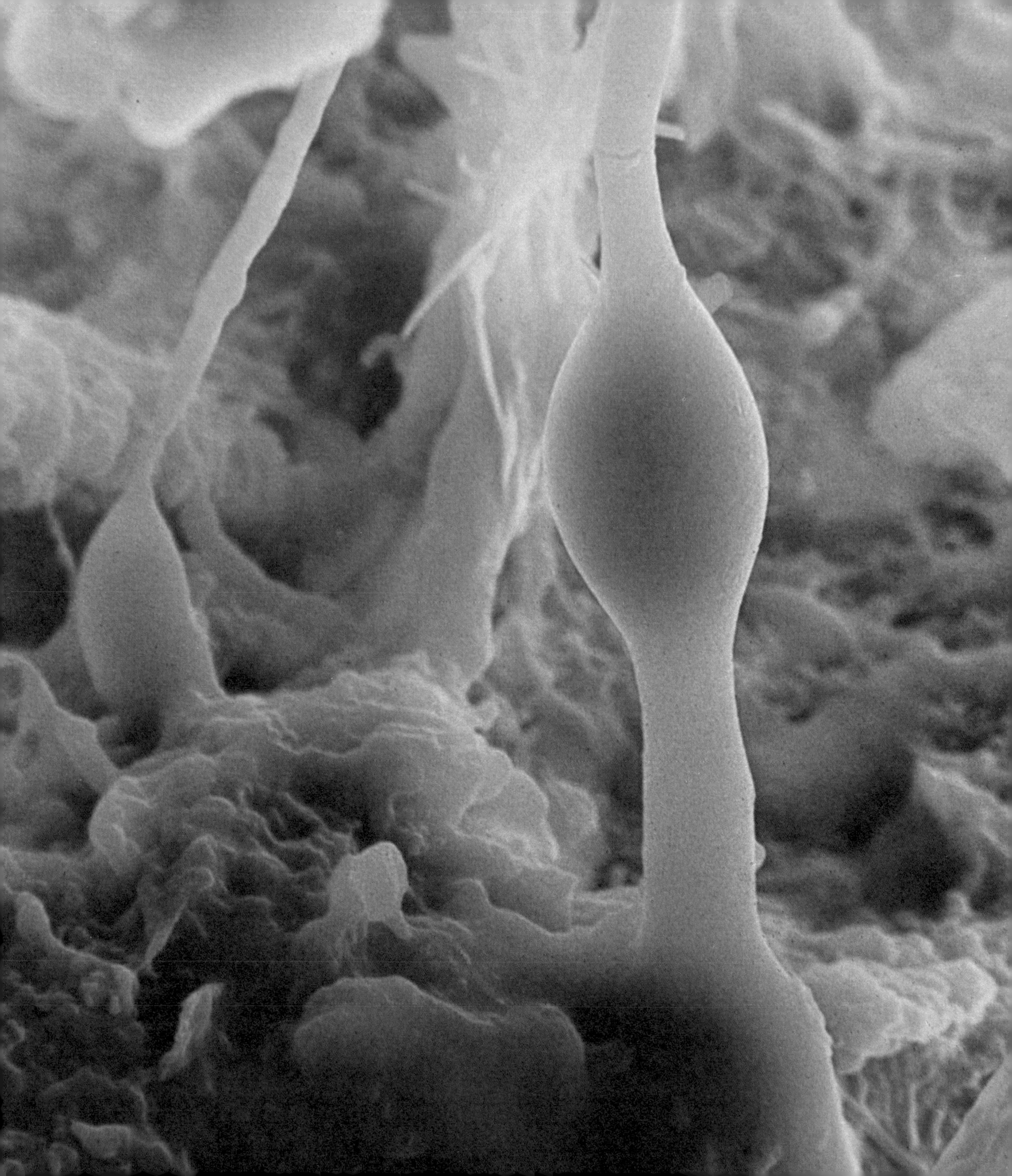

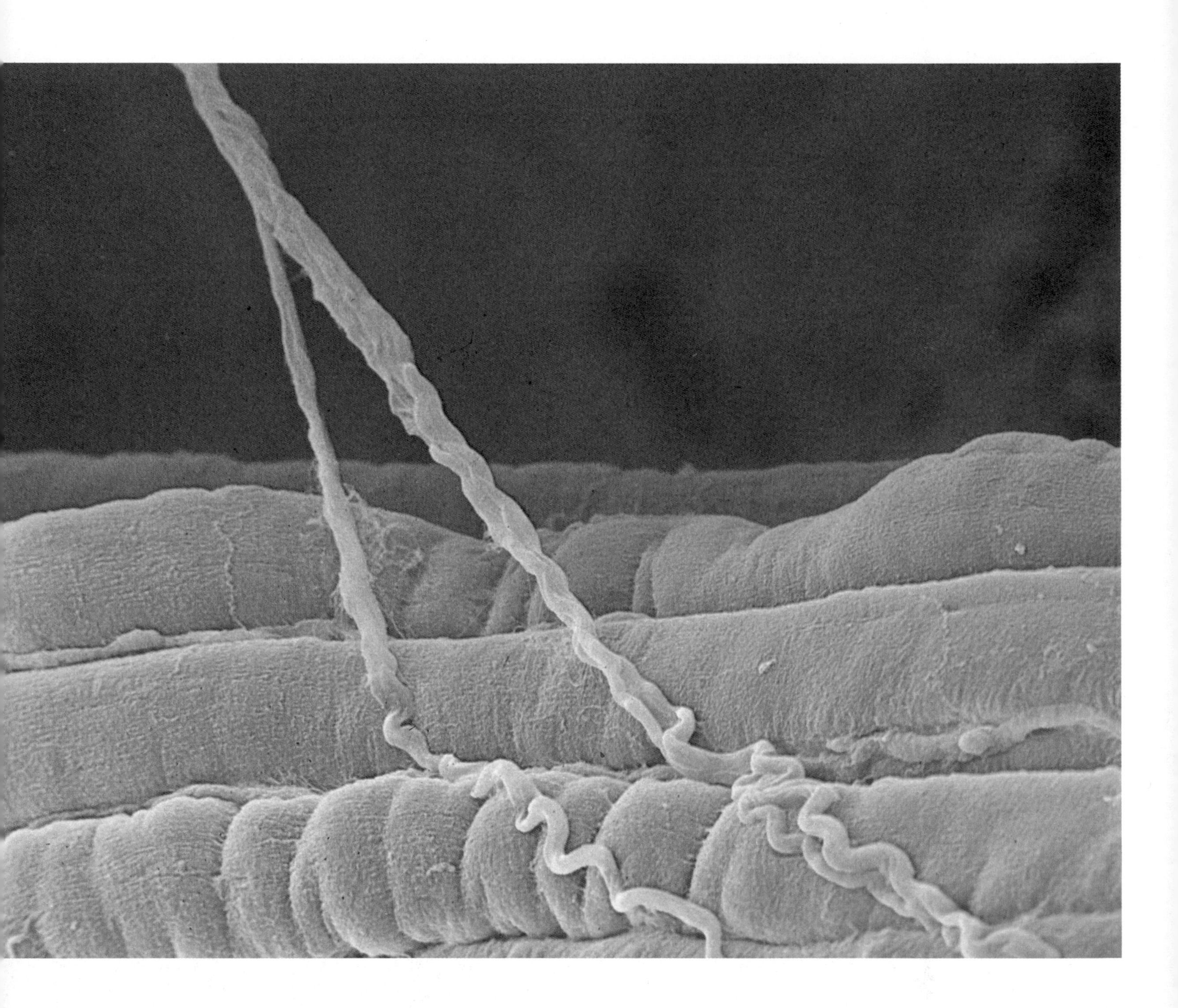

The nerve impulse,
conducted down a nerve fiber that ends in skeletal muscle, releases a small amount of the chemical acetylcholine. The action of acetylcholine at the motor end-plate initiates the chemical changes that cause the muscle to contract. The photo shows a number of nerve fibers leading to and crossing several striated muscle cells. (See also pages 118 and 119.)

The Senses

We experience the world around us through our sense organs. They are the channels through which we receive continuous and detailed information about sounds and smells, tastes, touches, and temperature. The nervous system uses this information to regulate behavior. In principle, behavior should always be biologically "correct," that is, appropriate to the given situation, so that the individual, and ultimately the species, can avoid danger and survive.

The sense organs are particularly alert to changes in the environment. Very simple experiments show this. For example, try putting one hand into cold water and the other into hot. When the one hand has cooled and the other warmed up, put both hands in lukewarm water. Now the cold hand will signal "hot" and the warm hand "cold." This kind of relative judgment is of fundamental interest. The sense organs are primed to note changes rather than absolute or static conditions. Indeed, the body tends to adapt to reasonable environmental conditions — we "tune out" until something new or different occurs. Then we notice it. After a while, the lukewarm water does not feel especially cold or hot to either hand. And even if the hands were kept in the first bowls they would gradually adapt.

If you hold a letter weighing one and a half ounces in your hand, you can distinguish it from one weighing one and a quarter ounces. But if you were carrying something weighing ten pounds, the addition or removal of a fraction of an ounce would not be felt. You'd need a change of at least three ounces to judge whether the new weight was lighter or heavier. Generally speaking, the sensory systems can perceive quantitative differences if the amount of change is a little over 2 percent of the original quantity.

In the eyes, light energy is converted into nerve impulses. Before light rays strike the retina at the back of the eye they are refracted or bent by the cornea, the lens, and the vitreous body, the gelatinous material behind the lens. All these media serve to focus the rays so that an image is formed on the retina where the visual receptor cells, the rods and cones, are located. The cones are essential in color vision. They are very dense in a central region of the retina, the macula lutea or yellow spot. The rods, which greatly outnumber the cones — 120 million compared to 7 million — are used for black-and-white vision and for seeing in dim light and darkness. The nerve impulses triggered by the visual receptors travel in the optic nerves to the visual centers of the brain, enabling us to perceive form and color, depth and motion in the world around us.

The ear contains the receptors for two sensory systems. In addition to the receptors for hearing, the ear contains the receptors of the vestibular apparatus. These organs are involved in maintaining body balance and in monitoring the position and movements of the head.

The ear is divided into three parts, the external, middle, and inner ear. The external ear funnels sound waves and, by way of the eardrum and the three auditory bones in the middle ear, transmits them to the organ of hearing, the cochlea, a snaillike structure in the inner ear. There the pressure of the sound waves causes movements in the fluid and membranes of the cochlea which ultimately stimulate the auditory receptor cells. The cells themselves are well protected in the middle partition of three which divide the cochlea along its spiraling length. (The partitions are called scalae — staircases — because they spiral up and down.) The nerve fibers stimulated by the receptor cells travel in the acoustic nerve to the auditory centers of the brain. The processing of the electrical impulses en route ultimately gives rise to the sensations of pitch, loudness, and other qualities of sound we perceive.

The receptor cells of the vestibular organs are located in two thin membranous sacs and three semicircular canals, the planes of which are perpendicular to one another (like the corner of a room). When the head moves, fluid in the semicircular canals is set in motion, pushing against tiny hairs (cilia) which fringe the receptor cells. Nearby nerve fibers are then excited and convey impulses to vestibular centers in the brain. As the head changes position with respect to gravity, the changes will also be felt by tiny calcium crystals, otoliths, found in the two membranous sacs. The crystals, obeying the law of gravity, will come to rest on whichever hair cells are closest to the ground. Signals from the sacs and the canals travel through the acoustic nerve to the brain, and give rise to an awareness of the body's position in space and to the many subtle and auto-

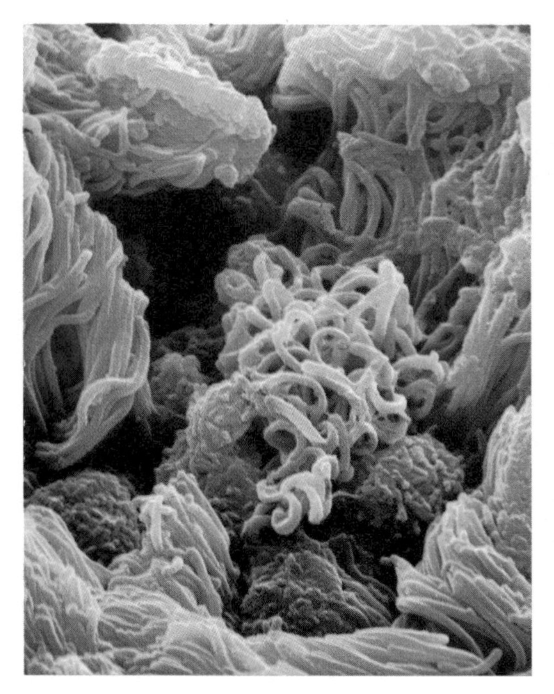

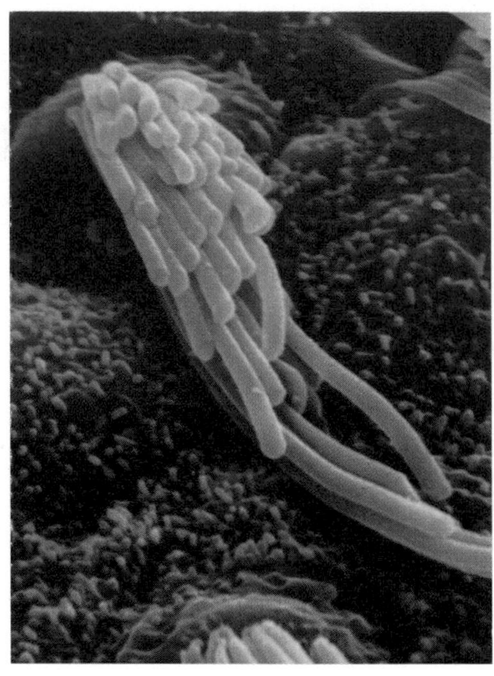

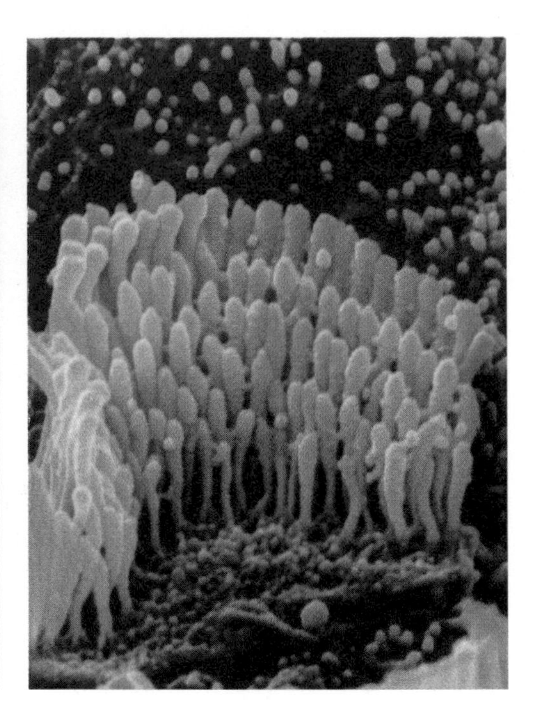

Receptors in many sense organs are much alike
The sensory receptor cells in a number of sense organs are similar in structure. These are the cells that respond to appropriate stimulation — sound energy or molecules of smellable substances — and initiate the process which converts the stimulation into electrical energy transmitted to the brain in the form of nerve impulses. The photo on the left shows hair cells in the olfactory mucosa, the center photo shows hair cells from the vestibular organs, and on the right are auditory hair cells. Ultimately, the nerve impulses are conveyed to different parts of the brain. We actually see, hear, feel, smell air, or taste food as a result of the processing of electrical impulses in the central nervous system. However, we refer the sensations to the regions in the body where they are received.

matic adjustments needed to maintain balance.

The receptors for taste are embedded in taste buds, which are found on the tongue as well as in the palate and the pharynx. The taste receptors are slender hair cells surrounded by supporting cells in the bud. Several nerves conduct gustatory impulses to the brain. While there appear to be only four basic tastes — sweet, salt, sour, and bitter — their combinations are limitless and are enriched by the sense of smell.

The olfactory organ is located in the roof of the nasal cavity. It covers an area the size of a small postage stamp in each nostril. Olfactory receptors are highly sensitive, reacting to minute quantities of volatile substances in air.

"Feeling" may mean how we feel as well as what we feel. Sense organs in the viscera convey information from interior organs and the muscles lining them. Cutaneous sense organs lie in the skin. Nerve endings and sense organs in the subcutaneous layer respond to pain, touch, cold, pressure, and heat. The distribution of sense organs varies over the body surface. There are fewer sense organs on the back, many more on the fingertips. The varied impulses are transmitted through the nervous system and give rise to sensations of pain, touch, pressure, vibration, heat, cold, et cetera. The cutaneous sense organs not only inform us of the world around us, but also function as an alarm system, alerting us to dangers to be avoided.

The total sensory input from the interior of the body is monitored in the brain and serves as an indicator of how we are "feeling" or "doing." This interior sensation greatly influences our behavior and conduct. Most activities in the body take place without thought. Impulses from the external environment are often consciously perceived, yet they too can evoke automatic or involuntary responses. Sometimes the total sensory input is so overwhelming that the body responds by feeling sick. In a heavy sea, for example, the information conveyed by the vestibular organs is unusual to most of us and may be in contradiction to a calmer report from the eyes surveying the interior space of a cabin. If you try to pick up a newspaper from the cabin floor and find it suddenly approaching you rapidly, your stomach too may get out of control.

Whether we are awake or asleep, there is a continuous transmission of impulses from sense organs to the central nervous system. These inputs contribute to the mechanisms by which we experience various stages of consciousness, but we understand

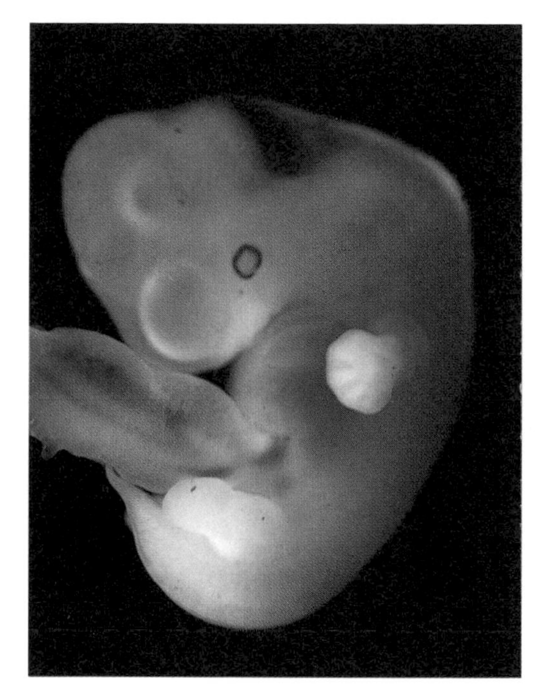

Primordial eye in a five-week-old embryo
The development of the eyes starts early, in the first month of gestation. At five weeks it appears as an incompletely closed black ring. This is the edge of the back of the eye, the fundus.

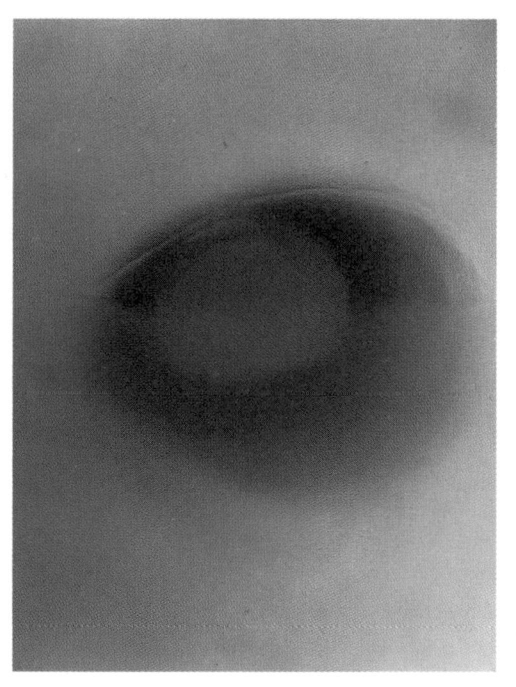

Transparent eyelids
are unmistakable at the end of the second month. They are half closed here. In a few days they will close completely and remain closed until the end of the seventh month.

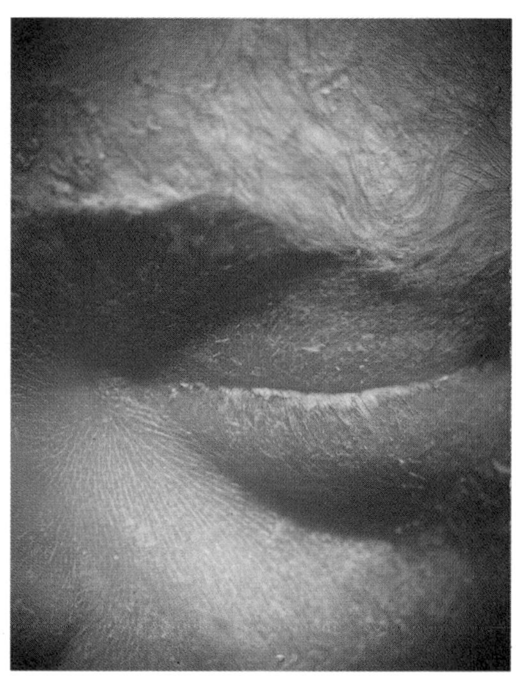

In the fifth month of gestation,
while the eyelids are completely closed, the skin is covered by a protective oily substance, the vernix caseosa. It is composed of epithelial cells that have been shed from the skin and sebum from the many sebaceous glands. Fetal hair keeps this oil on the skin.

very little of how this process takes place. It is known that incoming sensory impulses are relayed not only to the primary projection areas of the cerebral cortex, but also to nerve centers in the brain stem. These centers reflect the general state of arousal of the system and help maintain an appropriate degree of awareness or consciousness. With decreasing sensory input — in a condition of solitary confinement, for example — the degree of consciousness usually decreases. It is also known that lesions in certain centers in the brain stem may result in long-term coma.

Evolution of the Sense Organs

The evolution of higher forms of life was accompanied by the evolution of more complex sense organs. The survival of plants and animals depended on the ability to adapt to new conditions, an evolutionary principle still in effect. Existing plants and animals offer a variety of examples of sense organs at different stages of evolution.

The euglenas are monocellular flagellate organisms (possessed of taillike extensions). They have an "eye spot," a region in the cell sensitive to light. Many worms have light-sensitive cells spread all over the skin; in some the cells are concentrated in smaller regions. Light-sensitive cells are more protected when they are embedded in cavities in the skin, as in snails, and the animal has better opportunities of discovering a moving shadow, which often means danger. The smaller the opening of the cavity, the greater is the possibility of estimating the direction of the light.

A grain of sand or other foreign matter can easily lodge in an open cavity and completely block the light. In many animal species, a transparent membrane covering the eye cavity evolved, probably long ago. In the scorpion, for example, the protective membrane has thickened so that it forms an external lens. In many animals complex eyes have evolved. The many-faceted eyes of arthropods (the order that includes insects, spiders, and crustaceans) produce images in each of the hundreds of tiny prisms; the image-forming eyes of the cephalopods (octopus and squid), and our own eyes, all have different structures, but all satisfy the requirement of meeting the varied needs of the seer.

The eyes of the human fetus develop partly from brain tissue, partly from skin. Fetal development illustrates some of the earlier stages of evolution of the human eye.

Vision

Vision is probably the most important of the human senses. The most essential features of the external environment come to us through visual stimuli. About one-tenth of the cerebral cortex is preempted to serve vision. In the retina, visual stimuli are converted to electrical impulses and conducted through the optic nerves. Current psychological and neurophysiological studies point to a number of complex processes in the eye and brain which operate to sharpen outlines, fill in figures, or close gaps so that we are able to perceive patterns or images in color and in depth. The interpretation and understanding of visual information is aided by visual memory, preserved images of things previously seen.

Light

Visible light is a part of the electromagnetic wave spectrum that ranges from longwave radio and infrared radiation through visible and ultraviolet light up to shortwave X rays, gamma and cosmic rays, all of which differ in their effects on plants and animals. Visible light has wavelengths of between 7,000 Å and 4,000 Å (Ångstrom units — one Ångstrom unit equals 0.0000001 mm; that is, one-millionth of a millimeter).

According to quantum theory, light is propagated in packets of energy — quanta of light or photons. The sensitivity of visual receptor cells in the human eye is such that only a few quanta are needed to generate nerve impulses. The sensitive elements are the rods and cones. Only 10 percent of the light that enters the eyes reaches the receptor cells. Most of it is reflected or absorbed in other parts of the eye.

The Eye

The diameter of the eye is approximately 25 mm. The cornea, the lens, and the vitreous body form the lens system, the optical parts of the eye. The retina is the light-sensitive film. When parallel light rays from an object pass through the cornea, the lens, and the vitreous body, they are refracted so that an inverted image is formed on the retina. In good light, the rays are focused on the macula lutea, the retinal region where cones are numerous. Their name derives from their flasklike shape. Cones contain pigments which make them sensitive to color. In dim light, we depend more on rods for vision. These cylindrically shaped receptors are distributed throughout most of the retina. Both rods and cones contain photosensitive pigments whose chemical structure alters in the presence of light. The changes occurring in rods and cones in turn trigger electrical impulses in nerve cells in the retina, which are then transmitted in the optic nerves to the brain.

The Anterior Part of the Eye

The outer layer of the eye is the sclera. It is composed of tough, thick, gray-white connective tissue. The anterior part of the sclerotic coat constitutes the white of the eye. At the center, the white is modified to form the transparent cornea, which bulges out slightly. It is one of the few tissues of the body lacking blood vessels. It is supplied with nourishment through the aqueous humor, the fluid in the anterior chamber of the eye. For all practical purposes, the cornea is thus isolated from the rest of the body. This makes it possible to transplant a cornea from one

The eye in profile
Behind the cornea is the lens. It appears blue here because of the strong light. The crystallike dot in the center of the lens is a reflection of light.

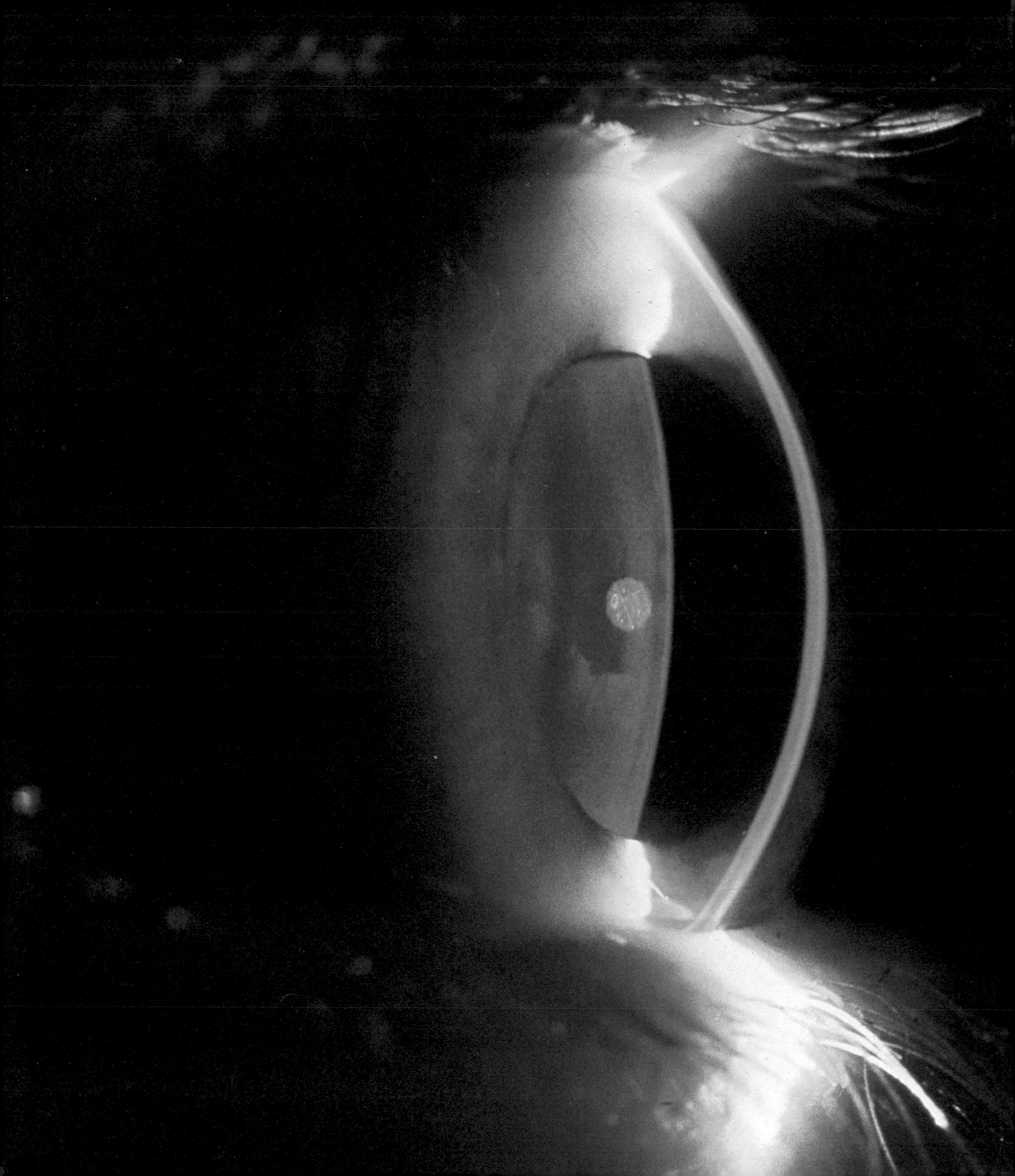

The Eye

In the interior of the eye,
light is refracted by the cornea and the lens to form a spot on the retina. This unusual photo was made using an extremely powerful wide-angle lens with a visual field greater than 180 degrees.

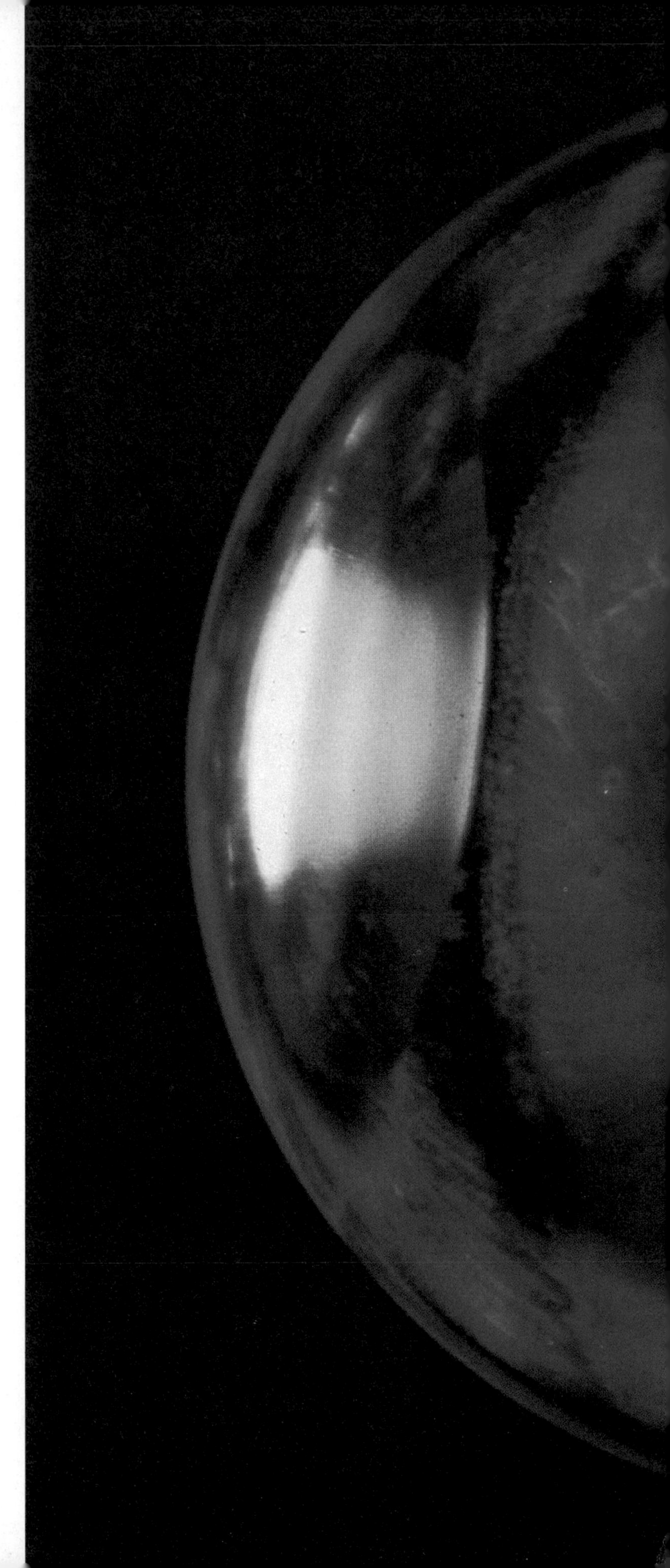

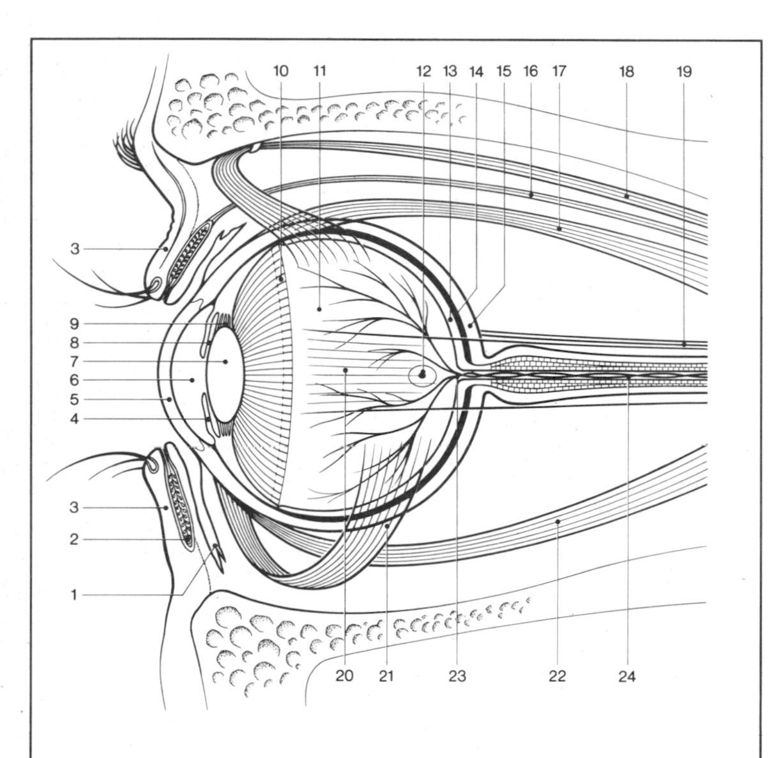

The eye

1 conjunctiva
2 tarsal gland
3 eyelid
4 iris
5 cornea
6 anterior chamber of the eye
7 lens
8 posterior chamber of the eye
9 ciliary ligament
10 anterior margin of the retina
11 vitreous body
12 macula lutea (on retina)
13 retina
14 choroid
15 sclera
16 levator muscle of upper eyelid
17 superior rectus muscle
18 superior oblique muscle
19 medial rectus muscle
20 lateral rectus muscle
21 inferior oblique muscle
22 inferior rectus muscle
23 blind spot (optic disc)
24 optic nerve

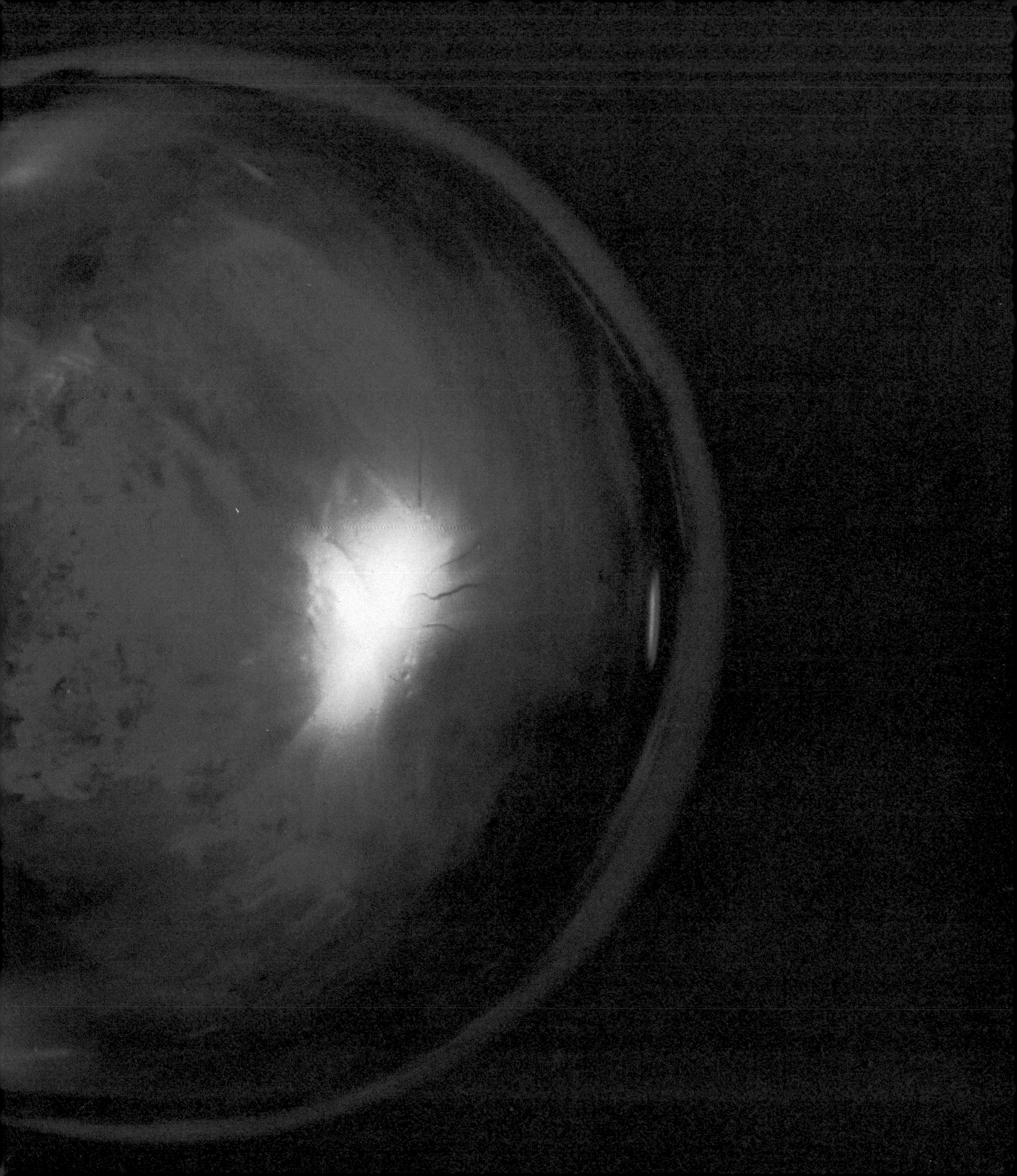

individual to another. The implanted tissue will not be rejected as foreign by the recipient, since blood-borne antibodies cannot reach it to attack it.

The fluid in the anterior chamber of the eye, the cavity between the cornea and the iris, is continuously secreted and absorbed, with a turnover rate of only four hours. Consequently, it is rare for sight to be disturbed by impurities in the aqueous humor.

It is generally believed that the cornea is transparent throughout. It is not completely so, since it contains nerves and lymph vessels, but these do not impair vision. The cornea protects the interior, more sensitive portions of the eye against injuries. It has a remarkable healing capacity and minor lesions are easily repaired.

Beginning at the point where the cornea passes into the sclera is a region where the choroid, the second layer covering the eye, is thickened to form the ciliary body. This body is composed of smooth muscles which hold the lens in place. The anterior part of the choroid forms the iris, the colored portion of the eye. In the center of the iris is a round hole, the pupil. The iris contains

The cornea is transparent
Although nerves and lymphatic vessels are present in the cornea, they do not impair vision. The white dot in the pupil is a reflection of light. The eyelashes on the eyelid help shield the eye from dirt or other harmful substances.

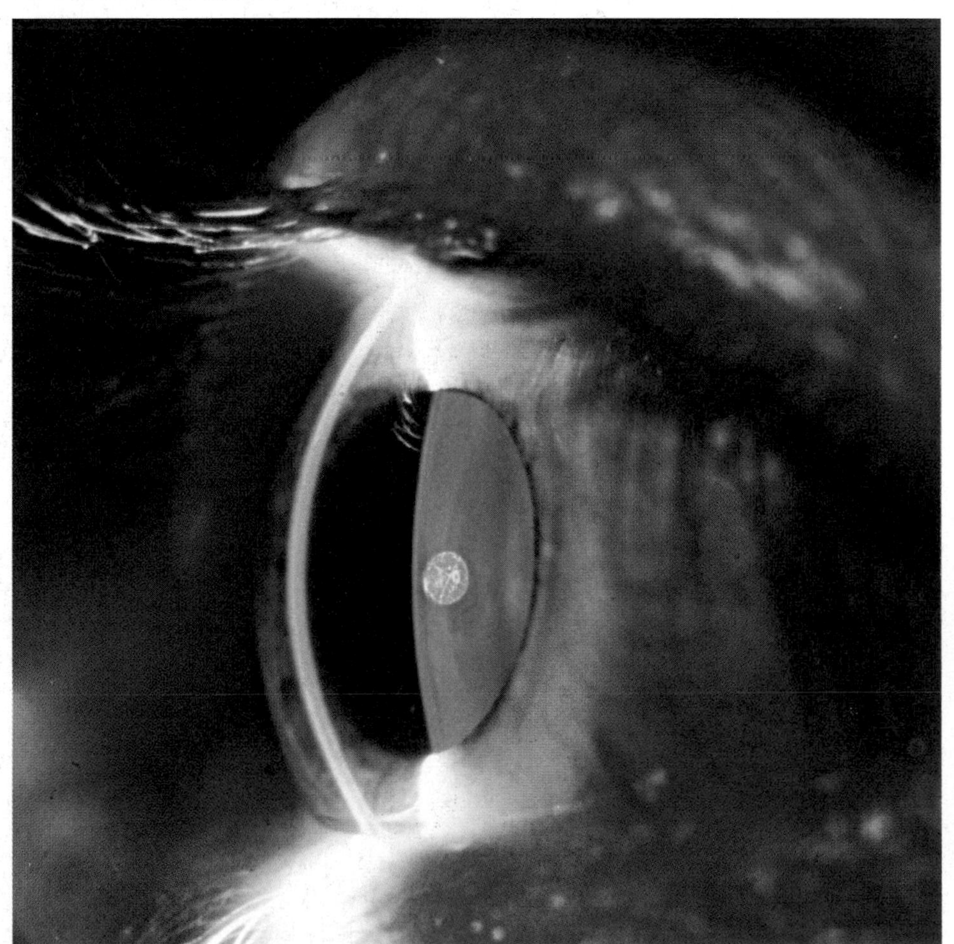

Longitudinal section through the cornea
as seen in a microscopic preparation. The cornea, which bulges out slightly from the eye, consists of several cell layers covered by a protective membrane.

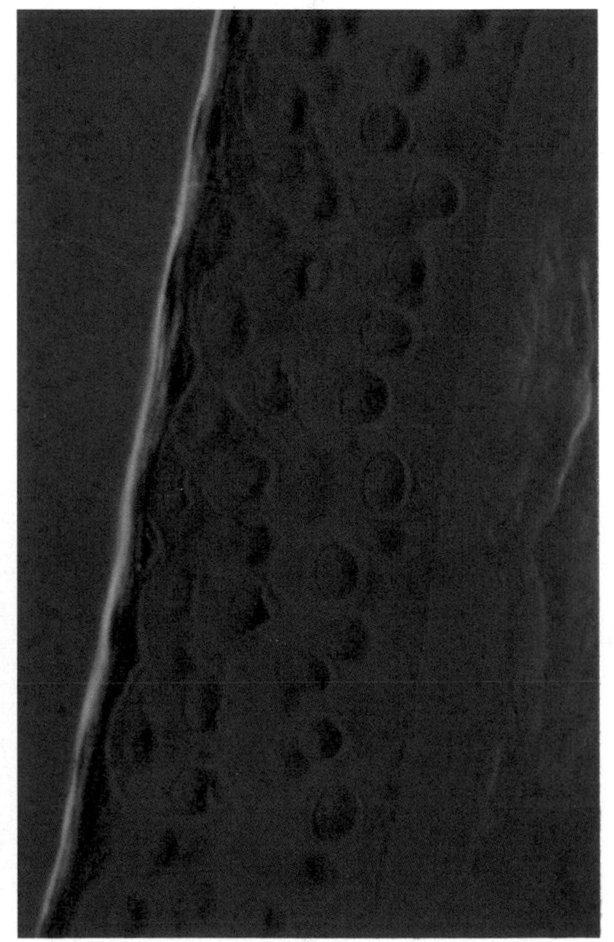

The Pupil

The pupil constricts (above) and dilates (below)
to admit the appropriate amount of light to the eye. For this reason, it is often compared with the diaphragm of a camera. Pupil adjustments occur automatically. When maximally dilated the pupil opening is sixteen times its minimum size. In bright light, the pupil constricts as in the top photo. In darkness the pupil dilates. Pupil size is also affected by emotion, becoming dilated in fear, for example. It narrows for vision at close distances. The pupil always looks black because of the refraction of light in the eye. This makes it impossible to see the interior of someone else's eye.

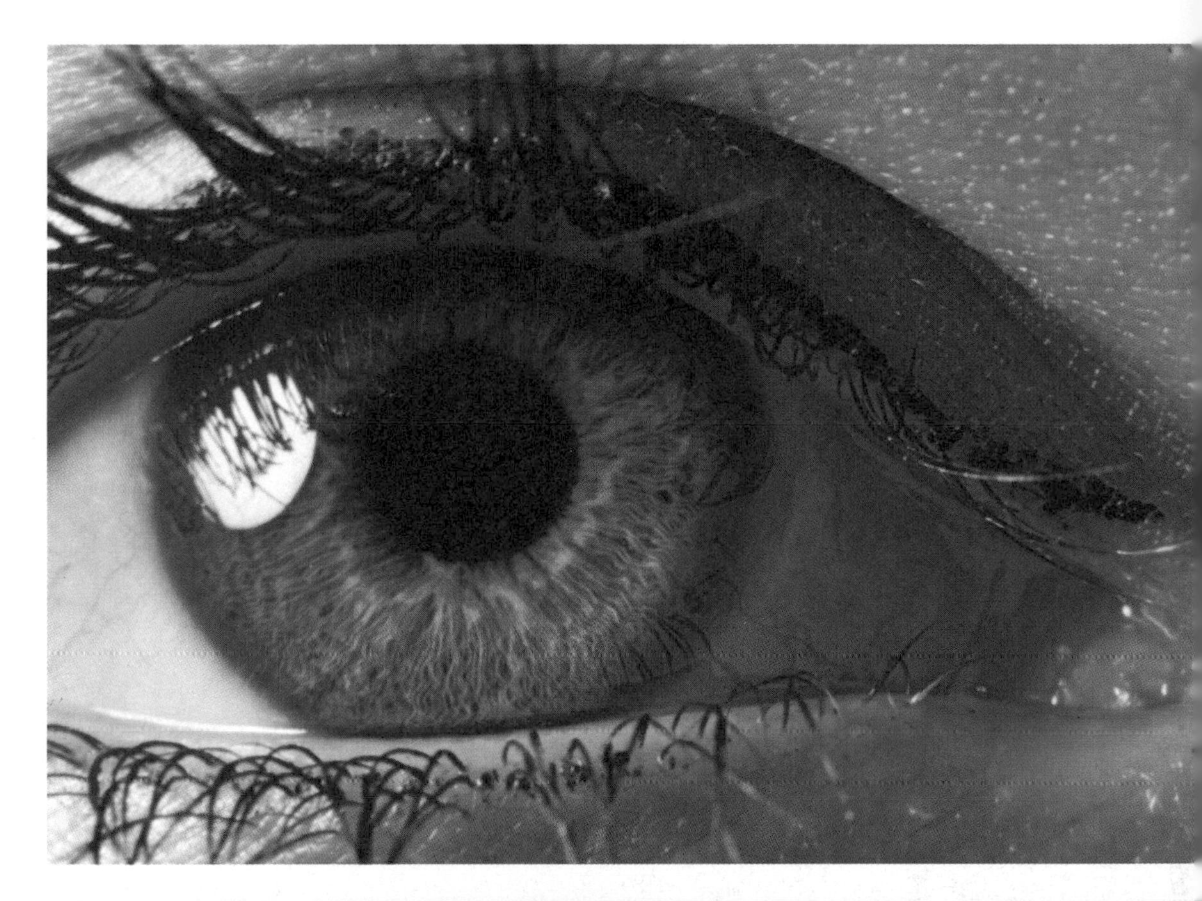

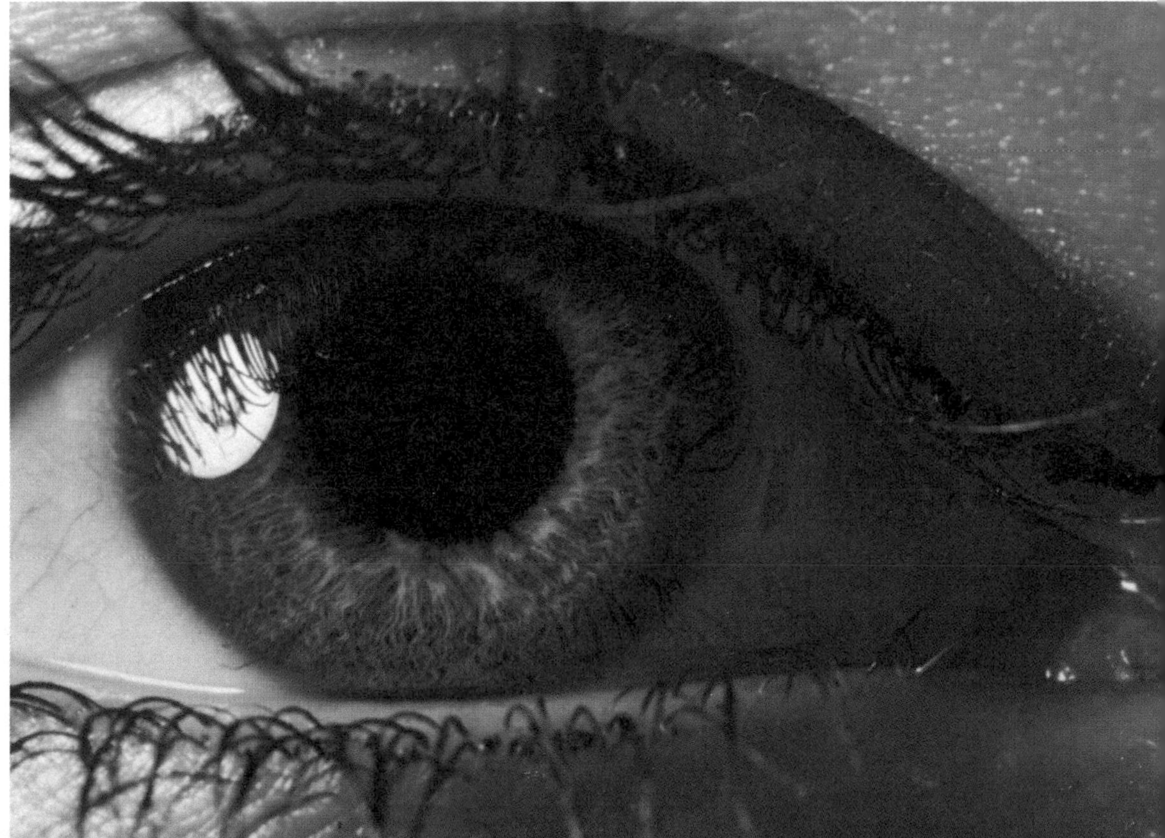

3 The outside world seen from the interior of an eye through the window framed by the iris

Eye color and the appearance of the iris change (1)
In many newborn infants, the iris is largely unpigmented. The eye then has a bluish color. Later the iris develops streaks, spots, or circles of various colors. The pattern is unique, in much the way that fingerprints are unique. Most often both eyes have the same pattern and color composition, but in some individuals the two eyes may be quite different. Eye color has no functional meaning as long as the iris is completely opaque. In albinos, who lack pigment, the eyes do not function well in strong light.

A brown eye magnified about twelve times (2)
The pigments look like spots. A child with brown eyes was born with a fair amount of pigment. If one parent has brown eyes, the chances are at least fifty-fifty that the child's eyes will be brown too, since the gene for brown eyes dominates those for other colors.

A view through the pupil (3)
The photo on the right-hand page is not a stained glass window with the sun shining through. It is a view through the pupil showing the way light is admitted to the eye to stimulate the visual receptor cells in the retina. The suspensory apparatus of the lens, the ciliary ligament, appears black in the faint light.

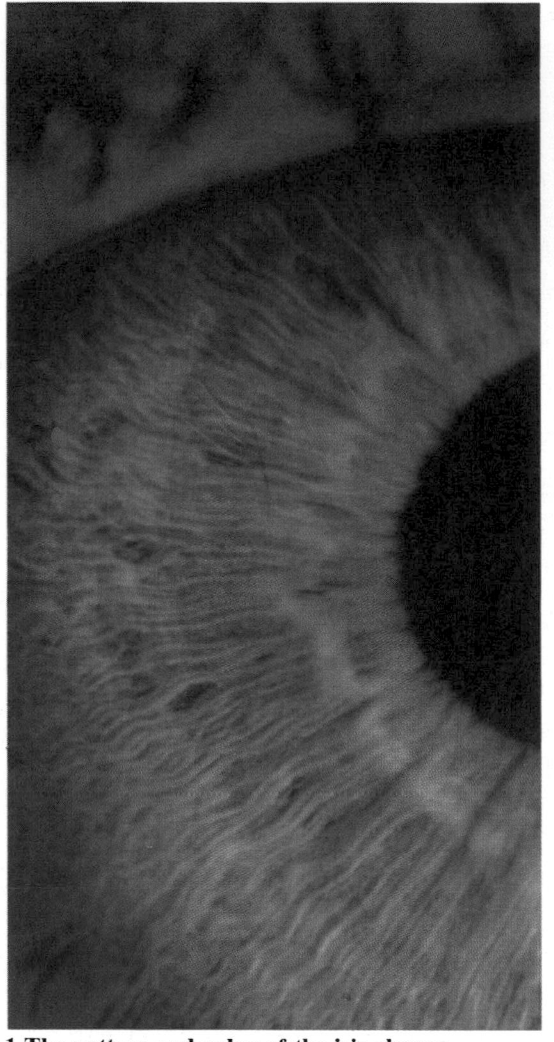

1 The pattern and color of the iris change

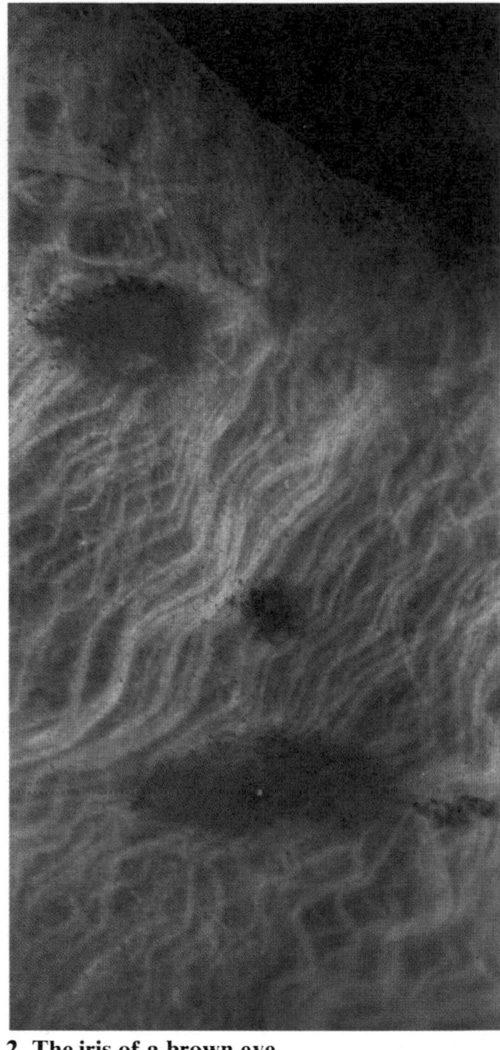

2 The iris of a brown eye

circular and radial smooth muscles by which the pupil is constricted and dilated. Their motions adjust the amount of light entering the eye.

The pupils of each eye are synchronized, or arranged in parallel, so that eye movements are coordinated. The diameter of the pupil is between 2 to 4 mm. The size is influenced mainly by the intensity of light, but it is also affected by emotions or interests. When a person is terror-struck or highly interested in something, the pupils dilate.

The posterior chamber of the eye — the space between the rear of the iris and the lens — is connected with the anterior chamber by the pupil. Like the cornea, the lens has no blood vessels and gets its nourishment from the aqueous humor. The main function of the lens is to bring to a focus at the retina the image of objects viewed at different distances. The lens accomplishes this by changing its own curvature. This is called accommodation. At rest, the normal eye is adjusted so that fairly distant objects can be focused sharply on the retina. The lens is then rather thin and flat. For vision at closer distances, the lens becomes thicker and more curved.

The lens is a transparent, solid, elastic body, yellowish in color. Its posterior part is more curved than the anterior. It is

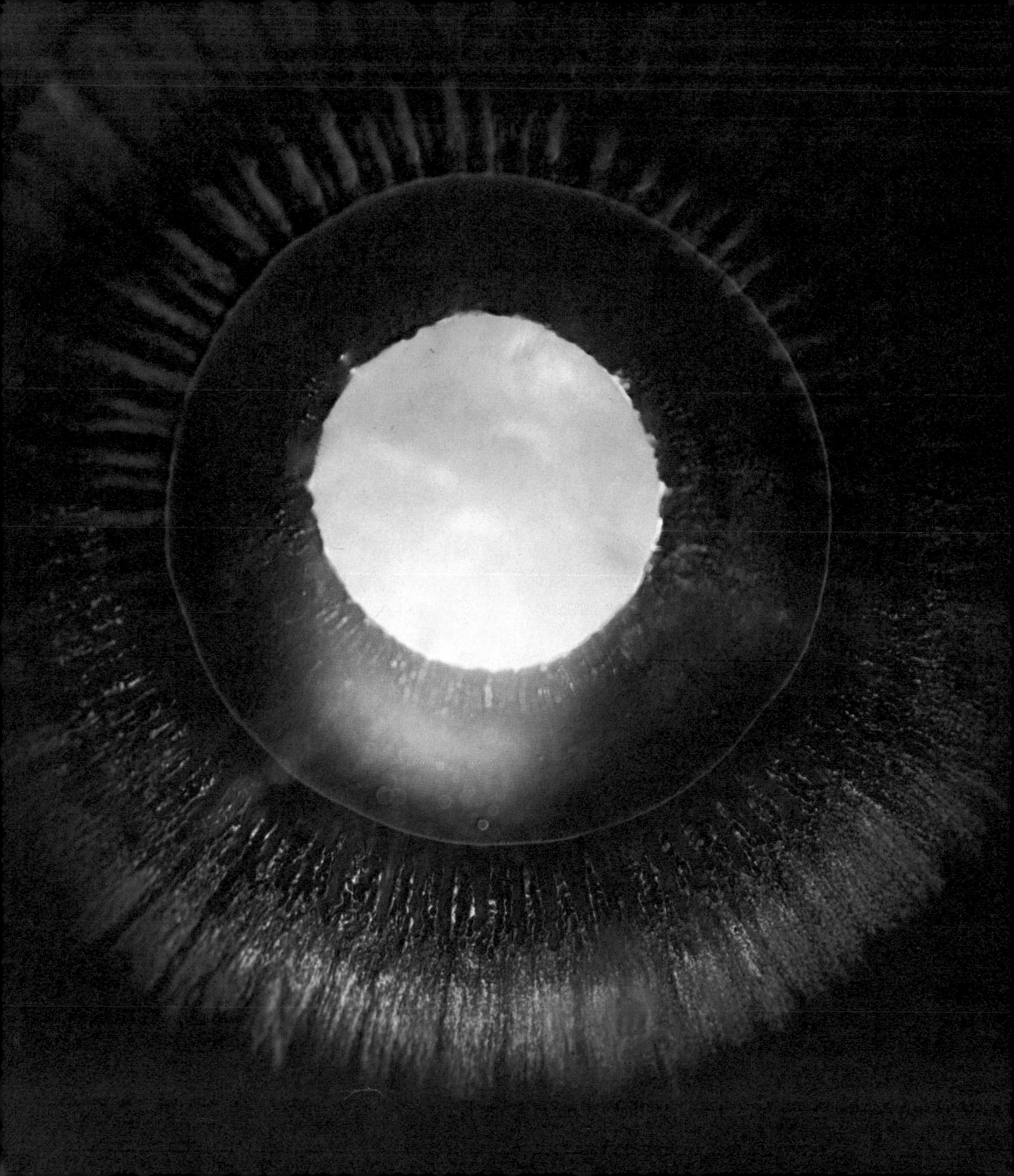

The convex lens of the eye ▷
seen from the side, penetrated by light rays.

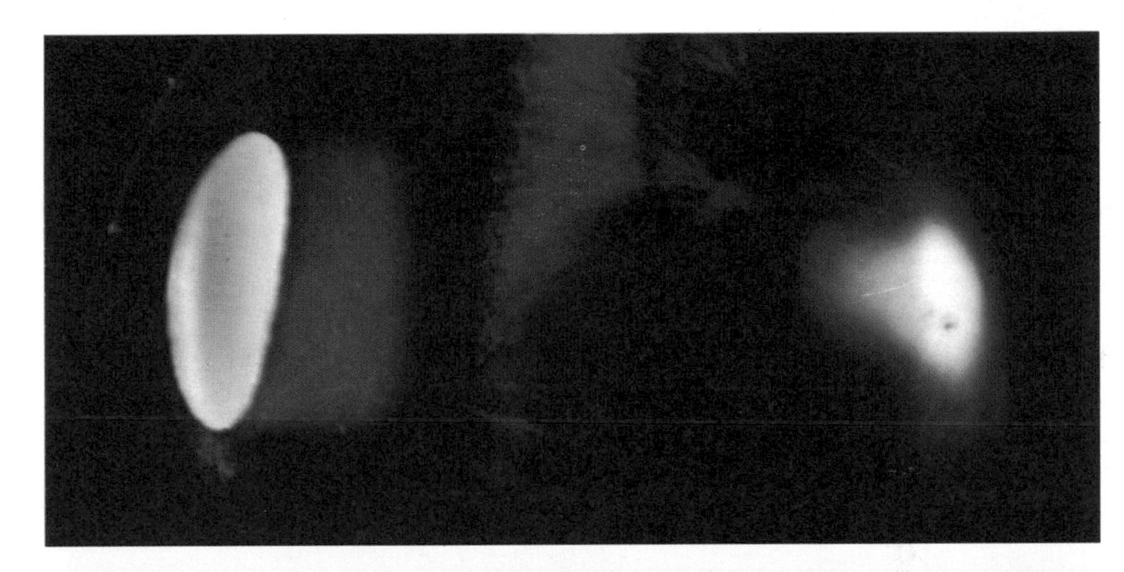

The lens is kept in place and in shape by thin fibers, ▷
which again are kept taut by being connected to the surrounding circular ciliary muscle like the spokes of a wheel. At rest and when viewing distant objects, the lens is relatively thin. To view near objects the lens thickens. This process, called accommodation, occurs when the ciliary muscle contracts, producing a change similar to what happens when we purse our lips. The tension in the fibers holding the lens lessens and the lens, which is elastic tissue, assumes a more balloonlike shape. The greater degree of curvature allows greater refraction of light rays so that the image of a nearby object will come to focus sharply on the retina.

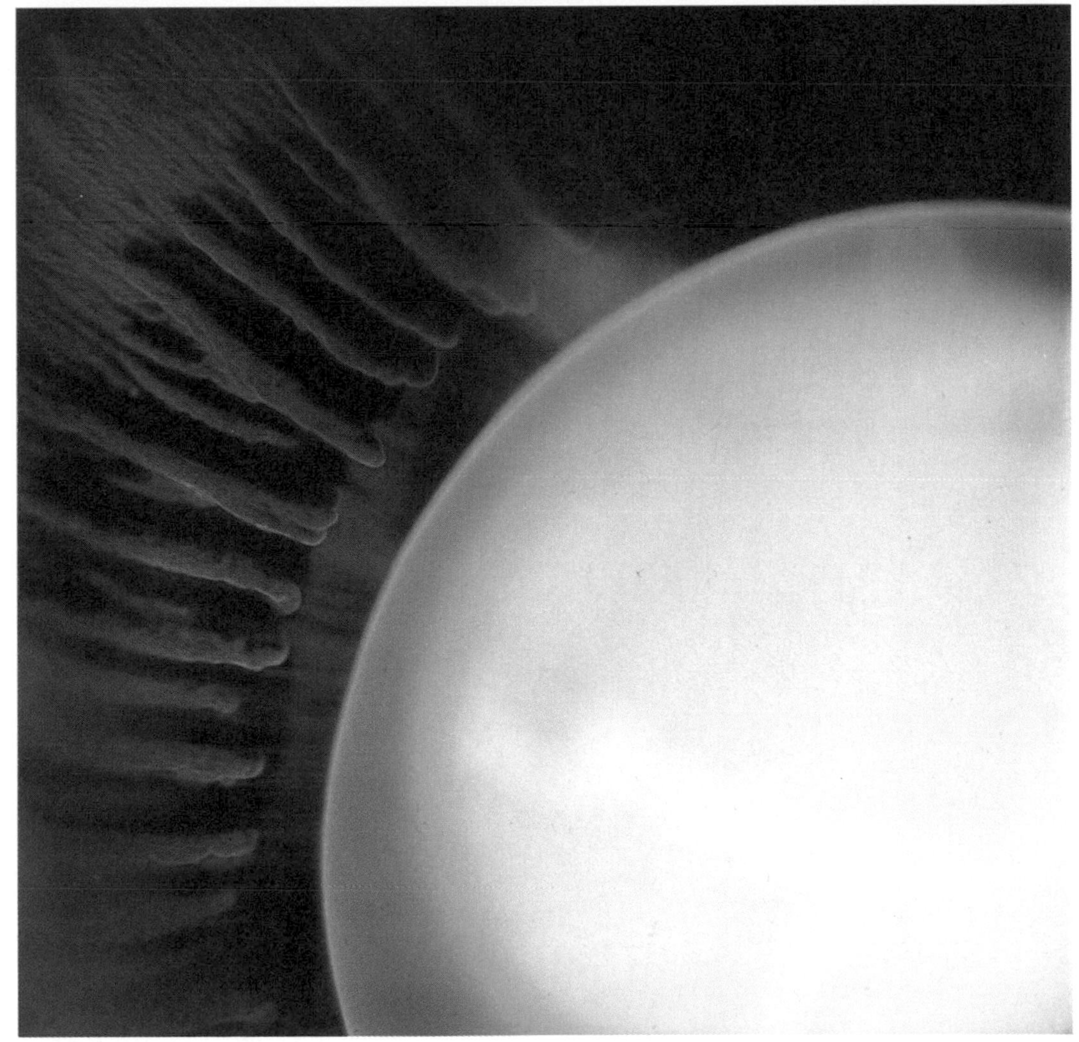

◁ **The ciliary ligament**
viewed from within the eye, in natural color and magnified about 400 times.

The retina reflecting an image of some flowers
The image has been projected into the eye and appears at the back. The optical arrangements within the eye make it comparable to a camera, but the analogy stops there. Visual perception depends on the electrical signals coded in the optic fibers. The position of the retinal image here suggests that the viewer is looking at flowers at a dark time of day, perhaps twilight. In bright light the image of an object under focus will generally be concentrated at the macula lutea.

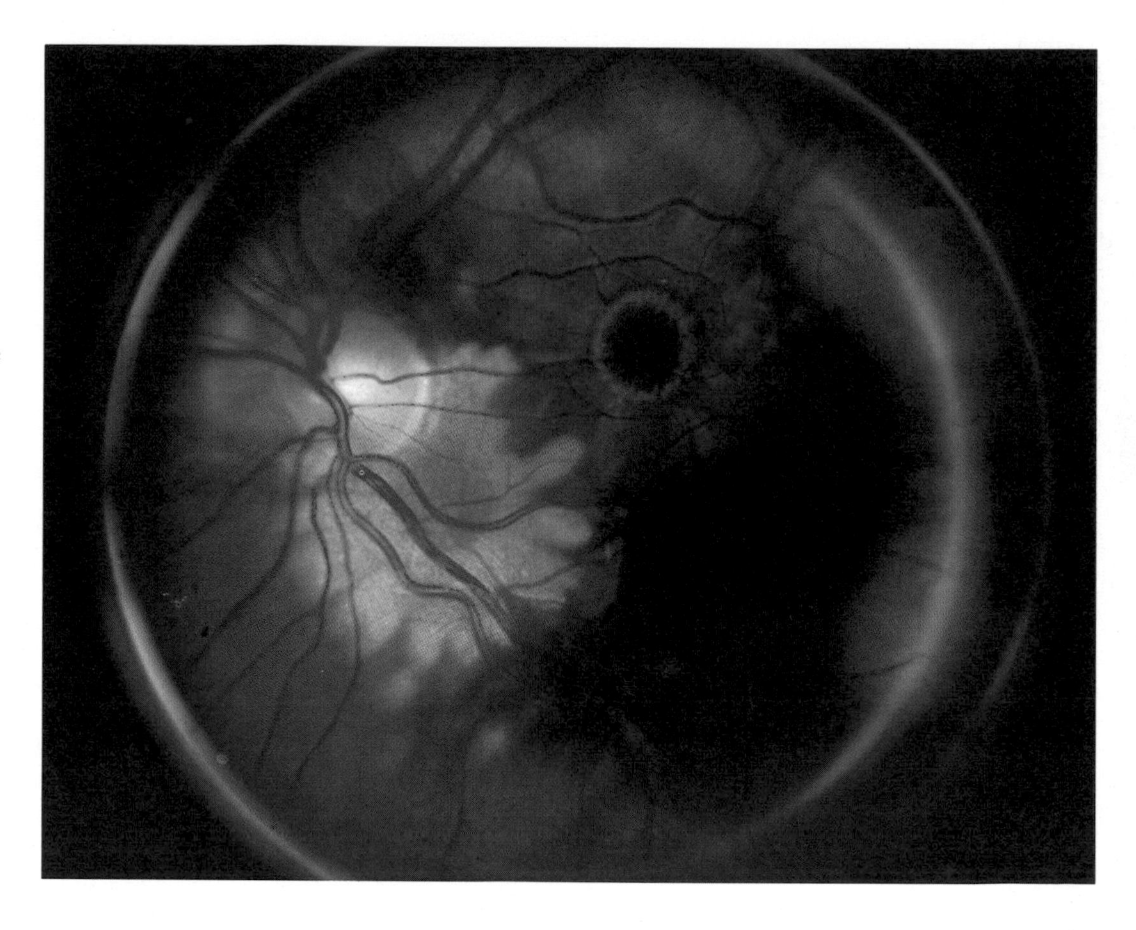

built up of thin concentric layers like an onion. The lens continues to grow throughout life, but at slower and slower rates. In the oldest parts, the cellular layers become increasingly isolated and cut off from a sufficient supply of nourishment or oxygen. Eventually they die. They then harden, and it becomes difficult for the lens to adjust its curvature for viewing near objects. That is why older people often cannot see well at close distances.

The vitreous body fills the eyeball behind the lens and the ciliary body. It consists of a clear gelatinous material and is the largest of the refractory components in the eye.

Eye Movements

Each eye is moved by six ocular muscles. The superior, the inferior, the lateral, and the medial rectus ocular muscles move the eyeball up, down, to the side, or toward the nose, respectively. The superior and the inferior oblique ocular muscles enable eyes to be rotated right or left.

It is easy to understand that we can see an object moving sideways across the field of vision. However, we can also judge the distance between the back and front of the object, that is, we can define its depth in space. In order to do this we normally make use of both eyes. The images seen by the two eyes are not identical but supplement each other. We perceive and interpret the small differences between the two images so that our vision is stereoscopic. Our stereo vision becomes evident from a very easy experiment: look at the branches of a tree with both eyes first. Then cover one eye for a little while without moving the eyes. On uncovering the eye again, the branches will stand out in depth.

It takes time for the eyes to be stimulated by the light rays from an object — a painting, for example — and it takes some additional time for the visual processing which enables us to perceive what that light stimulation was all about. Vision thus involves a certain time lag. This property is of importance in our perception of motion. Also, when we perceive an object in motion, memory processes collaborate by very briefly retaining the image of the movement of a moment before so that it can be compared with the present one. This visual time lag and immediate memory is central in the perception of movies. A strip of film is, after all, only a series of still pictures. At the rate at which it is shown, however, the difference between two consecutive pictures is so small, and the time between successive frames so short, that we enjoy the illusion of movement.

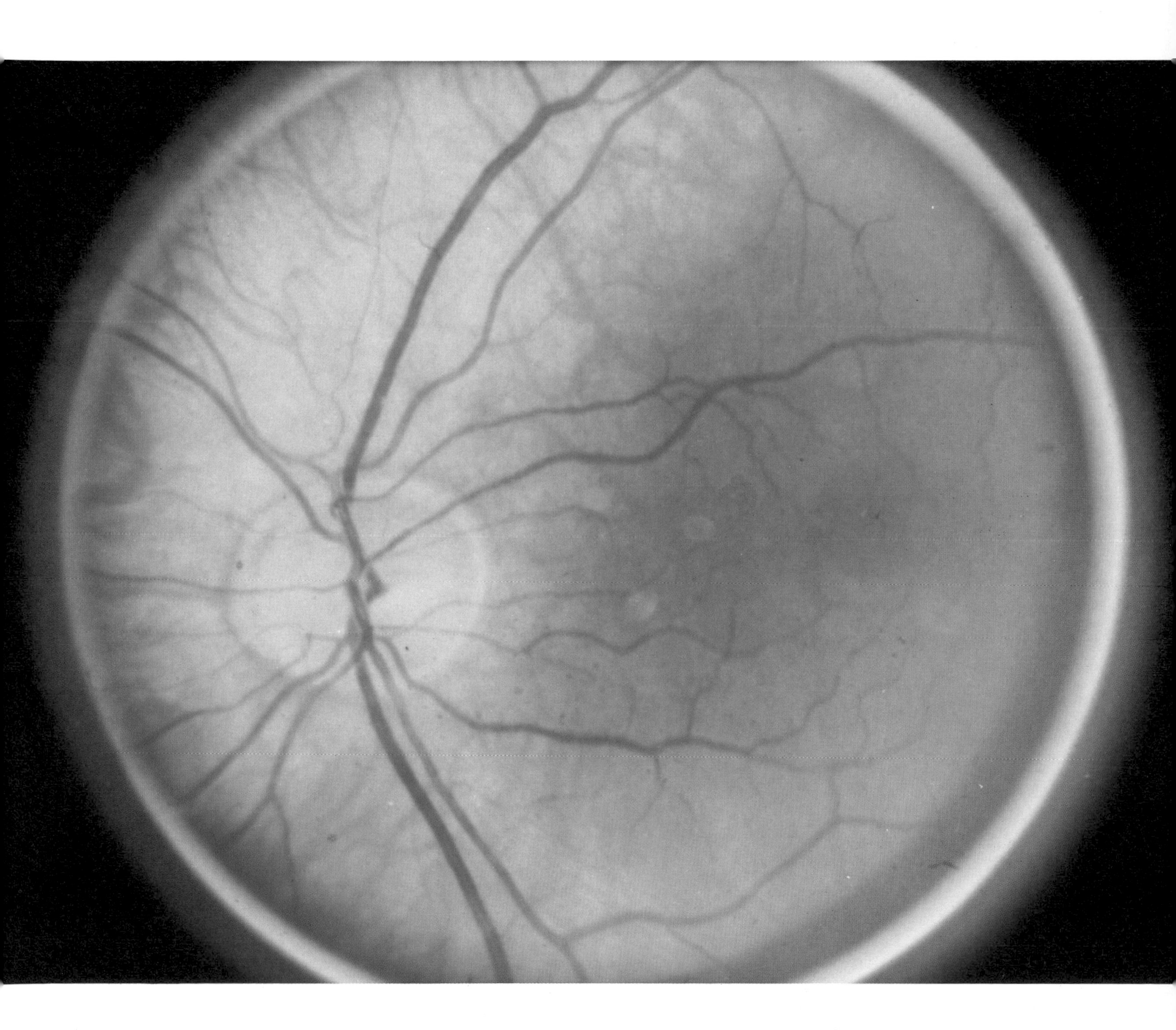

The fundus of an intact eye
It is routine for a doctor to examine the back (the fundus) of a patient's eye with a special mirror called an ophthalmoscope. The photo shows the fundus with retinal blood vessels and the blind spot where vessels and nerves exit from the eye. Near the blind spot is the macula lutea, the region in the eye that is especially rich in cones, the color-sensitive receptor cells.

The Macula Lutea

Macula lutea, magnified ▷ 25,000 times

Section through the macula lutea photographed in a light microscope. The slide was oriented in such a way that light entered the eye from the left as in the accompanying diagram.

The macula lutea, magnified 25,000 times The photo of the macula on the right-hand page was made with a scanning electron microscope. The macula is not circular but elliptical. The long axis of the ellipse is about 1.5 mm and is positioned horizontally in the eye. Current theory suggests that some cones are sensitive to longwave light (red), others to shortwave light (blue), and still others to light with wavelengths in between (green). Combination of these three colors (red, green, and blue) can produce all other colors seen in nature. The center of the macula is the most sensitive part of all, where vision is most acute. There are about two thousand cones there and no rods. The smallest are no more than a millionth of a millimeter thick. Each cone in this center is connected with a single nerve fiber. Further away from the macula, on the retina, the ratio of receptor cells to nerve fibers increases, and vision is less acute. The most peripheral rods enable us to perceive passing shadows or other less distinct occurrences in the visual field.

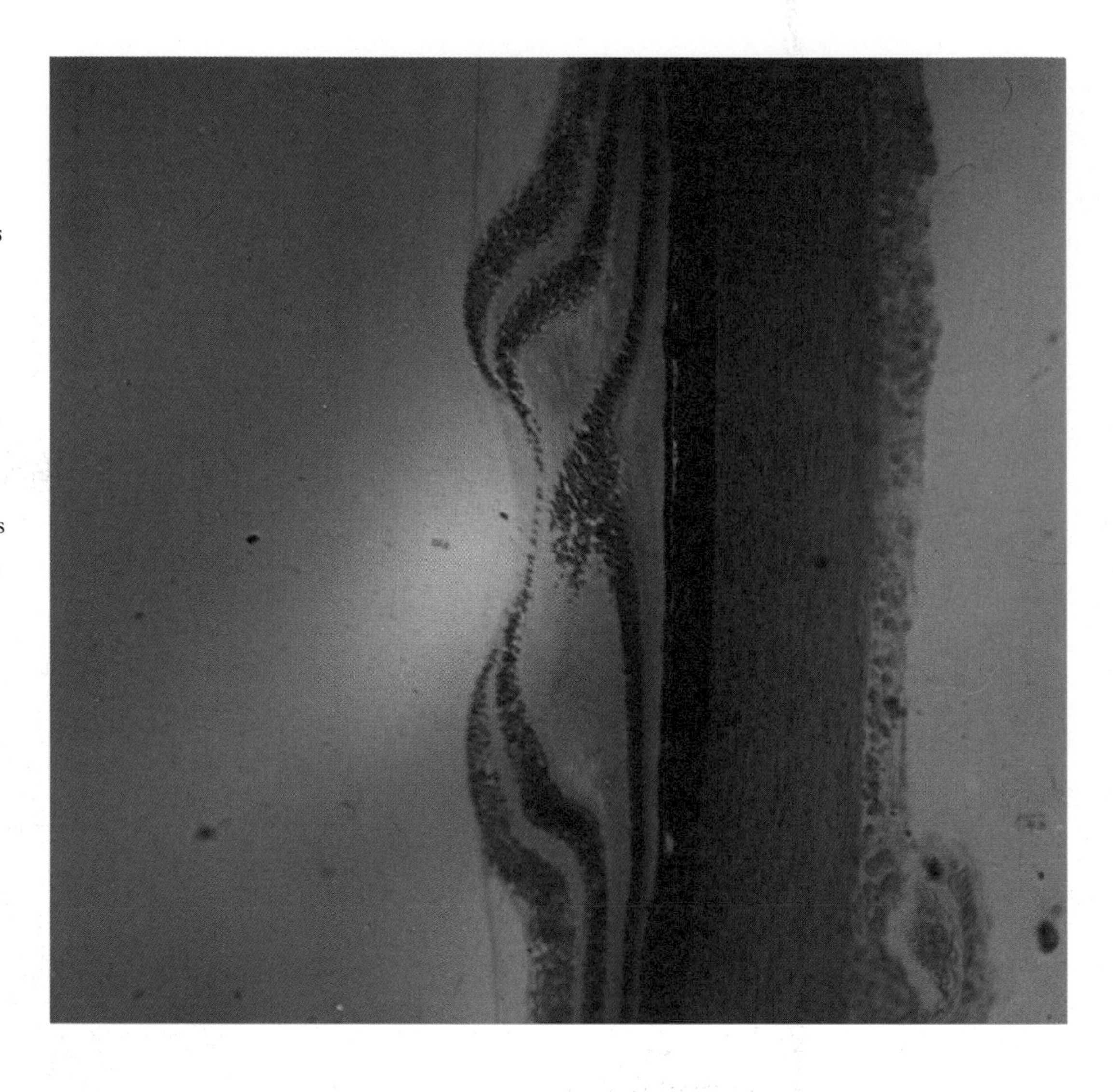

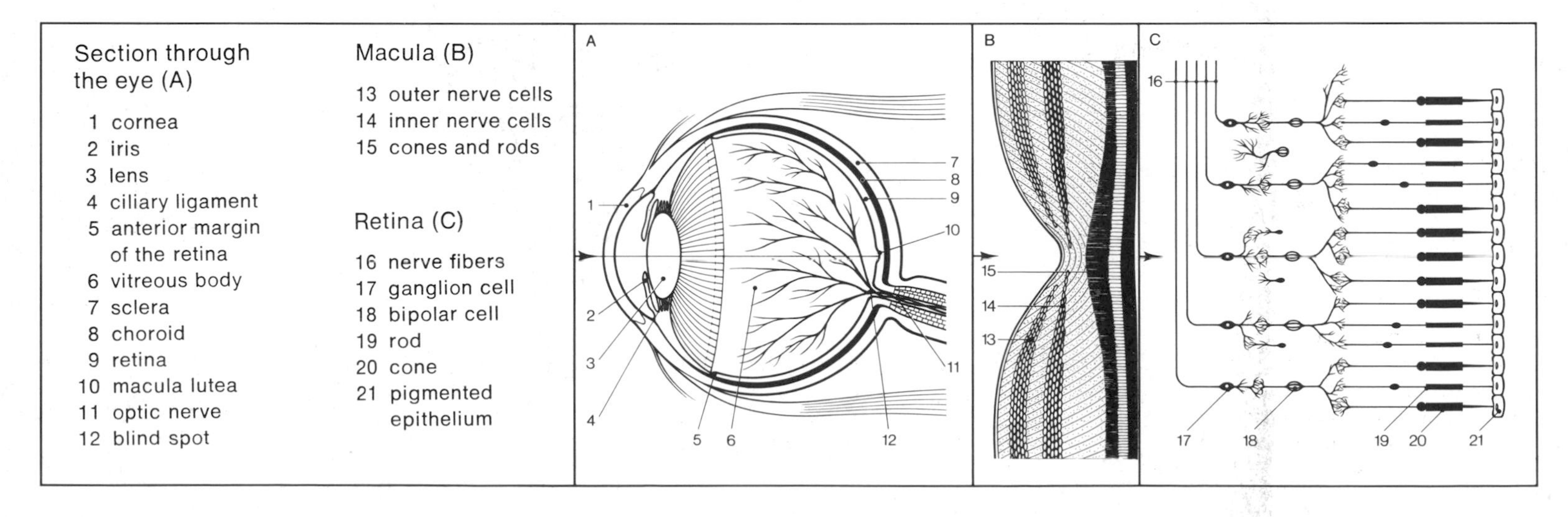

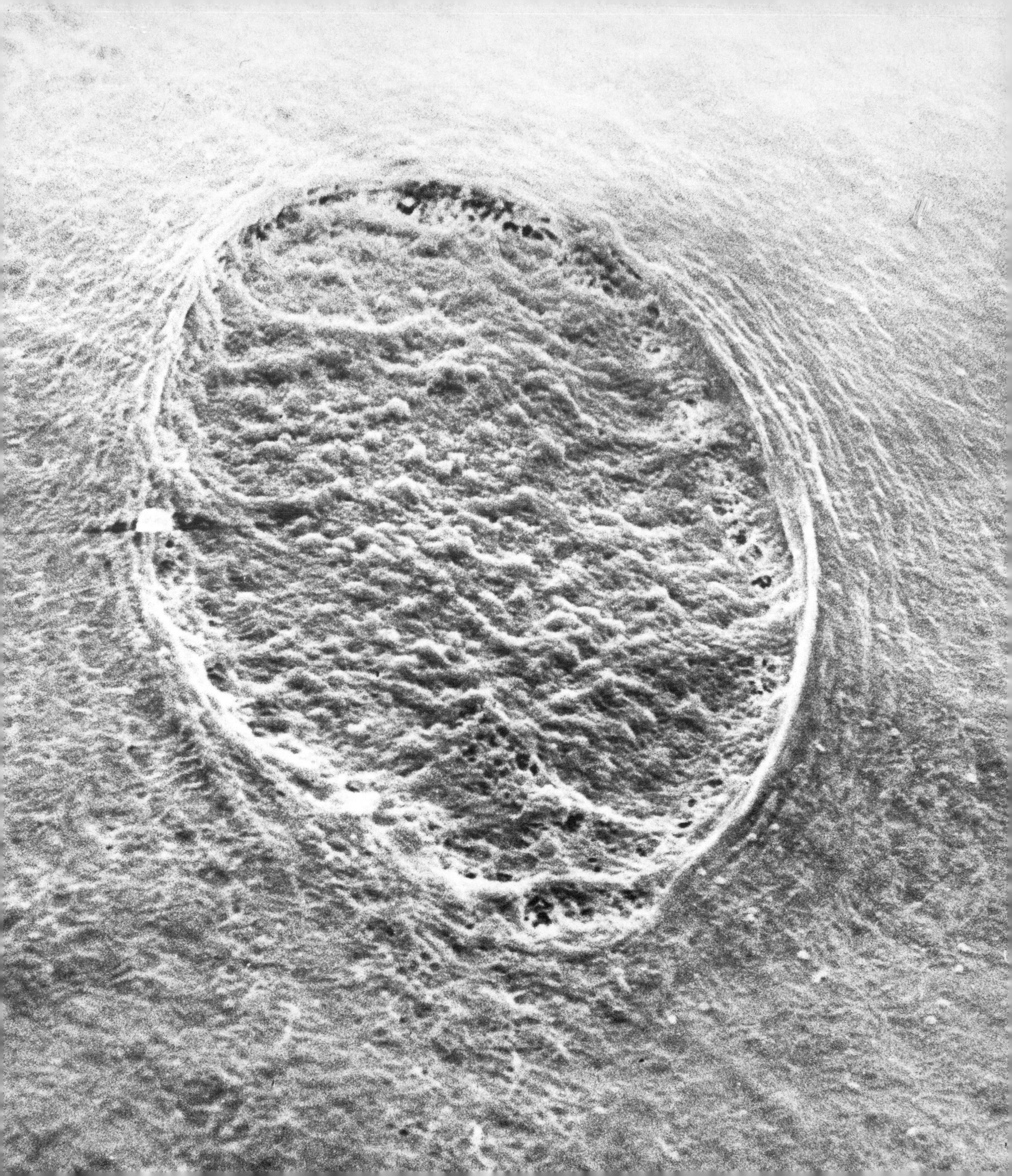

The Retina

Light reaching the retina from above △
The pigmented layer and a part of the thick sclera can be seen at the bottom of the photo.

The macula seen in a microscopic preparation ▷
The section is the same as that on page 192, but in this case the photo was taken in different light with different color effects. The photo was also made with light entering from above as in the previous photo. Seen from the bottom up are the sclera, the pigmented choroid, and the retina. The bottom layer of the retina consists of cones and rods. Above them are the layers of nerve cells. Note that these layers are particularly thin in the center of the photo. It is here, at the center of the macula, that visual acuity is best.

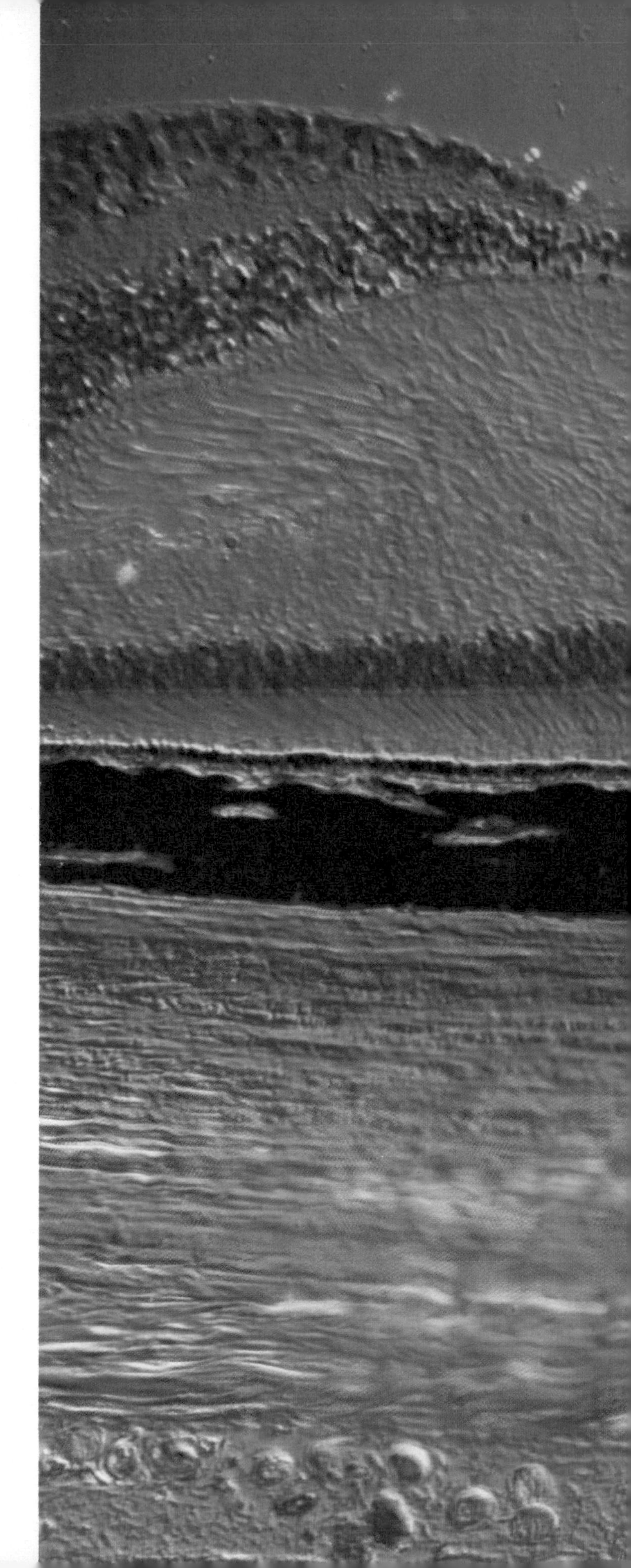

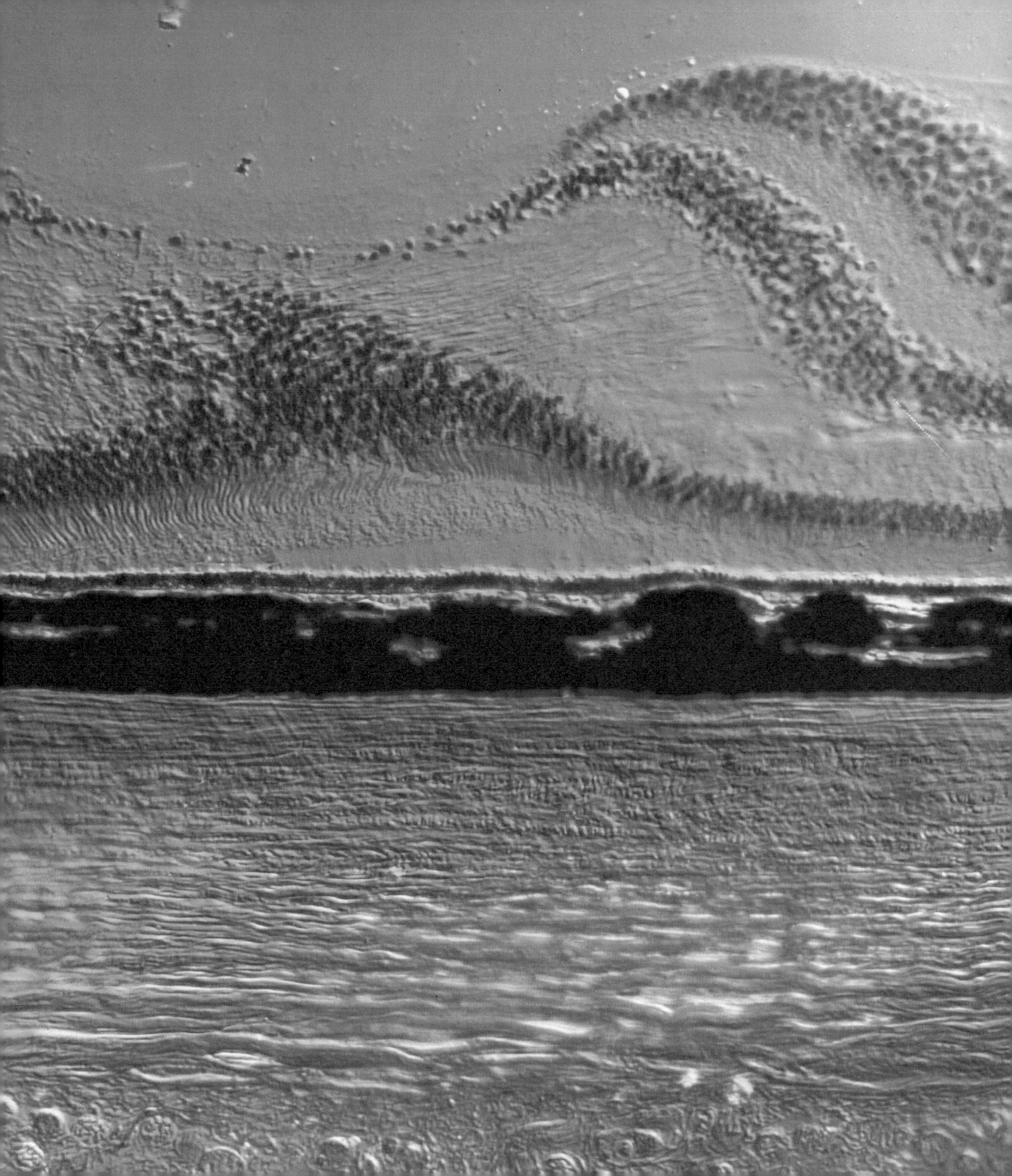

The Retina

Fat cones and slender rods,
magnified approximately 45,000 times. The left end of the photo illustrates how cones and rods are packed in the retina, with numerous rods squeezed in and around a single cone or a pair of cones. The cones and the rods are the visual receptor cells which convert light into electrical energy. There are about 120 million rods and 7 million cones in the human eye.

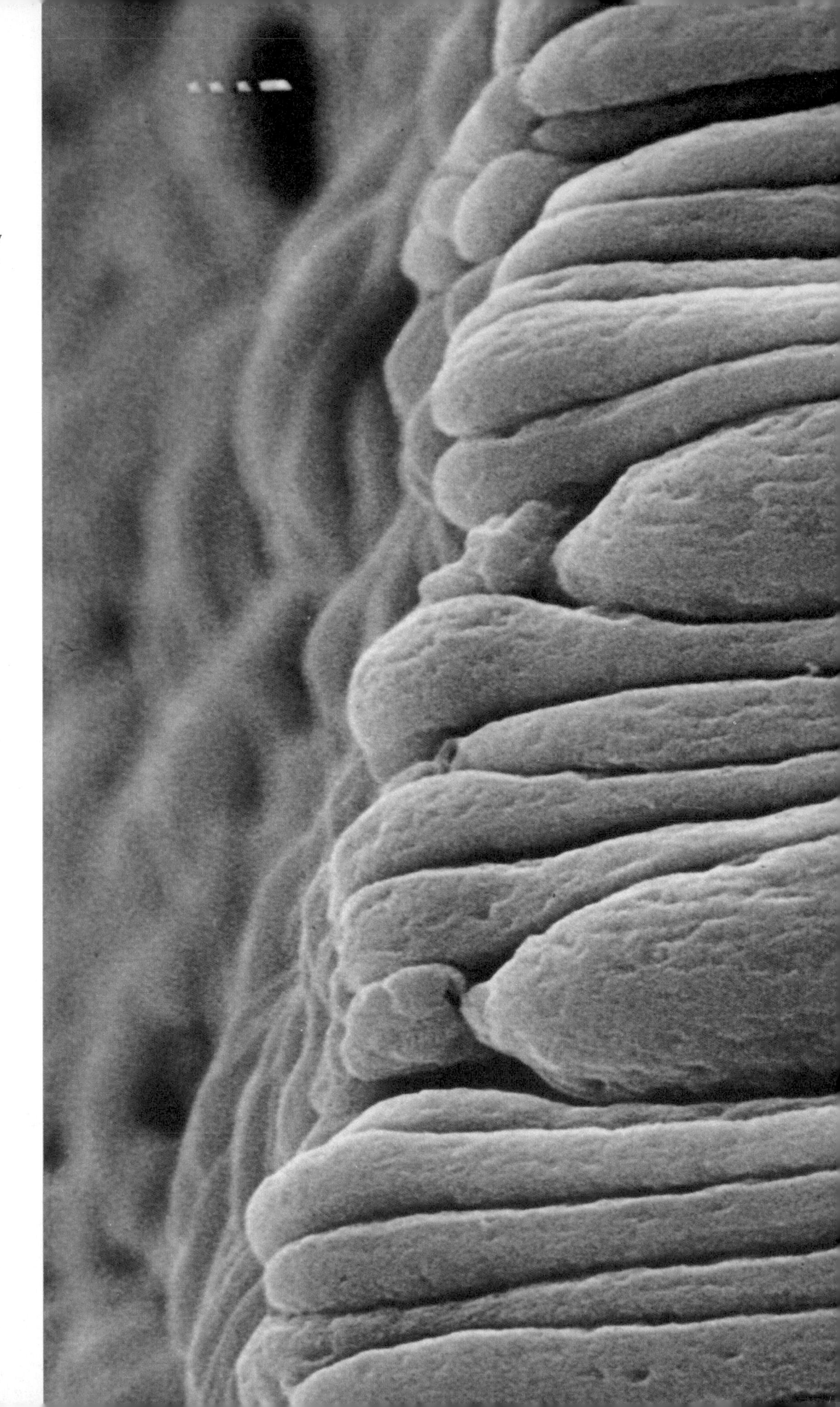

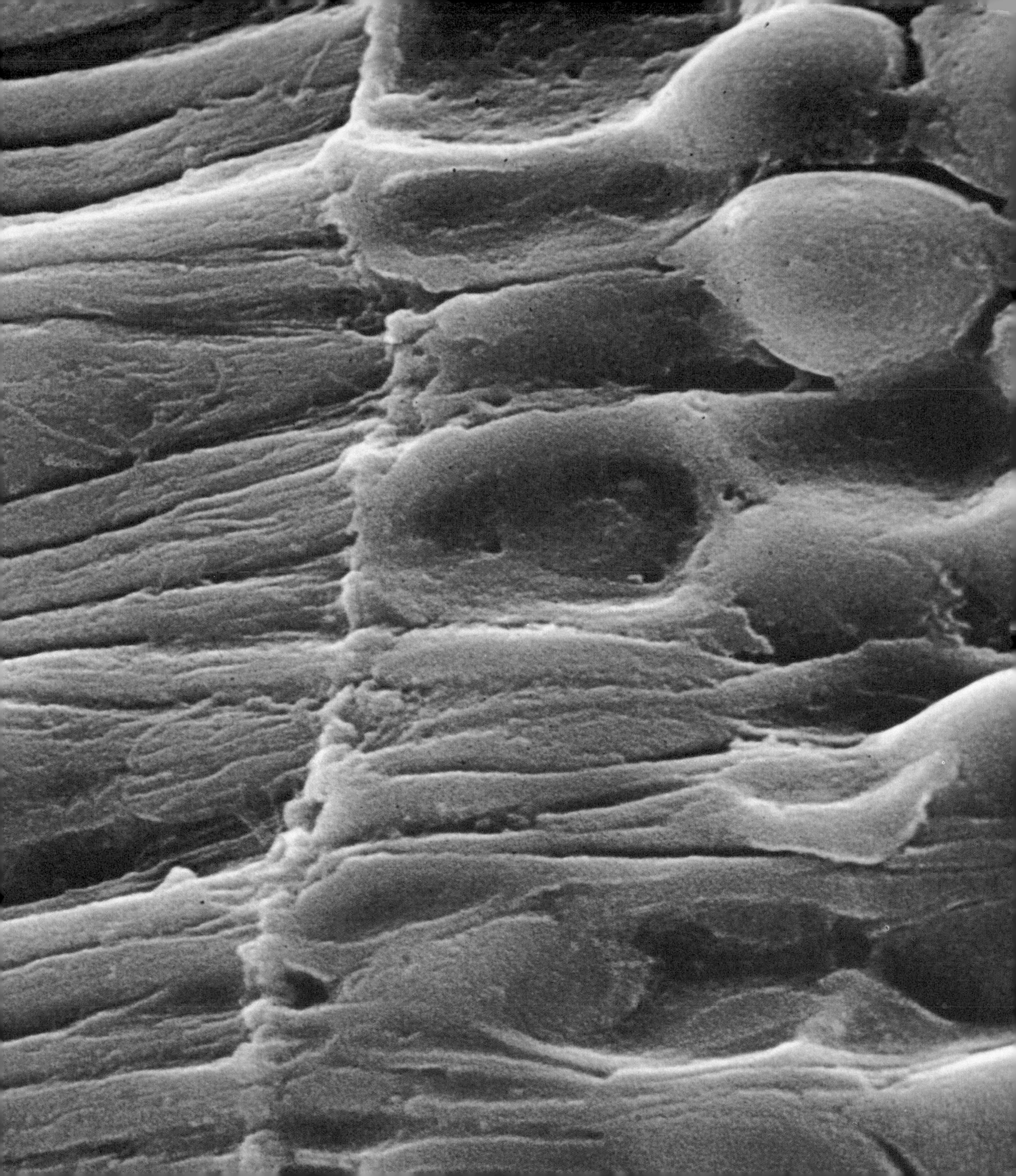

Tears

The eyelids are thin folds of connective tissue covered by thin skin. At the edge of the eyelid, the skin becomes a thin membrane, the conjunctiva. The function of the eyelashes is to prevent impurities from coming under the eyelids and irritating the eyeball.

Each eye has a lacrimal, or tear, gland located in the upper anterior portion of the orbit. Tears keep the surface of the eye constantly moistened. The salt-containing lacrimal fluid empties into small ducts in the nasal cavity. A grain of sand under the eyelid or emotional events of great sadness or joy cause tears to be shed so rapidly that they overflow. In the inner corner of the eye is a small pink piece of tissue which is a remnant of the nictitating membrane found in other species. In birds, this membrane is well developed and can be drawn across the eye like a curtain.

The Retina

The innermost of the three layers covering the eyeball is the retina. Its name comes from the Latin *rete*, a net, and refers to the network formed by the retinal vessels.

Since the retina contains the light-sensitive rods and cones, it is the most essential portion of the visual organ. Light must pass through the innermost layers of the retina — the nerve cells, blood vessels, and supporting cells — before it can reach the rods and cones. Through the action of the photosensitive pigment, visual purple, or rhodopsin, light energy is converted into electrical impulses. These impulses are conducted in nerve fibers in the retina which are bundled into the optic nerve. The point where the optic nerve leaves each eye is known as the blind spot, the optic disk. The optic nerve is not a "nerve" in the sense that it is a part of the peripheral nervous system in the body, but, like the retina, it is a part of the central nervous system itself.

The retina, magnified 20,000 times

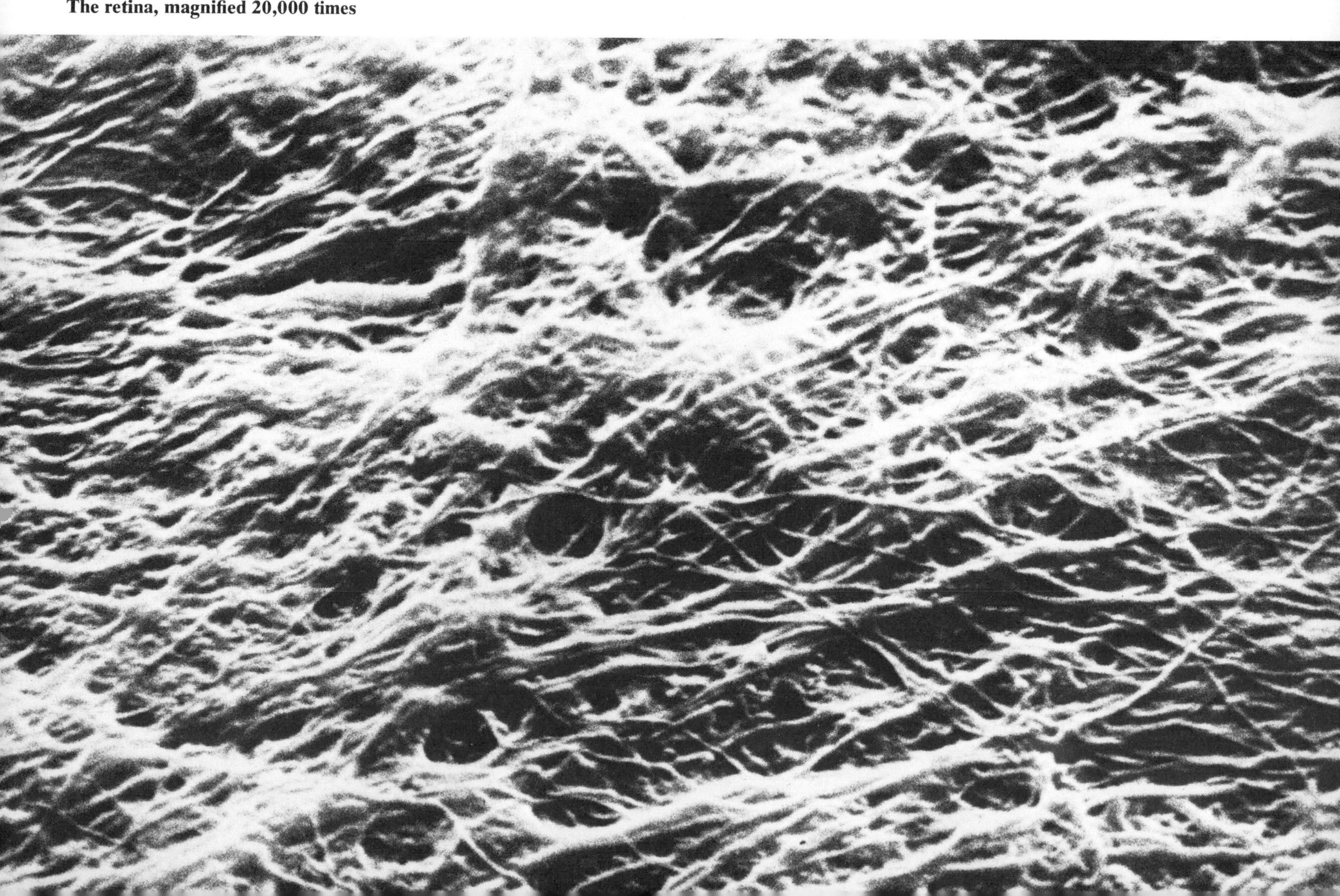

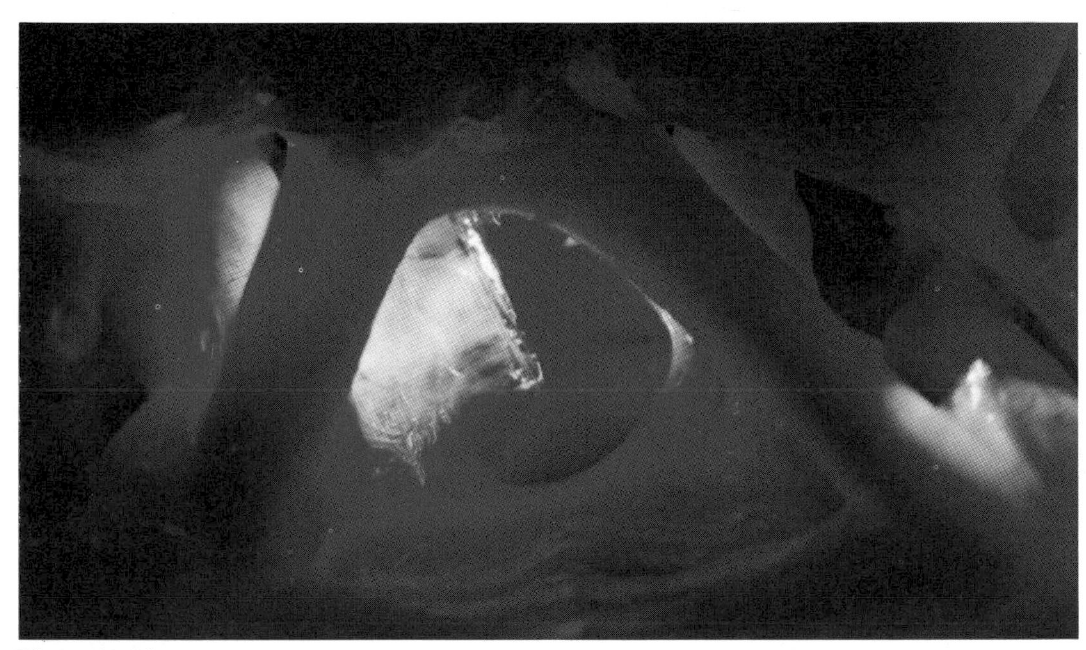

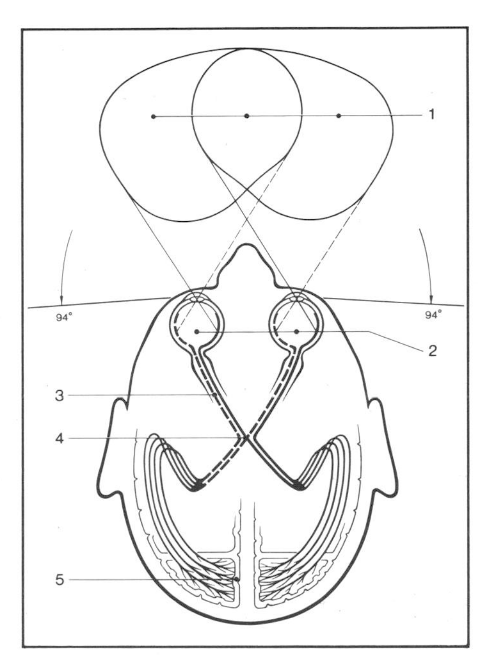

The optic chiasma,
where parts of each optic nerve cross to the opposite hemisphere on their way from the eyes to the brain. The optic chiasma is located in the midline, just below the cerebral hemispheres.

The diagram on the right shows that the visual fields (1) of the two eyes overlap to some extent. Objects directly in front of the eyes are seen distinctly, objects toward the sides, less clearly. Two different images are projected on the retinas (2), one from each visual field. Nerve impulses travel in the optic nerves (3) and continue across the optic chiasma (4) to the visual centers in the brain (5).

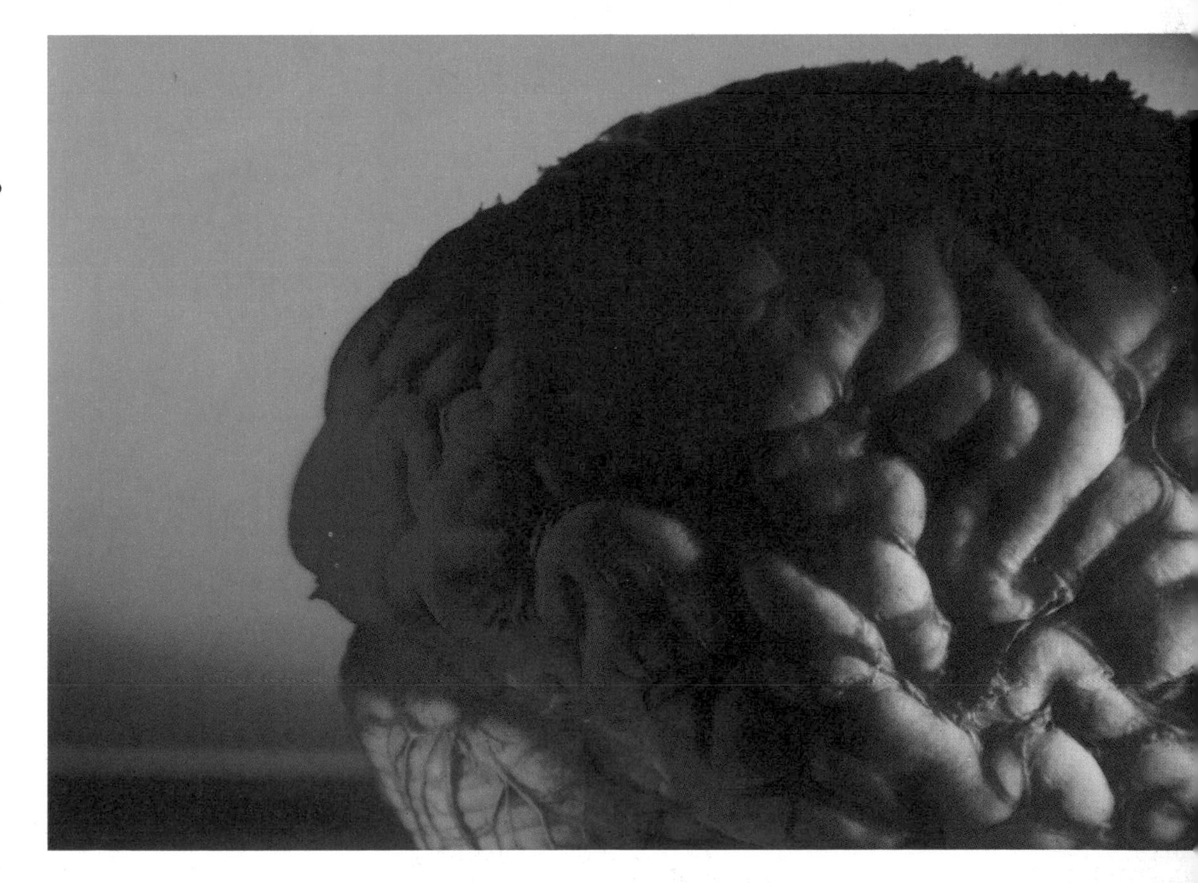

The visual center ▷
is located at the rear of the brain in the occipital lobes. In the photo, a red spotlight is shining on this region.

◁ **The retina, magnified 20,000 times**
in the scanning electron microscope. The surface of the retina is a dense underbrush of nerve fibers. The optic fibers transmit electrical impulses to the brain.

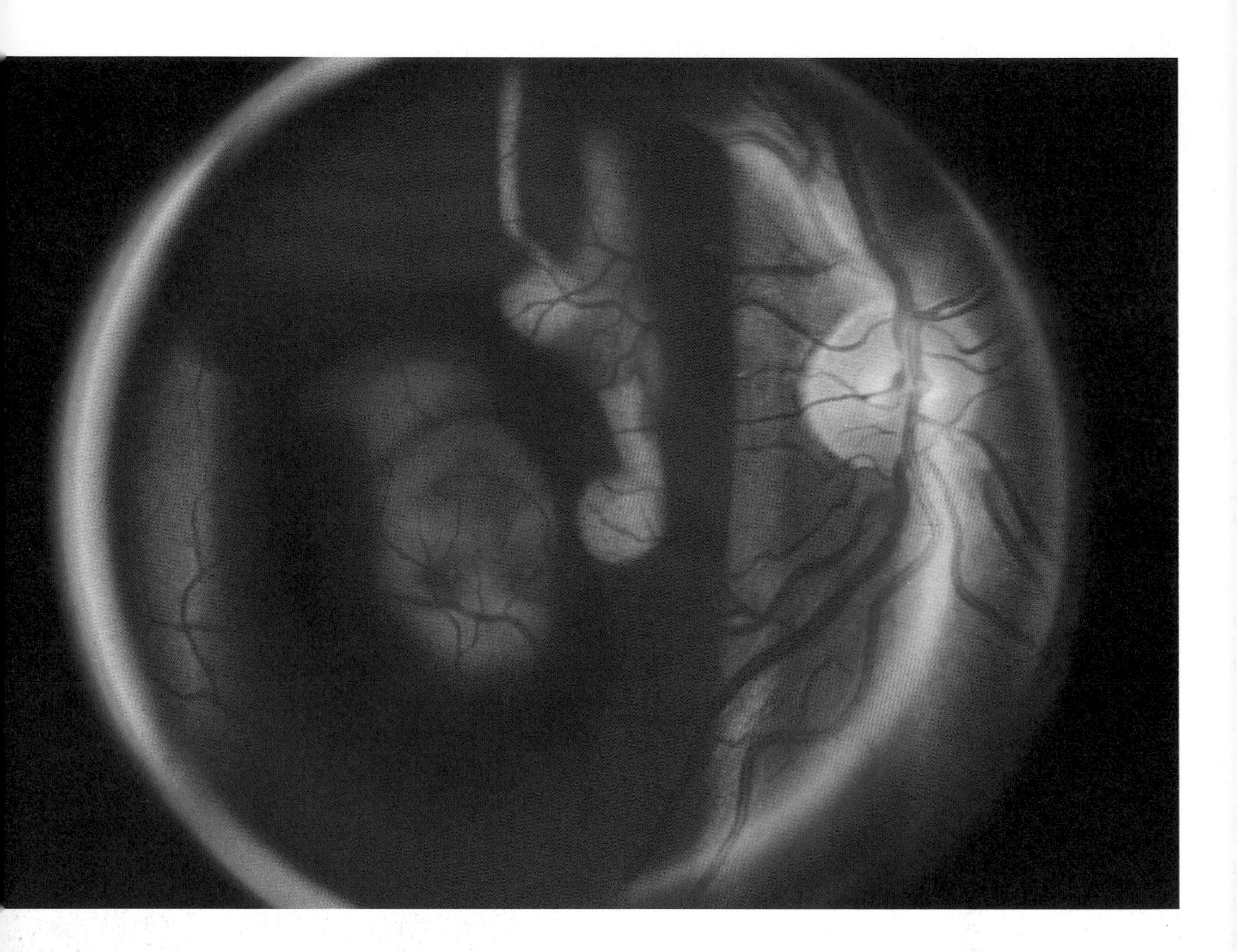

The camera-like parts of the eye
Here the inverted retinal image of a girl using a telephone illustrates that the eye in some ways behaves like a camera with lens, diaphragm, and retinal "film." The "photographer" was focusing on the girl's lips, consequently her lips and chin are seen in the macula of his eye, which appears as a relatively dark area. The image is projected onto the retina when the light from the object is refracted at the cornea and the lens. The picture was taken by means of a specially constructed camera, which could be directed into the eye through the pupil. The retinal image is inverted as the laws of optics dictate. At the right in the photo is a yellow region where blood vessels and nerve fibers exit. Since there are no cones and rods present here, it is called the blind spot.

Hearing

Information from the external environment reaches us through many channels. But only when the messages from the sense organs are combined with the experience we gain in living — especially during the early years — do we have the understanding and perception that enable us to act effectively. The infinitesimal waves that make up visible light are picked up by the eye. The pressure waves vibrating in the air as sounds are picked up by the ears. Chemical stimuli excite the organs of smell and touch.

In some ways the human ear is not as sensitive as the ear of many animals. Moreover, some animals have movable external ears. These help to funnel sound as well as aid in determining where it comes from.

It is often stated that, from a social point of view, hearing is our most important sense. Language, the special human communication system, depends on our ability to perceive particular sounds transmitted to the ear. Formally created combinations of sounds other than the language — music, for example — are also expressions of our higher mental life.

Sound is generated by a vibrating object. The sound waves are propagated in all directions from the sound source, generally reaching a human ear through air. The velocity of sound through air is 1,080 feet per second (at 68°F). At the same temperature, sound travels about 3 miles per second in iron. Like light or other forms of wave motion, sound waves can be reflected, absorbed, or refracted. Sound waves lose energy the farther they travel from the source; we perceive this as a decrease in loudness.

Sound waves are collected by the external ear and conducted to the eardrum, which starts vibrating when sound reaches it. From the eardrum, the sound waves pass to the three auditory bones in the middle ear and from there to the fluid-filled bony cochlea in the inner ear. At this stage the sound energy is ready to be converted to nerve impulses. This is the function of the auditory receptor cells, some 24,000 in each ear. The excited receptors stimulate auditory nerve fibers, which conduct signals to the various auditory centers in the brain. The processing of the signals gives rise to the sensation of sound and the meanings sounds have for us.

If you put a finger in each ear, you close off air-conducted sounds. It is then possible to hear a pulsing, buzzing sound. This is mainly due to the contractions of the jaw muscles.

The eardrum
is set into vibration by sound waves in the air. These vibrations initiate the process by which sound energy reaches the organ of hearing with little loss of energy. The photos illustrate the mobility of the eardrum — it can literally bulge in and out. The small bone attached to the center of the eardrum is the malleus (hammer). The movements are very exaggerated here to illustrate how the eardrum functions.

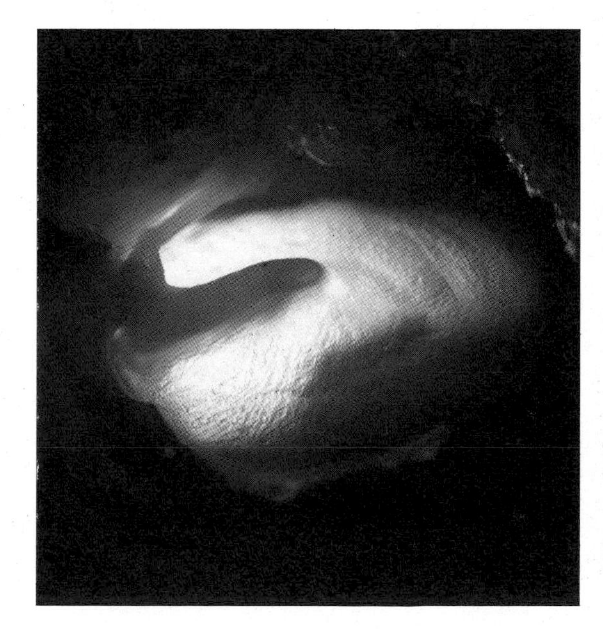

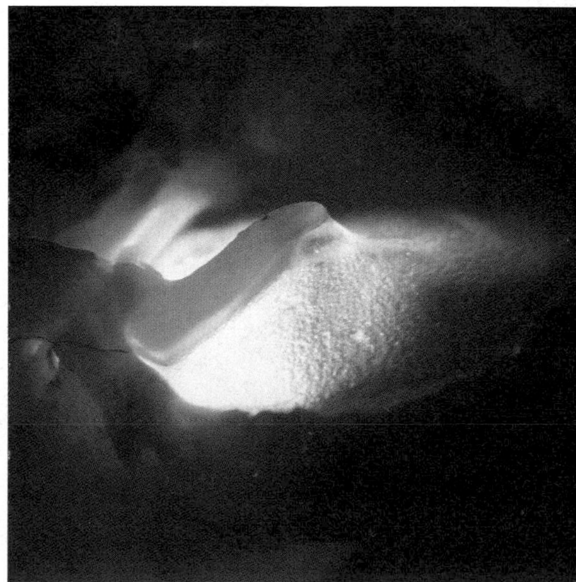

A view inside the external auditory canal
The eardrum is at the end of the canal.

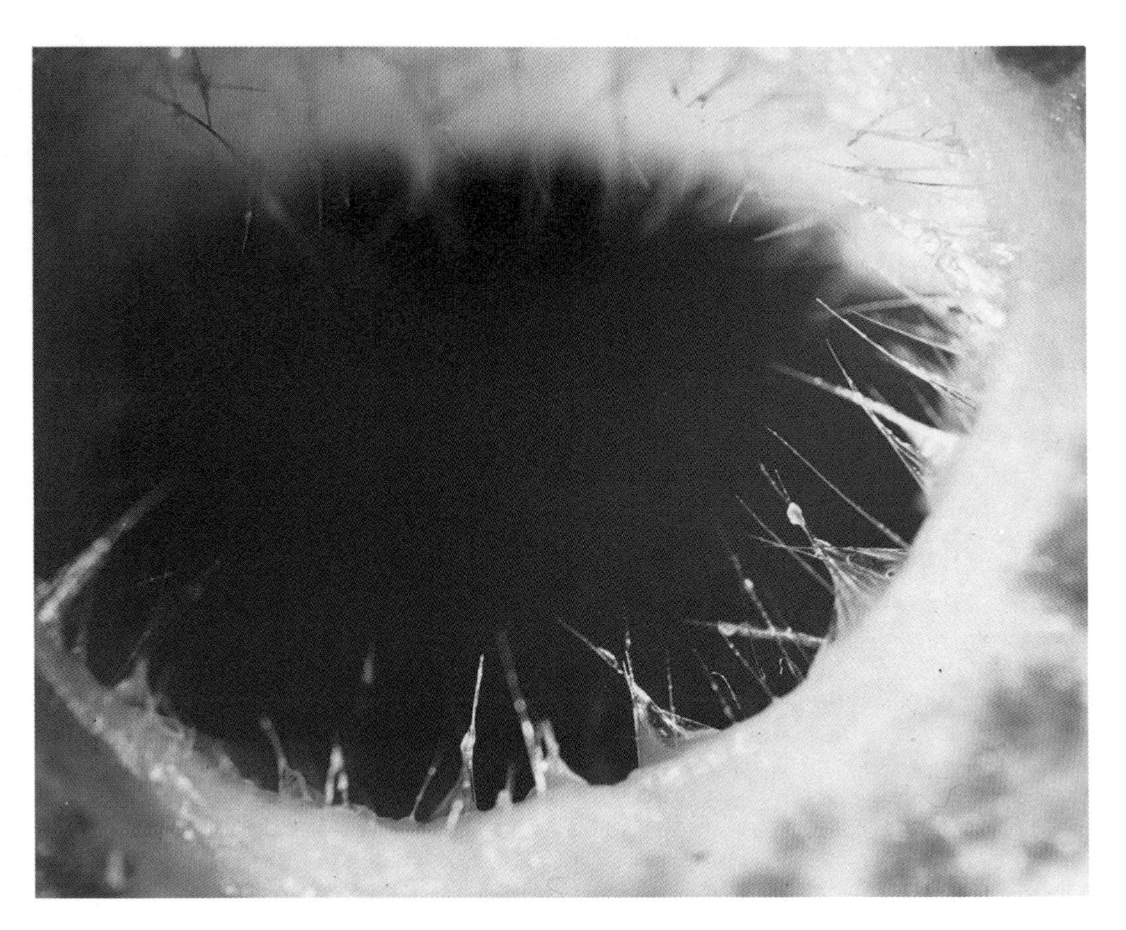

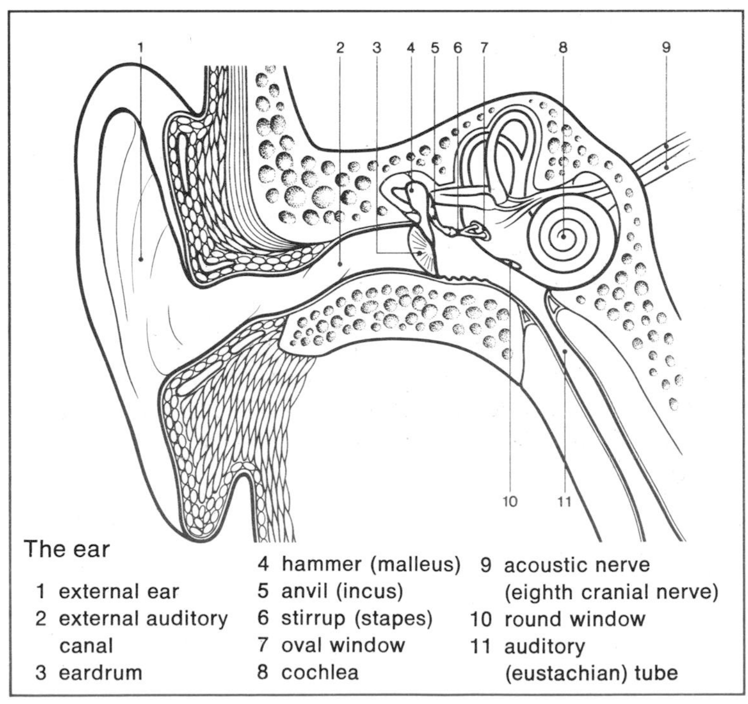

The ear

1 external ear
2 external auditory canal
3 eardrum
4 hammer (malleus)
5 anvil (incus)
6 stirrup (stapes)
7 oval window
8 cochlea
9 acoustic nerve (eighth cranial nerve)
10 round window
11 auditory (eustachian) tube

The External Ear

Many mammals have large and easily movable external ears. The human external ear is relatively small, flat, and usually cannot be moved — except by those who can wiggle their ears. The way we localize the source of a sound depends largely on the slight differences in the time of arrival of a sound stimulus at each ear. We actively cooperate in this nervous-system process by turning the head to see what effect this will have. Our ability to localize sounds is quite good — it is within a few degrees (considering the head as the center of a circle).

The external ear is essentially a fold of skin supported by funnel-shaped, elastic cartilage. The lower portion, the earlobe, does not contain cartilage but is filled with soft fatty tissue. The earlobe is well supplied with nerve endings and is sensitive to touch. The central inner portion of the external ear is the external auditory canal. It forms a slightly curved passage, less than an inch long, leading to the middle ear. The outer part of the canal is made up of connective tissue and cartilage; the deeper part is surrounded by bone. The curvature of the canal

Earwax in the external auditory canal
Droplets of wax are seen on the skin and hairs. Sometimes too much wax is present, or it becomes hardened. This may result in the formation of a wax plug, which blocks the canal and inhibits the movements of the eardrum, so that hearing is reduced. Gentle flushing of the canal with lukewarm water dissolves the wax plug, and restores hearing. As there is some risk of damaging the delicate eardrum, the removal of a plug should be left to a doctor or nurse.

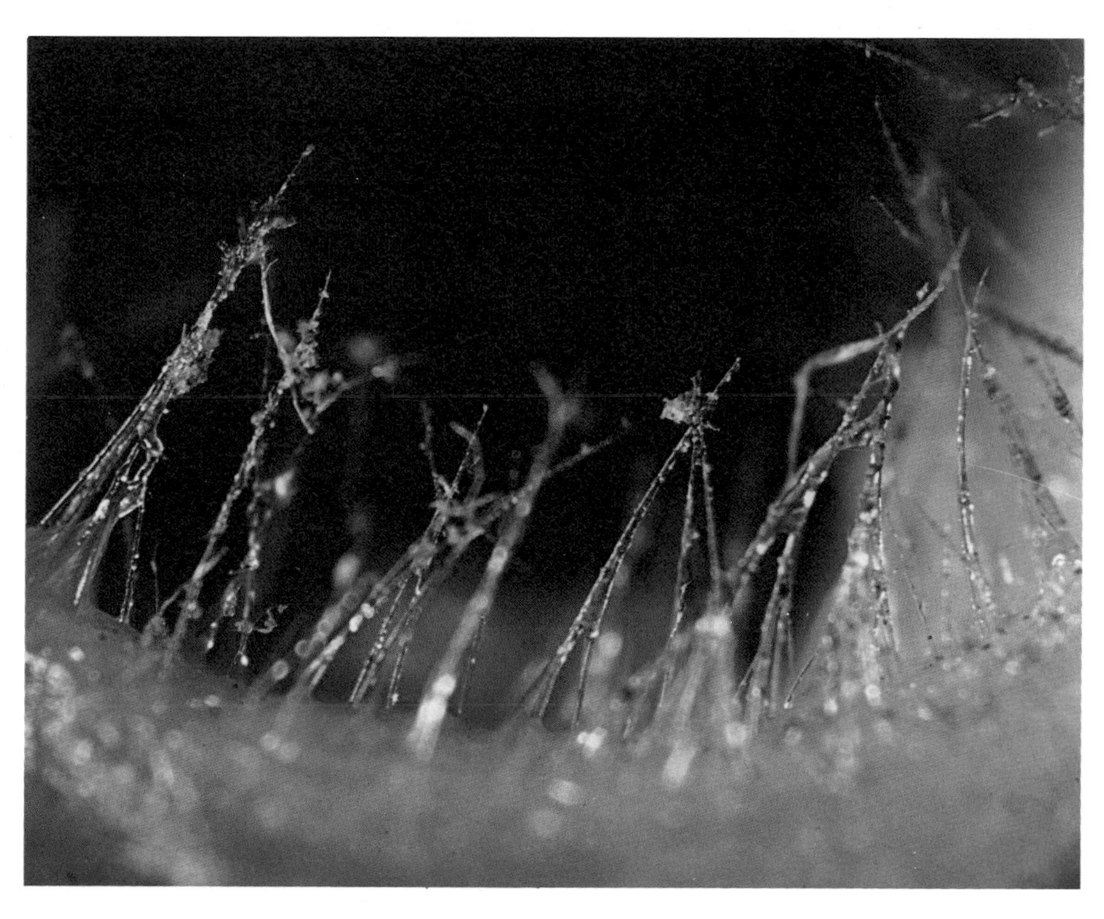

The eardrum behind a yellow wax plug

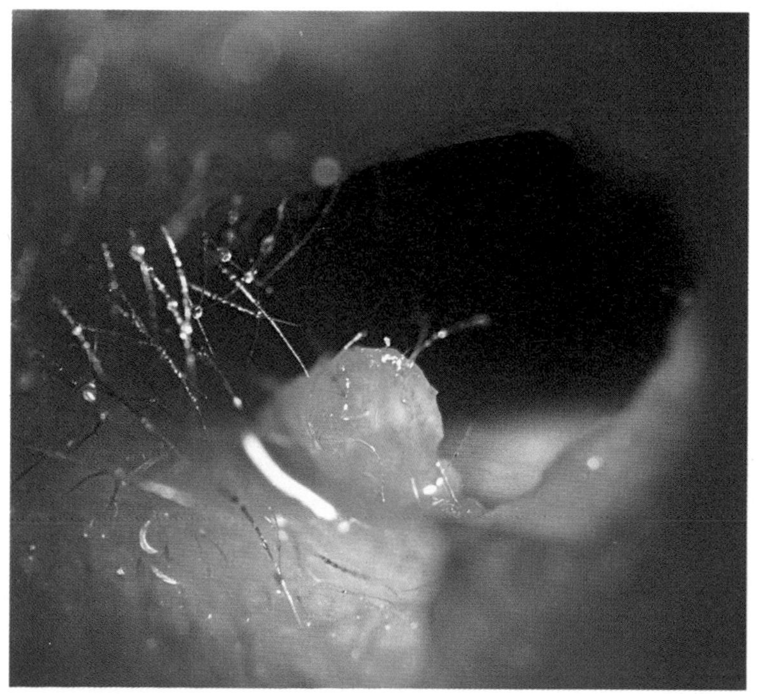

makes it relatively inaccessible to large objects. Small dust particles or insects, on the other hand, can enter, but usually become embedded in the earwax produced by glands in the skin lining the canal. The structure of these glands is similar to that of sweat glands, such as those in the armpits.

The external auditory canal is closed by the eardrum, the tympanic membrane. The eardrum is an oblique, almost circular, and very thin membrane of connective tissue. The midportion curves inward a little so that the resting eardrum is slightly conical in shape. The eardrum forms a natural boundary between the external and the middle ear.

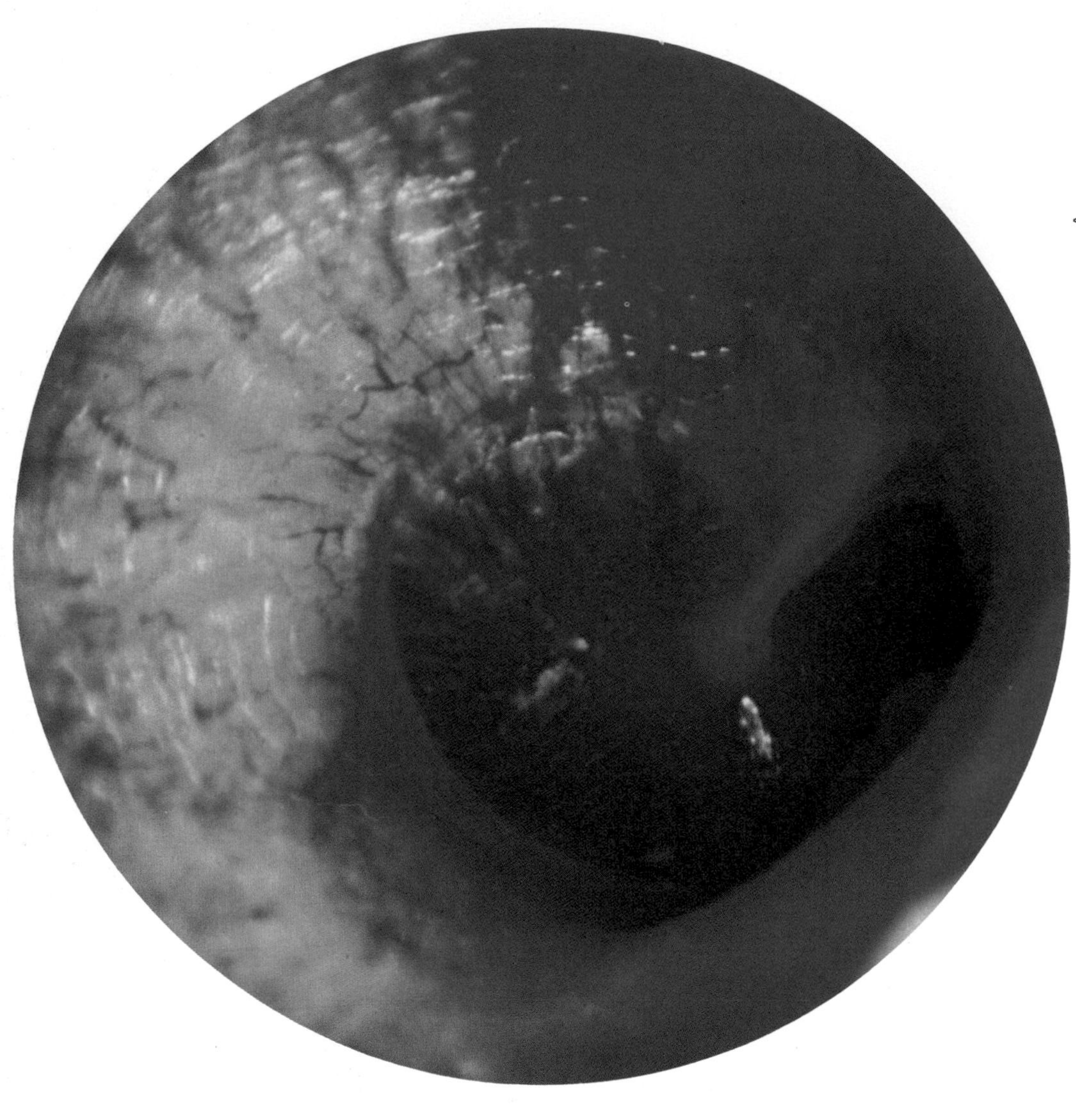

◁ **The eardrum from outside**
The diameter of the eardrum is approximately 15 mm. While it is composed of several layers of connective tissue, it is so thin that the attachment of the malleus, the first of the bones of the middle ear, can clearly be seen through the tissue.

Exterior view of the eardrum, under higher magnification ▷
The handle of the malleus is again seen through the membrane. When the lowest perceptible sound vibrations reach the eardrum, it moves only about a millionth of a millimeter back and forth. This minute movement is further dampened by the auditory bones. If a small insect should fly into the auditory canal, the flapping of its wings may sound like the roar of a jet engine.

The Middle Ear

The middle ear consists of a system of small cavities in the temporal bone of the skull. The walls are lined with mucosa and the cavities are air filled. The three auditory bones are situated in the largest cavity, the tympanic cavity. This cavity is approximately 4 mm deep and 8 mm in diameter. The outside wall consists primarily of the eardrum. The inside wall, separating the middle and inner ear, consists of solid bone in which there are two small openings. The upper one is the oval window, the lower, the round window.

The auditory bones or ossicles — the hammer (malleus), the anvil (incus), and the stirrup (stapes) — convey airborne sound waves from the eardrum to the oval window. The auditory bones are the only skeletal elements that do not increase in size as the rest of the body grows. They are fully developed in childhood. Part of the hammer is attached to the eardrum. Its other end articulates with the anvil, which in turn is in contact with the stirrup, whose footplate covers the oval window. When intense vibrations — very loud sounds — strike the ear, two small muscles, the tensor tympani and the stapedius, contract. These restrain the bones' movements, to prevent damage to the delicate structures of the inner ear. The effect of the bones' action is to deliver the sound pressure originally applied to the eardrum to the much smaller surface of the oval window. The result is a twenty-two-fold increase in pressure.

The tympanic cavity opens into the upper portion of the pharynx through the auditory, or eustachian, tube. The one-and-a-half-inch long tube is usually closed and its walls collapsed. But it opens during swallowing, allowing air to pass through. This equalizes the air pressure on each side of the eardrum. If the external pressure alters without a compensatory change in the tympanic cavity, we suffer a partial deafness. This happens

The Middle Ear

sometimes when riding in a fast elevator, driving a car up and down steep hills, or when flying. The deafness and uncomfortable feeling disappear if you swallow a few times. Some discomfort is also experienced when the auditory tube becomes stopped up, as a result of a cold and swelling in the throat. In that case it doesn't help to swallow, for air can't get in to equalize the pressure difference. This can make flying when you have a cold very unpleasant.

Young adults can generally hear sounds with frequencies between 20 and 20,000 cycles per second. Infants may be able to perceive sounds up to 40,000 cycles per second. With increasing years, the eardrum thickens, the tissues in the middle ear become more rigid, and we experience hearing losses. The ability to perceive high sounds usually diminishes first. Some tests have shown that men over age forty experience an average monthly loss in sensitivity in the upper frequency range of about 13 cycles per second.

An unusual photo of the middle ear
This photo of the middle ear was made possible by the use of advanced fiber optic techniques and a round lens slightly over a millimeter in diameter. At the left is the eardrum, curving inward a little. The black spot at the far right is the round window. Just left of it is the footplate of the stapes, which covers the oval window. The distance between the eardrum and the oval window is actually less than a centimeter.

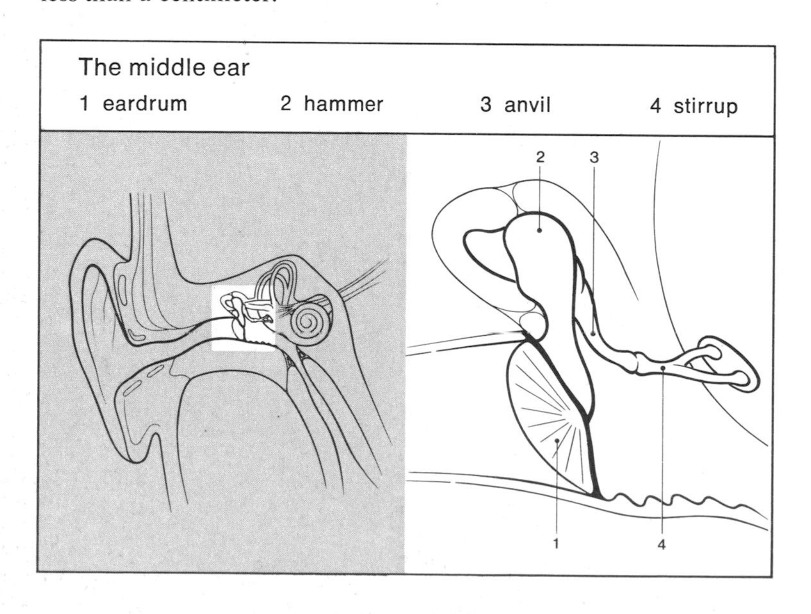

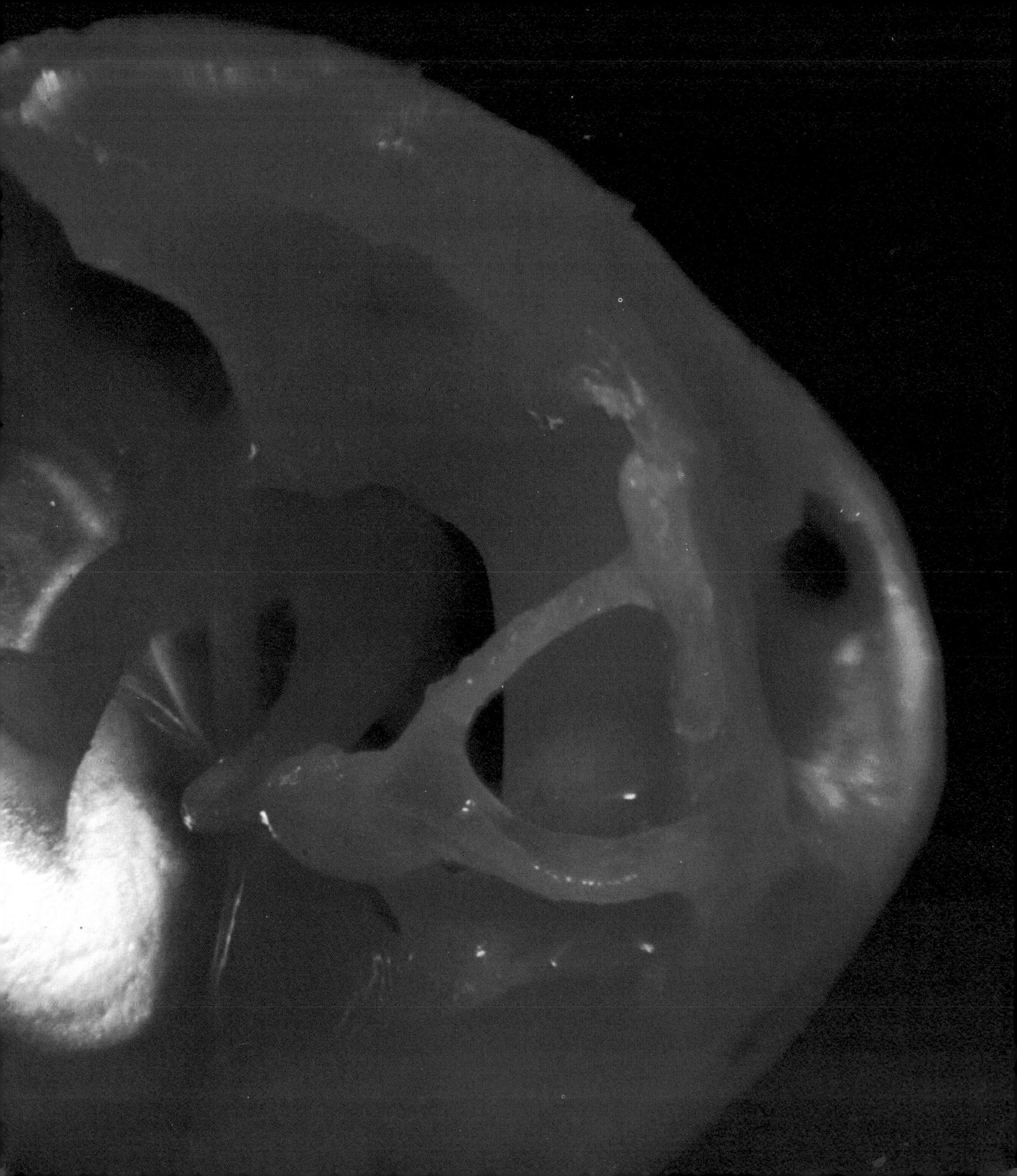

Inside the Tympanic Cavity

Details of the tympanic cavity, the largest of the cavities in the middle ear. It is about four millimeters deep and eight millimeters in diameter. The two photos on this page and the upper one on the right-hand page show the characteristic structure of the stapes from different angles. The thick nerve running below it in the upper picture on the right-hand page is a division of the facial nerve, the seventh cranial nerve. Some of its fibers convey impulses from the gustatory receptors in the tongue. In its circuit to the brain it passes through the middle ear.

The bottom picture on the right-hand page is a detail of the bony wall of the tympanic cavity.

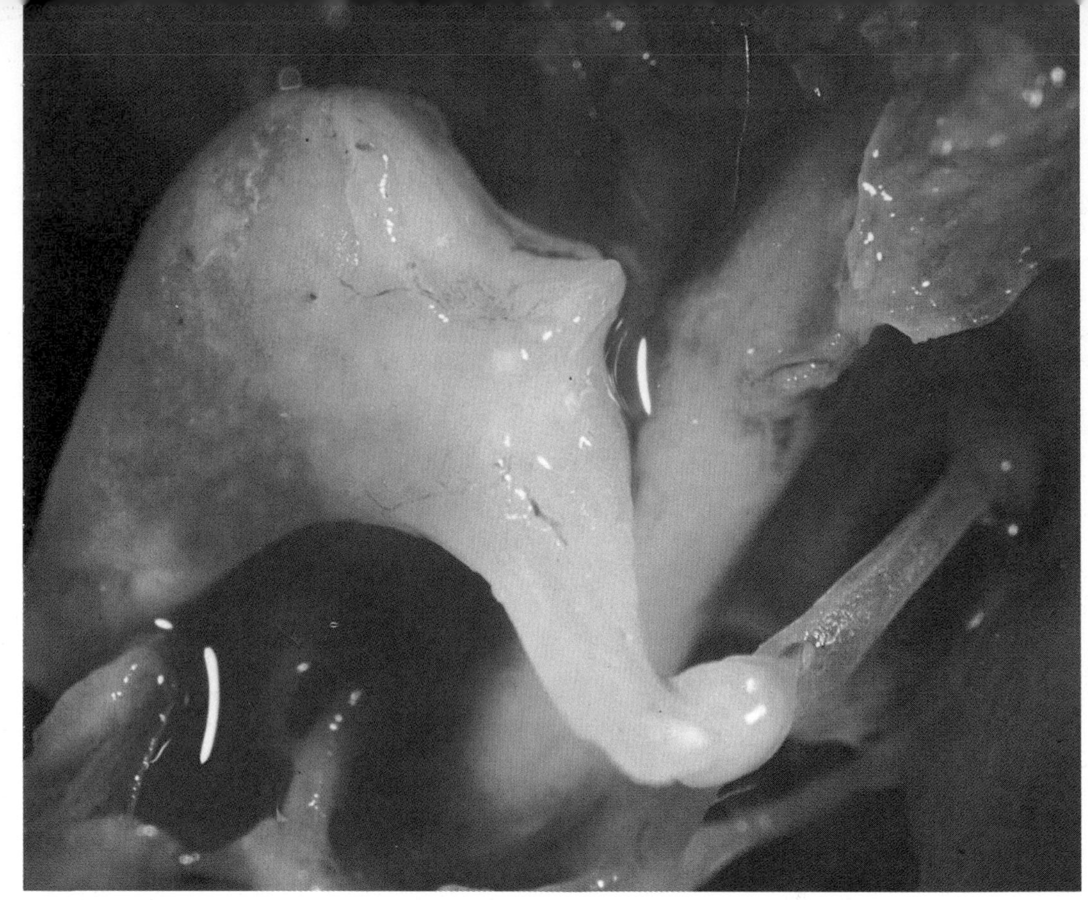

The anvil and the stirrup, two of the three bones that transmit sound waves to the inner ear

The photo illustrates why this small bone is called the stirrup

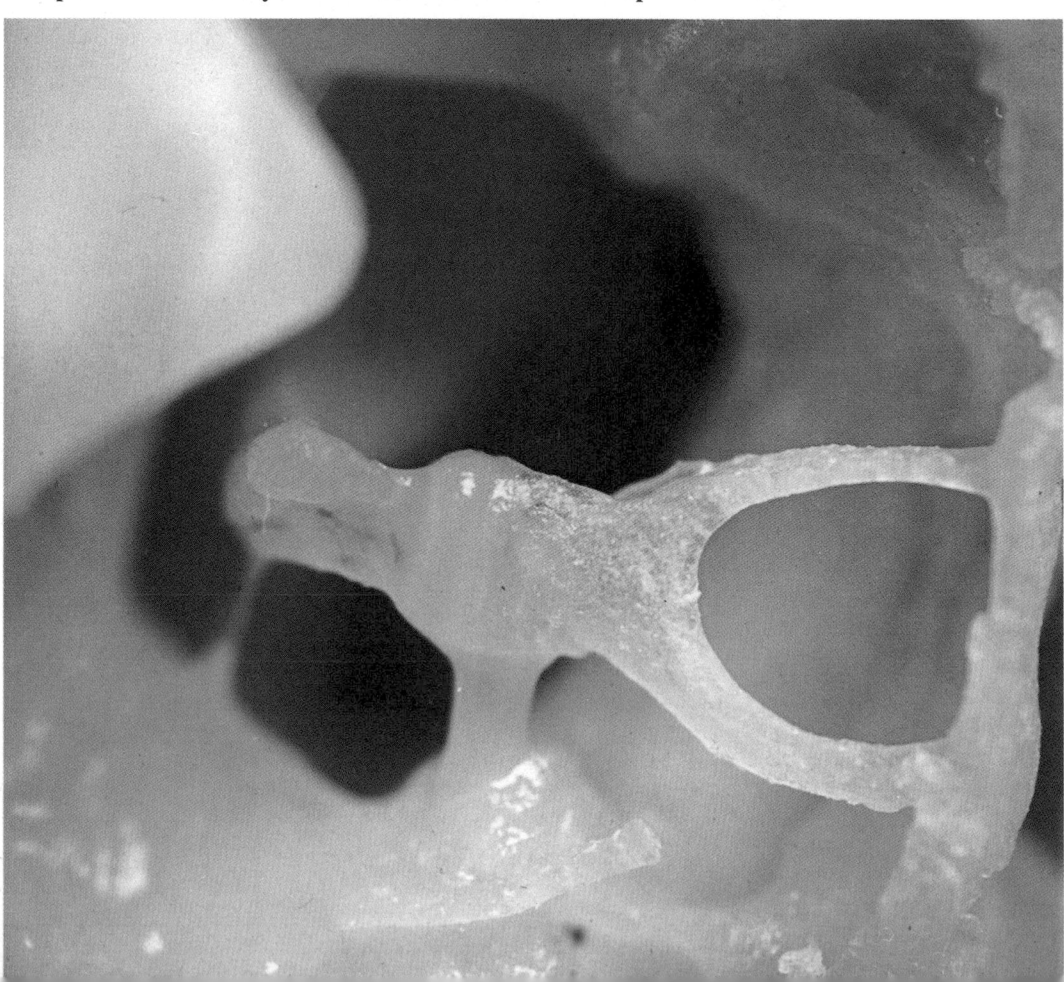

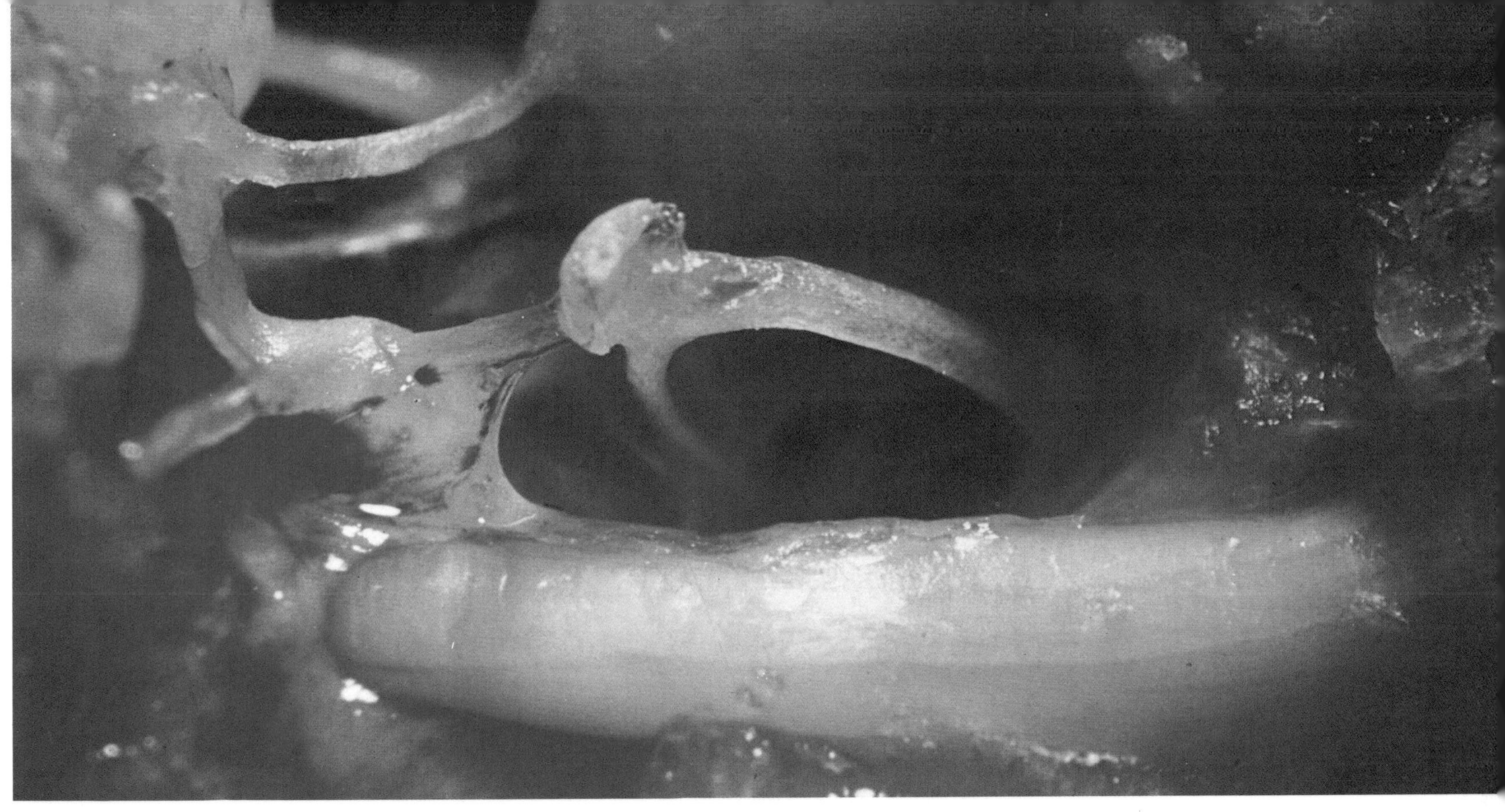

Below the stirrup is the seventh cranial nerve, which passes through the tympanic cavity on its way to the brain.

The walls of the tympanic cavity are covered by projections of bone tissue

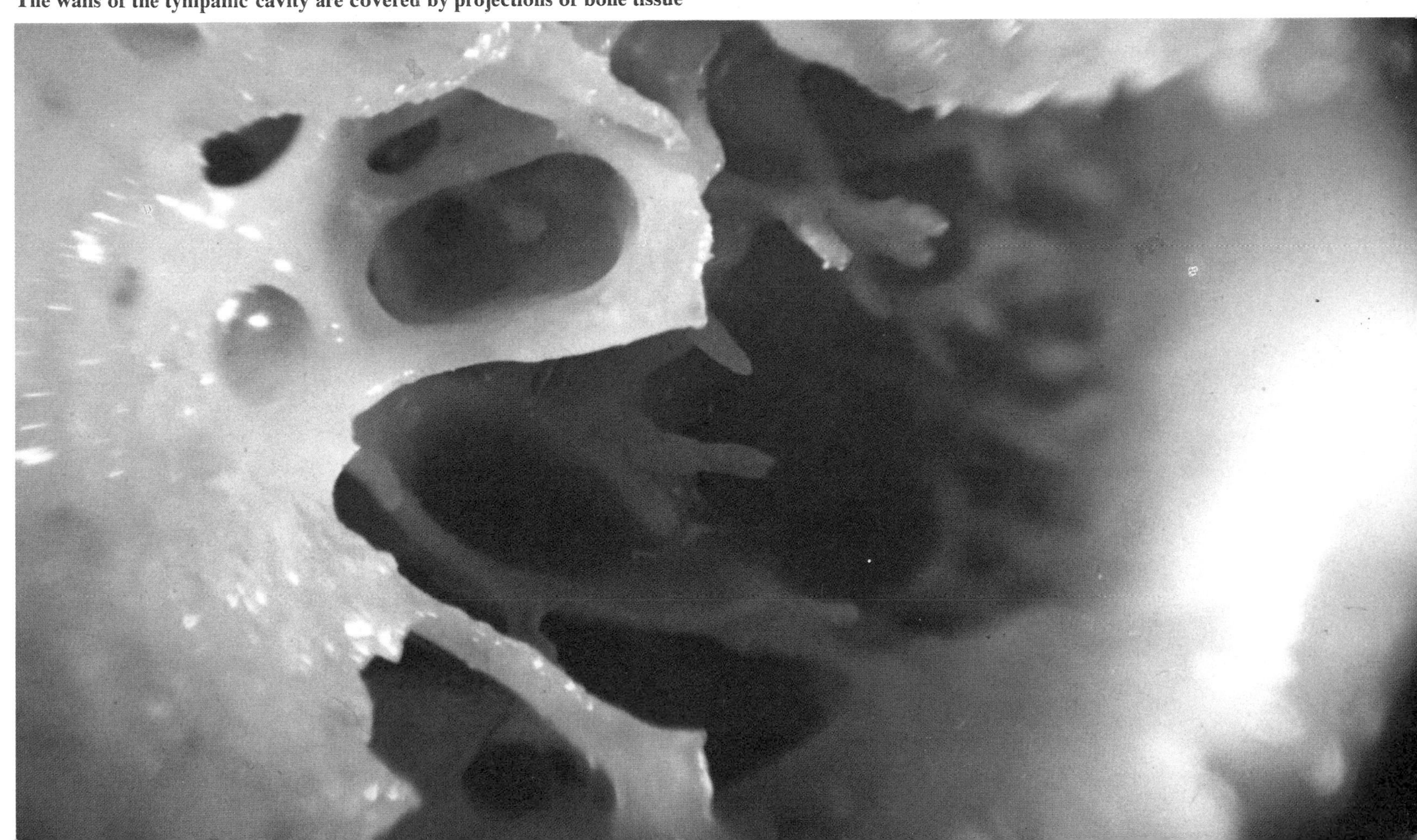

The Inner Ear

The inner ear

1. footplate of the stirrup against the oval window
2. saccule
3. endolymph duct
4. round window
5. cochlear duct
6. the cochlear duct ends before the top of the cochlea
7. cochlea
8. scala tympani
9. scala vestibuli
10. acoustic nerve
11. bony wall
12. band of connective tissue, the spiral ligament
13. epithelium
14. basilar membrane
15. organ of Corti with auditory receptor cells

The cochlea in the inner ear
lies well protected in the hard temporal bone. The auditory receptor cells are located in the cochlea.

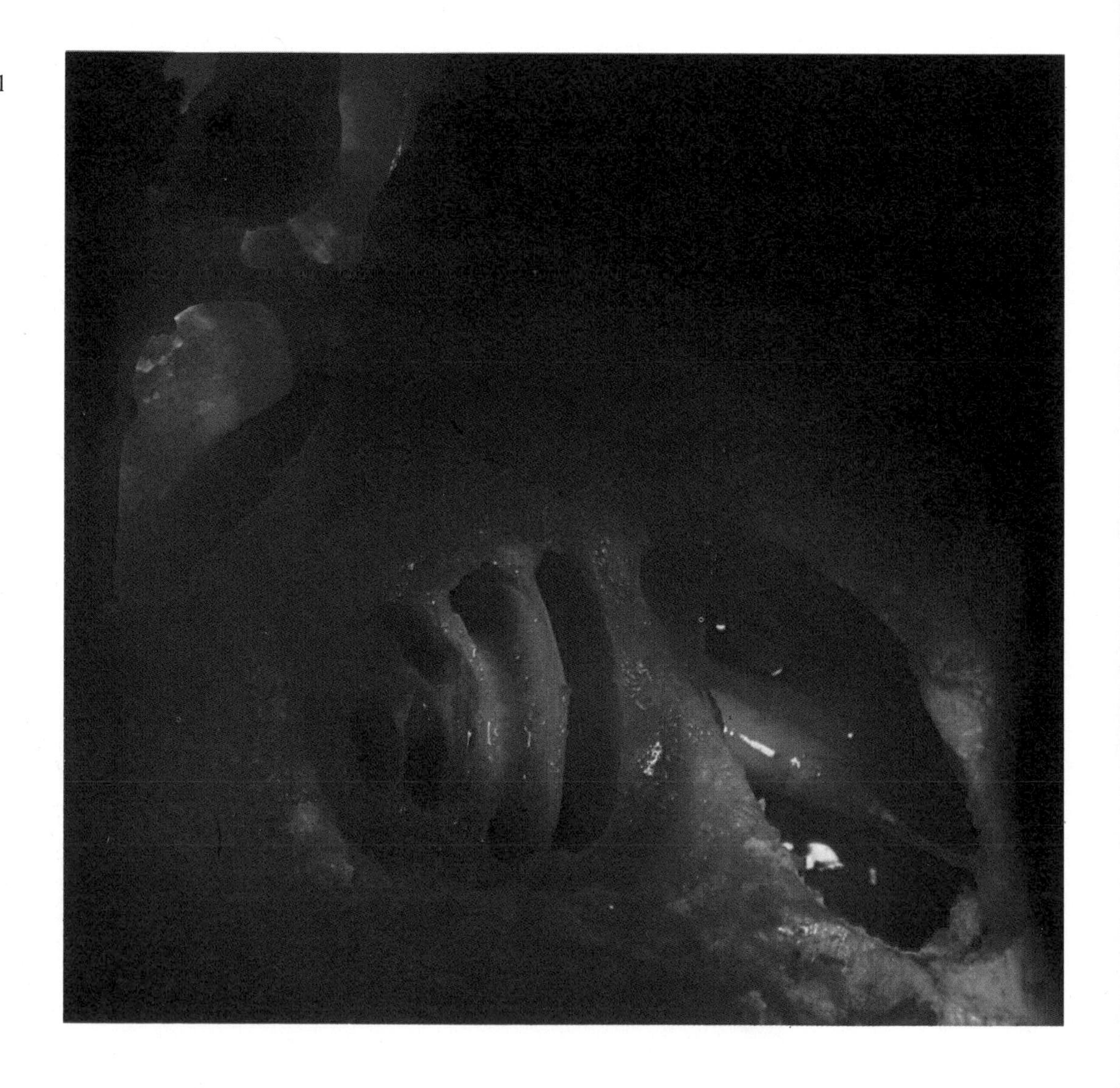

The Inner Ear

The inner ear is a cavity in the very hard temporal bone. The cavity is irregularly shaped and is difficult to describe. It is called the bony labyrinth. Its walls consist of bone covered by epithelium. The bony labyrinth contains a watery fluid called the perilymph. However, the fluid does not completely fill the labyrinth. A great part of the space is occupied by a system of thin-walled channels and sacs, the membranous labyrinth. The membranous labyrinth more or less duplicates the bony labyrinth, fitting it like a hand in a glove. Its walls consist of soft connective tissue. The membranous labyrinth also contains a watery fluid, the endolymph, which is similar to amniotic fluid. The portion of the bony labyrinth just inside the oval window is an irregularly shaped chamber called the vestibule. The footplate of the stirrup covers the oval window, while a thin membrane of connective tissue covers the round window.

The cochlea begins in the anterior part of the vestibule. The name cochlea, meaning snail shell, refers to the spiral structure of this cavity.

The central part of the membranous labyrinth consists of two sacs located in the vestibule. One is round, the saccule, the other oval, the utricle. Both sacs have systems of receptor cells and calcium crystals that respond to gravitational and accelerating forces and so relay information concerning the position of the head. Thus, in addition to the organ of hearing, the inner ear also contains the balancing organs (also called vestibular organs, because of their location in the vestibule).

The delicate and complex parts of the inner ear that form the hearing organ will be described first. A narrow channel leads from the saccule to the cochlear duct, which lies inside the cochlea. The cochlear duct is a narrow, wedge-shaped tube of

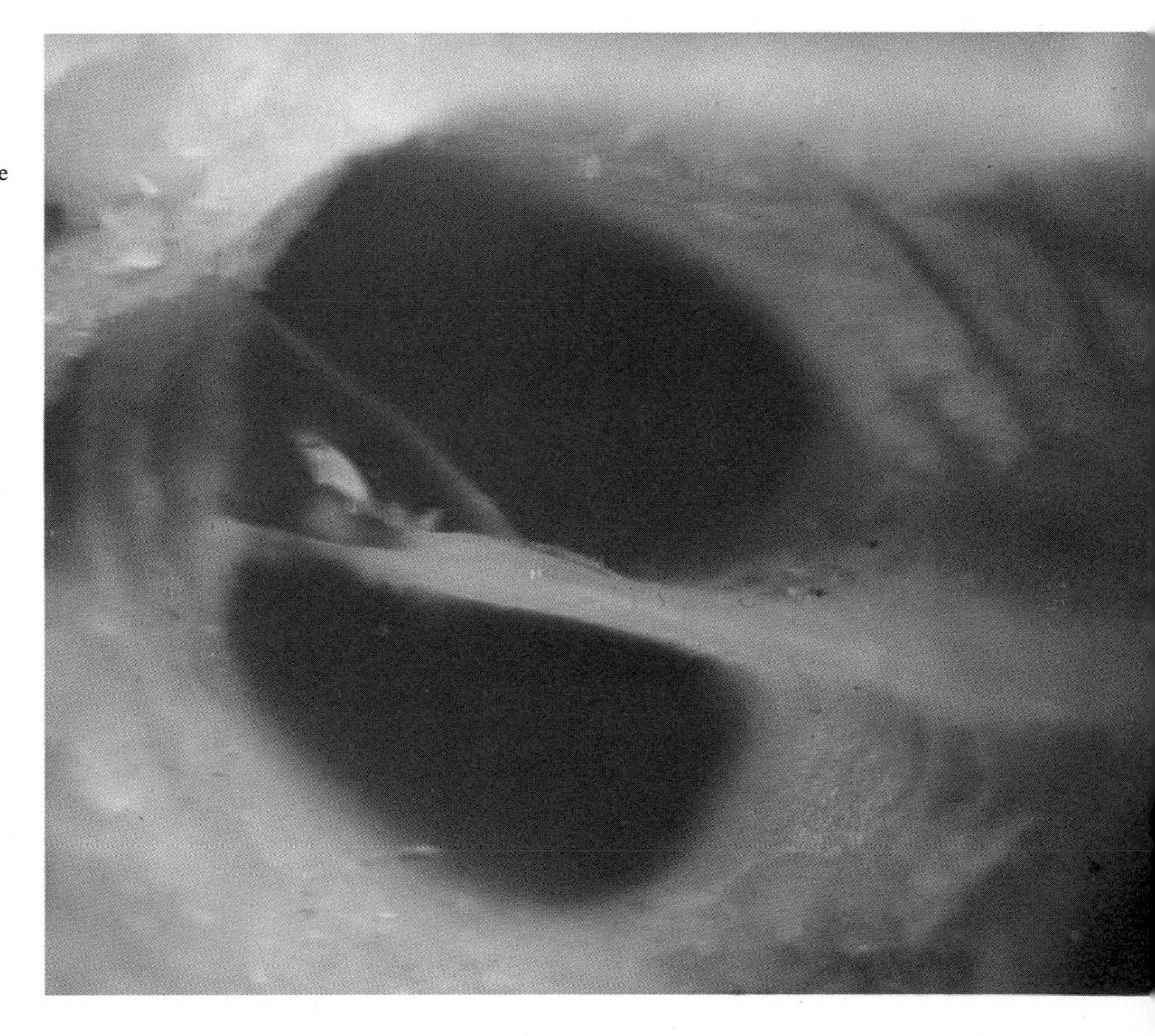

A slice through the cochlea shows a cross section of one of its turns. The diameter here is less than a millimeter. The three cochlear chambers — the scala vestibuli, the scala media, and the scala tympani — are clearly seen. The scala media contains the organ of Corti.

The neurons that are in contact with the receptor cells
The auditory receptor cells (fanlike structures in the center) are stimulated by vibrations in the endolymph. The stimuli are converted to electrical impulses conducted to the auditory centers of the brain. The thirty thousand cell bodies of the neurons that mediate these impulses lie well protected in the bony core of the cochlea and communicate with the receptor cells through their fiber processes. The axons of these nerve cells form the cochlear division of the acoustic nerve, the eighth cranial nerve. Another division carries nerve impulses from the equilibrium organs.

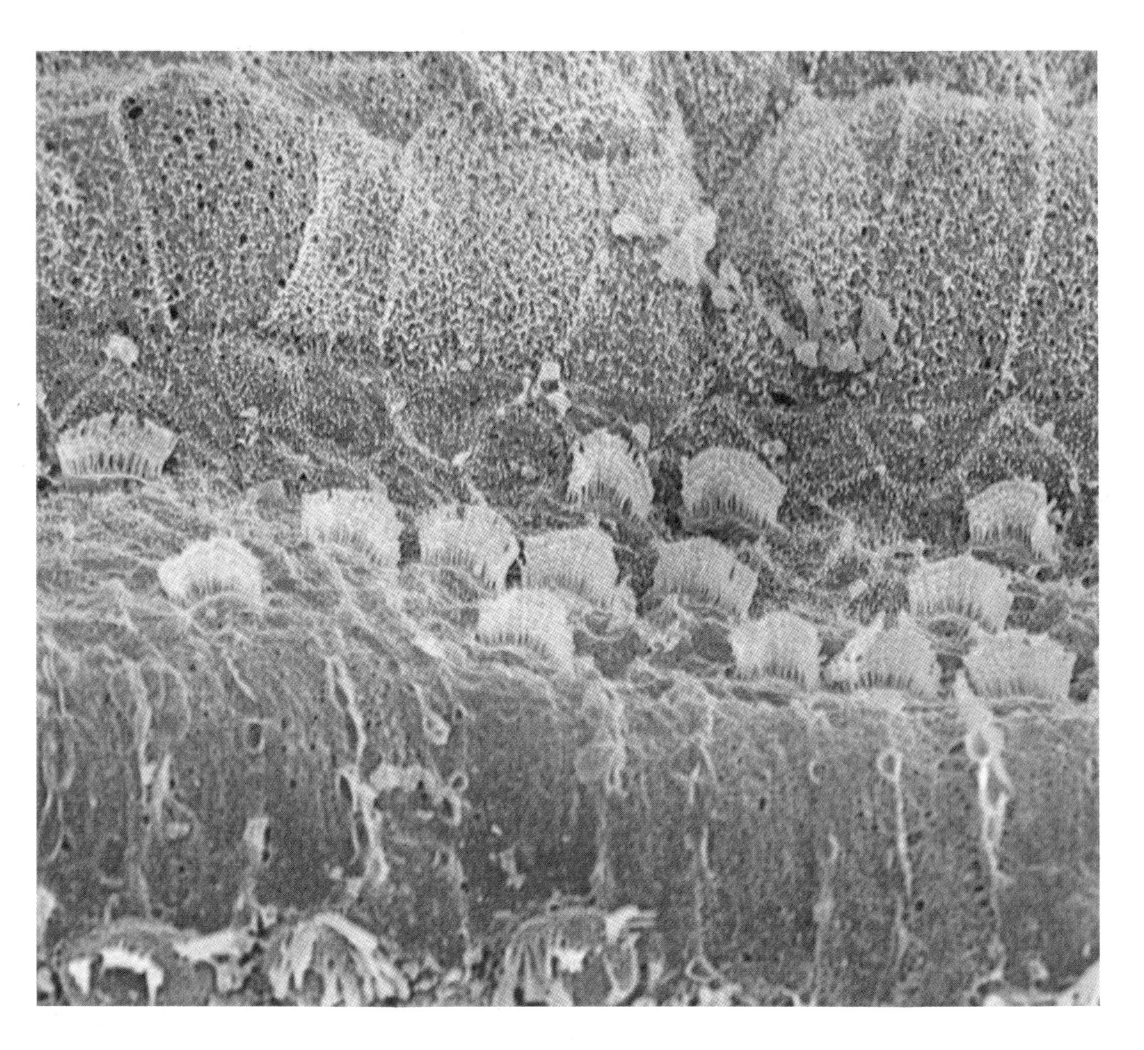

connective tissue, which spirals two and a half times up the cochlea following its snail shape. It is also called the membranous cochlea. The dimensions of these parts of the inner ear are exceedingly small. The cochlear duct rises less than a quarter of an inch from base to apex. The cochlea itself is less than half an inch across at its base. The diameter of the bony coil whose spirals form the cochlea is only 1/25 of an inch at the base turn, decreasing steadily as it winds to the top. The cochlear duct inside is smaller yet, and actually makes up a very minor portion of the total volume of the cochlea: partitions divide the bony coil into upper and lower parts with the duct in the middle. The base of the duct consists partly of bone, and partly of soft tissue, the basilar membrane.

The basilar membrane makes up the floor of the cochlear duct. It extends from the bony part of the partition to the outer cochlear wall, where it is attached to a band of connective tissue rich in blood vessels, the spiral ligament. Another thin membrane forms the ceiling of the cochlear duct. It too extends from the bony shelf of the partition to the upper edge of the spiral ligament. Thus, a cross section of the coil shows it to be divided into three chambers. Two are large and almost semicircular in section. The cochlear duct is the small, wedge-shaped tube sandwiched in between. The upper chamber starts at the vestibule and winds upward toward the top of the cochlea. It is called the scala vestibuli. The cochlear duct (also called the scala media) is a little shorter than the bony spiral and ends short of the top. Hence, the two chambers on either side of the cochlear duct are in contact with each other at the top. The lower chamber winds down from the top of the cochlea toward its base. It ends at the round window, which is located in the bony wall facing the tympanic cavity. It is therefore called the scala tympani.

The basilar membrane, the floor of the cochlear duct, is a little over an inch long when extended. While the diameter of the bony coil is largest at its basal turn and decreases toward the top, the width of the basilar membrane increases as it rises, widening from 1/250 inch at the base to about 1/50 inch at the top. The auditory receptor cells are situated on the basilar membrane. Together with a large population of supporting cells they form the organ of Corti, also called the spiral organ. Another component of the organ of Corti is a thin, gelatinous membrane, the tectorial membrane, which gently roofs the receptor cells. These are hair cells whose tips are embedded in the gelatinous membrane. When the vibrations of the auditory bones set the perilymph in motion, the cochlear duct vibrates. This stimulates the auditory receptor cells and in some way initiates a flow of electricity in the nerve fibers clustered at the base of the receptor cells.

The vibrations of the duct reach a maximum at different parts of the basilar membrane, depending on frequency. High-pitched sounds have a maximal effect on the lower parts of the membrane. Low-pitched sounds have vibratory peaks near the top of the cochlea. A measure of intensity of vibrations in the cochlear duct is also transmitted to the brain. This makes it possible for us to distinguish sounds according to loudness. The vibrations themselves always occur in the same way, no matter whether the sound waves are airborne and conducted by the eardrum and the auditory bones, or reach the inner ear by conduction through the bones of the skull itself.

The nerve endings at the base of the hair cells are from neurons which make up the cochlear division of the acoustic nerve. The cell bodies are located in the modiolus, the core of the cochlea. Surrounded by bony tissue, these "spiral ganglion" neurons are well protected. When the auditory receptor cells are stimulated, the impulses travel through the fine nerve endings to the spiral ganglion. From here, the impulses are transmitted through the acoustic nerve, the eighth cranial nerve, to various stations in the brain stem, ultimately reaching the cortex. On its way to the brain, the acoustic nerve passes through a canal in the temporal bone, called the internal auditory meatus. Nerve fibers from the equilibrium organs join the acoustic nerve so that it has a vestibular as well as a cochlear division.

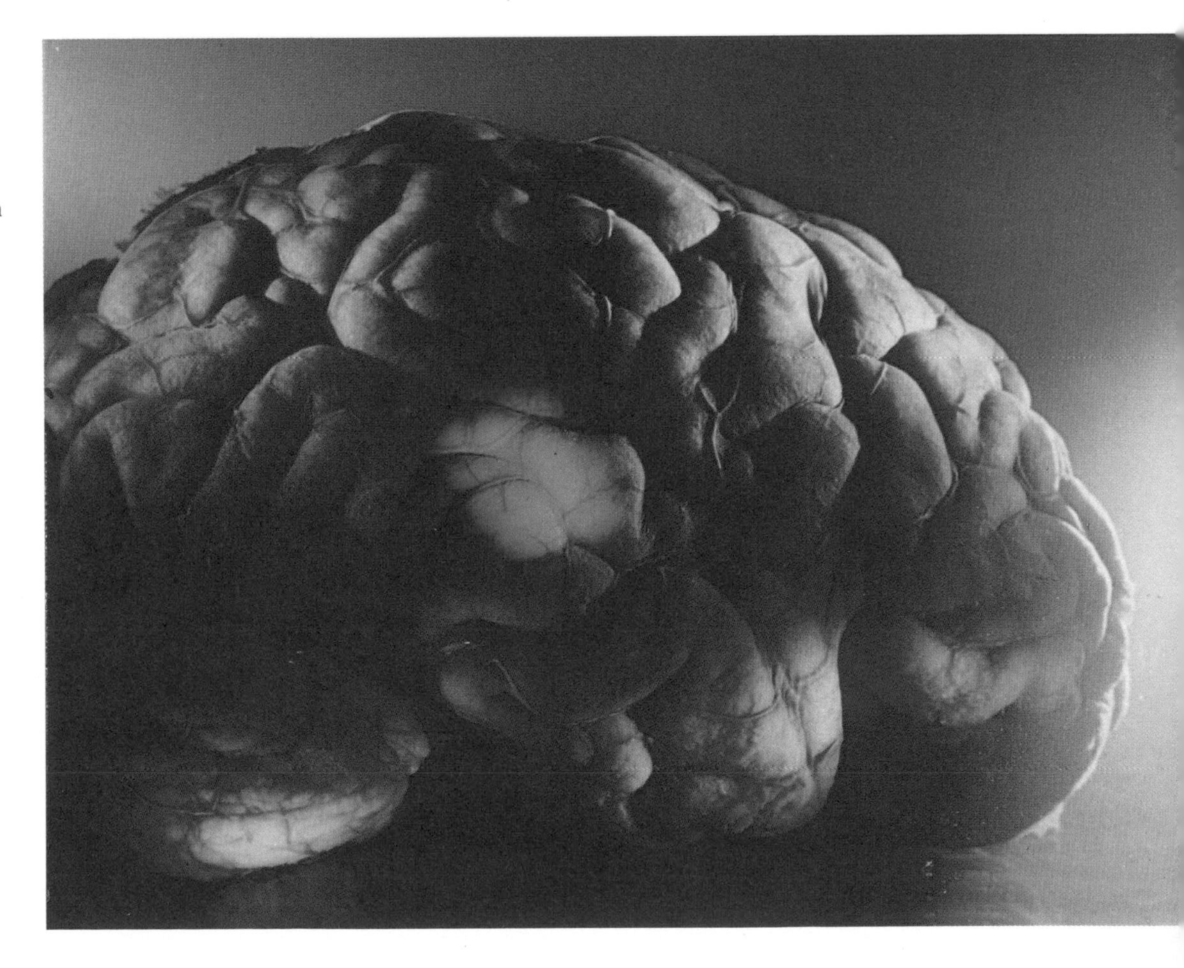

An auditory center in the brain
Impulses from the cochlea are conveyed through the acoustic nerve to the brain stem and to higher parts of the cortex. A major auditory center in the brain is located in the temporal lobe of the cerebral cortex, below the region marked with blue light in this photo.

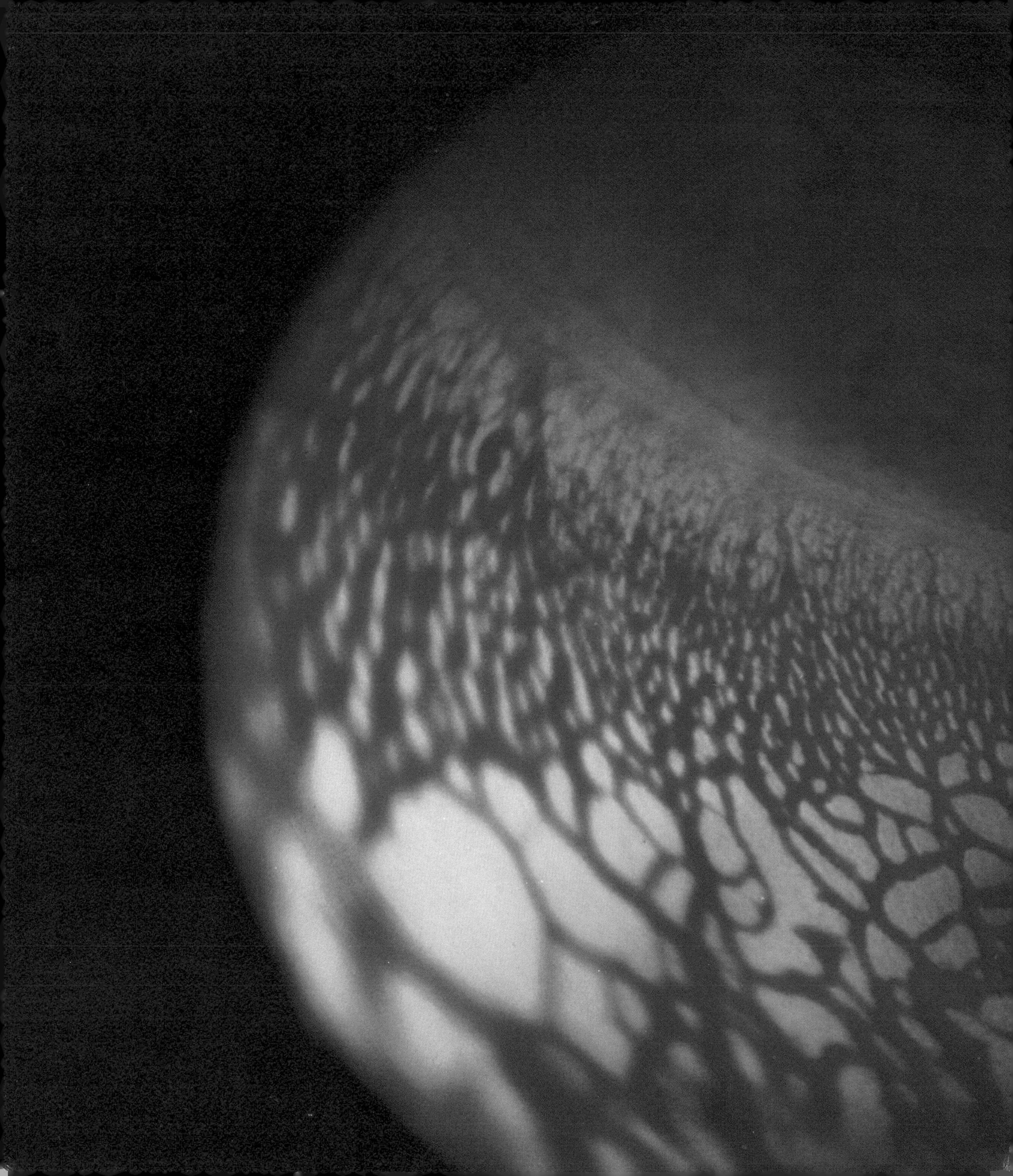

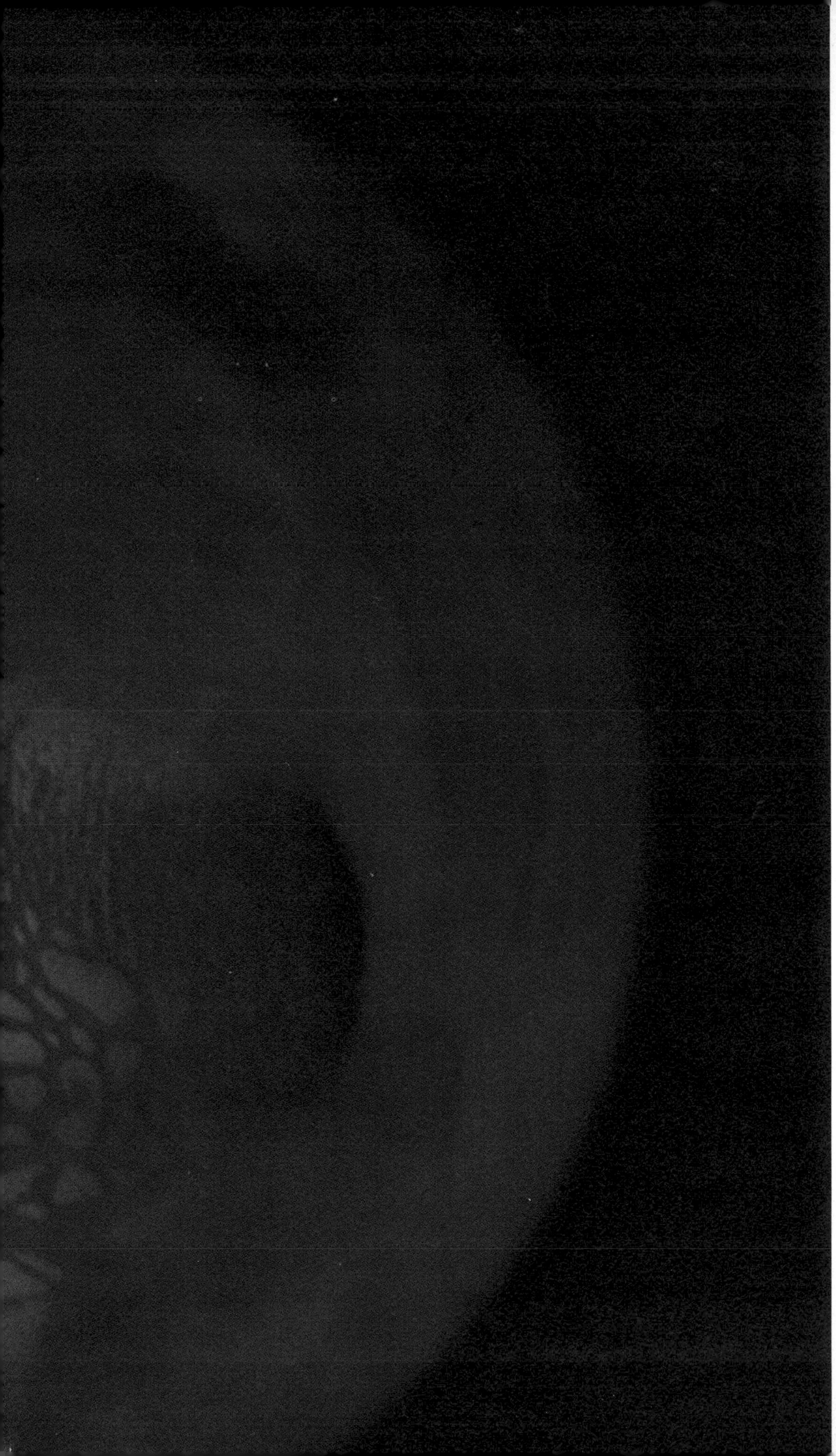

Inside the Inner Ear

The scala tympani,
one of the outer spiral chambers in the bony cochlea, where sound waves move after passing the upper part of the cochlea. The basilar membrane separates the scala tympani from the cochlear duct. The picture is of a fetus; that is why the walls are so thin and the blood vessels so clearly visible.

The cochlear duct winds from under the left portion of the base of the scala vestibuli. At the bottom of the duct are the auditory receptor cells. They and their supporting cells are nourished by the fluid present in the duct. A similar situation exists in the cornea and lens. There a clear fluid also provides nourishment so that blood vessels do not block the path of light.

Auditory Receptor Cells

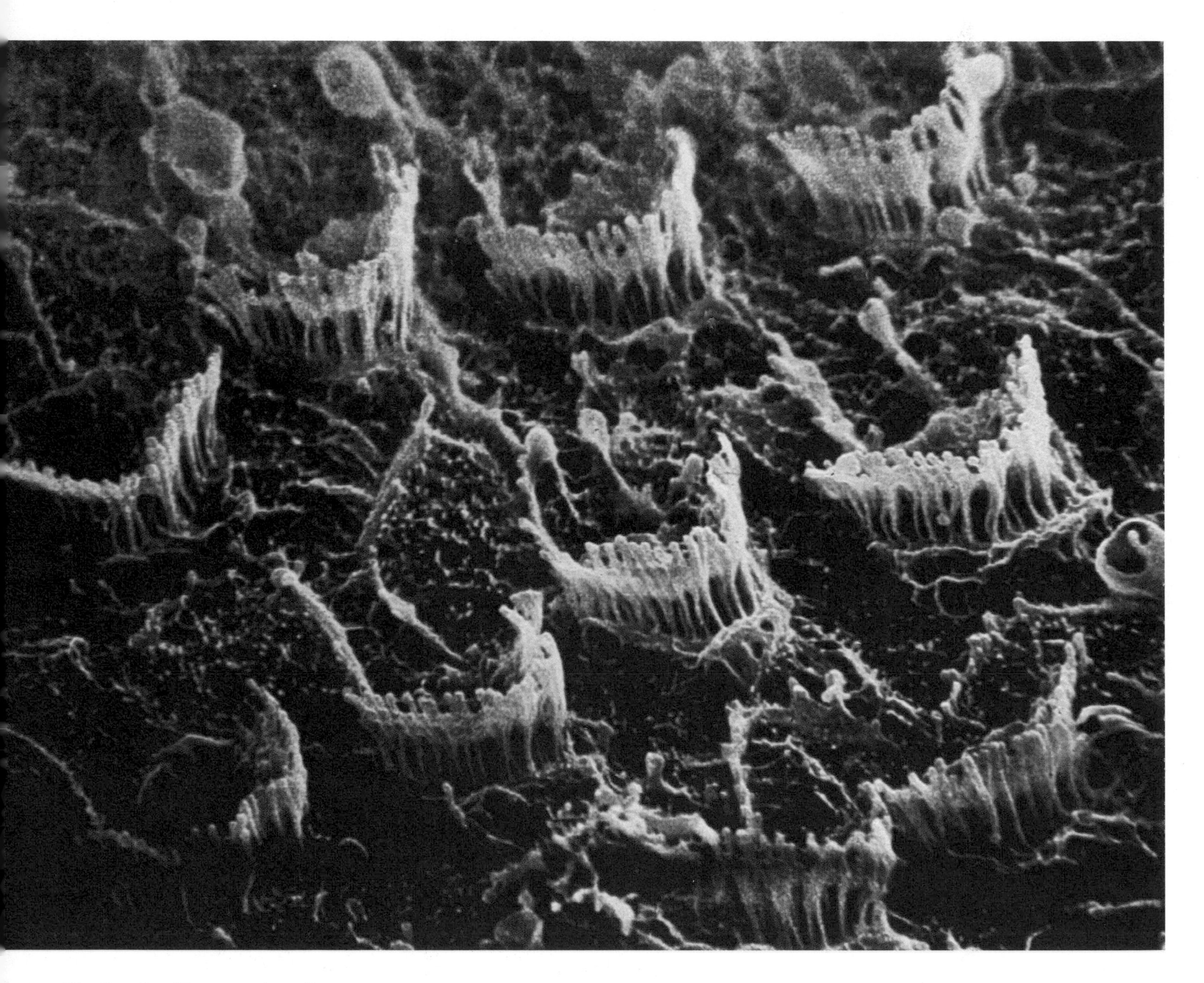

The tips of auditory receptor cells
When magnified 3,500 times, the basilar membrane looks like a landscape with regularly placed groups of picket fences. These are the ciliated tips of auditory receptor cells which, on close inspection, form W-like patterns. The cell groups are viewed from behind, which means that the oval window is located diagonally upward to the left.

Auditory Receptor Cells

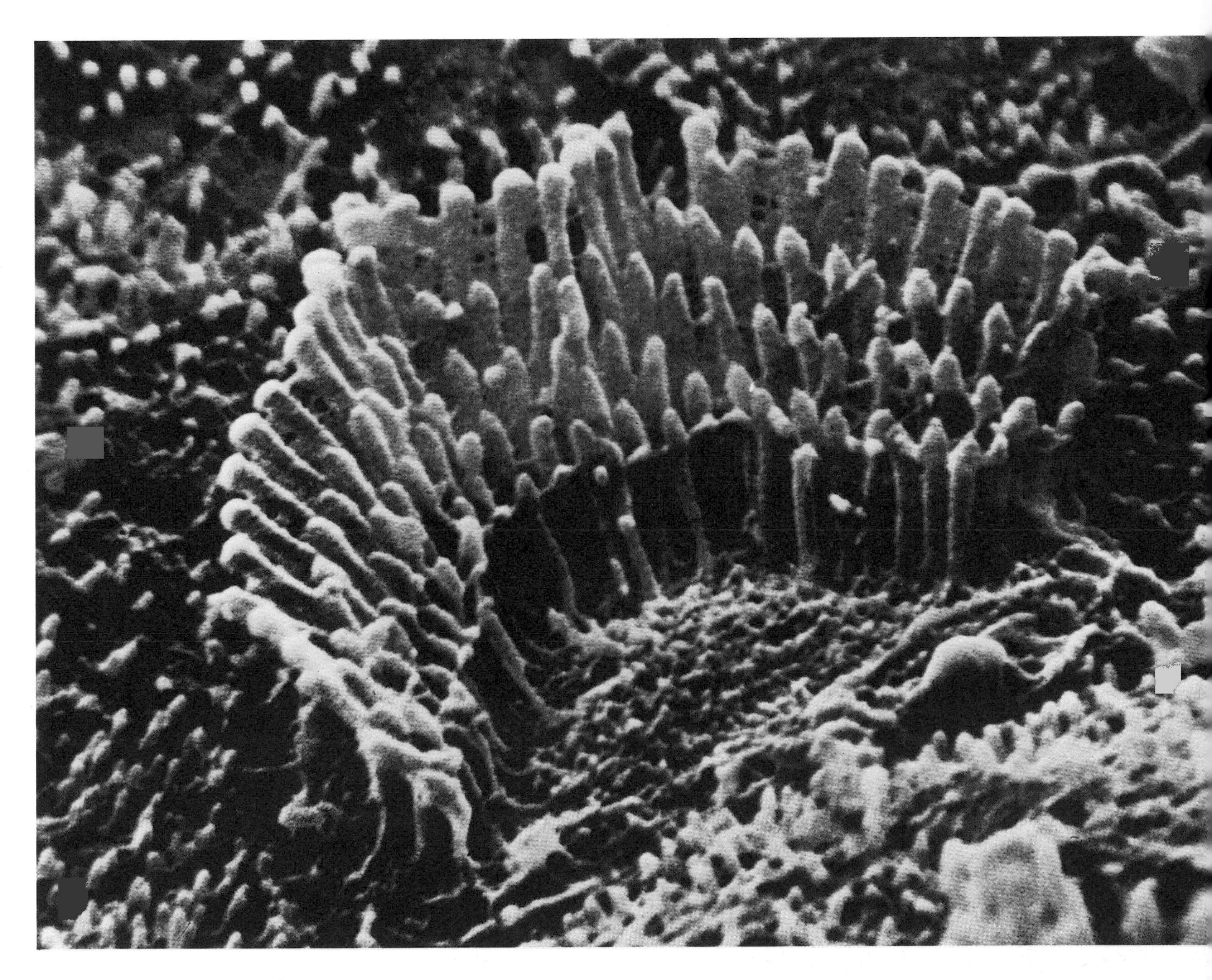

Detail of an auditory receptor cell
The movements of sound waves in the air are eventually felt by the hairs of the auditory receptor cells. When the fluid in the inner ear vibrates, it disturbs the basilar membrane and jostles the hairs. The particular pattern of jostling becomes converted into electrical impulses, which are transmitted to the brain. The hairs, which are not of equal length, are thought to be stimulated differently by different volumes of sound. Very loud sounds may cause such strong vibrations that the hairs of the receptor cells are destroyed and hearing impaired.

Hearing gradually diminishes in aging. It is possible that receptor cells die off one by one, so that we finally lack whole groups of cells.

The external ear in the fifth month, ▷ when it is almost fully developed.

The Development of the Ear

The embryo is only a few weeks old when the ear — the external ear as well as the hearing and equilibrium organs — starts to develop. All three germ layers of the embryo participate in their formation.

During the sixth week, the external ear begins to develop. At first it looks like a series of six small tissue protuberances around one of the neck folds at the side of the embryo. Gradually, the six small offshoots move upward to the region behind the joint of the jaw. Here the formation of the external ear with its cartilage structures surrounding the external auditory canal takes place. The external auditory canal itself is formed by an inversion of ectoderm. At its base, the eardrum develops.

An outpocketing of entoderm at the same time forms a cavity in the side wall of the pharynx where the middle ear and the eustachian tube will eventually develop.

The auditory bones — the hammer, the anvil, and the stirrup — are products of mesoderm. When the embryo is only a few weeks old, a series of cell proliferations takes place in the mesoderm which develops into the primordia of these important small bones.

The organs of hearing and equilibrium themselves, with their sensory epithelium, the hair cells, originate in the ectoderm. The hearing and equilibrium apparatus forms in the same way as the eyeballs, by a turning in, or invagination, of the ectoderm. With further development the tissue becomes closed off from the surface of the body and forms a complicated system of branches and bulges which eventually become the cochlea, the semicircular canals, et cetera. Portions of the epithelium are transformed into the special receptor cells, which communicate with the fibers of the acoustic nerve. Thus, it is by the action of cells which originally formed part of the exterior of the body that we ultimately hear and perceive changes in posture.

Fetal development of the ear
The external ear starts developing at five or six weeks, when the primitive structures of the eyes and nose are also visible.

The photos below show the development of the external ear in the third and fourth months of gestation.

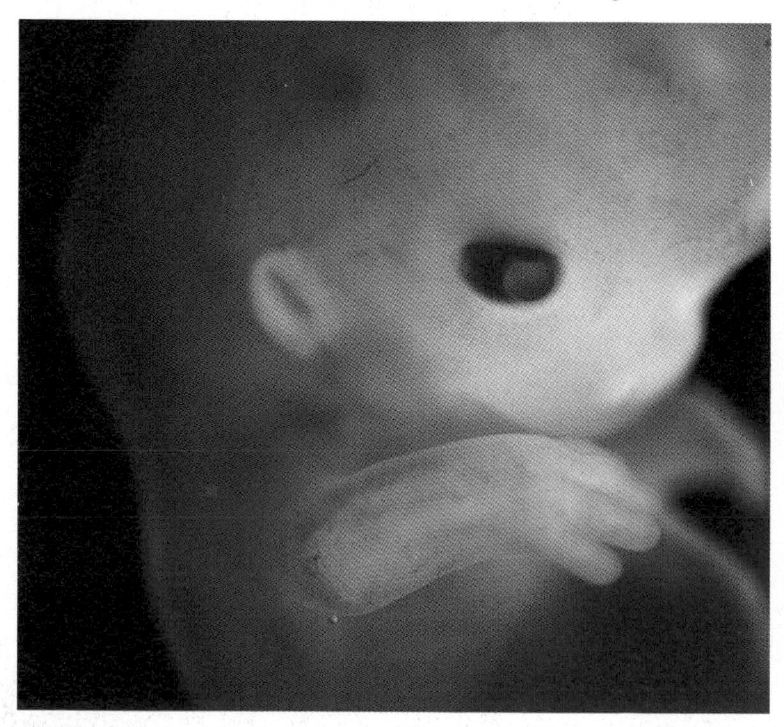

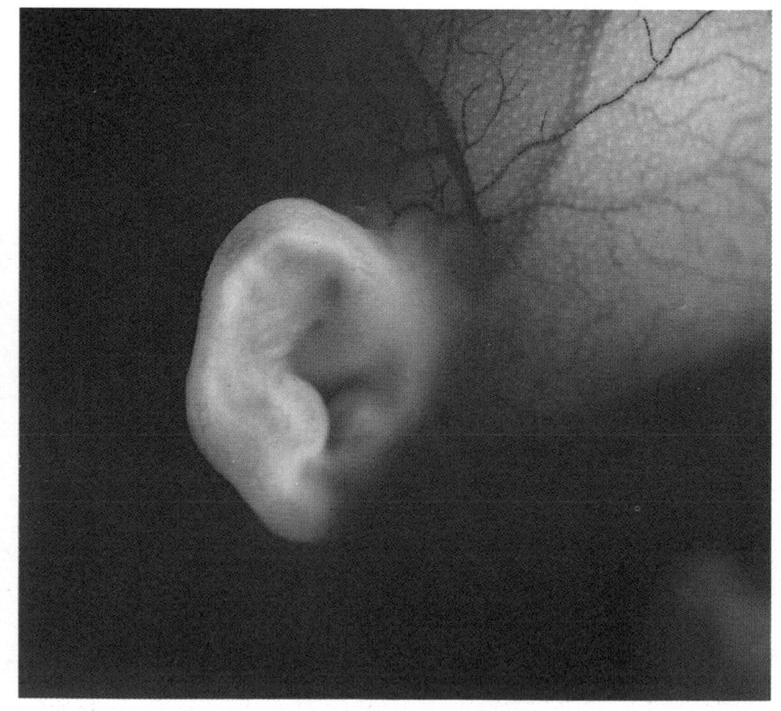

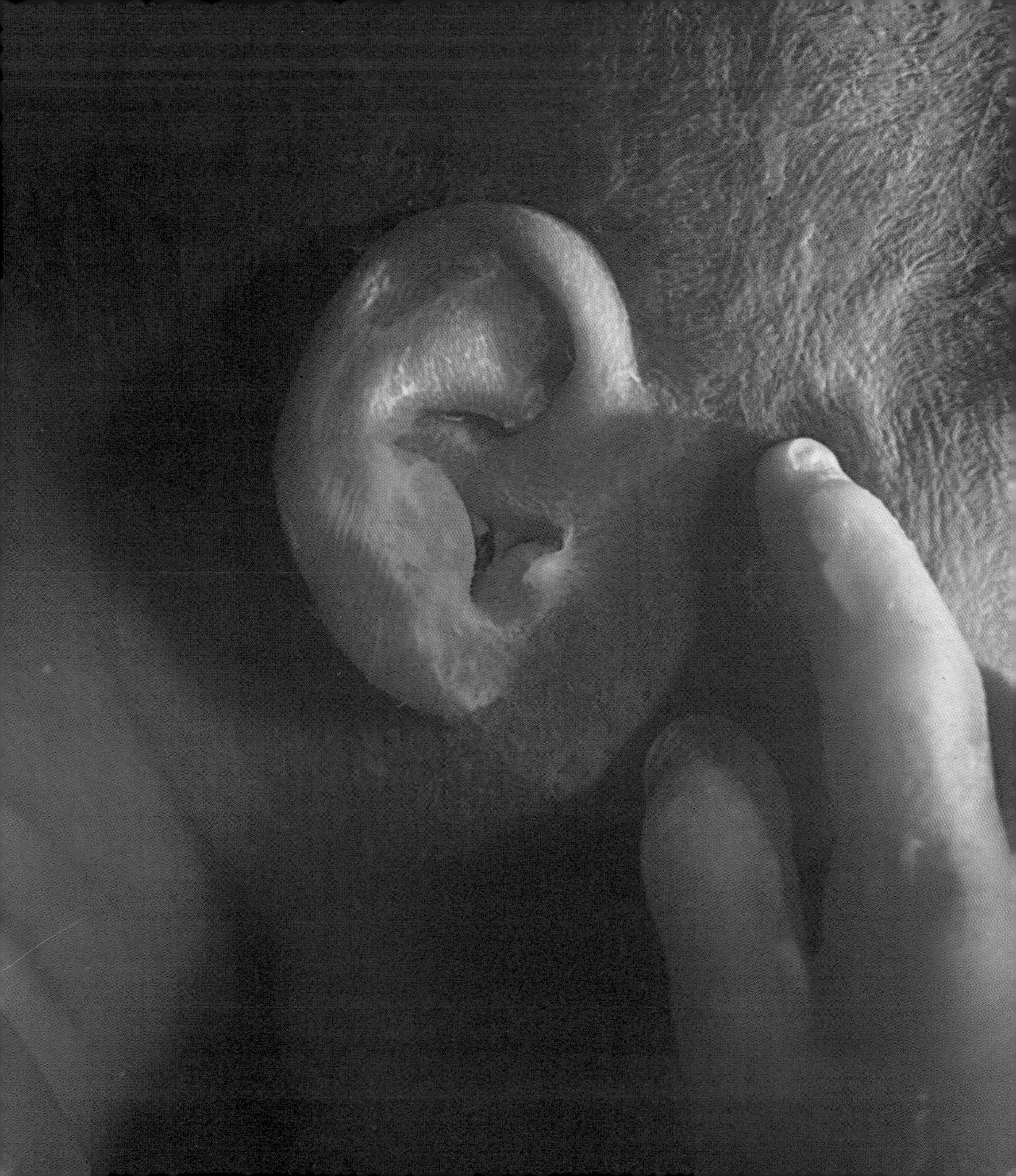

The Organs of Equilibrium

The inner ear contains receptor cells for the sense of equilibrium, or balance, as well as those for hearing. The function of the equilibrium sense is to make us aware of the position of the body. As soon as the head moves forward, backward, or sideways, we notice it because the motion stimulates receptors in the inner ear.

The auditory portion of the internal ear was described earlier as part of the contents of the bony labyrinth. The labyrinth is filled with the watery fluid, perilymph, bathing the membranous labyrinth which contains endolymph. Two large sacs, the saccule and the utricle, form the central portion of the membranous labyrinth. Both sacs contain receptor cells that respond to gravitational or accelerating forces. The hairs of the receptor cells are surmounted by a gelatinous film in which are embedded crystals of calcium salts. They are usually called statoliths, meaning standing stones, or otoliths. The gelatinous mass and its crystals have a greater weight per unit volume than the endolymph. When the head is bent, this "inert mass" does not follow the movement, but slides over the receptor cells like miniscule plumb bobs seeking the center of the earth. In this way different hair cells are deflected by even minor changes in positions. The hair cells in turn excite nerve endings which convey impulses to the brain. Among other centers, the signals are relayed to the cerebellum, which automatically regulates muscle contractions so that the body maintains its balance. Normally, you do not notice such corrections. If you want to bend your head in some direction, signals originating in the motor cortex are relayed to the appropriate muscles. Because the receptor cells immediately inform the brain about every change in position, you very quickly perceive what is happening in the situation in question. The responses take place so fast and so effectively that you have the sensation of bending your head to look at the ground virtually at the same time you perform the movement.

The three semicircular canals of the bony labyrinth contain membranous canals that branch off from the utricle. The canals are oriented in such a way that their planes are mutually perpendicular, like three-dimensional axes for length, width, and height. When standing, the lateral canal is roughly horizontal, the anterior roughly parallel to the side of the head, and the posterior roughly parallel to the facial plane. A spool-shaped enlargement, the ampulla, is located where each of the membranous canals ends at the utricle. Each ampulla contains receptor cells that register the rotational movements of the head. For example, when the head is turned horizontally, receptor cells in the ampulla of the lateral semicircular canal are stimulated.

The receptor cells are located along and atop a ridgelike structure in the interior of the ampulla. The cells are ciliated like those in the saccule and the utricle, and the tips of the hairs are also embedded in a gelatinous, flat conical structure, the cupula (without otoliths, however). When the head moves, the motion is imparted to the endolymph in the semicircular canals, which pushes the cupula in the direction of the flow. This causes the

The semicircular canals are parts of the vestibular system
The three canals are roughly perpendicular to each other in the three planes of space. Any conceivable angle of rotation of the head will be registered by hair cells in one or more semicircular canals in both ears.

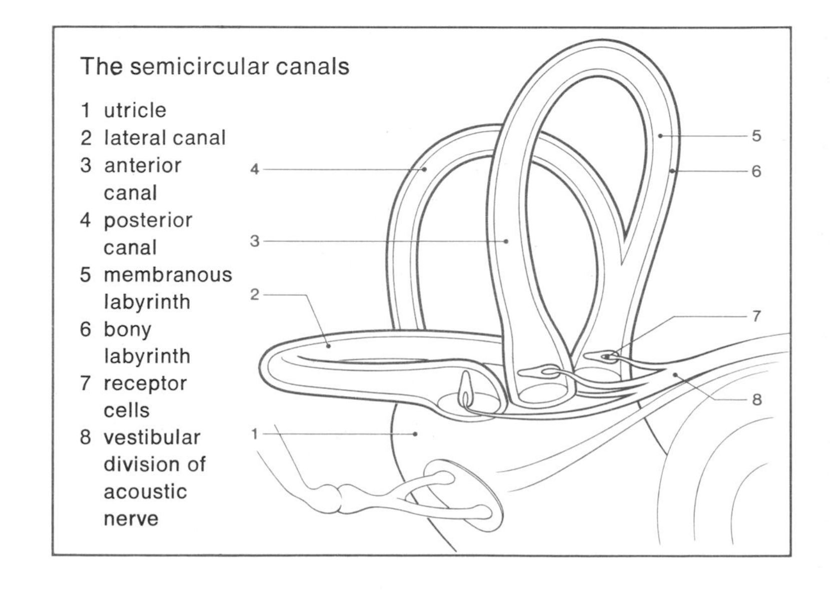

Calcium crystals
in the vestibular organ ▷

hairs to bend; the receptor cells are stimulated, and electrical impulses are sent to the brain. Nerve fibers from the canals join those from the saccule and utricle to form the vestibular nerve. The vestibular nerve forms a division of the eighth cranial nerve, which contains cochlear fibers as well. The vestibular pathways in the brain give rise to adjustments not only in limb muscles but in the eyes as well. In this way we can keep balanced and look fixedly at a point even if the position of the head is changed. Sometimes the remarkable coordination of sense organs and muscles, which usually takes place unconsciously, is temporarily overwhelmed and lost. Thus, if you whirl around very fast, the information about body posture from the balancing organs and from the eyes differs. You then become dizzy and may no longer be able to keep your balance.

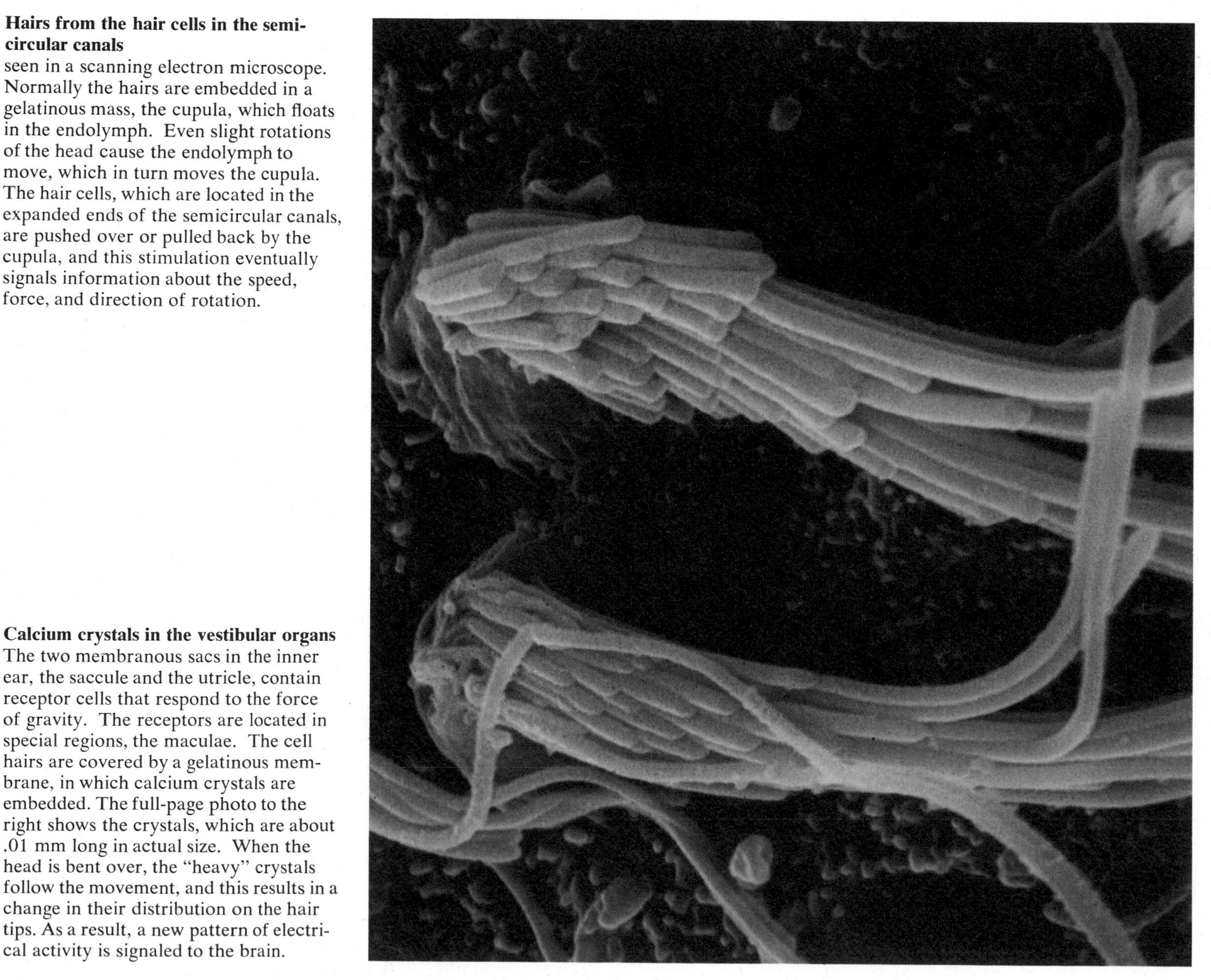

Hairs from the hair cells in the semicircular canals
seen in a scanning electron microscope. Normally the hairs are embedded in a gelatinous mass, the cupula, which floats in the endolymph. Even slight rotations of the head cause the endolymph to move, which in turn moves the cupula. The hair cells, which are located in the expanded ends of the semicircular canals, are pushed over or pulled back by the cupula, and this stimulation eventually signals information about the speed, force, and direction of rotation.

Calcium crystals in the vestibular organs
The two membranous sacs in the inner ear, the saccule and the utricle, contain receptor cells that respond to the force of gravity. The receptors are located in special regions, the maculae. The cell hairs are covered by a gelatinous membrane, in which calcium crystals are embedded. The full-page photo to the right shows the crystals, which are about .01 mm long in actual size. When the head is bent over, the "heavy" crystals follow the movement, and this results in a change in their distribution on the hair tips. As a result, a new pattern of electrical activity is signaled to the brain.

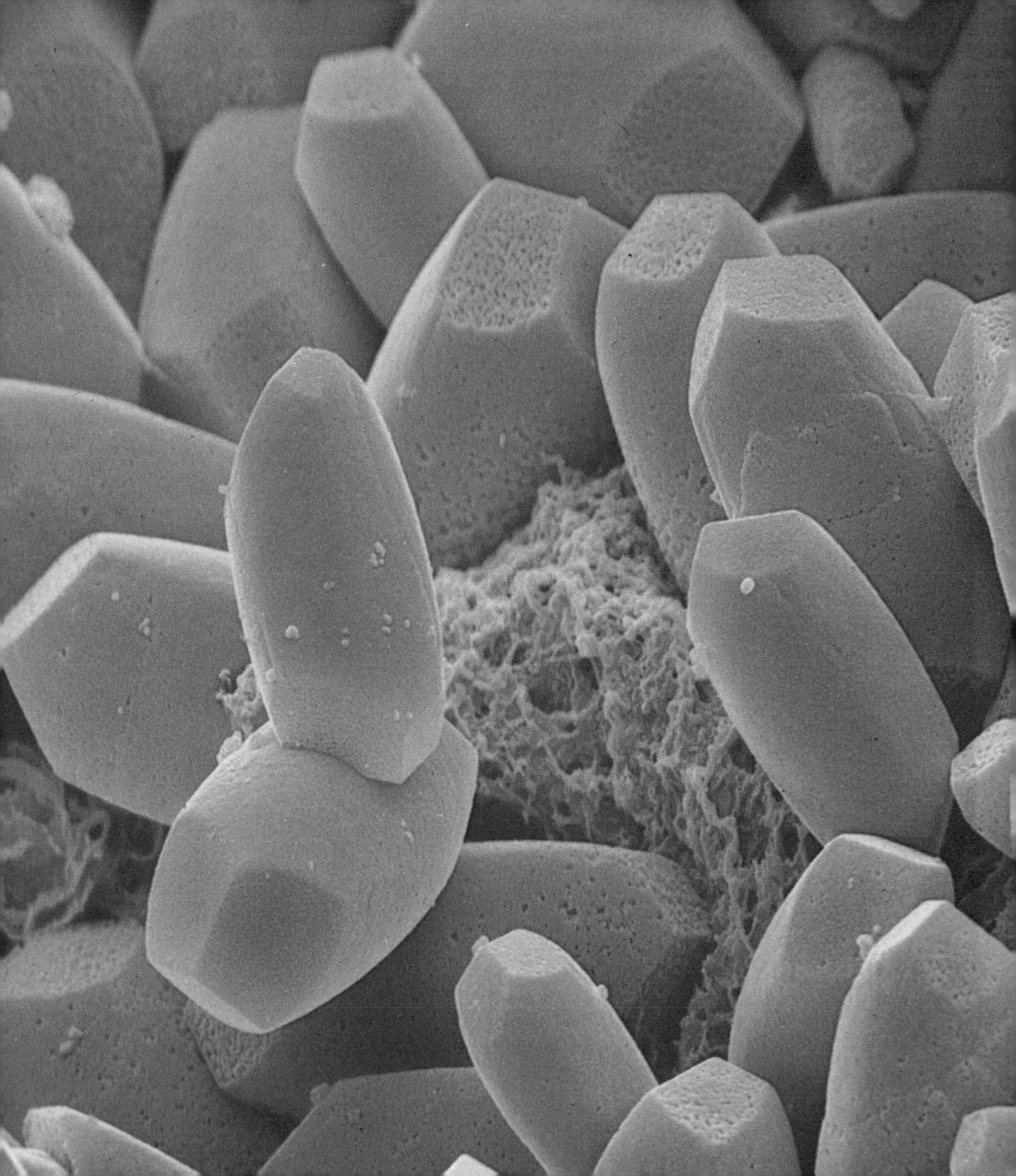

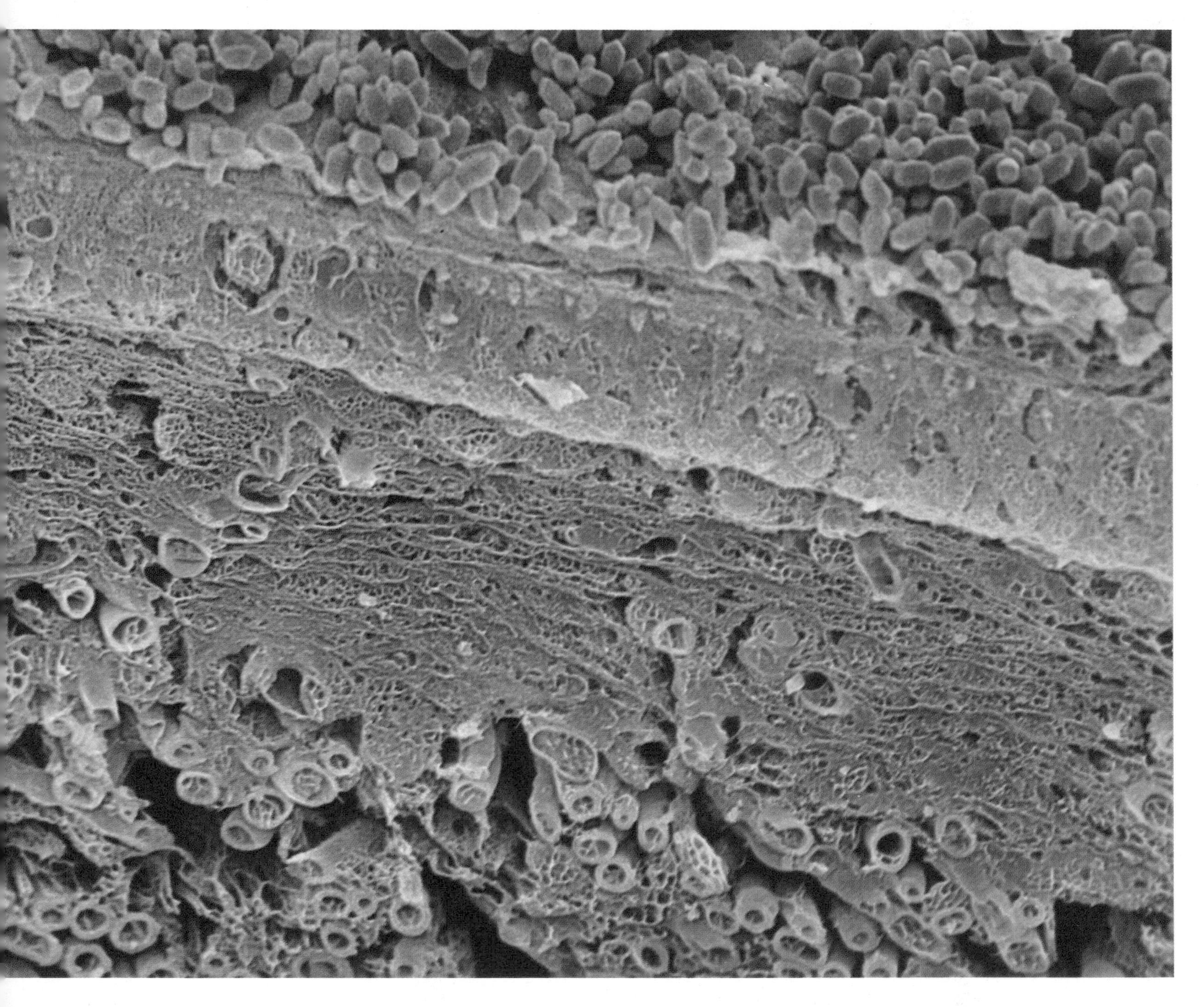

Nerve fibers in contact with the receptor cells
in the saccule and the utricle come together in the vestibular division of the eighth cranial nerve. The oblong structures in the upper part of the photo are calcium crystals pressing on the hairs of the receptor cells. In the center are hair cells and supporting cells. At the bottom, a large number of nerve fibers are seen, cut across their length. The individual nerve fibers are insulated by myelin sheaths. In the photo the sheaths look blue-green.

Touch and Pain

The literal ability to "feel" includes a number of different tactile qualities which provide information about the environment. We can touch or be touched by objects, experiencing a light touch or gentle tickle or very rough scrapes, pushes, or pressures. Such feelings are monitored by touch (tactile) receptors throughout the body. Other sense organs respond to mechanical vibrations, or to warm or cold temperatures. Muscle and joint senses, which inform the nervous system of the degree of stretch in skeletal muscles or of the position of joints, are also at times considered part of the general "feeling" senses. Finally, pain impulses are included in the large stream of sensory input from the body to the brain.

Many impulses reporting these qualities of feeling never reach consciousness. Yet the constant stream of routine information from the skin, the muscles, and the interior organs is a necessary condition for the normal functioning of the nervous system. What we say we feel is only what we become conscious of, what our attention turns to. Since we are especially sensitive to pain, impulses from pain receptors generally take priority and are dealt with immediately.

The impulses related to the varied qualities of feeling are initiated in peripheral end-organs, which may be naked nerve terminals or specially designed and encapsulated receptor organs. Naked nerve endings have been associated with pain perception, but they may also respond to other sensory qualities as well. The endings are usually finely branched and found in the lower layers of the skin. They are also present in the outer tissue layers of hollow organs such as the intestines. The theory is that stimulation by stretch or pressure causes changes in the electrical potential of the membrane of the nerve fibers with the result that pain impulses travel toward the spinal cord and the brain. Special sense organs as well as fine networks of nerve fibers are also found around hair roots. The individual hair acts as a lever, so that even a very gentle touch at the hair tip — without touching the skin — can trip signals in nearby nerve fibers.

The skin on the fingers, toes, the palms of the hands, the soles of the feet, and other parts of the body contain numerous Pacinian corpuscles. These sense organs consist of nerve endings surrounded by many fine layers of connective tissue so that they resemble miniature onions. Pacinian corpuscles are pressure receptors which respond to mechanical forces or stretches applied to the skin. In hairless regions of skin, particularly on the fingertips and lips, are touch receptors called Meissner's corpuscles. These corpuscles are entirely encapsulated and consist of spiral nerve endings with thin cellular layers between the windings. A sense organ that is sensitive to cold has been named after Krause, a German anatomist. Krause's corpuscles are situated immediately under the epidermis; they are bulb-shaped and contain a coil of nerves. Warmth is believed to be registered by Ruffini's corpuscles, which also consist of a fine network of nerve endings surrounded by a capsule.

The sense organs in the skin are not evenly distributed. Thus, the distance between cold-sensitive spots on the back may be an inch or more while it is only a fraction of an inch on the tip of the nose. The fingertips, lips, and tongue are areas of great sensitivity, well supplied with sense organs. Pain receptors are more evenly distributed in the skin.

When a particular sense organ or a group of naked nerve endings in skin is stimulated, the impulses travel through afferent nerves to the spinal ganglions, enter the cord, and ascend to the thalamus, the main relay center for sensory impulses in the middle of the brain. From here, the impulses are relayed

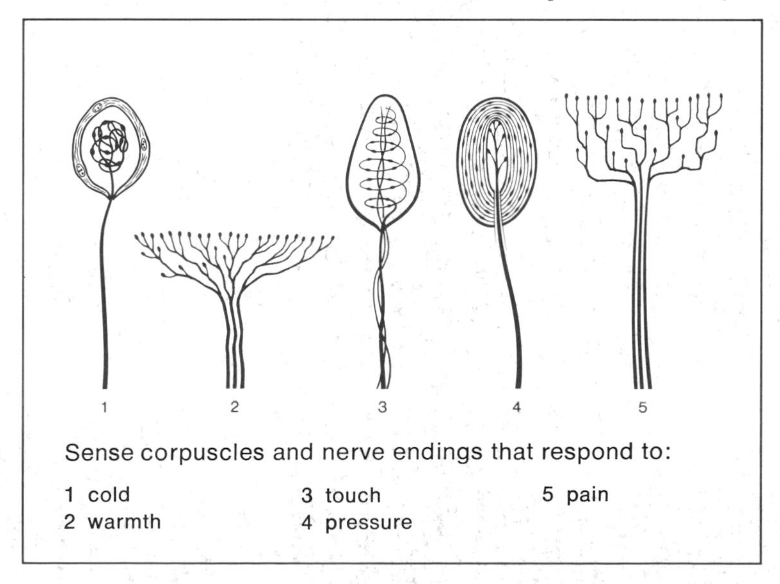

Sense corpuscles and nerve endings that respond to:

1 cold
2 warmth
3 touch
4 pressure
5 pain

3 An end-organ sensitive to touch

to the cerebral cortex. Branching and crossing over occur at various levels, and in the cord itself the sensory impulse may trigger a reflex response. For example, if you touch a hot object, the sensory signals in the cord may immediately relay with motor cells which will trigger contractions in the hand muscles so that the hand is withdrawn instantly. Higher up in the brain stem, the same burst of impulses from the hand may lead to a change in respiration, a faster pulse rate, or a rise in blood pressure. Somewhat later, perhaps not more than a few hundredths of a second, we become conscious of what has happened and can interpret the event.

The impulses conveyed to the spinal cord and the brain through the various sensory pathways are always of the same type. They consist of electrical nerve impulses. The nerve cell, as well as the naked nerve ending or the fiber inside a sense corpuscle, is covered by a membrane. A potential difference of about sixty-thousandths of a volt exists between the outside and the inside of this membrane. When the nerve cell body or its

The fingertips (1)
are extremely sensitive. To some extent they substitute for eyes in the blind. Nerve endings and sense organs are numerous at the fingertips. When they are stimulated, nerve impulses travel through the nerve trunks of the fingers to the spinal cord and on to higher relay stations in the brain stem and cortex. At the cortex the various parts of the body are represented by a map which is a projection of the sensory endings in the skin. The skin contains sense receptors responsive to touch and pressure, tickling, vibrations, et cetera.

The pattern on the fingertip (2)
The skin on the palm of the hand and at the fingertips has a characteristic papillary pattern, unique for every individual. The elements of the pattern are ridges and furrows formed by connective tissue fibers in the dermis. Sweat glands have openings on the ridges.

A Meissner's corpuscle (3)
consists of an encapsulated nerve fiber, the distal end of which forms a spiral. It is designed to respond to touch, in contrast to other end-organs which may be stimulated by cold, warmth, or pain. The corpuscle lies in the dermis, immediately under the epidermis, and is surrounded by disk-shaped epithelial cells. A touch on the skin overlying a Meissner's corpuscle deforms the capsule and this leads to the production of nerve impulses in the nerve fiber extending out from it. A deeper touch produces greater deformation and leads to an increase in the frequency of the impulses. Thus, impulse frequency is one of the cues the nervous system uses to measure the intensity of a stimulus.

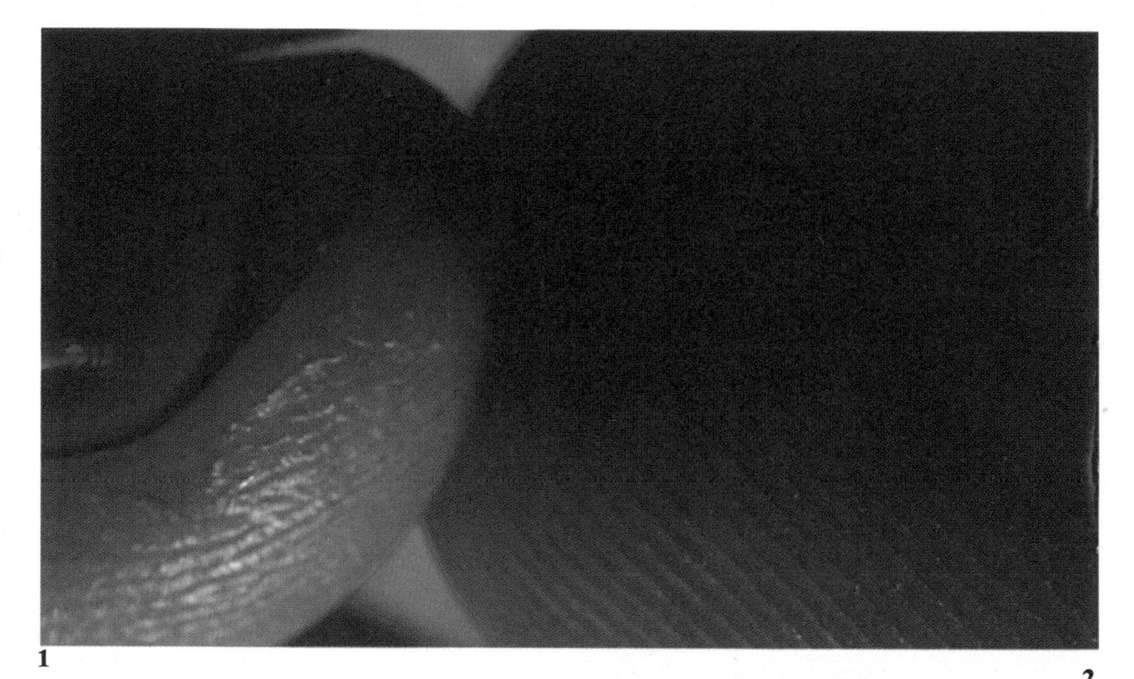

1

2

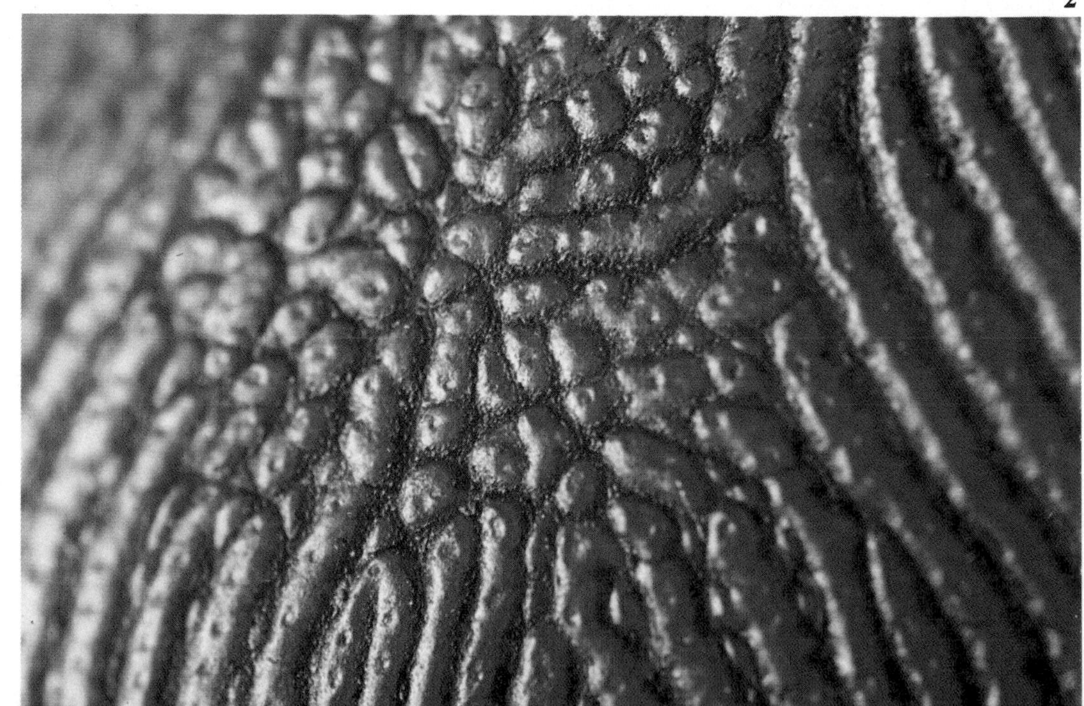

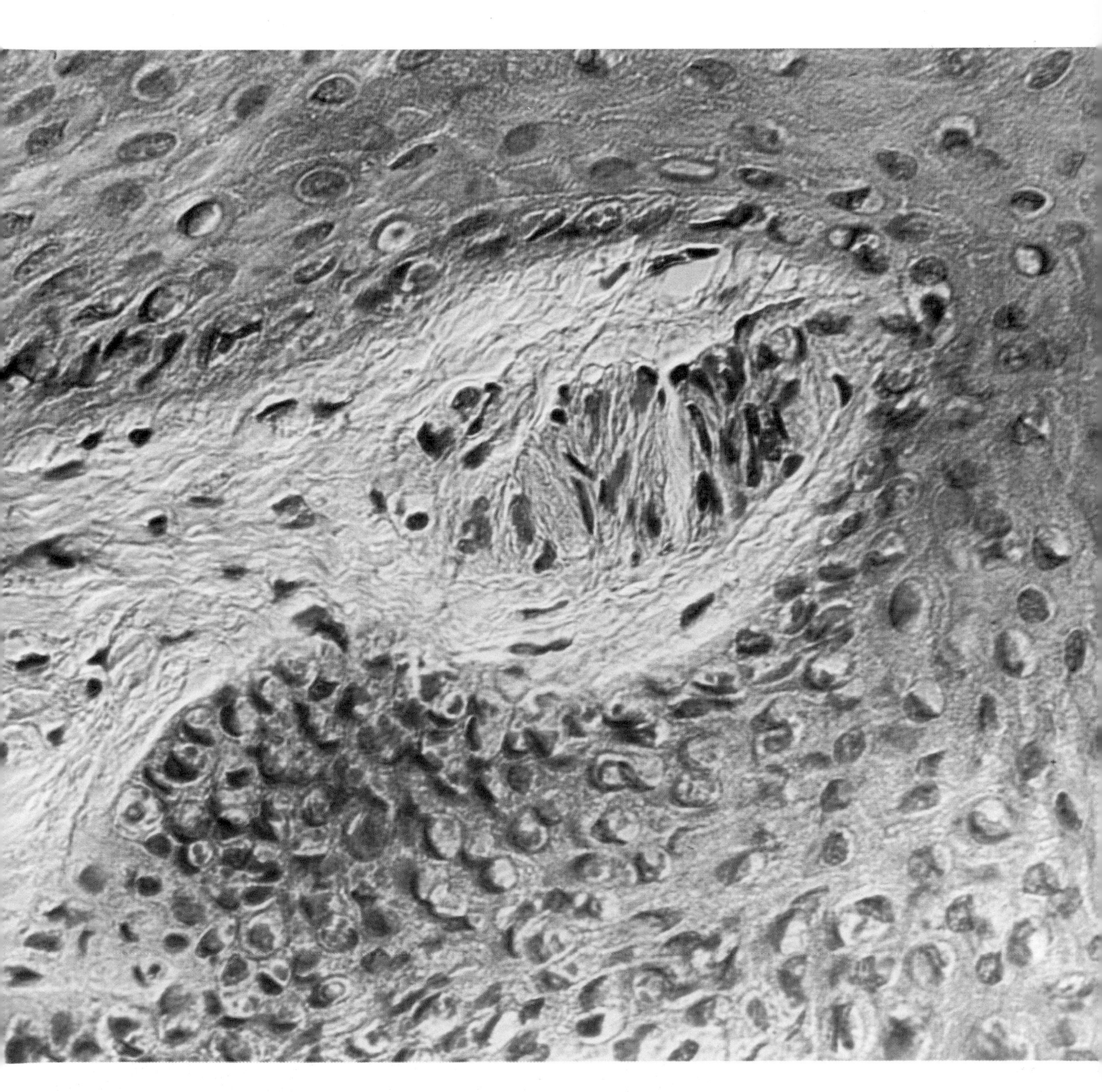

The Pacinian corpuscle
is one of the largest sensory receptors in the body. It is specialized to respond to pressure. Its diameter ranges from 1 to 4 mm, so it can be seen with the naked eye. Pacinian corpuscles are numerous in the skin, the heart, the cornea of the eye, the pancreas, loose connective tissue, and elsewhere in the body. The photo shows how the nerve ending is surrounded by cells arranged concentrically. The nuclei of the cells appear as dark dots in the bands. The corpuscle in this micrograph was from the pancreas.

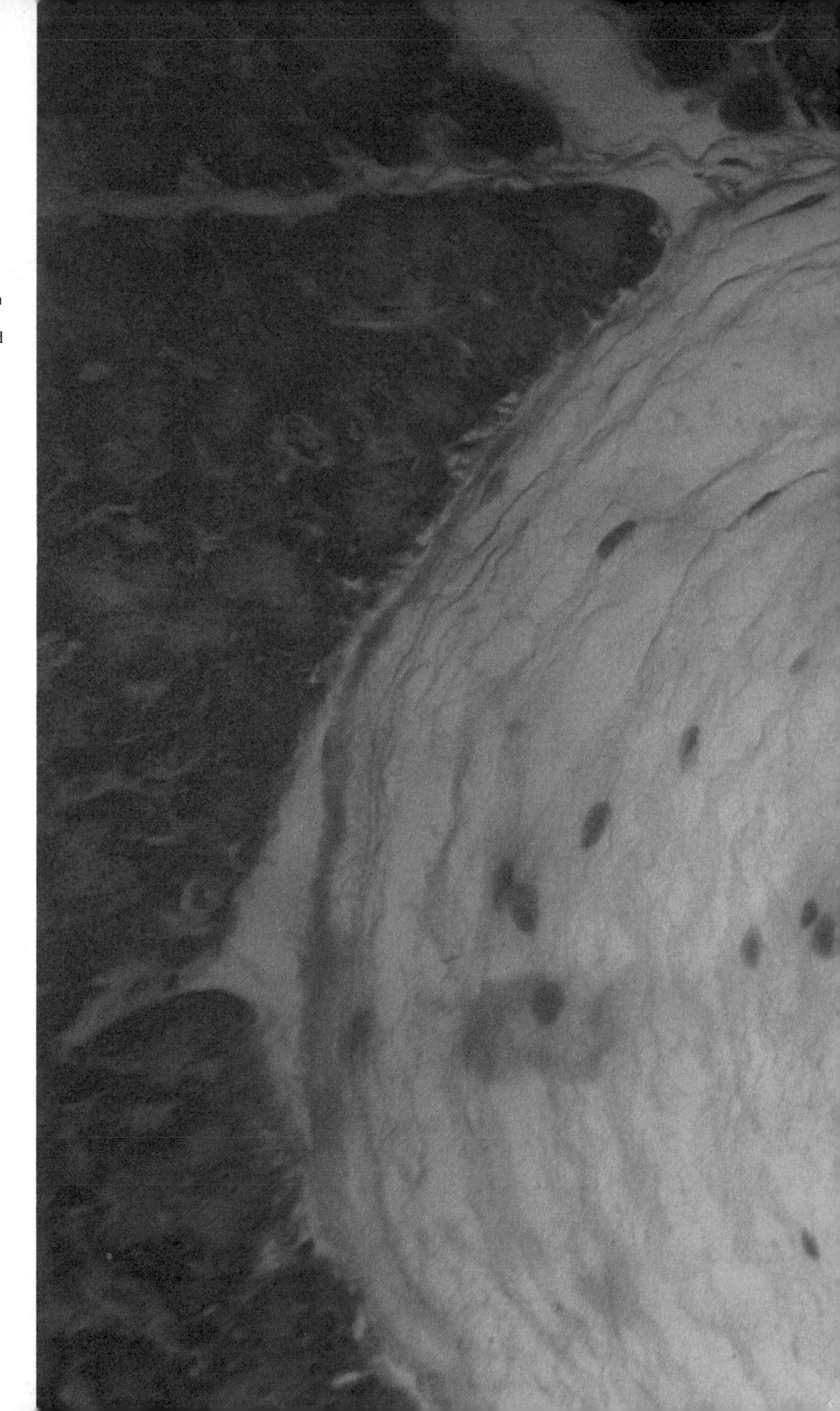

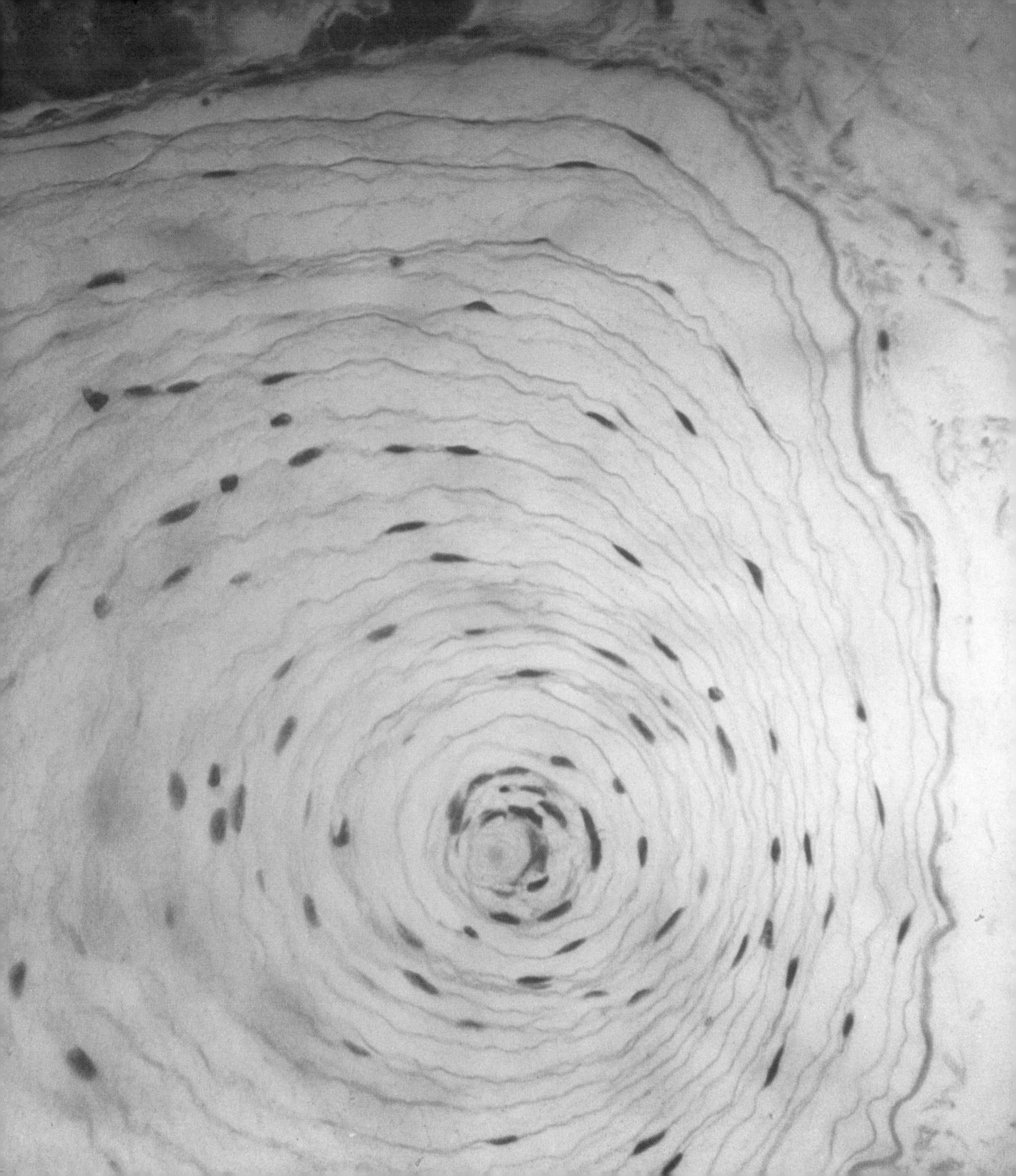

fiber processes is sufficiently stimulated, it is depolarized, and a weak current is generated that can be propagated along the membrane. This current is the actual nerve impulse. The velocity at which the impulse is conducted depends on the thickness of the nerve fiber. The speed of conduction varies between several feet to a few hundred feet per second. Generally, conduction velocity is most rapid in fibers conveying impulses from muscles and from some sensory endings in skin. Fast-conducting fibers are sheathed in a fatty substance (myelin), which is interrupted at periodic intervals. The nerve impulse in effect "jumps" from one interval to the next. A photo of such a myelinated nerve fiber is shown on page 156.

It used to be thought that beyond a certain threshold, all nerve endings give rise to sensations of pain. In all probability this is not true. The evidence points to specific pain receptors present in the skin and in the interior organs. The function of these end-organs is to alert the central nervous system to circumstances threatening the integrity of the body. These would include disturbances in blood circulation, excessive heat or cold, chemical irritants, or strong pressures.

The perception of pain often involves two distinct qualities. Stubbing your toe against a table leg produces an immediate sharp pain. After this there is a pause — short, but clearly perceptible. This is followed by a more dull, aching pain in the toe. Neurophysiologists explain that the reason this occurs is that pain impulses can travel in two kinds of nerve fiber, one which conducts at a fast rate, the other more slowly.

A mixture of fast and slow impulses may also be involved in certain compound sensations such as a tickle, which has elements of touch and pain. The ratio of fast to slow impulses may also figure significantly in how we judge the intensity of sensations. Injuries to nerve trunks in the extremities may affect this ratio and lead to conditions of prolonged pain or other anomalies of pain sensation which are little understood.

Our sensibility to pain is not constant. We are likely to be more sensitive to pain when tired than when awake and alert. Anxiety tends to lower the threshold to pain, while a relaxed mood raises it. The more we think about pain, the more intense it may become. Distraction, on the other hand, by suggestion or hypnosis, can diminish pain to the point where it may disappear completely. However, the function of pain generally is to make us aware: its signal means that the body is being threatened. We must respond by taking action to protect the body against the danger to which it has been exposed.

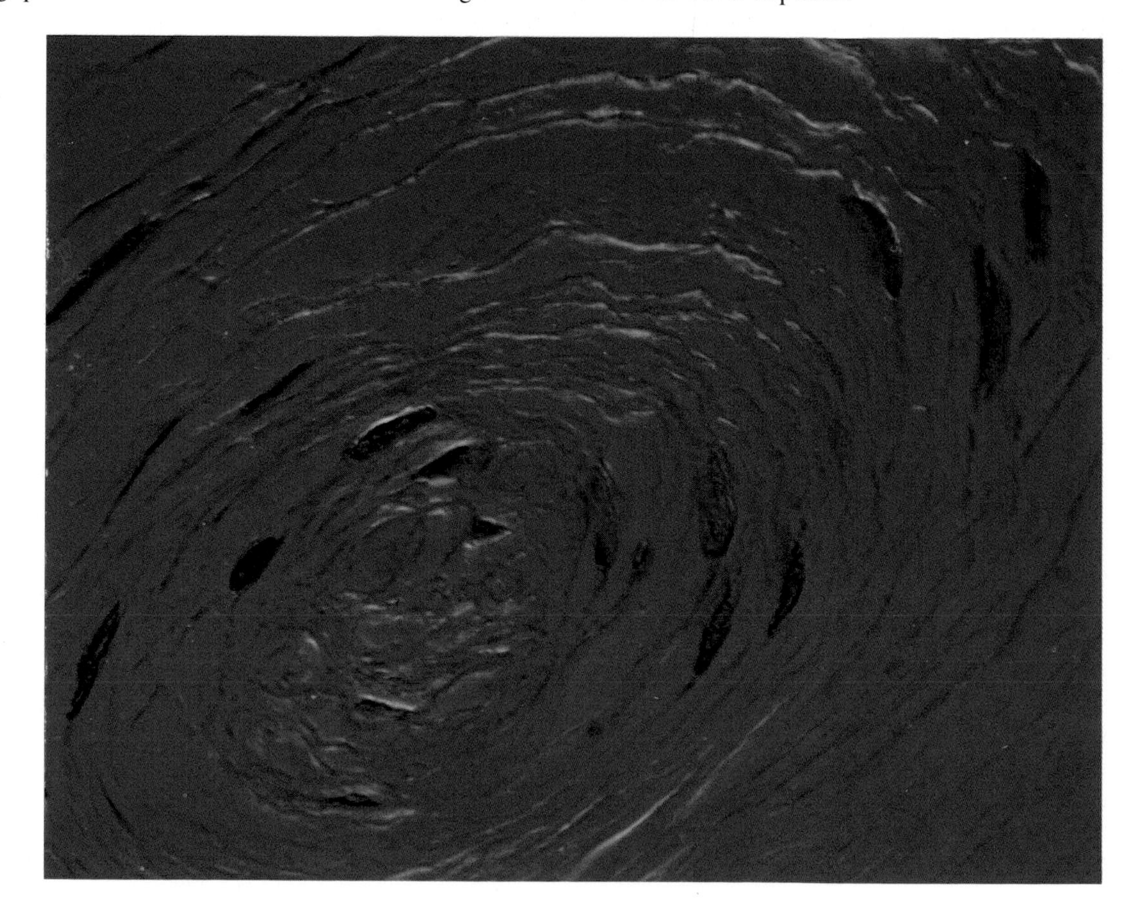

A Pacinian corpuscle,
seen in an interference microscope. This type of encapsulated corpuscle is numerous in skin and other organs where pressure information is important. Thus, Pacinian corpuscles in the walls of the intestine measure the degree of their distention or stretch in the process of digesting food.

Olfaction

Man's senses are dominated by vision and hearing. In other animals, the sense of smell (olfaction) plays the greatest role. The dog is an excellent example. The huge number of receptors in the dog's olfactory area — much greater than in the human nose — makes its sense of smell one of the most acute in the animal kingdom.

Generally speaking, the dominance of a particular sense in an animal is related to its evolutionary importance. The survival of man, whose ancestors once lived in trees, depends more on stereoscopic vision, to provide good depth perception, than on a fine sense of smell. When early man began to walk upright, the importance of vision to detect, distinguish, and estimate objects at a distance increased still more. In the course of time the sense of smell became less and less important.

Nevertheless, the sense of smell serves several important functions. It warns us against potential dangers — of food gone bad, for example, or of poisonous exhausts from cars, or smoke from fire. It can help in the search for food, given more primitive circumstances, as well as enormously contribute to the pleasures of eating (aiding digestion, in fact). It is also of importance in sexual life. Recent physiological research confirms that particular odors produced by glandular secretions in many animal species are sexually arousing. They are the natural "perfumes of love."

The olfactory organ is located in the roof of the nasal cavity and covers an area of about one square inch in the epithelium of either nostril. The olfactory epithelium is called the olfactory mucosa. It is slightly yellowish in color because of the presence of a substance related to carotene, the orange pigment found in carrots which is the source of vitamin A.

The olfactory mucosa contains the olfactory receptors and supporting cells. The receptor cell is a thin double-ended (bipolar) nerve cell, rounded in the middle where the nucleus is. One slender ending sticks up into the mucosa, its tip fringed with hairs (cilia) pointing up into the air passageway through the nostrils. The cilia are normally bathed by mucus. The opposite end of the cell is the nerve axon. This fiber joins others passing through the cribriform plate of the ethmoid bone in the skull to form the olfactory nerve. After several relays in a region called the olfactory bulb, the fiber pathways continue to centers on the underside of the brain. The processes by which we recognize smells, recall memories associated with them, or react emotionally with feelings of happy delight or utter disgust are little understood.

In addition to the olfactory mucosa, the nasal cavity also contains certain naked nerve endings that respond to airborne substances. Irritating gases such as ammonia and chlorine stimulate the naked nerve endings as well as the olfactory receptors, producing fast and painful smell sensations.

The precise way an olfactory receptor cell responds to stimulation is not yet known. Clearly, an odor-producing substance must be volatile — that is, its molecules must be present in the air — in order to react with the receptor cells. Moreover, its molecular structure appears to influence its odor quality. A small change in the configuration of a group of atoms in one part of an odoriferous molecule may increase or lower its ability to stimulate an olfactory receptor cell or change its odor entirely. The fact that the substance and the receptor are in intimate contact may also mean that the odoriferous molecules are absorbed at certain parts of the receptor cell membrane. A number of experiments indicate that we have different types of receptors that respond to different types of odor-producing substances, and that the impulses from groups of similar receptors are integrated and analyzed at higher relay centers.

The stimulation of olfactory receptor cells is the initial step in the journey from nose to brain, but it is no guarantee that we will actually be aware of a smell sensation. Our ability to experience odors decreases fairly rapidly. A possible explanation of this adaptation is that the impulses reaching the higher brain centers begin to decline in frequency, because they are suppressed by other brain centers. Thus, even though the olfactory receptors may continue to report an odor over a long period, we may cease to smell it because the brain inhibits the signals coming in from the periphery.

While there are many theories about how the sense of smell works, none of them can explain how we can distinguish among so many odors. Most people can recognize several hundred odors; a skilled chemist may distinguish among several thousand.

Olfaction

The nostrils
The photo on the right shows the nostrils in a four-and-a-half-week-old embryo, as compared with a seven-week-old embryo below. The olfactory mucous membrane develops from a thickening in the ectoderm, the olfactory plate. This plate gradually descends and forms an olfactory pit, the primordium of the nasal cavity. The cells of the olfactory plate thus become located in the upper portion of the nasal cavity, where they develop into the olfactory mucous membrane. In the lower portion of the nasal cavity are the nostrils.

The photo on the right shows the nose magnified about 170 times; that below, about 180 times.

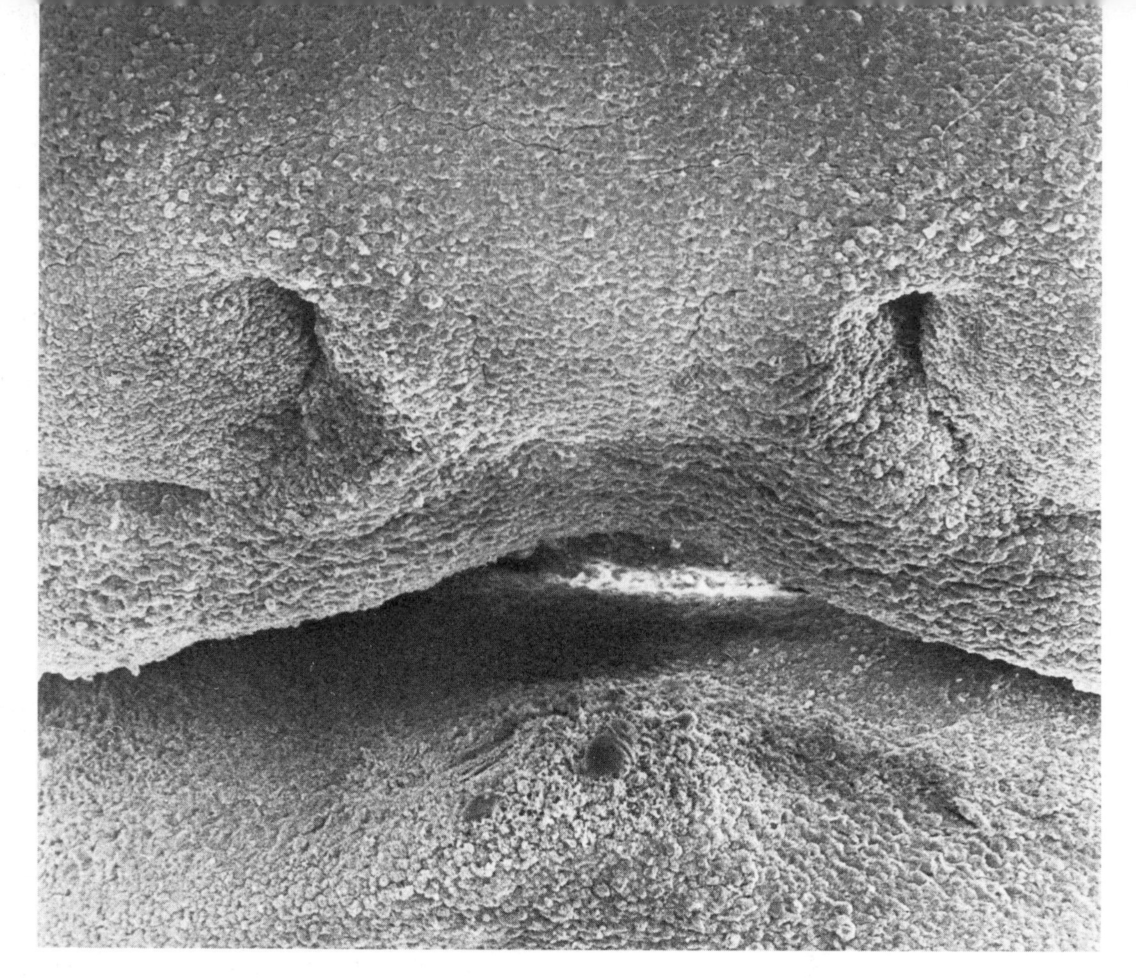

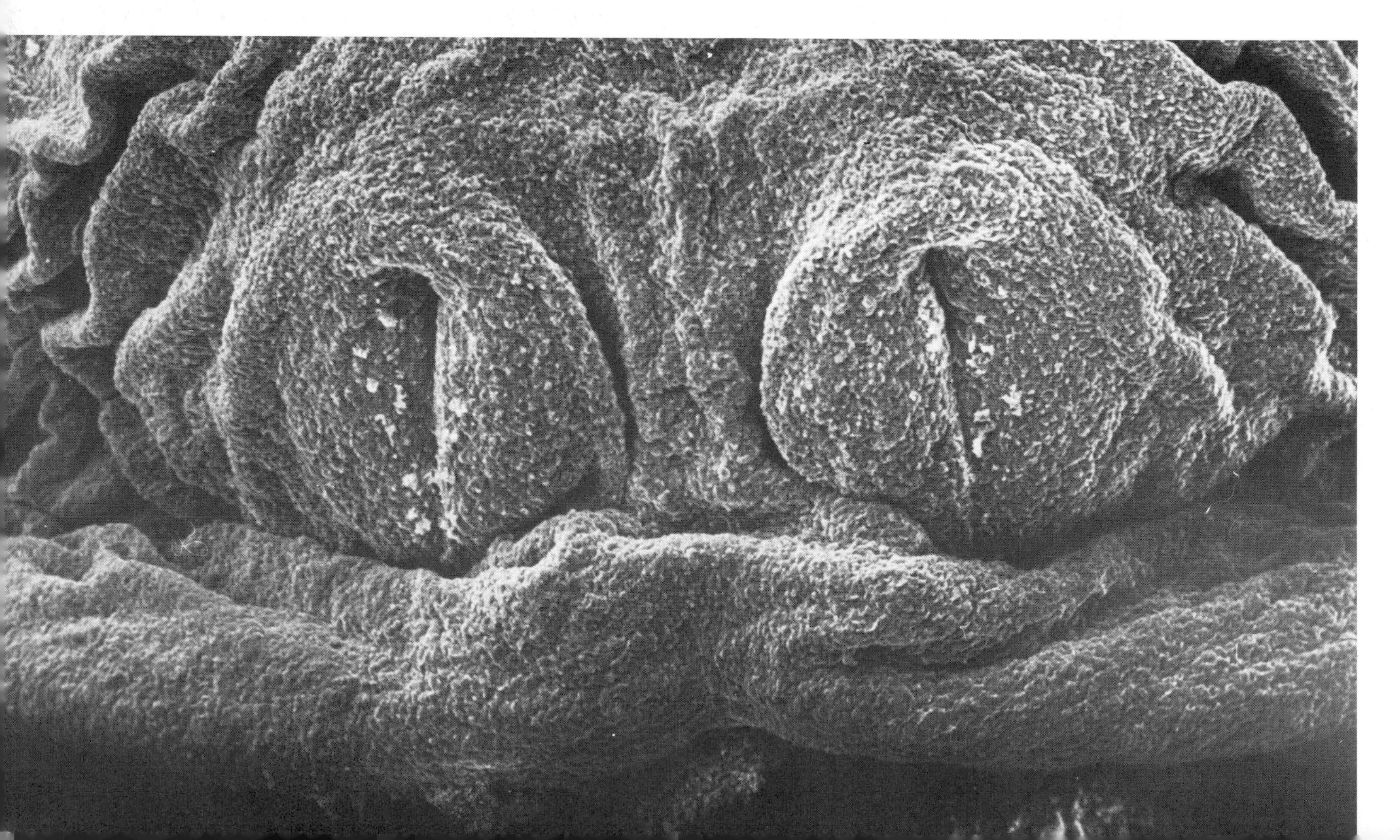

Attempts to classify odors have failed because, among other reasons, the relation between the chemical composition of a substance and its odor is still not clear. One theory divides smell sensations into four main types: pleasant, fiery (smoky), cheese-like, or rotten. In other experiments where individuals were exposed to hundreds of odors, a "smell prism" was constructed consisting of six basic qualities: fruity, flowery, decaying, spicy, resinous, and fiery.

Among the strongest odors we encounter is that of mercaptan, an unpleasant smelling sulfurous substance chemically related to alcohol. It can be perceived at the minute concentration of one part mercaptan to 460 million parts air. Sensitivity to sulfurous smells is generally high, which presents great problems to cellulose and paper manufacturers. In fact, experiments have shown that the human nose is far more sensitive to a sulfurous smell than any machine designed to measure smells. In

The nose
still looks snubbed at about the fourth month of gestation. The white is cartilage and the surrounding structures are skin and muscle tissue.

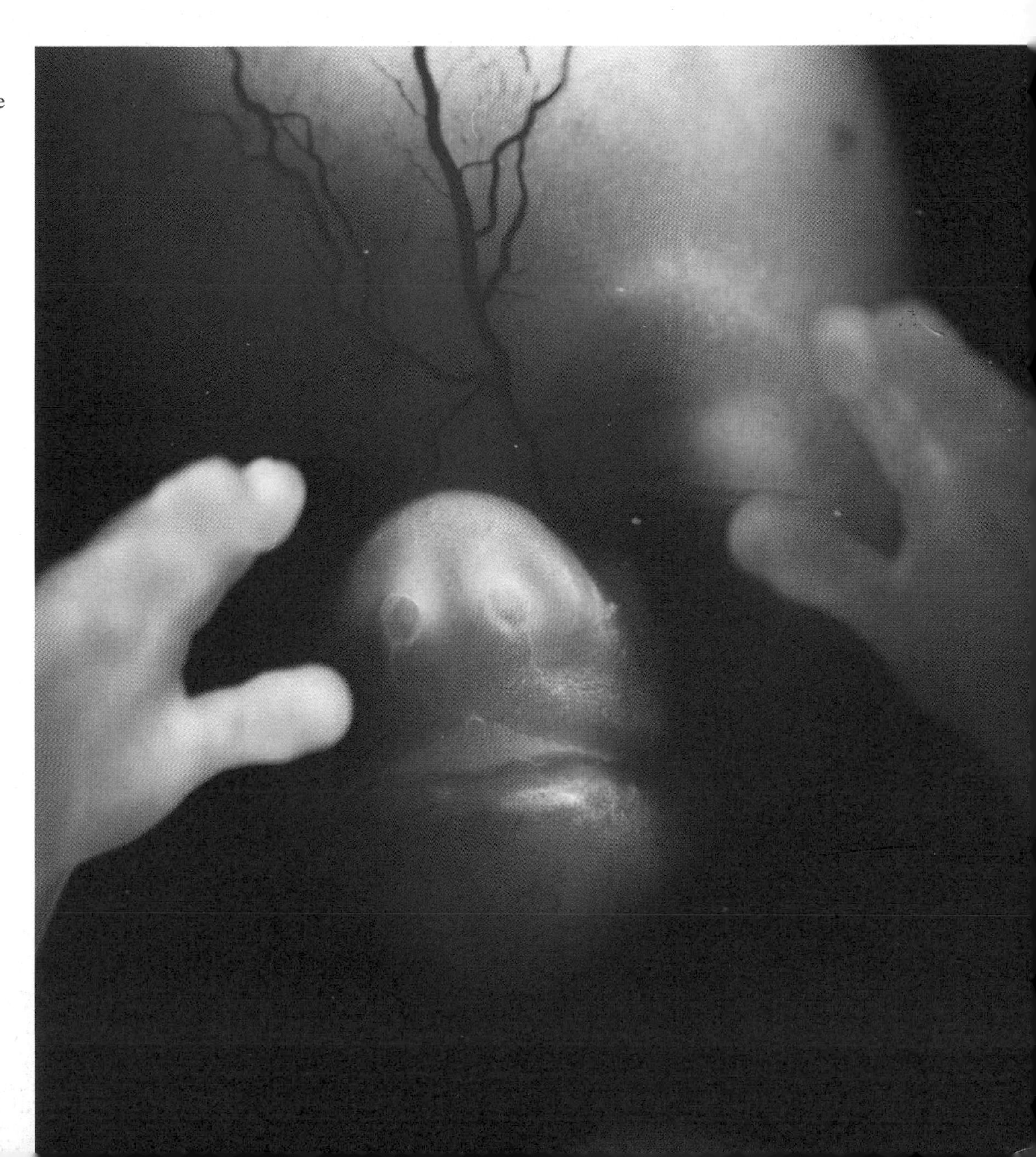

their pure form, only seven basic elements can be smelled by man: fluorine, chlorine, bromine, iodine, oxygen (in the form of ozone), phosphorus, and arsenic. Otherwise, the olfactory receptors are stimulated chiefly by organic (carbon-containing) compounds.

In addition to being volatile, a substance which excites the olfactory receptor may also have to be soluble in fat and water. Substances which are soluble in both fat and water generally have strong smells. The principal ingredients in the scent of flowers are volatile oils — for example, rose oil and lily of the valley oil. These are mixtures which belong to the group of hydrocarbons called terpenes. Also belonging to this group are menthol from peppermint oil, camphor from the camphor tree, and caoutchouc from the rubber tree.

When we speak about the "taste" of food, we generally mean the total sensation we get from food in the mouth. But often a cold and a stopped-up nose can prevent food from having any

The olfactory organ is located in the roof of the nasal cavity where it covers a small-postage-stamp-sized area. About twenty-five bundles of nerve fibers lead from the olfactory receptors to the olfactory bulb, the next relay center. Impulses from the bulb in each nostril are conducted to the brain in the olfactory tracts.

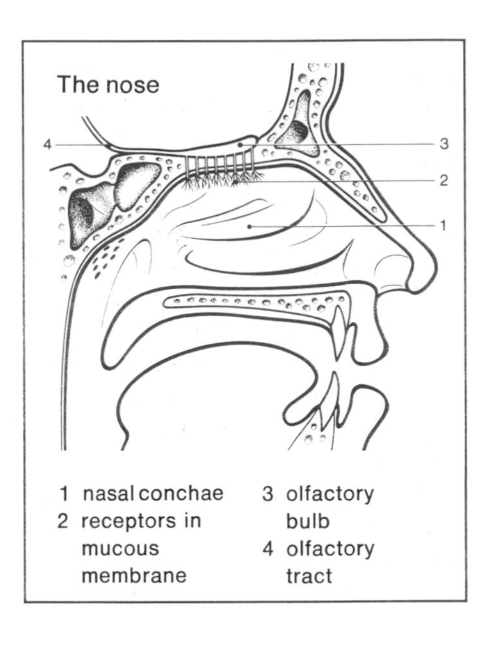

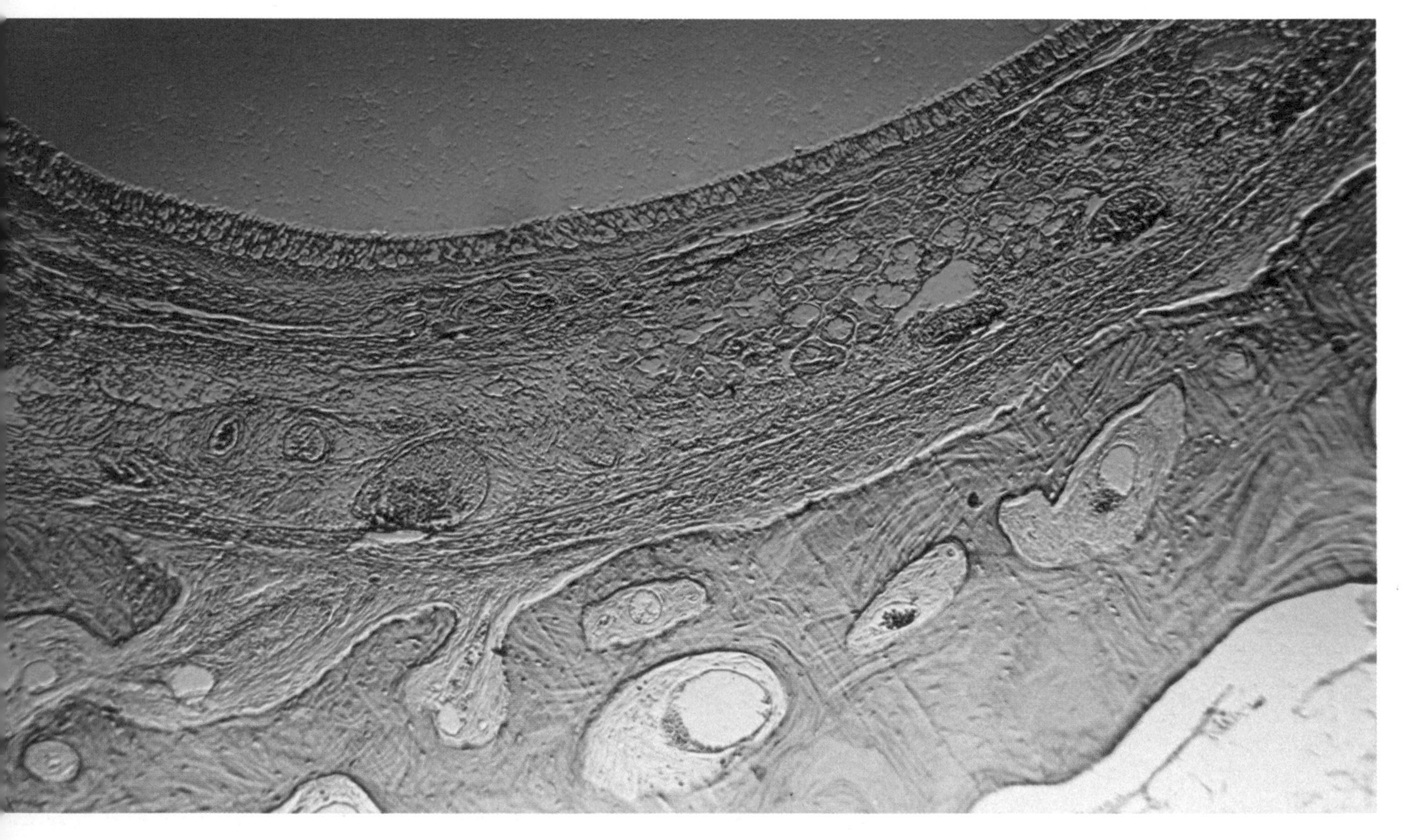

taste at all. Thus, the sense of smell plays an important role in our experience of food. Strictly speaking, taste in the mouth relates to sensations and combinations of sweet, salt, sour, and bitter perceived by the taste buds. The other "taste sensations" are more closely linked to the sense of smell. Looked at this way, it becomes obvious that the common seasonings, except salt, sugar, and sour substances, are used in cooking largely because of their smell. Smell then collaborates with taste in eating. If you hold your nose (and are blindfolded) it may be impossible to distinguish between mashed apple and mashed onion. Gourmets, of course, would acknowledge that the bouquet of the wine and the flavor of excellent roast beef are almost entirely perceived by the nose.

Nowadays, the production of scents and odor-neutralizing agents is a large industry. Artificial essences of all kinds are added to foods to make them more appetizing, or perhaps even to camouflage less salable smells. Perfume, like alcohol, is one of the earliest inventions of man. Incense and fragrant oils are mentioned in the oldest writings. Incense was also burned to appease or influence the gods. Today, the deodorant industry dominates the scent market. Movies, newspapers, magazines, subway advertisements, and commercial television programs are indefatigable crusaders for man to smell like anything but man. The volatile oils, from pine needle to periwinkle, are the principal smell-camouflaging agents in the more simple deodorants. More complex ones contain agents that combine with and neutralize some of the odor-producing substances.

The individual odor of a person varies, and this variation is not just a matter of hygiene. The smell of a baby differs from that of an adult, and those of elderly people differ from those of younger ones. Diet is believed to influence body odor, probably because the composition of sweat and other odors is affected by diet. As a common example, you can always tell when a person has eaten garlic.

◁ **A section through the roof of the nasal cavity**
At the top is porous bone tissue with the hollows covered by a well-vascularized mucosa. The olfactory mucous membrane is located in the lower portion of this mucosa. It is rather thick, slightly yellowish in color, and composed of receptors and supporting cells. The receptors are double-ended nerve cells with cilia at one end. The other end forms the fine nerve fiber which joins with others into the bundles which make up the olfactory nerve.

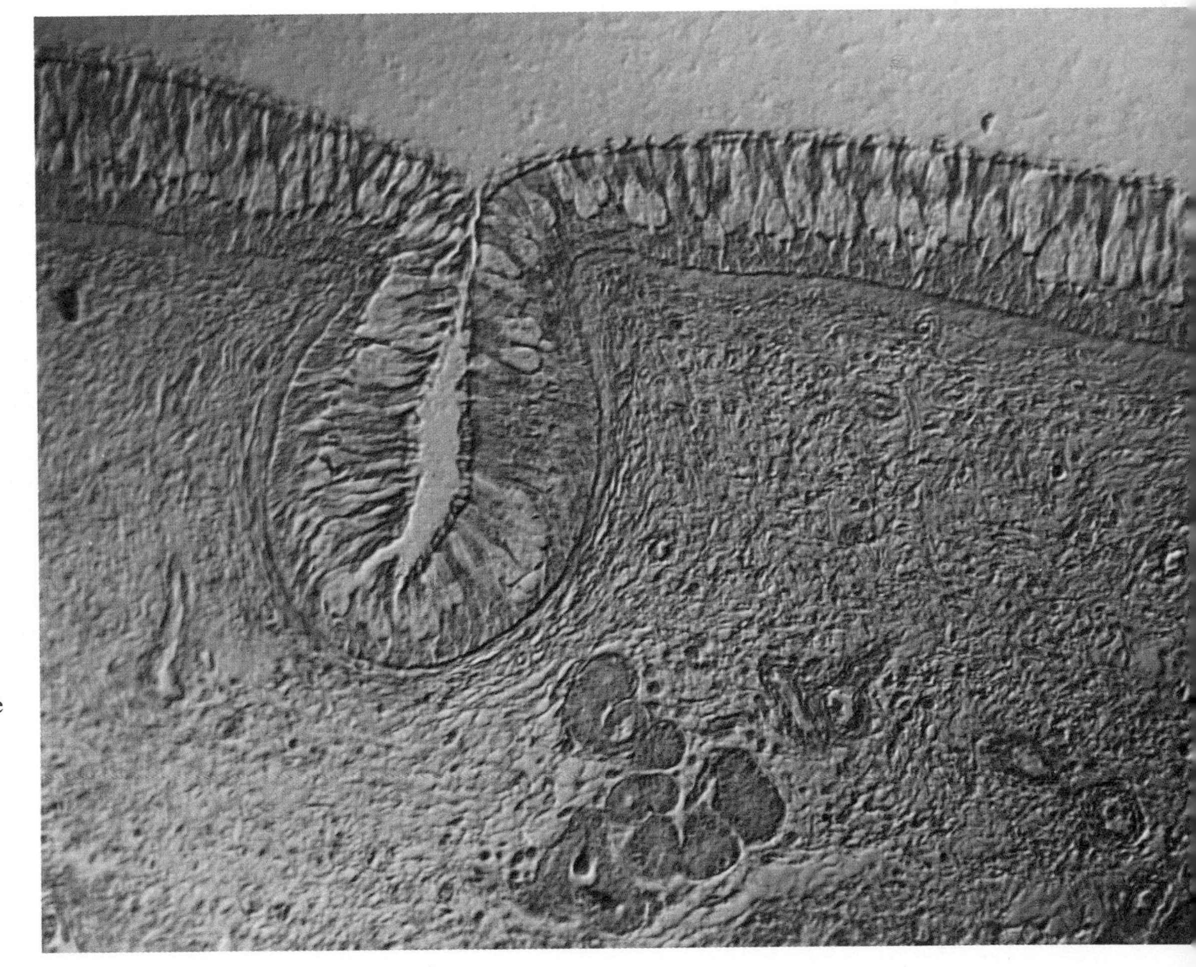

The olfactory mucous membrane ▷
in man is certainly small, but being convoluted, as shown in the photo, it has a relatively large surface. The olfactory receptors appear as a narow band in the upper parts of the photo. The receptors respond to stimulation by odiferous substances by generating electrical impulses. The very fine processes at the surface of the olfactory receptors are cilia. They are surrounded by mucous secretions. The cilia play a crucial role in picking up odor-producing molecules from the air as it passes through the nose.

Olfaction

The olfactory mucous membrane
consists of columnar epithelium. It covers an area of approximately five hundred square millimeters and is about six-hundredths of a millimeter thick. It is composed of three different types of cells. The fine hairlike processes projecting from the surface of the receptor cells are cilia.

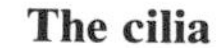

The cilia
are approximately two-thousandths of a millimeter long. They project from the surface of the receptor cell and are covered by a thin layer of mucus, which traps chemical substances. The appropriate substances stimulate the cilia so that electrical impulses are generated in the nerve fibers.

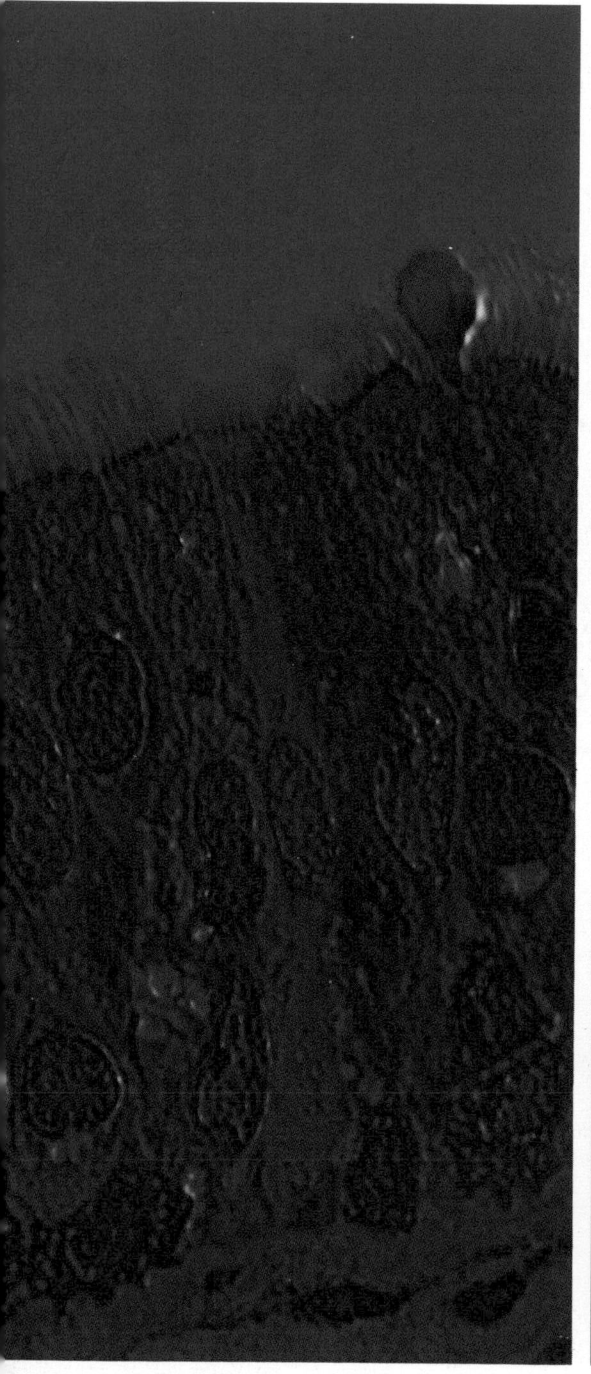

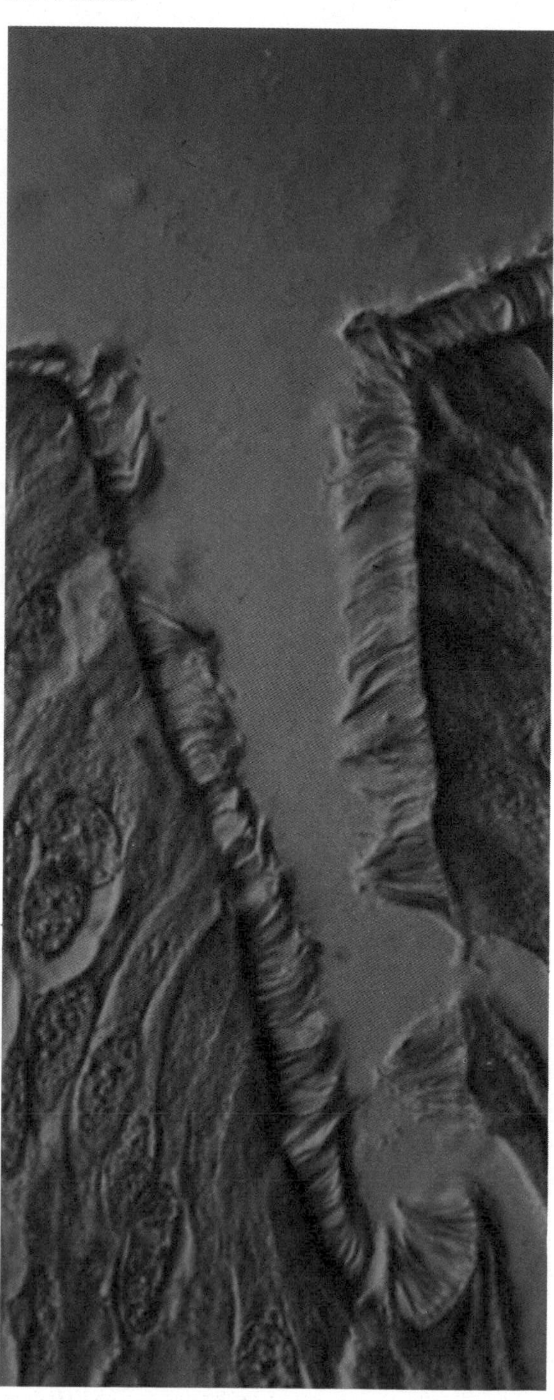

In the upper air passages, ▷
the mucosa is covered by ciliated respiratory epithelium. The constant upward swaying of the cilia carries mucus from the pulmonary bronchi up the trachea and to the nose. The respiratory epithelium is the site of the olfactory mucous membrane. The cilia are clearly visible in this electron micrograph because the secretion normally surrounding the cilia has been removed.

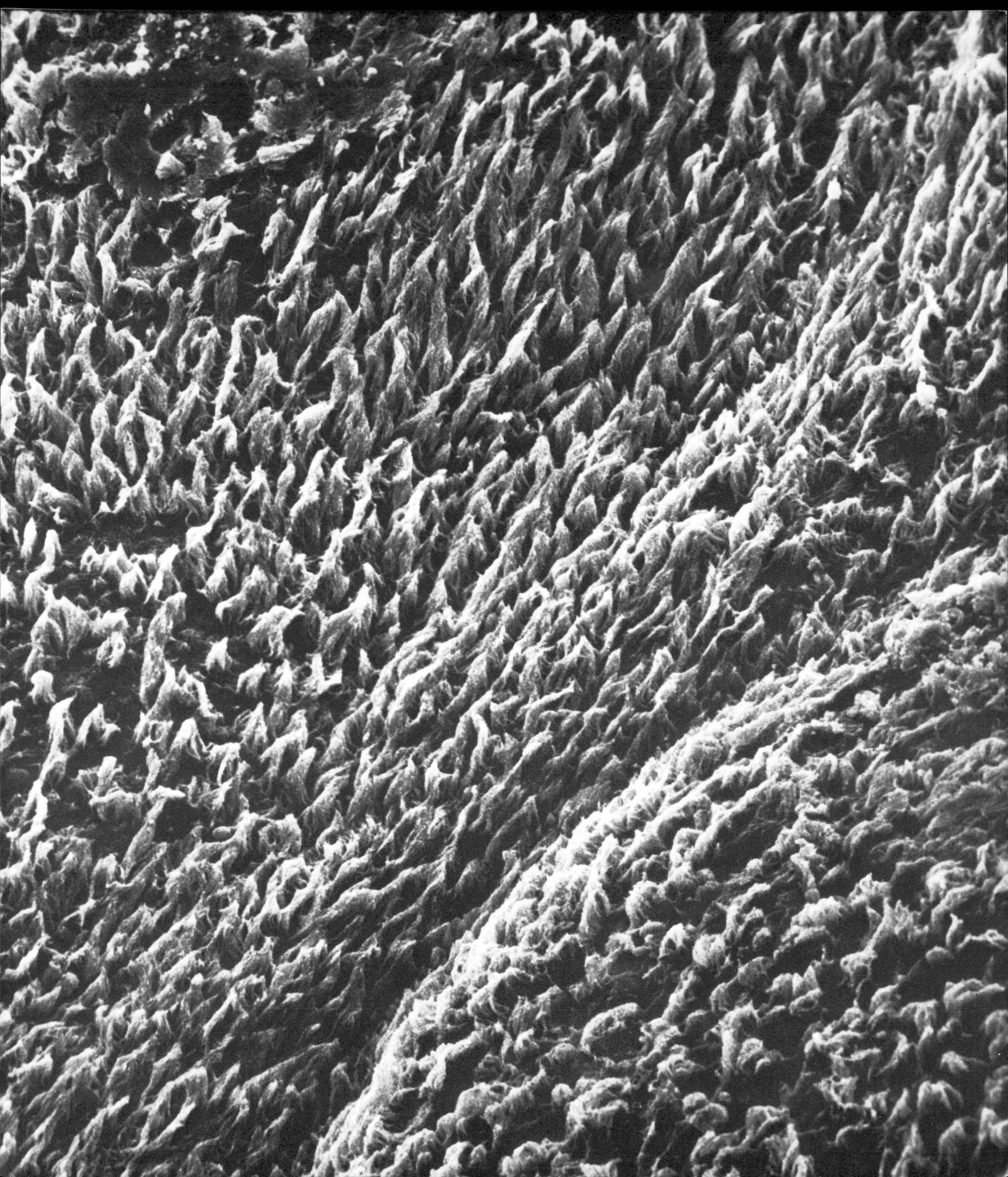

Olfaction

Olfactory receptor cells, greatly magnified
The cilia fringing the cell are embedded in a mucous secretion. Appropriate odor-producing substances caught in the mucus can excite the receptor so that electrical impulses are sent to the brain. It is commonly held that the individual olfactory receptor specializes in trapping certain types or classes of substances, and that impulses from cells responding to the same odors are integrated in the olfactory bulb, the first relay center for olfactory signals. A probable explanation for the fact that the nose seems to adapt rapidly to odors is that nerve cells in the brain inhibit incoming signals from the olfactory bulb. The sense of smell in man is not perfect. A number of poisonous gases, among them carbon monoxide, are not detected by the olfactory receptors, which means they are odorless.

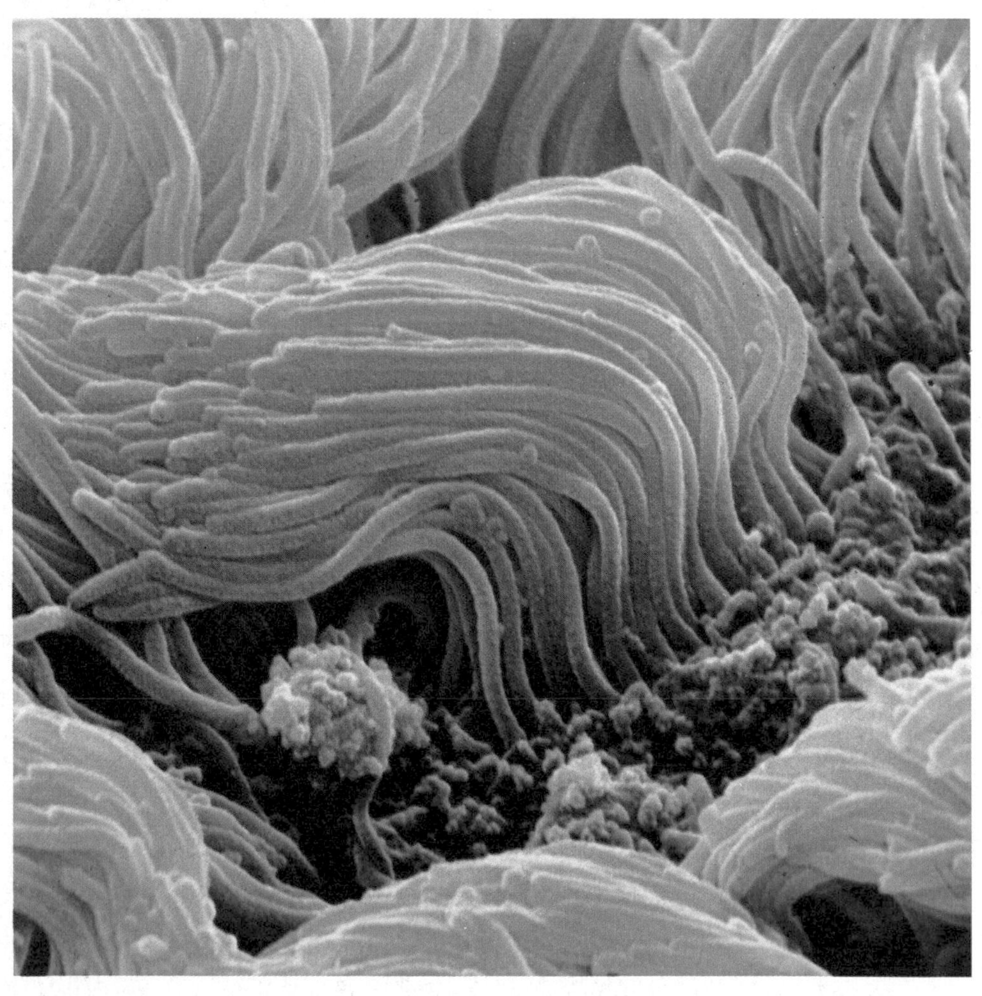

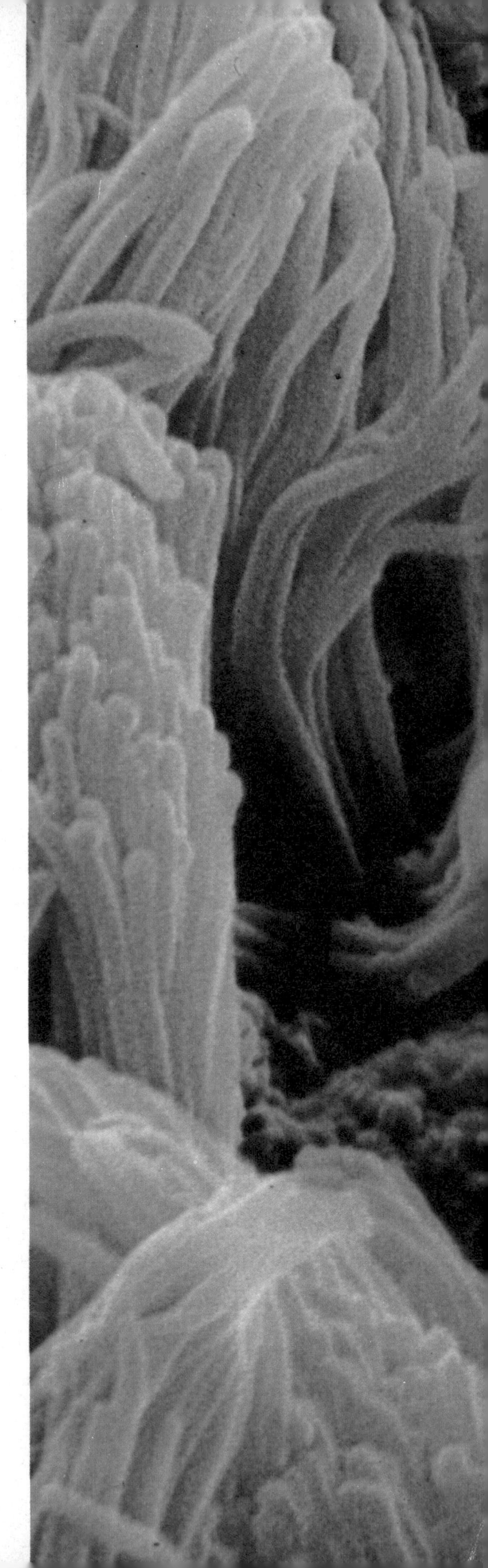

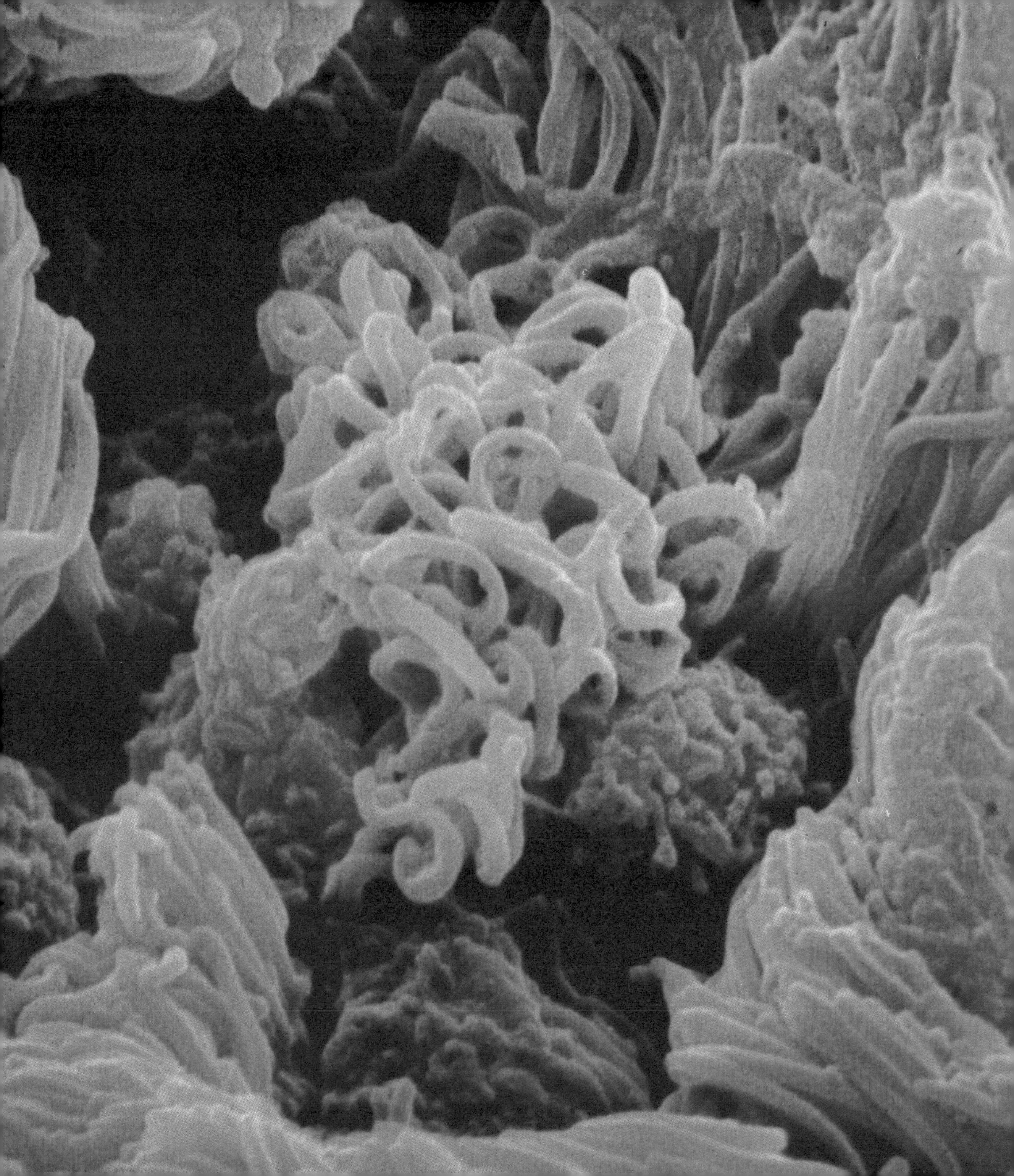

Taste

The word "taste" has several meanings. In the description of smell it was pointed out that much of what we call the taste of the food depends on smell sensations. Biologically, the human sense of taste has a relatively limited range. The several thousand taste buds in the mouth respond only to salt, sour, sweet, and bitter flavors, yet there is no denying that taste is an effective and sensitive sense, working as it does in close collaboration with the sense of smell. The taste buds are especially sensitive to bitter tastes. This may be a consequence of the fact that many plant toxins are alkaloids, which have a bitter taste. A most important function of the sense of taste, then, is to warn us against harmful substances.

The sense of taste is derived from the responses of the taste receptors in the taste buds. The buds are located in papillae on the tongue but also in other parts of the mouth. Each taste bud is made up of taste cells and supporting cells. Hairs at the tip of the receptor cells are in direct contact with chemical substances present in the mouth. The taste buds respond to water-soluble substances only.

A number of fine nerve fibers arise from each taste bud. They combine and travel part of the way with the nerve innervating the tongue. Later they follow the facial nerve to the brain stem. Since the nerve fibers are bundled closely together, injuries to nerves in the face may affect the sense of taste, too. Impulses in sensory nerve fibers from taste buds may trigger reflexes such as the secretion of saliva. A bite into a sour apple or into a slice of lemon can thus be felt as far back as the ears, by virtue of the large salivary glands nearby.

The papillae are located chiefly on the tip of the tongue, its sides, and its posterior portion. There are several types of papilla. The largest ones are the vallate papillae, which are arranged in a V on the back of the tongue. A vallate papilla consists of a mound surrounded by a moat. The taste buds are located on the banks of the moat. Other types are the foliate, the filiform, and the fungiform papillae. Small glands open around the taste buds, and their watery secretions dissolve the substances in food so that they can gain access to the hairs of the taste receptors.

The way the taste cells operate is similar to the process by

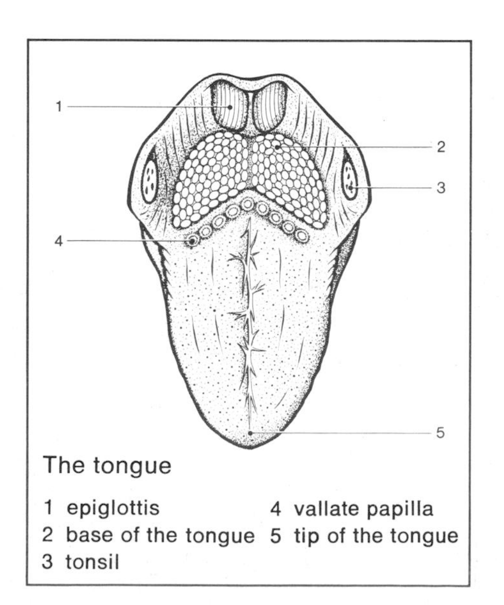

The tongue

1 epiglottis
2 base of the tongue
3 tonsil
4 vallate papilla
5 tip of the tongue

Salt crystals
on a papilla on the tongue. In solid form, salt has no taste. The taste buds are not stimulated until the crystals dissolve in the saliva — i.e., in water — and come into contact with the receptors. Salt crystals are rapidly dissolved, however, so when we get some salt on the tongue, the taste sensation occurs almost immediately.

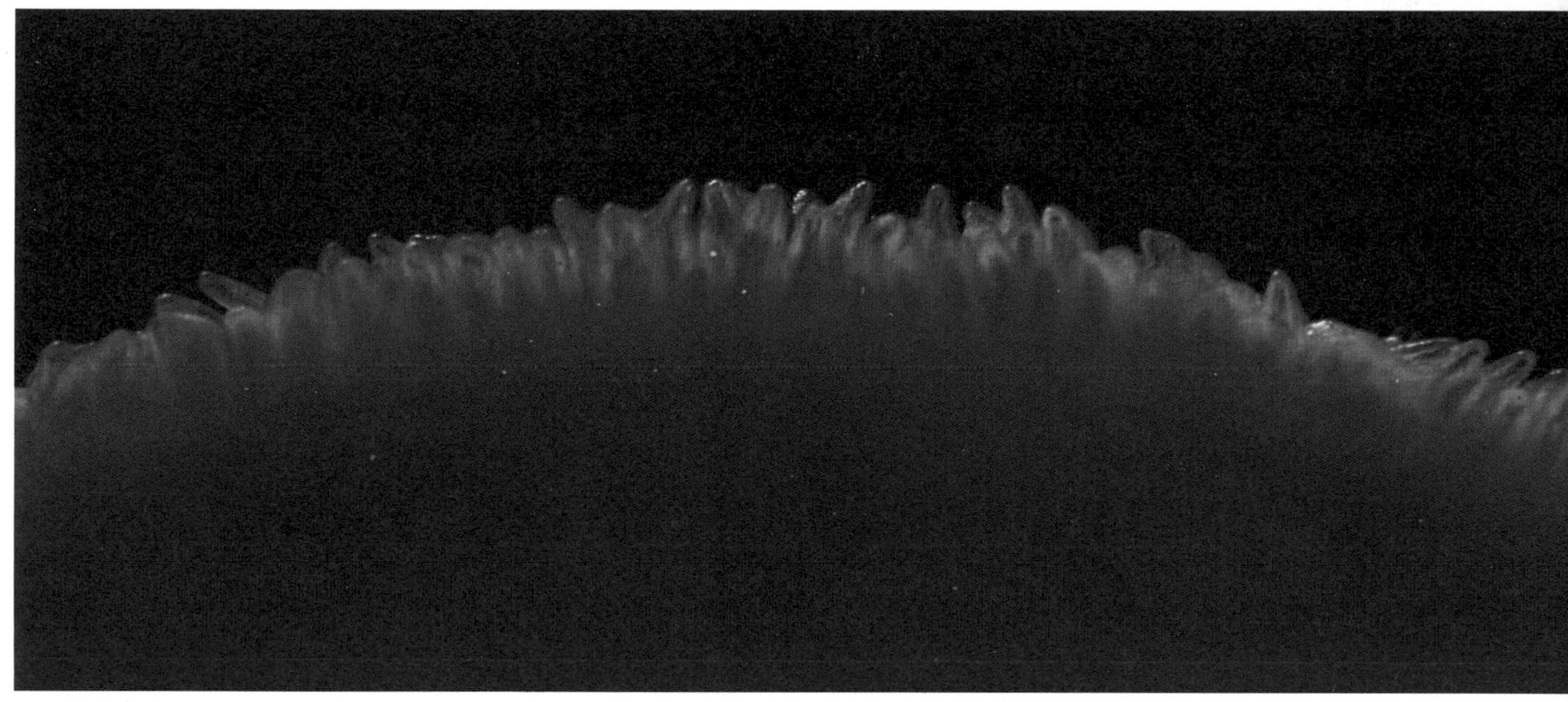

The tip of the tongue
in a small child shows the numerous threadlike papillae that give the surface a velvety look. The photo also shows the fine blood vessels in the center of each papilla. The tongue is essentially a muscle inside the oral cavity. Its mucosal covering is the principal site of the organ of taste. The musculature of the tongue consists of muscle bundles arising from the base of the skull, the lower jaw, the hyoid bone, and the pharyngeal wall. Because of this rich network of muscles, the tongue is extremely flexible, a fact of major importance in chewing, swallowing, and speaking.

which different types of stimulus energy become converted into nerve impulses. A recent theory is that each taste bud responds not only to one of the four tastes, but also to combinations of them such as sour-salt or sour-salt-sweet. This would imply that the taste buds contain individual taste cells that are stimulated by different substances. Studies of taste have also shown that an individual taste bud often "changes taste," and this phenomenon may express variations in the sensitivity of different types of receptors.

When the hairs of the taste cell come into contact with whatever type of molecule excites them, electrical impulses are generated in nearby nerve endings and pass into the nervous system. The intensity or strength of a taste depends generally on the number of cells excited and the frequency of impulses generated. A strong taste is associated with a high frequency of impulses.

The tongue is not taste sensitive over its entire surface. Central portions on its upper side have no papillae with taste buds. The sensitivity to salt is highest on the tip and along the edges of the tongue. Sweet is best perceived on the tip, sour along the edges, and bitter best in the vallate papillae in their V-shaped arrangement far back on the tongue.

Various factors influence the sensation of taste. Very cold food may seem completely tasteless. Generally, the sensitivity of taste cells seems to peak — that is, the taste threshold is lowest — when the stimulus is at body temperature. A sour taste is chemically related to the concentration of hydrogen ions. The more hydrogen ions — the lower the pH — the more sour. Salts with a low molecular weight taste salty, whereas those with higher weight may taste bitter. Mercury salts and salts of certain other heavy metals have a metallic tang, whereas salts of lead and beryllium have a treacherously sweet taste. The sweet taste, otherwise, is generally a feature of sugars and alcohols, which are organic compounds, but also of synthetic sweetening agents such as saccharin and cyclamate. The threshold concentration to

Taste

Taste

On the upper side of the tongue are the papillae, commonly divided into four types: foliate, fungiform, filiform, and vallate. The vallate and the foliate papillae are located on the posterior portion of the tongue.

In the large photo at the left, the root of the tongue is seen most clearly. This portion contains much lymphatic tissue. At the border between the root and the body of the tongue, the vallate papillae form a V. Around the sides of the papillae are the taste buds. Anterior to the vallate papillae are fungiform and filiform papillae. The epithelium covering the fungiform papillae often, but not always, contains taste buds.

The photo on the right is a detail of the surface of the body of the tongue. It shows fungiform and filiform papillae in a region just before the vallate papillae.

which the taste buds respond is seven hundred times lower for a solution of saccharin than for a solution of sugar. The structure of the sugar molecule, too, influences the taste threshold — even small changes in its molecular structure may cause the particular type of sugar to have either no taste or even a bitter taste.

Sensations of bitterness are associated with a number of substances, such as the aforementioned alkaloids, which include quinine, strychnine, and caffeine. They are perceptible even in extremely low concentrations, which suggests that the body is especially on guard against bitter substances. Since a number of alkaloids are strong nerve poisons, the existence of the human species may, at earlier stages of development, have been dependent on the ability to sort out the bitter substances from food. A remarkable finding of modern science is that some people have lost the ability to taste certain bitter compounds. This inability is genetic and assumed to be due to a recessive gene which affects taste at the receptor level.

Unfortunately, not all poisons taste bitter. A dangerous substance such as lead acetate tastes sweet, and was used as a sweetening until its toxic effect was established. In many countries, the use of cyclamate as an artificial sweetener has been prohibited because of its possible role as a cancer-producing agent. Authorities in many countries are increasingly attentive to the problems arising from synthetic additives which can "fool" the taste buds or smell receptors but whose physiological activity is little known. The sense of taste evolved in concert with the natural environment. With the advent of man-made substitutes in the diet, the tongue has become a less reliable judge.

Vallate Papillae

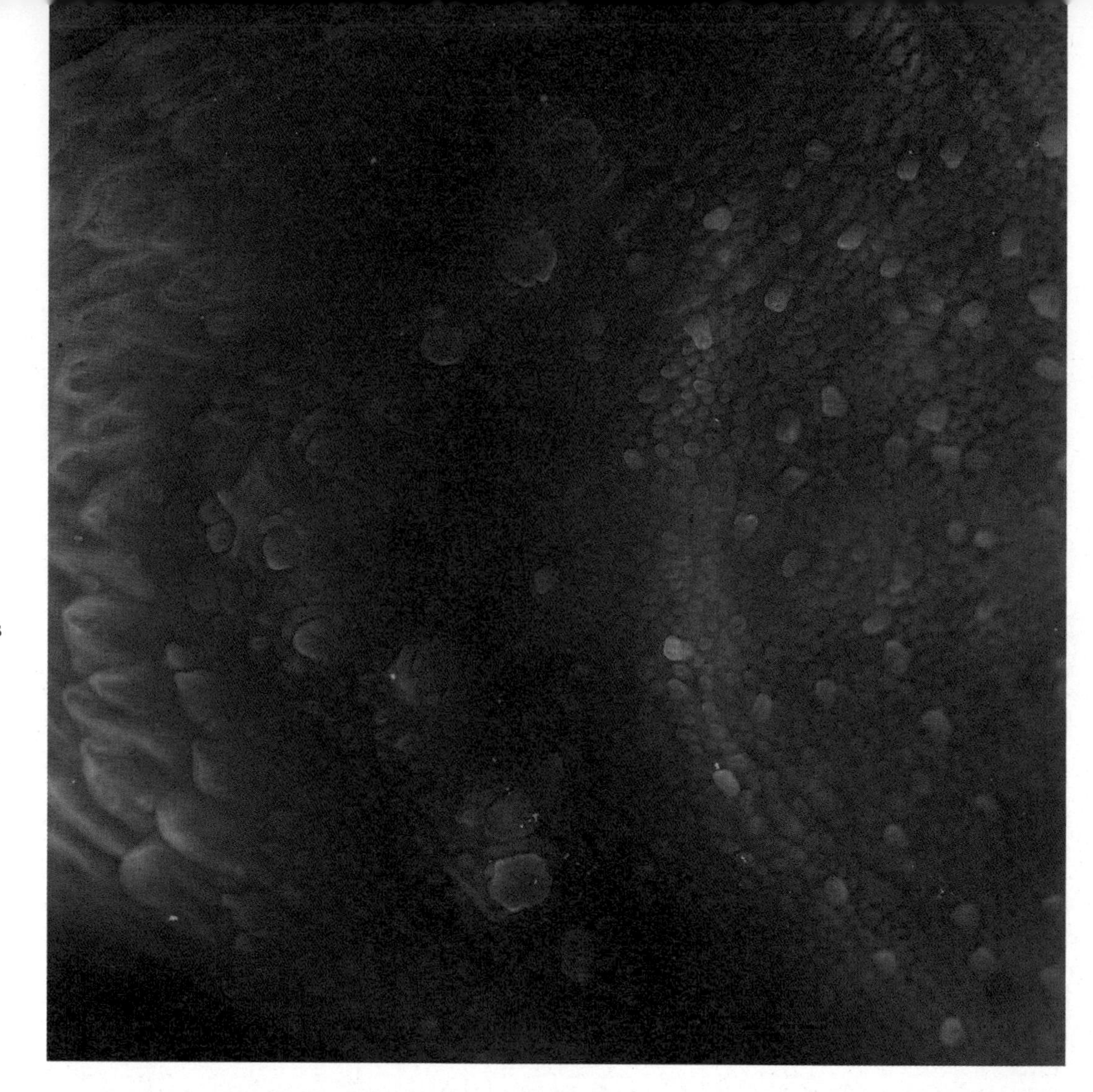

V-shaped arrangement of papillae on the tongue
At the back of the tongue, the small photo on the right shows the V-shaped system of vallate papillae. Taste buds here are especially sensitive to bitter flavors. Fungiform papillae are at the right in front. They are more sensitive to sour tastes. Furthest to the left is the pharyngeal portion of the tongue. There is an accumulation of lymphatic tissue here with lymphocytes, specific kinds of white blood cells.

The photo on the opposite page shows vallate papillae more highly magnified (about 70 times).

The papillae in a child
look more fresh and less worn. The large vallate papillae in their V formation appear clearly. Lymphatic tissue is visible at the left. Filiform and fungiform papillae are seen very clearly defined at the right. Under high magnification some fungiform papillae resemble cactus flowers.

Taste

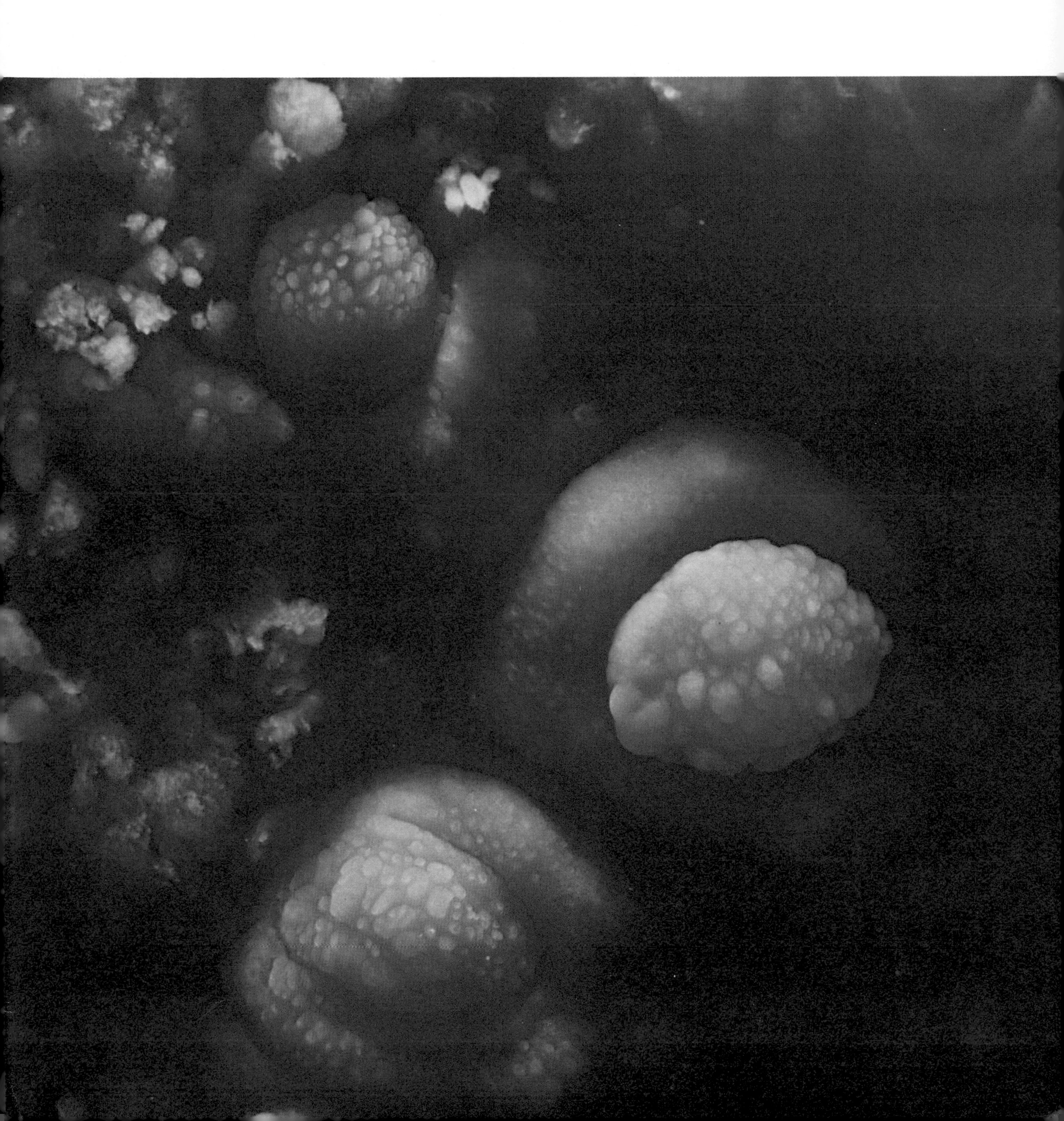

Taste

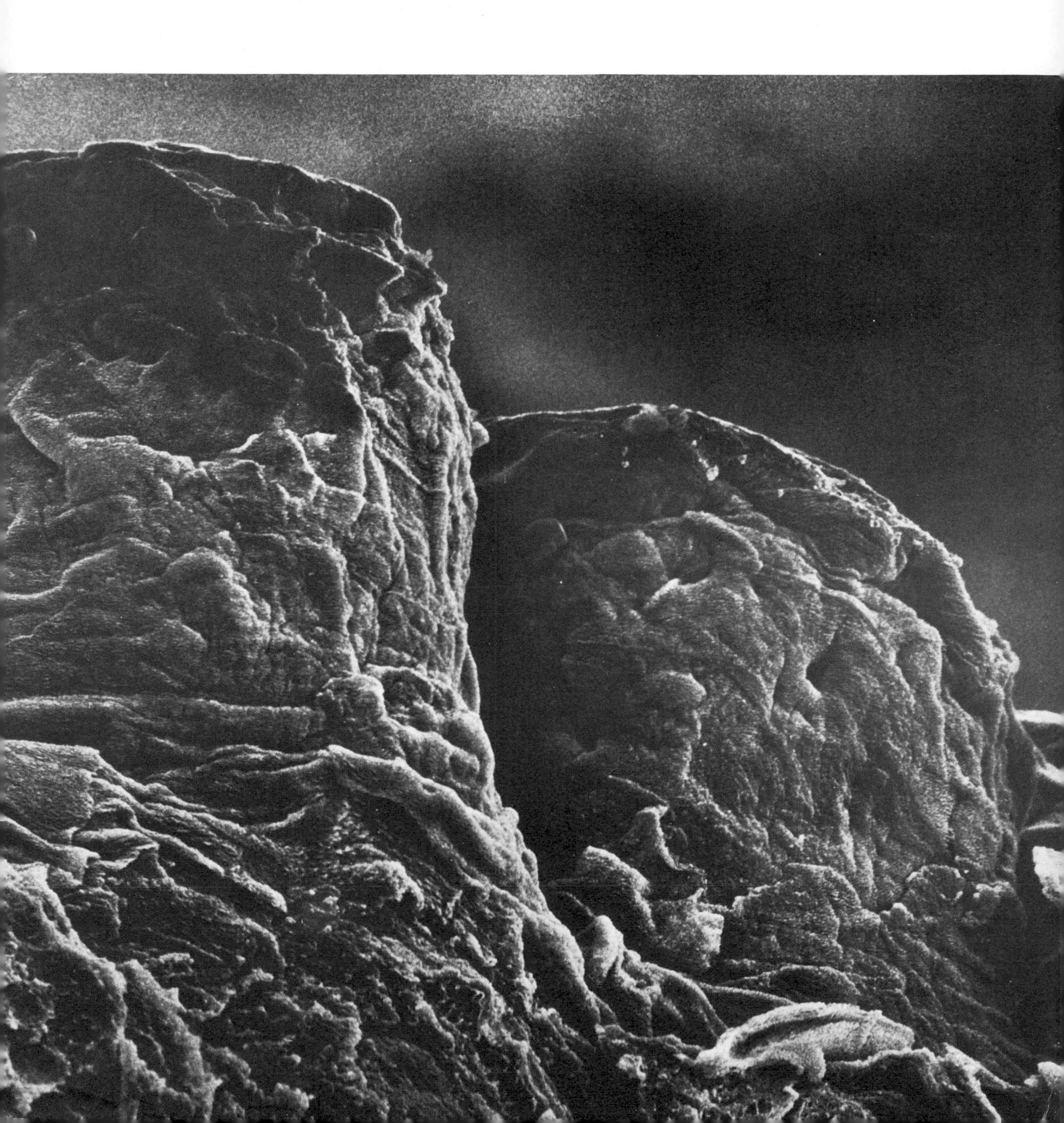

Taste

Fungiform papillae on the tongue, magnified 1,600 times, are shown in the photo at the left. Under higher magnification (18,000 times), the upper photo on the right shows the structure of the surface of the papillae. All the photos on these pages were taken in a scanning electron microscope. The oblong or elliptic hollows visible in the papillae are probably taste pores, openings that lead to the lower parts of the taste buds. The hairs of the receptor cell project into these pores and can be seen dimly in the close-up on the right.

The left-hand photo at the bottom of this page is a photo of the surface of the tongue magnified 38,000 times. The angular structures are individual cell surfaces with their many processes, the microvilli. Oval and round bacteria, which are numerous in the oral cavity, are also visible on the surface. The lower right-hand photo shows the surface of the edge of a papilla on the tongue, magnified 26,000 times.

As previously mentioned, the sense of taste often involves a close collaboration with the sense of smell. In addition, the sense of touch is also of great importance. It responds to the texture of food: some foods are exciting because they are granular and hard; others, because they are smooth or soft. In addition we can perceive when something is wet, and measure the degree of viscosity of a fluid. Finally, the temperature sense greatly influences taste sensations.

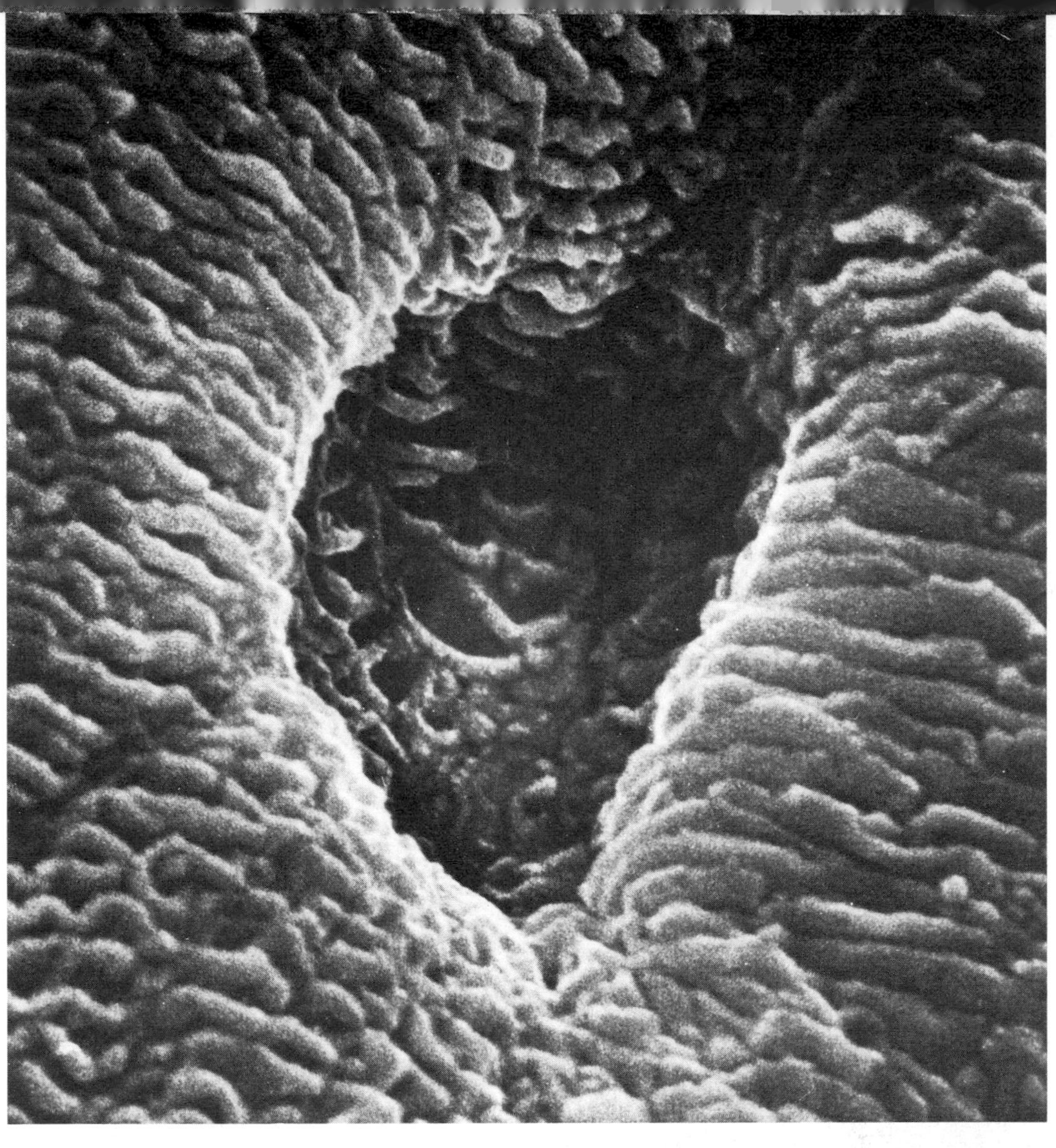

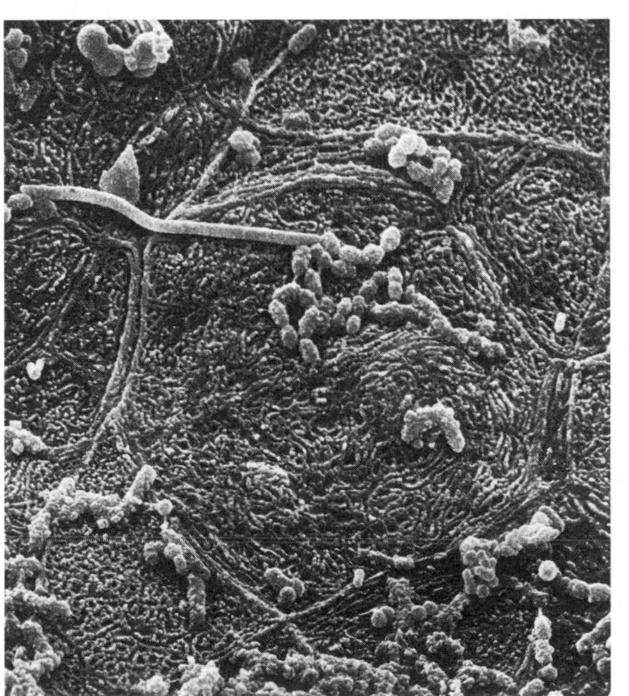

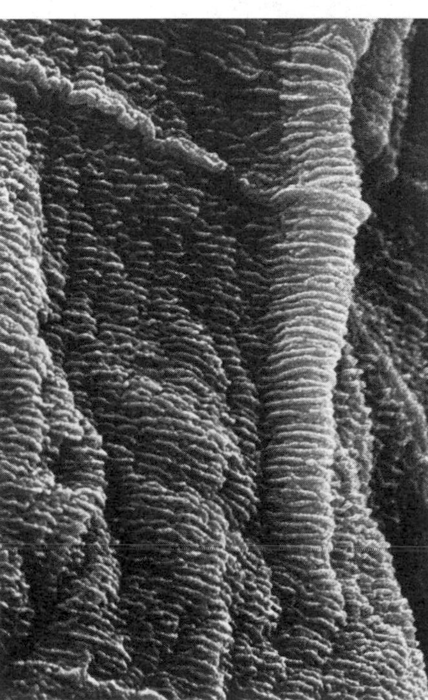

Index

Abdominal cavity, 34, 36, 86, 96
Acetylcholine, 118, 129, 176
Actin, 115, 116
Actomyosin, 115
Adam's apple, 86
Adenosine diphosphate (ADP), 152
Adenosine triphosphate (ATP), 116, 152
Adrenal cortex, 31
Adrenal glands, 30
Adrenaline, 30, 32, 34, 70, 72
Adrenal medulla, 32, 34
Agglutinogens, 74
Aging, 11, 73, 81, 190, 217
Albinos, 186
Alveoli, pulmonary, 82, 83, 88, 89
Amino acids, 138
Amniotic fluid, 52, 56, 211
Amniotic sac, 52, 54
Ampullae, 220
Amylase, 130, 142
Anaerobes, 81
Androsterone, 35
Anemia, pernicious, 152
Anesthesia, 164
Antibodies, 59, 74, 78, 180
Antigen, 74
Anus, 22, 38, 42, 129, 152
Anvil (incus), 120, 201, 202, 204, 206, 207, 208, 209, 218
Aorta, 63, 64, 65, 66, 68, 72, 76, 96, 116, 143
Apatite crystals, 124
Appendicitis, 148
Appendix, 129, 148
Aqueduct, cerebral. *See* Cerebral aqueduct
Aqueous humor, 180, 184, 186, 214
Arachnoid membrane, 172
Arbor vitae ("tree of life"), 162, 174
Arm, primordial, 52, 55, 160, 161. *See also* Hand.
Arteries, 38, 63, 64, 66, 68, 71, 108, 162, 164, 165; carotid, 64, 66, 72, 156; coronary, 65, 116; femoral, 64; hepatic, 143; iliac, 68; intercostal, 83; pulmonary, 68, 70, 76, 82; renal, 68, 98, 100; subclavian, 64, 66
Arteriosclerosis, 66, 68
Arytenoid cartilages. *See* Cartilage.
Atlas (first cervical vertebra), 122
ATP. *See* Adenosine triphosphate
Atrium, 63, 64, 68, 70, 71, 72, 73, 76, 82, 116, 138. *See also* Heart
Auditory bones (ossicles). *See* Anvil, Hammer, Stirrup
Auditory canal, external, 114, 201, 202, 203, 204, 218; internal, 202, 203, 213
Auditory centers, 201, 212, 213
Autonomic nervous system, 153, 154, 177
Axon, 12, 156, 166, 212, 231

Backbone. *See* Spinal column
Balance. *See* Equilibrium
Baldness, 112
Basilar membrane, 210, 212, 213, 214, 216, 217
Beard. *See* Hair
Behavior, 26. *See also* Emotions
Bile, 22, 96, 103, 137, 142, 143, 144, 148, 150
Birth control, 49
Bladder, 38, 42, 96, 97, 103, 104
Blastocyst, 51, 52
Blind spot, 182, 191, 192, 200
Blood, 23, 48, 49, 57-77, 78, 81, 83, 96, 115, 126, 127
Blood cells, red, 10, 23, 57, 59, 60, 63, 71, 74, 76, 82, 89, 92, 93, 122, 124
Blood cells, white, 20, 59, 60, 61, 80, 82, 92, 124
Blood circulation, fetal, 64, 76, 79, 82, 116, 156, 159, 164, 230
Blood clotting, 32, 57, 59, 62, 108, 152
Blood corpuscles. *See* Blood cells
Blood platelets, 57, 59, 62
Blood pressure, 26, 63, 68, 72, 156, 226
Blood transfusion, 74
Blood types, 74
Blood vessels, circulation, 15, 52, 57-64, 65-71, 72, 76, 84, 94, 99, 120, 129, 139, 144, 159, 165, 214-215
"Blue baby," 76
Bone, 52; skeletal, 14, 24, 25, 76, 120-125, 202; solid (dense), 122, 124; spongy (porous), 122, 123, 124
Bone cells, 14, 25, 121, 122
Bone marrow, 57, 59, 60, 62, 76, 122, 124
Bouton. *See* End-feet
Bowman's capsule, 96, 98, 99, 248
Brachiocephalic trunk vein, 64
Brain, 162-165, 213, 217, 225. *See also* Cerebellum, Cerebrum, Nervous system
Brain stem, 26, 153, 154, 156, 165, 171, 179, 213, 226, 240
Brain ventricles. *See* Ventricles of the cerebrum
Brain vesicles, 52, 53, 161
Breastbone. *See* Sternum
Bronchi, 82, 88, 89, 93

Calcium, 120, 121, 122, 124, 126, 211. *See also* Statolith
Cancer, 78, 93, 243
Capillaries, 23, 59, 63, 64, 71, 74, 78, 82, 83, 89, 96, 99, 105, 138, 248
Carbon dioxide, 23, 26, 59, 60, 64, 71, 76, 81-83, 89, 96, 129, 156, 164
Carbon monoxide, 93
Cardia ventriculi, 133
Caries, dental, 128
Carotene, 152, 231
Carpal bones, 120
Cartilage, cartilage cells, 14, 21, 25, 52, 88, 120, 121, 122, 124, 125, 202, 218, 233; arytenoid, 86; cricoid, 86, 88; thyroid, 86
Cauda equina, 155
Cecum, 148
Cell, 9, 10-20, 21-29, 49, 50, 51, 52, 59, 78, 112, 115
Cell division, 10, 11, 21, 28, 29, 30, 40, 50, 51, 52
Cell membrane, 10, 11, 17
Cementum, 126
Centriole, 10
Centrosome, 10
Cerebellum, 12, 26, 153, 162, 166, 172, 174, 220
Cerebral aqueduct, 171
Cerebral blood vessels, 164, 165
Cerebral cortex, 26, 83, 86, 156, 159, 163, 164, 165, 166, 179, 213, 226
Cerebral meninges, 162, 163, 172
Cerebrospinal fluid, 165, 171, 172
Cerebrum, 26, 153, 156, 162, 163, 164, 165, 172
Cervix, 18, 45, 48, 49
Chest, 82, 83, 120, 122, 125
Cholesterol, 66, 144
Cholinesterase, 118
Choroid, 182, 184, 192, 194
Choroid plexus, 171, 172
Chromosomes, 11, 28, 29, 30, 34, 50, 51, 52, 152
Chyme, 137, 138, 148
Chymotrypsin, 142
Cilia, 18, 40, 45, 82, 93, 177, 231, 235, 236, 237, 238
Ciliary body, 184, 189, 190, 192
Citric acid cycle, 152
Clavicle. *See* Collarbone
Climacteric. *See* Menopause
Clitoris, 38, 42, 45
Coagulation. *See* Blood clotting
Coccyx, 120, 122, 125
Cochlea, 177, 201, 202, 210, 211, 212, 213, 214-215, 218
Cochlear duct, 210, 211, 212, 213, 214-215
Collarbone, 120
Colon, 22, 129, 136, 140, 148, 150, 152
Coma, 179
Conchae, nasal, 82, 234
Cones (visual), 26, 177, 180, 191, 192, 194, 196, 198, 200
Conjunctiva, 182, 198
Connective tissue, 12, 20, 21, 24-25, 37, 38, 52, 115, 120, 121, 122, 124, 126, 157, 202, 210, 211, 212, 225, 228
Consciousness, 26, 82, 94, 156, 165, 178, 179, 225
Cornea, 177, 180, 182, 184, 186, 192, 200, 214, 228
Coronary arteries. *See* Arteries
Corpora quadrigemina, 162, 171
Corpus callosum, 162, 171
Corpus hemorrhagium, 42, 43
Corpus luteum, 42, 48, 49, 51
Corti, organ of, 210, 211, 213
Cortisone, 31
Cranial sutures, 158, 159
Cranium. *See* Skull
Creatinine, 96
Cribriform plate (of the ethmoid bone), 231
Cricoid cartilage. *See* Cartilage
Cupula, 220, 222
Cytoplasm, 10, 11, 61

Defecation, 26, 150
Dendrites, 12, 166
Dentin, 126, 127, 128
Dermis, 105-106, 108, 110, 112, 225, 226
Diabetes mellitus, 31, 142
Diaphragm (midriff), 83, 86
Diastole, diastolic pressure, 72. *See also* Blood pressure
Digestion, 21, 22, 52, 126, 129-142, 156, 230
Disk, embryonic, 52
Disk, spinal, 122
DNA (deoxyribonucleic acid), 17
Duodenum, 133, 136, 137, 142
Dura mater, 172. *See also* Spinal meninges

Ear, 177, 201-224
Ear, external, 14, 52, 114, 177, 201, 202, 206, 218, 219
Ear, inner, 120, 177, 202, 210-217
Ear, middle, 177, 202, 203, 204, 206, 207, 208, 209, 218
Eardrum (tympanic membrane), 120, 177, 201, 202, 203, 204, 205, 206, 207, 213, 218
Earwax, 203
Ectoderm, 21, 52, 56, 161, 218, 232
EEG (electroencephalogram), 165
Egg, egg cell. *See* Oocyte, Ovum
Einthoven, Willem, 64
Ejaculation, 18, 34, 36, 38
EKG (electrocardiogram), 64
Embryo. *See* Gestation
Emotions, 26, 156, 185, 186, 231
Enamel, dental, 52, 126, 127, 128
End-feet, 168
Endocrine glands, 21, 30-32, 69, 172
Endolymph, endolymphatic duct, 210, 211, 212, 214, 220, 222
Endoplasmic reticulum, 10
End-organs, peripheral, 225, 226, 227, 230
End-plate, motor, 115, 116, 118
Enterogastrone, 137
Entoderm, 21, 52, 56, 218
Enzymes, 11, 22, 60, 94, 116, 126, 129, 137, 152
Epidermis, 17, 105-106, 108, 112, 114, 225, 226
Epididymis, 35, 36, 37, 38, 39
Epiglottis, 86, 129, 130, 240
Epilepsy, 165
Epiphysis cerebri. *See* Pineal gland
Epithelial cells, epithelium, 17, 105, 106, 108, 110, 112, 210, 211, 218, 226, 231, 236, 237, 242, 243
Equilibrium, 21, 177, 178, 211, 212,

213, 218, 220 - 224
Erectile tissue, 36, 38, 106, 112
Erection, 36, 38
Erythrocytes. *See* Blood cells, red
Esophagus, 129, 130, 133
Estradiol, 40, 41, 48
Estriol, 40, 41, 48
Estrogens, 35, 40, 48, 49, 51. *See also* Estradiol, Estrone, Progesterone, etc.
Estrone, 41, 48
Eustachian tube, 202, 204, 206, 218
Excretion, 21, 96 - 100, 103 - 104
Exercise, 116
Exhalation (expiration), 81, 82, 83, 86
Extracellular fluid. *See* Interstitial fluid
Extrauterine pregnancy. *See* Pregnancy, ectopic
Eye, 52, 179, 180 - 200, 214; primordial, 84, 161, 179
Eye, anterior chamber, 180, 182, 184, 186
Eye, posterior chamber, 182, 186
Eye spot, 179
Eyeground. *See* Fundus
Eyelashes, 184, 198
Eyelid, 179, 182, 184, 198
Eyes, faceted, of the arthropoda, 179

Fallopian tube, 18, 28, 40, 42, 43, 45, 46, 49, 50, 51, 52
Falx of cerebrum, 172
Fat cells, 12, 24, 105, 108
Femur. *See* Thighbone
Fertilization, 9, 38, 40, 41, 45, 46, 50, 51
Fetus. *See* Gestation
Fibrin, Fibrinogen, 62
Fibroblast, 80
Fibrocartilage, 124
Fibula, 120
Fimbriae, 45
Fingernails. *See* Nails
Fingerprint, 94, 106, 226
Follicle, hair. *See* Hair
Follicle, ovarian, 40, 41, 42, 43, 45, 48, 49, 50
Fontanelle, 52
Foramen magnum, 154, 159
Foramen ovale, 56
Foreskin, 38
Frontal bone, 158 - 159
Frontal lobe, 162
Fructose, 39
FSH. *See* Hormone, follicle-stimulating
Fundus, of eye, 179, 190, 191, 200

Gallbladder, 129, 137, 142, 143, 144
Gamete, 51
Gamma globulin, 59
Ganglion, 155, 225. *See also* Neuron
Ganglion, autonomic, 154
Ganglion, spiral, 213
Gastric juice (secretion), 22, 129, 133, 134, 136, 137, 148
Gastric mucosa, 133, 134, 136, 137
Gastrin, 133, 137
Genes, 10, 11, 21, 22, 28, 30, 34, 49, 51, 52, 243
Genitalia. *See* Reproductive systems
Germ cells, primordial. *See* Oocyte, Spermatocyte
Germ layer. *See* Ectoderm, Entoderm, Mesoderm
Gestation, 21, 28, 40, 45, 49 - 56, 76; 4th week, 52, 53, 56, 116; 5th week, 22, 52, 53, 56, 76, 84, 124 - 125, 160, 161, 232; 6th week, 22, 52, 53, 56, 65, 84, 124, 179, 218; 7th week, 52, 53, 84, 85; 8th week, 53, 108, 109, 125, 179, 232; 3rd month, 42, 54, 124 - 125, 218; 4th month, 36, 38, 42, 54, 114, 124 - 125, 218, 233; 5th month, 16, 17, 22, 23, 36, 38, 43, 55, 76, 77, 114, 179, 218, 219; 6th - 8th month, 111, 114; 7th month, 116
Gill slits in the embryo, 52, 53, 76, 84, 85
Glans, 36, 38, 45
Glia, glial cells, 164, 174
Glomerular filtrate, 96, 99, 103, 248
Glomerulus, 96, 98, 99, 100, 101, 248, 249
Glottis, 86, 87
Glucose, 138, 143, 152, 164
Glycogen, 116, 143
Glycolysis, 116, 152
Golgi apparatus (in cell), 10
"Goose pimples," 112
Graafian follicle. *See* Follicle, ovarian
Granulocytes, 60

Hair, 52, 105, 106, 110 - 114, 126, 225
Hair follicle, 106, 112
Hairs, gray, 111, 112, 114
Hairs, lanugo, 111, 114
Hammer (malleus), 120, 201, 202, 204, 205, 206, 207, 218
Hand, 22, 23, 109, 120, 125, 160, 161, 179
Harvey, William, 64
Hearing, 26, 159, 177, 201 - 219
Heart, 21, 23, 32, 52, 63, 64, 65, 66, 68, 70, 72, 73, 74, 75, 76, 78, 82, 84, 96, 115, 116, 146, 156, 161, 228
Hemoglobin, 57, 59, 60, 71, 89
Heredity. *See* Genes
Hipbones, 25, 120, 122
Hippocrates of Cos, 165
Hormones, 30, 31, 32, 34, 35, 37, 40, 41, 57, 69, 112, 124, 152; antidiuretic, 97, 100; follicle-stimulating (FSH), 48, 49; gonadotropic, 35; luteinizing (LH), 48, 49
Horny cells (epithelial), 108, 112, 114
Humerus, 120, 125
Hunger, 48, 133, 156
Hymen, 42
Hyoid bone, 84, 120, 122, 130, 241
Hypnosis, 230
Hypophysis cerebri. *See* pituitary gland
Hypothalamus, 48, 94, 95, 156, 162, 172

Ileum, 136
Induction, 52
Infundibulum, 42, 43, 45
Inhalation (inspiration), 81, 82, 83, 86
Insulin, 31, 142
Interstitial fluid, 21, 23, 83, 159
Intestinal blood vessels, 69, 146
Intestinal juices (secretions), 22, 129, 142, 148
Intestinal villi, 138, 139, 140
Intestines, 15, 22, 46, 78, 79, 129, 133, 136 - 141, 144, 148, 152, 225, 230
Iris, 182, 184, 186, 192

Jawbone, 84, 218, 241
Jejunum, 137
Joint capsule, 124, 125
Joints, 14, 124, 125, 225; ball-and-socket, 124, 125; gliding, 124; hinge, 124, 125; saddle, 124

Keratin, 17, 105, 106, 112. *See also* Horny cells
Kidneys, 22, 59, 96 - 101, 103 - 104, 248, 249
Kneecap, 120, 125, 153
Krause's corpuscles, 225

Labia, major and minor, 38, 42, 45
Labyrinth, bony, 211, 220, 221
Labyrinth, membranous, 211, 220, 221
Lactic acid, 108, 110, 116, 128
Lamellae, 122
Langerhans, islets of, 142
Lanugo hairs. *See* Hairs, lanugo
Large intestine. *See* Colon
Larynx, 84, 86, 87, 88, 129
Leg, primordial, 52
Lens, 177, 180, 182, 186, 189, 190, 192, 200, 214
Leukocytes. *See* Blood cells, white
LH. *See* Hormone, luteinizing
Ligaments, 122, 124, 125, 210, 211, 212
Light, 180, 201
Lingual follicles (in lymph nodes), 130
Lipase, 142
Liver, 32, 35, 52, 59, 60, 65, 69, 76, 116, 129, 133, 137, 138, 142, 143, 146, 148
Lobules (of testes), 34, 37, 39
Long bones (tubular bones), 122
Lungs, 20, 32, 52, 59, 60, 63, 64, 70, 71, 76, 81 - 93, 116
Lymph, 21, 23, 78 - 80, 138 - 139
Lymph nodes, 23, 78, 79, 130
Lymphatic vessels, 21, 23, 78 - 80, 138, 139, 184
Lymphocytes, 23, 59, 60, 61, 78, 244

Macrophages, 20, 80, 81, 92
Macula lutea, 177, 180, 182, 190, 191, 192, 194, 200
Maculae acusticae, 222
Malleus. *See* Hammer
Maltose, 130, 142
Mammary glands, 40
Mechanoreceptors, 26, 156. *See also* Receptors
Medulla oblongata, 83, 86, 162, 225
Medulla, spinal. *See* Spinal cord
Meiosis. *See* Cell division
Meissner's corpuscles, 225, 226, 227
Melanin, 105, 110, 111
Melanocytes, 106
Memory, 26, 156, 159, 164, 180, 190, 231
Menopause, 34, 40, 48, 124
Menstrual cycle, 30, 40, 41, 48, 49, 51
Mesentery, 79, 137, 146, 148
Mesoderm, 21, 52, 56, 218
Metabolism, 21, 22, 32, 59, 78, 81, 129, 150 - 152, 164, 165
Metastases, 78
Microscope, electron, 9, 11, 12, 84, 222, 247, 248, 249
Microscope, interference, 9, 11, 126 - 127, 230, 248, 249
Microscope, phase-contrast, 9, 12, 248, 249
Microvilli, 17, 247
Midriff. *See* Diaphragm
Mitochondria, 10, 11, 156
Mitral valve, 72
Modiolus, 213
Molars. *See* Teeth
Monocyte, 60
Mons veneris, 42
Mitosis. *See* Cell division
Morris, Desmond, 111
Morula, 51
Mother cell, egg. *See* Oocyte
Mother cell, sperm. *See* Spermatocyte
Mouth, 52, 82, 129, 140, 156. *See also* Oral cavity
Mucus, 18, 40, 41, 45, 46, 48, 49, 133, 134, 136, 148, 150, 152, 231, 236, 238
Muscles, 12, 15, 21, 24, 34, 52, 70, 83, 115 - 119, 166, 201, 226, 230, 233, 241; cross-striated (cardiac), 70, 115, 116, 156; smooth, 15, 25, 63, 68, 104, 105, 115, 129, 136, 138, 184, 186; striated, skeletal, 15, 25, 115, 116, 118, 130, 156, 157 - 158, 176, 225. *See also* Piloerectile muscles
Myelin, 154, 156, 157, 224, 230
Myofibrils, 115
Myofilaments, 115
Myosin, 115, 116

Nails, 22, 52, 105, 125
Nasal cavity, 82, 86, 87, 129, 130, 198, 231, 232, 234, 235
Nephron, 96, 99, 103
Nerve, nerves: 12, 13, 26, 115, 116, 117, 118, 120, 126, 127, 153 - 157, 166, 176, 201, 202, 208, 209, 224, 226, 240, 241; acoustic, 177, 202,

210, 212, 213, 218, 220, 221, 222, 224; facial, 209, 240; motor, 115, 116, 118, 153, 154, 155, 156, 157, 163, 165, 220; olfactory, 234-235; optic, 159, 180, 182, 192, 198, 199; vagus, 129. *See also* Neurons, Receptors
Nerve cell. *See* Neuron, Purkinje cell
Nervous system, 26, 30, 32, 52, 72, 81, 94, 153-176, 178, 225, 228; central, 15, 118, 153-176, 178, 230; peripheral, 153-176
Neural tube, 21, 154, 160, 161
Neurite. *See* Axon
Neurons, 11, 26, 153, 164, 165, 166, 168, 212, 213
Noradrenalin, 32, 72
Nose, 14, 82, 114, 163, 218, 231-239
Nostrils, 232, 234
Nuclear membrane, 10
Nucleolus, 10
Nucleus, 10, 11, 12, 14, 15, 17, 29, 30, 39, 49, 50, 60, 121, 122, 152, 156, 164, 166, 228, 231

Occipital bone, 158-159
Occipital lobes, 156, 159, 162, 172
Ocular muscles, 182, 190
Olfaction (sense of smell), 26, 86, 178, 201, 231-239, 240, 243, 247
Olfactory bulb, 231, 234, 238
Olfactory mucous membrane, 26, 178, 231, 232, 234, 235, 236, 237
Olfactory pit, 232
Olfactory plate, 232
Oocyte, 29, 30, 46, 50
Ophthalmoscope, 191
Optic chiasma, 199
Optic disk. *See* Blind spot
Oral cavity, 82, 86, 87, 129, 130, 241, 247
Organelles, 10, 11
Orgasm, 38, 45
Osteoblasts, 122, 124
Osteoclasts, 122
Osteocytes, 122
Otoliths, 177, 220
Oval window, in ear, 202, 204, 206, 207, 210, 211, 216
Ovaries, 34, 40, 41, 42, 43, 45, 48, 49
Ovulation, 41, 45, 48
Ovum (egg), 8, 9, 10, 11, 18, 21, 28, 29, 30, 34, 38, 40, 41, 42, 45, 46, 47, 48, 49, 50, 51, 52, 56
Oxygen, 23, 26, 57, 59, 60, 63, 65, 70, 71, 72, 76, 81-93, 116, 129, 152, 164, 190, 234
Oxytocin, 41

Pacinian corpuscles, 225, 228, 230
Pain, 26, 178, 225-230
Palate, hard, 130
Palate, soft, 129, 130
Pancreas, pancreatic juice, 22, 31, 129, 137, 142, 144, 148, 228, 229
Papilla, defined, 112
Papillae, oral (foliate, filiform, fungiform, vallate), 26, 130, 131, 240, 241, 242, 243, 244, 245, 246, 247
Papillae, renal, 97, 98, 99, 103
Papillary muscles, 70, 72
Papillary pattern. *See* Fingerprint
Parathyroid glands, 124
Parietal bones, 158-159
Parietal lobe, 162
Patella. *See* Kneecap
Pelvis, 25, 34, 120, 122
Penis, 35, 36, 38, 42, 45
Pepsin, 133
Perception of depth (stereoscopic vision), 156, 159, 190, 231
Perilymph, 211, 213, 220
Periosteum, 25, 120, 121, 124
Peristalsis, 40, 45, 129, 136, 146
Peritoneum, 134, 137
Phagocytosis, 61
Pharynx, 22, 26, 82, 86, 87, 129, 130, 150, 163, 178, 204, 218, 241
Phosphocreatine, 116
Photography, 9
Photosynthesis, 81
Pia mater, 172. *See also* Meninges
Pigmentation. *See* Skin, Melanin
Piloerectile muscles, 106, 112
Pineal gland, pineal body (epiphysis cerebri), 162, 171
Pituitary gland (hypophysis cerebri), 35, 40, 48, 49, 97, 100, 159, 162, 172
Placenta, 41, 45, 49, 51, 52, 53, 54, 56, 65, 76, 84, 160, 161
Plasma, blood, 23, 57, 59, 62, 69, 78, 99
Pleura, 82, 83
Polar body, 29, 46, 50
Pons, 162
Pores, 106, 110
Pregnancy, ectopic, 45
Primitive segments, 56
Primitive streak, 52, 56
Process, defined, 12
Progesterone, 41, 49, 51
Prostate, 35, 36, 38, 104
Protoplasm, 11, 61, 115
Puberty, 34, 37, 40, 48
Pubis, 120
Pulmonary alveoli. *See* Alveoli
Pulmonary arteries. *See* Arteries
Pulmonary veins. *See* Veins
Pulp, dental, 126, 127, 128
Pulse, 34, 63, 64, 68, 72, 94, 226
Pupil (of eye), 32, 184, 185, 186, 200
Purkinje, Johannes, 106
Purkinje's cells, 12
Pus, 61
Pylorus, 133, 134, 140, 142

Radius, 120
Ranvier, node of, 156, 157
Raphe, 38
Receptors, auditory, 177, 178, 201, 210, 212, 213, 214-215, 216, 217, 218, 220
Receptors, gustatory. *See* Taste buds
Receptors, olfactory, 178, 231, 234, 235, 236, 237, 238, 239
Receptors, tactile, 225
Receptors, visual. *See* Cones, Rods
Receptors, for equilibrium, 177-178, 220, 221, 223, 224
Receptors, for pain, 225
Rectum, 129, 136, 140, 148, 150, 151, 152
Reflex, 153
Renal cortex, 96, 98, 99, 100, 101, 248, 249
Renal medulla, 96, 98, 99
Renal pelvis, 96, 97, 98, 99, 103
Reproduction, 21, 28, 29, 33-56
Reproductive system, female, 40-49; male, 34-39
Respiration, 21, 23, 81-93, 94, 116, 156, 226
Respiratory center in the brain, 83, 154, 156
Resuscitation, 82
Reticulum, of cell, 11
Retina, 26, 156, 159, 177, 180, 182, 186, 190, 191, 192, 194, 196, 198, 199, 200
Rh factor, 74
Rhodopsin, 152, 198
Ribosomes, 10, 11
Ribs, 82, 83, 120
Rickets, 152
Rima glottidis. *See* Glottis
RNA (ribonucleic acid), 17, 164
Rods, retinal, 26, 177, 180, 192, 194, 196, 198, 200
Round window, in ear, 202, 204, 206, 207, 210, 211, 212
Ruffini's corpuscles, 225

Saccule (in ear), 210, 211, 220, 222, 224
Sacrum, 120, 122
Saliva, salivary glands, 22, 126, 129, 130, 142, 148
Scala media. *See* Cochlear duct
Scala tympani, 210, 211, 212, 214-215
Scala vestibuli, 210, 211, 212
Scanning electron microscope. *See* Microscope, electron
Sclerotic coat, sclera, 180, 182, 184, 192, 194
Scrotum, 34, 35, 36, 38
Sebaceous glands, 105-106, 112, 114
Sebum, 105, 112, 114, 179
Secretin, 142
Sella turcica, 159, 172
Semen, 36, 38, 39, 45
Semicircular canals, 177, 218, 220, 221, 222
Seminal vesicles, 35, 36, 38
Sense organs, 26, 153, 156, 177-179, 201-247
Septicemia (blood poisoning), 78
Sex chromosomes, 11, 28, 30
Sex organs. *See* Reproductive organs
Sexual cycle, female. *See* Menstrual cycle
Shinbone. *See* Tibia
Shoulder blade, 125
Sight, 156, 159, 177, 180-200, 201
Sinoatrial node, 70
Skeletal muscles. *See* Muscles
Skeleton, 11, 21, 24, 34, 120-125
Skin, 11, 21, 24, 52, 63, 105-114, 126, 152, 162, 225, 226, 228, 230, 233; pigmentation, 110, 111
Skull (cranium), 120, 121, 158, 159, 162, 164, 213, 231, 241
Sleep, 165
Small intestine, 133, 136, 137, 138, 139, 140, 146-147, 148
Smell. *See* Olfaction
Smoking, 93
Sodium, 159
Somites, 22, 57
Sound, 201, 202, 204, 206, 213, 217
Speech, speech apparatus, 86, 126, 201, 241
Spermatic cord, 36
Spermatic ducts, 35, 36, 37, 38
Spermatids, 29
Spermatocyte, 29, 30, 39, 50
Sperm cell (spermatozoon, spermatozoa), 8, 9, 18, 28, 29, 30, 34, 36, 37, 38, 39, 40, 45, 46, 48, 49, 50, 51
Spinal column (vertebral column), 115, 120, 122, 154
Spinal cord (spinal medulla), 26, 83, 115, 120, 122, 153, 154, 155, 157, 159, 161, 162, 171, 225, 226
Spinal ganglion, 155
Spinal meninges, 155
Spleen, 60, 63, 78, 148
Stapedius, 204
Stapes. *See* Stirrup
Statoliths, 220, 222, 223, 234
Stereoscopic vision. *See* Perception of depth
Sternum (breastbone), 83, 120
Stethoscope, 72
Stirrup (stapes), 120, 201, 202, 204, 206, 207, 208, 209, 210, 218
Stomach, 22, 52, 129, 133-137, 148
Subcutaneous tissue, 105-106, 108, 110
Sulcus, central, in brain, 156, 163
Sweat, sweat glands, 59, 94, 96, 105-111, 203, 226, 235
Symphysis pubis, 42
Synapse, 155, 168
Synovial fluid, 124
Systole, systolic pressure, 72. *See also* Blood pressure

Tail, in the fetus, 124-125
Taste, 234, 235, 240-247
Taste buds, 26, 130, 131, 178, 208, 231, 240, 241, 242, 243, 247
Taste pores, 247
Tears, tear glands, 198
Tectorial membrane, 213
Teeth, 52, 123, 126-128, 152
Temperature, body, 32, 48, 57, 59, 81, 94-95, 143, 152; regulation of, 94-

95, 105, 108, 110, 115, 143, 156; sense of, 177, 225, 247
Temporal bone, 120, 204, 210, 211, 213
Temporal lobe, of the brain, 86, 162
Tendons, 25, 121, 125
Tensor tympani, 204
Tentorium of cerebellum, 172
Testes, testicles, 34, 36, 37, 38, 39, 40, 48
Testosterone, 35, 39
Thalamus, 165, 225
Thighbone (femur), 120, 124, 125
Thirst, 48, 95, 156
Thoracic cavity, 86
Thoracic duct, 78, 79, 138
Thrombocytes. *See* Blood platelets
Thrombosis, 66
Thumb, 23
Thymus, 78
Thyroid cartilage. *See* Cartilage
Thyroid gland, 32
Thyroxine, 32
Tibia, 120, 125
Toenails. *See* Nails
Tongue, 129, 130, 131, 163, 178, 240-247
Tonsils, 78, 130, 240
Tooth bone. *See* Dentin
Tooth cement. *See* Cementum
Touch, 26, 156, 159, 201, 225-230, 247
Trabeculae, 122, 123, 124
Trachea (windpipe), 71, 82, 86, 87, 88, 93, 129, 130
Trypsin, 142
Tubular bones. *See* Long bones
Tubules: renal, 96, 97, 98, 99, 100; seminiferous, 34, 36, 37, 39
Tympanic cavity, 204, 208, 209, 212
Tympanic membrane. *See* Eardrum

Ulna, 120
Umbilical cord, 52, 53, 54, 56, 65, 161
Urea, 96, 103, 143
Ureter, 96, 97, 98, 103, 104
Urethra, 36, 38, 96, 97, 104
Uric acid, 96
Urinary organs, 52, 96-97, 103, 104
Urine, 22, 40, 41, 59, 60, 96, 97, 99, 100, 103, 104
Uterus (womb), 18, 28, 40, 41, 42, 43, 45, 46, 48, 49, 51, 52, 56
Utricle, of ear, 211, 220, 221, 222, 224

Vacuole, 10
Vagina, 36, 40, 42, 45
Vagus nerve. *See* Nerves
Valve, aortic, 68, 73; cuspid, 68, 70, 74, 75; mitral, 72, 73, 74, 75; pulmonary (semilunar), 68, 70; tricuspid, 70
Vascular system. *See* Blood vessels
Vein, 36, 38, 63, 64, 71, 78, 165; common iliac, 64; hypatic, 143; jugular, 64; portal, 138, 143, 146; pulmonary, 12, 82; renal, 98; subclavian, 64, 78, 79, 138; umbilical, 76
Vena cava, inferior and superior, 63, 64, 70, 96, 138
Ventricles, of heart, 64, 66, 68, 70, 72, 73, 74, 82. *See also* Heart
Ventricles, of cerebrum, 95, 161, 162, 165, 171, 172
Vernix caseosa, 108, 179
Vertebra, first cervical. *See* Atlas
Vertebrae, 14, 52, 60, 120, 122, 123, 154; caudal, 124-125; cervical, 122, 154, 159; lumbar, 120, 154; sacral, 154; thoracic, 83, 154
Vertebral column. *See* Spinal column
Vestibular organs, 211, 212, 213
Vestibule, of ear, 177, 178, 211, 212, 213, 222, 223
Vision. *See* Sight
Visual cortex, 156, 177, 199
Visual purple. *See* Rhodopsin
Vital capacity, 83
Vitamins, 111, 124, 152, 231
Vitreous body, 177, 180, 182, 190, 192
Vocal cords, voice, 86, 87

White of the eye, 180
Window. *See* Oval window, Round window
Windpipe. *See* Trachea
Wisdom teeth. *See* Teeth
Womb. *See* Uterus
Wounds, 12, 62
Wrist, 120, 124

X chromosome, 34, 40

Y chromosome, 34, 40
Yellow spot, on the retina. *See* Macula lutea
Yolk granules, 11
Yolk sac, 52, 53, 76, 160, 16

Göran Alsterborg, engineer, Allan Danielsson, Ph.D., and Anita Kajland, laboratory assistant, Analytica AB, Stockholm, have assisted in photographing the pictures on pages 58, 59, 61, 71, 86 top, 106 bottom right, 107, 110, 112 top, 113, 114 bottom, 193, 198, 216, and 217, with the Cambridge scanning electron microscope. The rest of the electron microscope pictures, 42 photographs, were taken by instruments made by Jeol.

We would like also to express our gratitude to the photographer Katarina Carlquist.

L. N.
J. L.